Nanostructure Science and Technology

Series Editor
David J. Lockwood

For further volumes:
http://www.springer.com/series/6331

Sonke Svenson • Robert K. Prud'homme
Editors

Multifunctional Nanoparticles for Drug Delivery Applications

Imaging, Targeting, and Delivery

Springer

Editors
Sonke Svenson, Ph.D.
Drug Delivery Solutions LLC
Arlington, MA, USA

Robert K. Prud'homme, Ph.D.
Department of Chemical and Biological Engineering
Princeton University
Princeton, NJ, USA

ISSN 1571-5744
ISBN 978-1-4614-2304-1 e-ISBN 978-1-4614-2305-8
DOI 10.1007/978-1-4614-2305-8
Springer New York Dordrecht Heidelberg London

Library of Congress Control Number: 2012930125

Printed on acid-free paper

Springer is part of Springer Science+Business Media (www.springer.com)

Contents

Contributors

Wafa' T. Al-Jamal Nanomedicine Laboratory, Centre for Drug Delivery Research, The School of Pharmacy, University of London, London, UK

Mansoor Amiji Department of Pharmaceutical Sciences, School of Pharmacy, Northeastern University, Boston, MA, USA

James R. Baker Jr. Michigan Nanotechnology Institute for Medicine and Biological Sciences, University of Michigan, Ann Arbor, MI, USA

Ravi V. Bellamkonda Wallace H Coulter Department of Biomedical Engineering, Georgia Institute of Technology and Emory School of Medicine, Atlanta, GA, USA

Erik A. Bey Departments of Pharmacology and Radiation Oncology, Head and Neck Surgery, Harold C. Simmons Comprehensive Cancer Center, University of Texas Southwestern Medical Center at Dallas, Dallas, TX, USA

David A. Boothman Departments of Pharmocology and Radiation Oncology, Head and Neck Surgery, Harold C. Simmons Comprehensive Cancer Center, University of Texas Southwestern Medical Center at Dallas, Dallas, TX, USA

Stephen G. Boyes Department of Chemistry and Geochemistry, Colorado School of Mines, Golden, CO, USA

Marcelo Calderón Department of Chemistry and Biochemistry, Organic and Macromolecular Chemistry, Freie Universität Berlin, Berlin, Germany

Chia-Chih Chang Department of Chemistry and Geochemistry, Colorado School of Mines, Golden, CO, USA

Jianjun Cheng Department of Materials Science and Engineering, University of Illinois at Urbana – Champaign, Urbana, IL, USA

David P. Cormode Translational and Molecular Imaging Institute, Mount Sinai School of Medicine, New York, NY, USA

Omid C. Farokhzad Department of Anesthesiology, Laboratory of Nanomedicine and Biomaterials, Brigham and Women's Hospital, and Harvard Medical School, Boston, MA, USA

Zahi A. Fayad Translational and Molecular Imaging Institute, Mount Sinai School of Medicine, New York, NY, USA

Mauro Ferrari The Methodist Hospital Research Institute, Department of Nanomedicine, Houston, TX, USA

Suresh Gadde Department of Anesthesiology, Laboratory of Nanomedicine and Biomaterials, Brigham and Women's Hospital, and Harvard Medical School, Boston, MA, USA

Jinming Gao Departments of Pharmacology, Head and Neck Surgery, Harold C. Simmons Comprehensive Cancer Center, University of Texas Southwestern Medical Center at Dallas, Dallas, TX, USA

Anita Gianella Translational and Molecular Imaging Institute, Mount Sinai School of Medicine, New York, NY, USA

Biana Godin The Methodist Hospital Research Institute, Department of Nanomedicine, Houston, TX, USA

Rainer Haag Department of Chemistry and Biochemistry, Organic and Macromolecular Chemistry, Freie Universität Berlin, Berlin, Germany

Wilasinee Hatakeyama Department of Chemistry and Geochemistry, Colorado School of Mines, Golden, CO, USA

Mark M. Banaszak Holl Michigan Nanotechnology Institute for Medicine and Biological Sciences, University of Michigan, Ann Arbor, MI, USA

Gang Huang Departments of Pharmacology, Head and Neck Surgery, Harold C. Simmons Comprehensive Cancer Center, University of Texas Southwestern Medical Center at Dallas, Dallas, TX, USA

Shardool Jain Department of Pharmaceutical Sciences, School of Pharmacy, Northeastern University, Boston, MA, USA

Chalermchai Khemtong Departments of Pharmacology, Head and Neck Surgery, Harold C. Simmons Comprehensive Cancer Center, University of Texas Southwestern Medical Center at Dallas, Dallas, TX, USA

Fernando J. Kim Department of Surgery – Urology, Denver Health Medical Center, Denver, CO, USA

Nagesh Kolishetti Department of Anesthesiology, Laboratory of Nanomedicine and Biomaterials, Brigham and Women's Hospital, and Harvard Medical School, Boston, MA, USA

Yong-Eun Koo Lee Department of Chemistry, The University of Michigan, Ann Arbor, MI, USA

Raoul Kopelman Department of Chemistry, The University of Michigan, Ann Arbor, MI, USA

Kostas Kostarelos Nanomedicine Laboratory, Centre for Drug Delivery Research, The School of Pharmacy, University of London, London, UK

Glen S. Kwon Pharmaceutical Sciences Division, School of Pharmacy, University of Wisconsin, Madison, WI, USA

Barry D. Liboiron Celator Pharmaceuticals Corp., Vancouver, BC, Canada

Kai Licha Mivenion GmbH, Berlin, Germany

Jongdoo Lim Department of Chemistry, Texas Christian University, Fort Worth, TX, USA

Xuewu Liu The Methodist Hospital Research Institute, Department of Nanomedicine, Houston, TX, USA

Lawrence D. Mayer Celator Pharmaceuticals Corp., Vancouver, BC, Canada

Kathleen McNeeley Wallace H Coulter Department of Biomedical Engineering, Georgia Institute of Technology and Emory School of Medicine, Atlanta, GA, USA

Daniel Q. McNerny Michigan Nanotechnology Institute for Medicine and Biological Sciences, University of Michigan, Ann Arbor, MI, USA

Willem J.M. Mulder Translational and Molecular Imaging Institute, Mount Sinai School of Medicine, New York, NY, USA

Douglas G. Mullen Michigan Nanotechnology Institute for Medicine and Biological Sciences, University of Michigan, Ann Arbor, MI, USA

Robert K. Prud'homme Department of Chemical and Biological Engineering, Princeton University, Princeton, NJ, USA

Joanna C. Read Translational and Molecular Imaging Institute, Mount Sinai School of Medicine, New York, NY, USA

Stephanie Reichert Department of Chemistry and Biochemistry, Organic and Macromolecular Chemistry, Freie Universität Berlin, Berlin, Germany

Benjamin T. Roller Wallace H Coulter Department of Biomedical Engineering, Georgia Institute of Technology and Emory School of Medicine, Atlanta, GA, USA

Misty D. Rowe Department of Chemistry and Geochemistry, Colorado School of Mines, Golden, CO, USA

Talia J. Sanchez Department of Chemistry and Geochemistry, Colorado School of Mines, Golden, CO, USA

Rita E. Serda The Methodist Hospital Research Institute, Department of Nanomedicine, Houston, TX, USA

Natalie J. Serkova Department of Anesthesiology, Biomedical MRI/MRS Cancer Center Core, University of Colorado Denver Anschutz Medical Campus, Aurora, CO, USA

Jinjun Shi Department of Anesthesiology, Laboratory of Nanomedicine and Biomaterials, Brigham and Women's Hospital, and Harvard Medical School, Boston, MA, USA

Eric E. Simanek Department of Chemistry, Texas Christian University, Fort Worth, TX, USA

Baran D. Sumer Departments of Pharmocology and Radiation Oncology, Head and Neck Surgery, Harold C. Simmons Comprehensive Cancer Center, University of Texas Southwestern Medical Center at Dallas, Dallas, TX, USA

Sonke Svenson Drug Delivery Solutions LLC, Arlington, MA, USA

Archana Swami Department of Anesthesiology, Laboratory of Nanomedicine and Biomaterials, Brigham and Women's Hospital, and Harvard Medical School, Boston, MA, USA

Li Tang Department of Materials Science and Engineering, University of Illinois at Urbana – Champaign, Urbana, IL, USA

Paul G. Tardi Celator Pharmaceuticals Corp., Vancouver, BC, Canada

Bowen Tian Nanomedicine Laboratory, Centre for Drug Delivery Research, The School of Pharmacy, University of London, London, UK

Rong Tong Department of Materials Science and Engineering, University of Illinois at Urbana – Champaign, Urbana, IL, USA

Jeroen Van den Bossche Nanomedicine Laboratory, Centre for Drug Delivery Research, University of London, London, UK

Alexander R. Votruba Department of Anesthesiology, Laboratory of Nanomedicine and Biomaterials, Brigham and Women's Hospital, and Harvard Medical School, Boston, MA, USA

Priya N. Werahera Department of Pathology, University of Colorado Denver Anschutz Medical Campus, Aurora, CO, USA

Chapter 1
Introduction: Benefits and Challenges for Multifunctional Nanoparticles in Medicine

Robert K. Prud'homme and Sonke Svenson

The US National Institutes of Health through the National Cancer Institute (NCI) has been charged with the goal of eliminating death and suffering from cancer by the year 2015. In order to achieve this very ambitious goal, the development of novel nanotechnology-based devices and therapeutics are envisioned that are capable of one or more clinically important functions, including detecting cancer at its earliest stages, pinpointing its location within the body, delivering anticancer drugs specifically to malignant cells, and determining if these drugs are killing malignant cells. The Cancer Nanotechnology Plan developed by the NCI has identified six major challenge areas, including "Multifunctional Therapeutics," that is, nanoscale devices that integrate diagnostic and therapeutic functions, target properties, and control the spatial and temporal release of therapeutic agents while monitoring the effectiveness of these agents.

The application of nanoparticles for combined targeting and delivery of diagnostic and therapeutic agents has received significant attention in recent years. Nanoparticles may be constructed from a wide range of organic and inorganic materials such as emulsions, solid lipid nanoparticles (SLN), nanocapsules, nanospheres, micelles, liposomes, dendrimers, quantum dots (QD), and fullerenes and carbon nanotubes (CNT). These materials are being used to encapsulate or solubilize chemotherapeutic agents for improved drug delivery in vivo or to provide unique optical, magnetic, and electrical properties for imaging and therapy. Several functional nanoparticles are being evaluated in preclinical and clinical studies and

R.K. Prud'homme
Department of Chemical and Biological Engineering, Princeton University, Princeton, NJ 08544, USA
e-mail: prudhomm@princeton.edu

S. Svenson (✉)
Drug Delivery Solutions LLC, 16 Temple Street, Arlington, MA 02476, USA
e-mail: ssvenson@drugdeliverysolution.com

S. Svenson and R.K. Prud'homme (eds.), *Multifunctional Nanoparticles for Drug Delivery Applications: Imaging, Targeting, and Delivery*, Nanostructure Science and Technology, DOI 10.1007/978-1-4614-2305-8_1,

some nanoparticle have advanced status including clinically approved liposome drug formulations and metallic imaging agents. The next generation of nanoparticle-based research is directed at the consolidation of functions into strategically engineered multifunctional devices, which may ultimately facilitate the realization of personalized therapy. These multiplexed nanoparticles may be capable of (1) improved delivery of poorly water-soluble drugs, (2) identifying malignant cells via molecular detection, (3) visualizing their location in the body by providing enhanced contrast in medical imaging techniques, (4) fostering transcytosis of drugs across tight epithelial and endothelial barriers, (5) targeting and killing diseased cells with minimal side effects through selective cell or tissue targeting, (6) temporal control of drug release, (7) monitoring treatment in real time, and (8) co-delivery of multiple drugs for combination therapy [2].

The total market for nanotechnology-enabled drug delivery is expected to rise to US$ 26 billion by 2012, with a compound annual rate of 37%. Besides overcoming bioavailability hurdles for newly developed drugs, that is, poor drug solubility or membrane permeability, nanoparticle-based drug delivery can enable reformulation of existing drug products to increase product lifecycle, expand intellectual property estate, and therefore, discourage competition during a drug's most valuable years [3]. One of the best known "classical" examples for reformulated, nanoparticle-based drug delivery is Doxil®. Doxil, approved in the USA in 1995, is the poly(ethylene glycol)-coated, liposome-encapsulated form of doxorubicin, a drug used in cancer chemotherapy. A more recent commercial product developed by Abraxis Bioscience, LLC., Abraxane®, consists of an albumin-based reformulation of the anticancer drug paclitaxel. Abraxane was approved in the USA in 2005. A docetaxel–albumin formulation, ABI-008, is in Phase 1 clinical trial. Examples of lipid formulations in completed or ongoing Phase 2 clinical trials are CPX-1 (irinotecan–floxuridine liposome) and CPX-351 (cytarabine–daunorubicin liposome), dual drug combinations from Celator Pharmaceuticals, Inc. Brakiva™, a liposomal formulation of topotecan from Talon Therapeutics, Inc. (formerly Hana Biosciences), is in Phase 1 trial. Micelle-based drug formulations, that is, formulations consisting of hydrophilic–hydrophobic block copolymers that self-assemble in aqueous solution into micellar nanoparticles with hydrophobic core and hydrophilic shell, are in clinical trials as well. NK012, a micellar formulation of the camptothecin derivative SN38 from Nippon Kayaku Co., Ltd., is in Phase 2 trial for the treatment of small-cell lung cancer, while NK105, a paclitaxel-incorporating micellar nanoparticle from the same company, is in Phase 2 trial as second-line treatment for advanced or recurrent gastric cancer.

Besides physical encapsulation, drugs have been chemically conjugated to polymers, forming prodrugs with improved solubility and reduced toxicity, which release the drug based on pH or enzymatic activation at the desired therapeutic site. Examples include CRLX101 from Cerulean Pharma Inc., a polymeric conjugate comprised of cyclodextrin–poly(ethylene glycol) copolymer conjugated to camptothecin [4]. CRLX101 is currently entering a randomized Phase 2 trial for patients with non-small-cell lung cancer (NSCLC). XMT-1001 from Mersana Therapeutics, Inc. is a camptothecin prodrug based on the carbohydrate-derived "Fleximer" polyacetal platform, which is currently evaluated in a Phase 1 study. Opaxio™, a prodrug of paclitaxel conjugated to poly(glutamic acid) from Cell Therapeutics, Inc., is in Phase 3 clinical trials for the treatment of ovarian cancer and, in combination with carboplatin, for the treatment of non-small-cell lung cancer. Other companies are engaged in the development of poly(ethylene glycol) (PEG)-based prodrugs. NKTR-102 and NKTR-105 are PEG prodrugs of irinotecan and docetaxel from Nektar Therapeutics, currently in Phase 2 clinical trials for patients with solid tumor malignancies including colorectal, breast, and ovarian cancers (NKTR-102) and Phase 1 trial in patients with certain types of solid tumors including hormone-refractory prostate cancer (NKTR-105). EZN-2208 from Enzon Pharmaceuticals, Inc. is a PEG prodrug of SN38. EZN-2208 is currently in two Phase 2 trials for patients diagnosed with metastatic colorectal cancer or metastatic breast cancer. These few examples clearly demonstrate the progression of nanoparticle therapeutics from basic research curiosities to viable business opportunities.

The chapters of this book highlight recent progress in the design and engineering of multifunctional nanodevices. Chapter topics include targeting and drug delivery, dual or multiple drug delivery (combination therapy), targeting and imaging, and diagnostic and therapy ("Theranostics"). Leading experts have been invited to summarize their research and comment on the progress, expected benefits, and challenges of this exciting and rapidly developing field. Targeting and drug delivery is being reviewed in four chapters. Farokhzad *et al.* provide an overview of nanoparticle platforms, design strategies, and targeting modalities. Cheng *et al.* focus on chemical aspects of nanomedicine development, including conjugation strategies for therapeutics and targeting ligands to the nanoparticles. Monocytes and macrophages form the basis of the human immune defense system and are distributed throughout the body and, therefore, can be utilized as "Trojan horses" to prevent and treat inflammatory diseases, cancer, and cardiovascular diseases. Amiji *et al.* provide a thorough review describing how nanoparticle delivery systems can target macrophages with the intent to distribute vaccines, imaging agents, or therapeutics. Dendrimers are synthetic, monomolecular polymeric nanoparticles with well-defined architecture and sizes below 10 nm, sometimes also referred to as "monomolecular micelles." Simanek *et al.* describe the development of a paclitaxel–triazine dendrimer nanoparticle, its efficacy, toxicity, and characterization, and the many hurdles that had to be and still have to be overcome in this process, including scalability, a challenge that is often undervalued in the development of successful nanomedicines. The topic of dual or multiple drug delivery (combination therapy) is being addressed in two chapters. Mayer *et al.* use their clinical candidate CPX-351 (cytarabine–daunorubicin liposome) as case study to discuss the importance of synergistic ratios between two anticancer drugs, how this ratio is being determined, and then maintained during drug delivery and released at the target site using either liposomes or polymeric nanoparticles. Glen Kwon reviews the use of polymeric micelles in drug delivery with special emphasis on the simultaneous delivery of multiple drugs and outlines the advantages compared to current cancer therapy, which requires sequential delivery of each drug in a combination therapy. The topic of targeting and imaging is being reviewed in two chapters. Mulder *et al.* provide an overview of nanoparticle classes used for imaging and delivery of therapeutics with special emphasis on inorganic and organic–inorganic hybrid nanoparticles. One interesting nanoparticle class is comprised of inorganic nanocrystals as core material, surrounded by a shell formed by lipoproteins. Multimodal imaging is achieved, for example, by combination of fluorescence and magnetic resonance imaging (MRI). The development of novel MRI contrast agents is the focus of the chapter provided by Boyes *et al.* This group presents a novel surface modification technique to produce a polymer-modified, positive contrast agent nanoparticle for targeted MRI. A range of hydrophilic and hydrophobic homopolymers, along with multifunctional copolymers, were synthesized using RAFT polymerization and subsequently used to modify the surface of gadolinium (Gd) metal-organic framework (MOF) nanoparticles.

The remaining seven chapters describe the various aspects and approaches to combined diagnostic and therapy ("theranostics"). Ferrari, Godin *et al.* discuss their multistage nanovector (MSV) system, designed to overcome sequential biological barriers. The first stage, comprised of porous silicon microparticles, targets tumor-associated endothelium. Once arrested at the disease site, second-stage nanoparticles carrying the therapeutic and imaging agents are released into the tumor microenvironment by time-dependent degradation of the silicon carrier. Details of MSV composition and fabrication as well as examples of its application are presented. Kopelman and Lee discuss the application of targeted, multifunctional hydrogel nanoparticles for imaging and treatment of cancer. Hydrogel nanoparticles made of natural, synthetic, or combinations of both polymers have been designed, prepared, and applied for treatment and imaging of cancer with various therapeutic and imaging modalities. The review describes the types of hydrogel nanoparticles developed for cancer applications and their preparation methods and analyzes their characteristics which make them suitable for cancer therapy and imaging. Gao *et al.* review the use of polymeric micelles for cancer imaging and therapy. This chapter focuses on recent advancements in the design of polymeric micelles and their application in the delivery of therapeutics and imaging using many different modalities and the combination of therapy and diagnostic. Bellamkonda *et al.* discuss the use of multifunctional nanoparticles in the diagnosis, monitoring, and treatment of cancer with specific focus on applications in personalized medicine. The many aspects of personalized medicine that need to be considered and have been studied are illustrated with numerous examples. Baker *et al.* summarize their research on the development of the dendrimer-based, multifunctional therapeutic agent G5-FA-MTX, which combines the targeting ligand folic acid with the antiproliferative agent methotrexate on the surface of a dendrimer nanoparticle. As needed, this dendrimer can also carry a fluorescence dye for imaging purposes. The chapter focuses on the challenges that appeared during the translation of G5-FA-MTX to the clinic, involving the variability and heterogeneity of the starting dendrimer, the development of analytical methods capable of defining the composition and assessing the reproducibility of the intermediate and final products, and the evolution of synthetic strategies that produce reproducible batches of the multifunctionalized dendrimer. Finally, the authors present a revised synthetic strategy that is expected to overcome these challenges. The challenges posed during scale-up of dendrimers have motivated Haag *et al.* to focus their research on hyperbranched, multivalent dendritic architectures. Hyperbranched polymers generally are not perfectly branched like dendrimers but can be prepared conveniently in one-step procedures via polyaddition, polycondensation, and radical polymerization on the kilogram scale. In their chapter, the authors review the medical applications of dendrimers and compare to hyperbranched polymers, specifically to their hyperbranched polyol platform, and discuss how this platform can be applied in the field of theranostics. In the final chapter, Kostarelos *et al.* review design and engineering of multifunctional, quantum dot-based nanoparticles for combined therapeutic and diagnostic applications. Quantum dots are semiconducting nanocrystals that are being used as fluorescent probes because they offer a significant increase in photostability and fluorescence lifetime compared to organic dye-based probes. Incorporation of quantum dots within previously established drug delivery systems or surface modification of quantum dots to allow interaction with therapeutic agents prepares this nanoparticle platform for theranostic applications. The Kostarelos group provides examples from many research groups to illustrate the potential of quantum dots in this field.

Lastly, the editors would like to thank all authors for their willingness to contribute to this book on "Multifunctional Nanoparticles for Drug Delivery Applications." We appreciate the time and hard work that went into the preparation of these chapters. We also thank the reviewers for their participation in this project as well as the editors and their staff at Springer for their invitation to prepare this book and their support and patience during the review and editing process. We certainly hope that this book will be informative and inspiring to readers in the many fields that participate in the development of multifunctional nanoparticles, including chemistry, engineering, materials science, pharmacology, medicine, and biology. It is our hope that the book will provide valuable information to experts in these fields and inspire students to participate in future endeavors developing a new generation of nanopharmaceuticals.

References

1. (2006) Cancer nanotechnology—going small for big advances. NIH Publication No. 04-5489
2. Farokhzad OC, Langer R (2009) Impact of nanotechnology on drug delivery. ACS Nano 3:16–20
3. Tim H (2007) The Nanotech Revolution in Drug Delivery. Report, Cientifica Ltd, October 10, 2007
4. Svenson S, Wolfgang M, Hwang J, Ryan J, Eliasof S (2011) Preclinical to clinical development of the novel camptothecin nanopharmaceutical CRLX101. J Control Release 153(1):49–55

Part I
Targeting and Drug Delivery

Chapter 2
Nanoparticles for Targeted and Temporally Controlled Drug Delivery

Archana Swami, Jinjun Shi, Suresh Gadde, Alexander R. Votruba, Nagesh Kolishetti, and Omid C. Farokhzad

2.1 Introduction

Therapeutic nanoparticle (NP) technologies have the potential to revolutionize the drug development process and change the landscape of the pharmaceutical industry [1–5]. By virtue of their unique physicochemical properties, nanoparticles have shown promise in delivering a range of molecules to desired sites in the body. Nanoparticle technologies may improve the therapeutic index of drugs by enhancing their efficacy and/or increasing their tolerability in the body. Nanoparticles could also improve the bioavailability of water-insoluble drugs, carry large payloads, protect the therapeutic agents from physiological barriers, as well as enable the development of novel classes of bioactive macromolecules (e.g., DNA and siRNA). Additionally, the incorporation of imaging contrast agents within nanoparticles can allow us to visualize the site of drug delivery or monitor the in vivo efficacy of the therapeutic agent [6, 7]. Thus far, over two-dozen nanotechnology products have been approved by the US Food and Drug Administration (FDA) for clinical use, and many are under clinic and preclinic development [2, 8, 9]. Interestingly, the majority of these clinically approved, first-generation nanotechnology products are comprised of liposomal drugs and polymer–drug conjugates, which are relatively simple and generally lack active targeting or controlled drug release components. To develop safer and more effective therapeutic nanoparticles, researchers have designed novel multifunctional nanoparticle platforms for cell/tissue-specific targeting, sustained or triggered drug delivery, co-delivery of synergistic drug combinations, etc. Among these functions, we believe that spatial and temporal controls in drug delivery may be critical for the successful development of next-generation nanotechnology products [5].

A. Swami • J. Shi • S. Gadde • A.R. Votruba • N. Kolishetti • O.C. Farokhzad (✉)
Department of Anesthesiology, Laboratory of Nanomedicine and Biomaterials,
Brigham and Women's Hospital, and Harvard Medical School, Boston, MA 02115, USA
e-mail: ofarokhzad@zeus.bwh.harvard.edu

S. Svenson and R.K. Prud'homme (eds.), *Multifunctional Nanoparticles for Drug Delivery Applications: Imaging, Targeting, and Delivery*, Nanostructure Science and Technology,
DOI 10.1007/978-1-4614-2305-8_2,

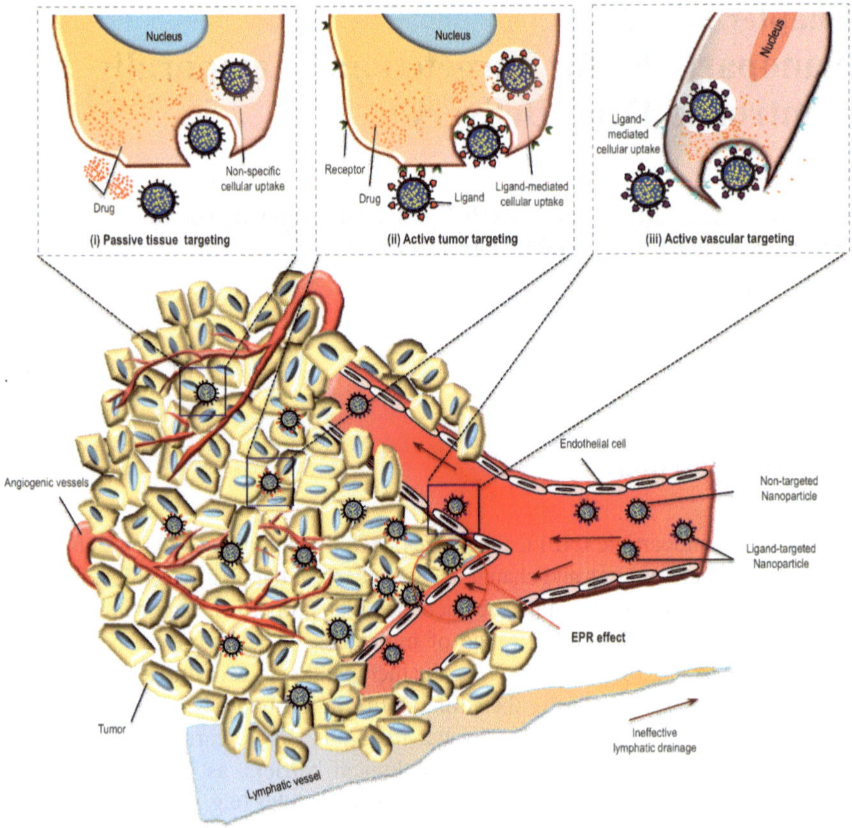

Fig. 2.1 Schematic presentation of passive *vs.* active targeting of nanoparticles. (**i**) Nanoparticles extravasate through the leaky vasculature and preferentially accumulate through the EPR effect. In this case of "passive targeting," the drugs may be released in the extracellular matrix and diffuse throughout the tissue for bioactivity. Some of these nanoparticles might also be taken up nonspecifically. (**ii**) After extravasation in the target tissue, the ligand-conjugated nanoparticles actively interact with the receptors present on target cell or tissue, resulting in cellular uptake through receptor-mediated endocytosis. This is referred as "active targeting." (**iii**) The targeted nanoparticles can be equipped for vascular targeting as well by incorporating ligands specific to endothelial cell surface receptors

Spatially controlled drug delivery can be obtained by conjugating drug-encapsulated nanoparticles with targeting ligands, which could facilitate the preferential delivery of nanotherapeutics to the sites of interest while reducing undesired side effects elsewhere. Since the first description of cell-specific targeted liposomes in 1980 [10, 11], targeted nanoparticles have shown some promising clinical and preclinical results in the treatment of different diseases. For tumor cell targeting, the presence of targeting ligands could enhance cellular uptake and retention of drugs via receptor-mediated endocytosis, although tumor accumulation through the enhanced permeability and retention (EPR) effect [12] is largely determined by the physicochemical properties of nanoparticles and long circulation half-life (Fig. 2.1) [3].

Active nanoparticle targeting is particularly essential for the delivery of biomacromolecules (e.g., DNA and siRNA) that require intracellular delivery for bioactivity [3]. In the case of vascular endothelial targeting for oncology or cardiovascular indications, ligand-mediated targeting may be critically important as nanoparticle localization is not a function of EPR (Fig. 2.1) [13–15]. In addition, efforts have been made to transport drugs across tight epithelial and endothelial barriers with nanotherapeutics (e.g., the blood–brain barrier) via ligand-mediated transcytosis [16]. More recently, targeted nanoparticles have been employed in solving the complex problems of multidrug resistance [8].

Controlled release polymer technology, resulting in the temporal control of drug exposure, has benefited virtually every branch of medicine over the past 4 decades. Many products utilizing this technology are now in clinical use, including Atridox®, Lupron Depot®, Gliadel®, Zoladex®, Trelstar® Depot, and Sandostatin® LAR [17]. Polymeric nanoparticles can encapsulate drugs and release them at sustained rates in the optimal range of drug concentration, thus enhancing the in vivo therapeutic efficacy, maximizing patient compliance, and facilitating the use of highly toxic, poorly soluble, or relatively unstable drugs [17, 18]. In general, drug release can be regulated by diffusion of the drug molecules through the polymer matrix or by differential surface and bulk erosion of the polymer [19]. Alternatively, drug release can be triggered by specific microenvironments in the body (e.g., changes in pH, temperature, and enzymatic activities) or manipulated by external events (e.g., electric field, magnetic field, and ultrasound) [20–22]. By further functionalization with targeting ligands, controlled release polymeric nanoparticles could deliver therapeutic agents in a spatiotemporally regulated fashion, which may be essential to many medical applications.

In this chapter, we focus on the major classes of organic nanoparticle platforms (*i.e.*, liposomes, polymeric nanoparticles, lipid–polymer hybrid nanoparticles, and dendrimers), their applications in targeted and/or temporally controlled delivery of therapeutic molecules, and their optimal design for escaping immune surveillances, as well as review the various available classes of ligands for targeted drug delivery applications.

2.2 Nanoparticle Platforms for Targeted and Temporal Delivery

Over the past few decades, different nanotechnology platforms were studied for their use in therapeutic applications [8, 9]. These nanoparticle platforms have been developed to enhance the pharmacological properties and therapeutic index of a myriad of drugs [1, 23]. Herein, we discuss four major classes of organic nanoparticle delivery systems, including liposomes, polymeric nanoparticles, lipid–polymer hybrid nanoparticles, and dendrimers (Fig. 2.2), which can encapsulate drugs with high loading efficiency and protect them from undesired effects of external conditions [3].

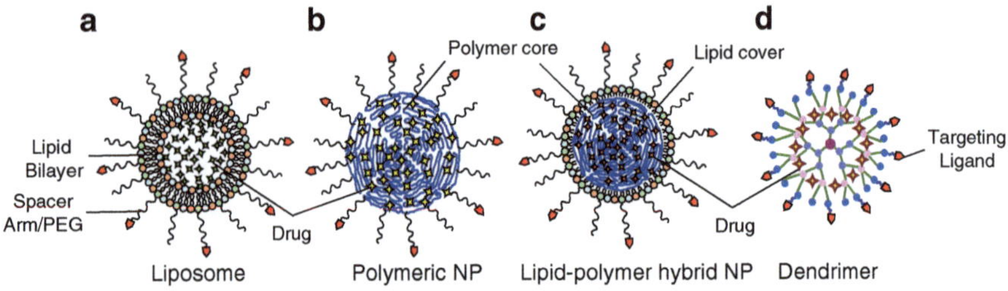

Fig. 2.2 Schematic representation of several major therapeutic nanoparticle platforms including liposome, polymeric nanoparticles, lipid–polymer hybrid nanoparticles, and dendrimers for the targeted and/or temporally controlled delivery of drugs

2.2.1 Liposomal Platforms

Liposomes are artificial, single, or multilaminar vesicles made with bilayered membrane structures, composed of natural or synthetic amphiphilic lipid molecules (Fig. 2.2a). As drug delivery carriers, liposomes exhibit several unique properties including favorable safety profiles, long systemic circulation half-life, and ease of surface modifications [24]. Among the clinically validated nanotechnology products, liposomal drugs were the first nanotherapeutics to get FDA approval for clinical use. Since the approval of DOXIL® (doxorubicin liposomes) for the treatment of AIDS associated with Kaposi's sarcoma in 1995 [9], several other liposomal drugs have been approved for clinical use, and many are in the various stages of clinical development, as shown in Tables 2.1 and 2.2. In addition to small molecule drugs, liposome systems also allow for the delivery of bioactive macromolecules (e.g., DNA) for therapeutic applications [25]. For example, Allovectin-7, composed of cationic lipid-based liposomes (DMRIE–DOPE), can carry plasmid DNA encoding HLA-B7 and β2 microglobulin that stimulate both innate and adaptive immune responses for cancer treatment [26].

Ligand-conjugated liposomes have also shown potential to enhance the therapeutic efficacy of drugs through targeted delivery. Three targeted liposomal systems have already entered clinical trials. MCC-465, a PEGylated liposome, is tagged with the $F(ab')_2$ fragment of the human monoclonal antibody GAH. Although this product does not appear to have progressed through development after a Phase I clinical trial, it has demonstrated superior cytotoxic activity against several human stomach cancer cells [27]. Recently, a novel *N*-glutaryl phosphatidylethanolamine (NGPE)-liposome formulation, MBP-426, which is conjugated to the human transferrin (Tf) ligand, improved the safety and efficacy of oxaliplatin through prolonging the drugs circulation time and by specifically targeting Tf receptors on tumor cells [28]. Another liposomal drug (SGT-53) also targets the Tf receptor on tumor cell surfaces by using the ligand of the anti-Tf receptor single-chain antibody fragment (Tf-R-scF_v) for the delivery of the tumor suppressor gene p53 [8].

The significance of targeted liposomes is further highlighted by extensive preclinical studies. For example, the monoclonal antibody 2C5 (mAb 2C5) recognizes intact

Table 2.1 Examples of nontargeted nanoparticles for drug delivery applications that are FDA approved or in clinical trials

Platform	Trade name	Drug	Clinical stage	Company
Liposomes				
PEGylated liposome	Doxil/Caelyx	Doxorubicin	Approved	Ortho Biotech, Schering-Plough
	SPI-77	Cisplatin	Phase II	Sequus Pharma.
Non-PEGylated liposome	Myocet	Doxorubicin	Approved	Zeneus
	DaunoXome	Daunorubicin	Approved	Gilead Sciences
	Onco TCS	Vincristine	Phase III (complete)	INEX Pharma.
	AmBisome	Amphotericin B	Approved	Gilead Sciences
	LE-SN38	SN-38	Phase II	NeoPharm
	CPX-1	CPT-11, floxuridine	Phase II	Celator Pharma.
	DepoCyt	Cytarabine	Phase I/II	SkyePharma
	NX211	Lurtotecan	Phase II	Gilead Sciences
	Nyotran	Nystatin	Phase I/II	Aronex Pharma.
	Atragen	All-trans retinoic acid	Phase II	Aronex Pharma.
	Aroplatin	NDDP	Phase II	Aronex Pharma.
	Annamycin	Annamycin	Phase I/II	Callisto
	L9NC	Camptothecin	Phase II	Verschraegen
Non-PEGylated, cationic liposome	Allovectin-7	DNA plasmid	Phase III	Vical Inc.
	EndoTAG	Paclitaxel	Phase II	MediGene
	E1A gene	E1A gene	Phase I	Targeted Genetics Corp.
Polymeric nanoparticles				
PEG-poly(L-lactic acid)	Genexol-PM	Paclitaxel	Approved in Korea	Samyang
PEG-poly(aspartic acid)	NK911	Doxorubicin	Phase I	Nippon Kayaku
PEG-poly(aspartic acid)	NK105	Paclitaxel	Phase II	Nano carrier
PEG-poly(glutamic acid)	NC-6004	Cisplatin	Phase I/II	Nano carrier
PEG-poly(glutamic acid)	NK012	SN-38	Phase II	Nippon Kayaku
PEG-cyclodextrin	CRLX101	Camptothecin	Phase II	Cerulean Pharma
Pluronic F127 and L61	SP1049C	Doxorubicin	Phase III	Supratek Pharma

Table 2.2 Targeted nanoparticles for drug delivery in different stages of clinical studies

Platform	Trade name	Targeting ligand	Drug	Clinical stage	Company
PEGylated liposome	MCC-465	F(ab')$_2$ fragment of human antibody GAH	Doxorubicin	Phase I	Mitsubishi Pharma.
NGPE liposome	MBP-426	Transferrin	Oxaliplatin	Phase I/II	Mebiopharm
Liposome	SGT-53	Single-chain antibody fragment	p53 gene	Phase I	SynerGene therapeutics
Cyclodextrin-based polymeric NP	CALAA-01	Transferrin	siRNA	Phase I	Calando Pharma.
PEGylated PLGA NP	BIND-014	Peptide	Docetaxel	Phase I	BIND Biosciences

nucleosomes (originating from apoptotically dying neighboring tumor cells), bound to the surface of live tumor cells. Conjugation of mAb 2C5 to DOXIL liposomes resulted in improved biodistribution and cell targeting and in increased drug efficacy [29]. Liposomes carrying poly(ethylene glycol) (PEG) chains on their surface and loaded with doxorubicin have been coupled with RGD peptides to target the integrins of tumor vasculature and have demonstrated increased efficacy against C26 colon carcinoma in murine model [30]. Folate-functionalized liposomes, encapsulating fluorescent calcein and doxorubicin, and targeting the folate receptor (type-β), have been used for the treatment of acute myelogenous leukemia [31]. However, while liposomes are commonly explored for drug delivery applications, they do not readily allow for sustained release of therapeutic molecules, which marks a significant shortcoming of this class of nanocarriers [32].

2.2.2 Polymeric Nanoparticles

Polymer–drug conjugates have made a significant clinical impact by improving the pharmaceutical efficacy and dosing of a variety of already approved drugs [8, 33]; however, their drug loading efficiency may be limited by the number of conjugation sites in the polymer, and most of them lack the ability of active targeting or controlling drug release. In order to further enhance the drug loading capacity and incorporate the spatial and/or temporal control over drug delivery, many biocompatible polymeric nanoparticle platforms have been developed [34–36].

Polymeric micelles have attracted substantial attention for their remarkable potential as therapeutic carriers [37]. Polymeric micelles can be formed by self-assembly of amphiphilic polymers with two or more polymer chains of different hydrophobicity. In aqueous environments, these block copolymers can spontaneously self-assemble into core-shell nanostructures, with a hydrophobic core and a hydrophilic shell (Fig. 2.2b) [35, 37]. To date several polymeric micelles have reached different stages of clinical development, and these systems have demonstrated enhanced accumulation of therapeutic agents at the target site and/or reduced adverse effects of therapeutic agents (Table 2.1) [9, 38]. Among them, NK911 [39] and NK105 [40] utilize PEG-poly(aspartic acid) copolymer to carry and protect the anticancer agents doxorubicin and paclitaxel, respectively. Notably, NK105 was shown to reduce the reported adverse effects of paclitaxel, which include neurotoxicity, myelosuppression, and allergic reactions [40]. A cisplatin-incorporated polymeric micelle-based system, NC-6004, is being examined in Phase I/II clinical trials and has demonstrated several distinct features, including sustained cisplatin release, promoted accumulation of cisplatin in cancer cells, and reduced nephrotoxicity and neurotoxicity associated with cisplatin [41]. Another PEG-poly(glutamic acid)-based polymeric micelle, NK012, loaded with 7-ethyl-10-hydroxycamptothecin (SN-38), has been shown to exert more potent antitumor activity against various human tumor xenografts than irinotecan (CPT-11), a water-soluble prodrug of SN-38 [42]. More impressively, the nontargeted polymeric micelle composed of poly(L-lactic acid) (PLA)-PEG (Genexol®-PM), for delivery of paclitaxel, was first approved for cancer therapy in Korea in 2007 [8] and is currently being evaluated in a clinical Phase II trial in the United States for the treatment of metastatic pancreatic cancer [43, 44].

The conjugation of polymeric nanoparticles with targeting ligands could also enable drug delivery in a spatially and temporally controlled manner, which may further enhance the therapeutic efficacy of drugs and reduce their toxic side effects. Our group has pioneered the development of aptamer-targeted polymeric nanoparticles and applied these nanoparticles to cancer therapy [14, 45–48]. For example, we have developed A10 RNA aptamer-conjugated poly(lactide-*co*-glycolide)-poly(ethylene glycol) (PLGA–PEG) nanoparticles that can recognize PSMA (prostate-specific membrane antigen), expressed on the cancer cell surface [49]. This PLGA–PEG-aptamer nanoparticle can substantially reduce tumor growth in a human prostate cancer tumor xenograft mouse model. More recently, we have reported a strategy for precisely engineering PLGA–PEG-aptamer nanoparticles with different biophysicochemical properties in a reproducible manner, whereby enabling the systematic screening of the targeted polymer nanoparticles for optimization [50]. Building on these efforts, BIND Biosciences has developed a self-assembled, targeted polymeric nanoparticle (BIND-014) and is currently evaluating this nanotherapeutic candidate in Phase I/II clinical trials for the treatment of solid tumors [51].

2.2.3 Lipid–Polymer Hybrid Nanoparticles

The success of polymeric nanoparticles and liposomes has also motivated the development of lipid–polymer hybrid nanoparticles (Fig. 2.2c), which could integrate the unique advantages of both polymeric nanoparticle and liposome systems, while overcoming some of their limitations. Thus far, several important lipid–polymer hybrid nanoparticles have been developed. For example, lipid-coated polymeric nanoparticles comprising a PLGA core, a PEG shell, and a lipid monolayer at the interface were recently described and characterized [13, 52–56]. The PLGA core is capable of carrying poorly water-soluble drugs, while the PEG shell helps to decrease biofouling and increase circulation half-life. The lipid monolayer that resides at the interface between PLGA core and PEG shell acts as a molecular fence, promoting drug retention and sustained release from the polymeric core [52, 56]. When compared to PLGA and PLGA–PEG nanoparticles, this lipid-coated PLGA nanoparticle allows for higher drug encapsulation, tunable and sustained drug release over a longer period of time, and excellent serum stability [56]. In another example, a liposome-enveloped PLGA nanoparticle, known as "nanocell," was developed in a multistep manner for the effective treatment of cancers. The nanocell has a PLGA core encapsulating the PLGA-conjugated anticancer drug doxorubicin, and a lipid multilayer shell containing the antiangiogenic agent, combretastatin. The synergistic effect of the two drugs is obtained through temporally controlled release, where combretastatin is first released to reduce vascularization, while the sustained release of doxorubicin from the nanocell directly kills the tumor cells [57].

Lipid-coated polymeric nanoparticles, developed by Zhang *et al.*, showed enhanced uptake in prostate cancer cells overexpressing PSMA antigens when conjugated with A10 aptamer, as compared to nontargeted hybrid nanoparticles [56]. More recently, Wang *et al.* have applied A10 aptamer-targeted lipid–polymer hybrid nanoparticles for the concurrent administration of a chemotherapeutic agent (docetaxel) and a radiotherapeutic agent (111-indium or 90-yttrium), which demonstrated higher level of cellular cytotoxicity, as compared to targeted nanoparticles containing only a single agent or nontargeted nanoparticles [55]. For the treatment of injured vasculature, Chan *et al.* developed a "nanoburr" system by conjugating the lipid-coated PLGA hybrid nanoparticle with a novel peptide ligand, screened from a combinatorial library of heptapeptide ligands against human collagen IV, which represents 50% of the vascular basement membrane [13]. The peptide-conjugated "nanoburr" demonstrated efficient targeting toward vascular basement membrane, high nanoparticle accumulation in the region of injured vasculature in a rat model, and sustained drug release over 2 weeks.

2.2.4 Dendrimers

Dendrimers are synthetic, branched macromolecules with a well-defined chemical structure (Fig. 2.2d), consisting of an initiator core and multiple layers with active terminal groups [9, 58]. Their specific molecular structure enables dendrimers to carry various drugs via covalent conjugation to the multivalent surfaces or encapsulation in the cavities of the cores through hydrophobic interaction, hydrogen bond, or chemical linkage [59, 60]. Besides, dendrimers can also carry bioactive macromolecules such as DNA by condensing them through electrostatic interactions [61]. The rigidity and the density of the branched units of dendrimers affect drug release kinetics. By use of pH- or enzyme-sensitive linkages, stimulus-responsive dendrimers can be generated [62].

Dendrimers are emerging as an important class of nanoparticle carriers for therapeutic delivery. For example, SPL7013 (L-lysine-based dendrimer) can be used in delivering microbicide for prevention of HIV and other sexually transmitted infections (STI) [63]. Frechet *et al.* have developed a biodegradable polyester dentritic drug delivery system with different architectures and molecular weights, for the delivery of doxorubicin [64]. Dendrimers composed of poly(amidoamine) (PAMAM) polymers have also been extensively investigated for the effective delivery of small molecular drugs [65]. Besides, the cationic nature of PAMAM, dendrimers allow them to effectively deliver macromolecular drugs such as DNA across cellular and subcellular barriers (*e.g.*, cell membrane and endosome) [66]. Attaching targeting ligands to their surface could further enhance the potential of PAMAM dendrimers in drug delivery. A case in hand is a folate-conjugated, methotrexate-loaded PAMAM(G5) dendrimer, which has demonstrated a tenfold reduction in tumor size and exhibited less systemic toxicity, compared to free methotrexate [67].

2.3 Optimal Design of Nanoparticles

One significant challenge for the successful development of therapeutic nanoparticles is rapid clearance during systemic delivery. When nanoparticles enter the bloodstream, the particle surface may experience nonspecific protein adsorption (opsonization), thereby making them more visible to phagocytic cells [67–69]. After opsonization, nanoparticles could be rapidly cleared from the bloodstream through phagocytosis by the mononuclear phagocyte system (MPS) in the liver and by spleen filtration [70, 71]. Therefore, the factors that could affect the clearance and biodistribution of nanoparticles, such as particle physicochemical properties and targeting ligand functionalization [68], should be carefully considered for the optimal design of therapeutic nanoparticles.

2.3.1 Size

On the basis of physiological parameters such as hepatic filtration, tissue extravasation/diffusion, and kidney excretion, it is clear that particle size plays a key factor in the long circulation and biodistribution of nanoparticles. Nanoparticles smaller than 10 nm can be rapidly cleared by the kidneys or through extravasation, while larger nanoparticles may have higher tendency to be cleared by cells of the mononuclear phagocyte system (MPS also referred to as reticuloendothelial system, RES) [4]. For example, in vivo biodistribution results of polystyrene nanoparticles with consistent composition and varying particle size of 50 and 500 nm showed higher level of agglomeration of the larger nanoparticles in the liver [72]. Another study compared different size ranges of PEGylated spherical nanoparticles (<100 nm, 100–200 nm, and >200 nm) for protein absorption, nanoparticle uptake by murine macrophages, and blood clearance kinetics [73]. It was observed that nanoparticles <100 nm have a higher potential to circulate in the blood for long periods of time and experience reduced hepatic filtration. Nanoparticle size also plays a key role in tumor accumulation through the EPR effect. Several studies have tried to determine the gap size in the leaky vasculature. For example, sterically stabilized liposomes of 100–600 nm were used for transvascular transport, and the cutoff size of the pores was estimated to be 400–600 nm in diameter [74]. In another study, the pore cutoff size was estimated to be between 7 and 100 nm at 34°C and was increased to >400 nm at 42°C, allowing all nanoparticles tested (~7 nm albumin, and 100, 200, and 400 nm liposomes) to be delivered to the tumor interstitium to some degree [75]. Therefore, to capitalize on the EPR effect and to efficiently escape from the physiological barriers, many studies advocate the optimal nanoparticle size range of approximately 10–250 nm [68].

2.3.2 Surface Charge

It has been established that the surface charge of nanoparticles also could affect their uptake by the MPS cells. Neutrally charged particles have demonstrated much lower opsonization rates than charged particles [76, 77]. It was found that positively charged nanoparticles generate a higher immune response (complement activation and conjugate activation) compared to neutral or negatively charged nanoparticle formulations [53]. For example, nanoparticles with a primary amine at the surface promote higher rates of phagocytic uptake when compared to those having sulfate, hydroxyl, or carboxyl groups at the surface [53, 68]. In a review study, Davis *et al.* have proposed that the optimal range of nanoparticle surface charge should be between −10 and +10 mV for reduced phagocytosis and minimized nonspecific interactions of nanoparticles [78].

2.3.3 PEGylation

Surface modification of nanoparticles with PEG, which has favorable intrinsic physicochemical properties (*e.g.*, high flexibility and hydrophilicity, and low toxicity and immunogenicity), was found to reduce nanoparticle accumulation in off-target organs such as liver and spleen [79]. A PEG shell on the nanoparticle surface shields hydrophobic or charged particles from attachment by blood proteins, leading to prolonged circulation half-life compared to non-PEGylated nanoparticles [25, 80]. The length, shape, and density of PEG chains on the nanoparticle surface largely affect its surface hydrophilicity and phagocytosis [81]. For example, at low PEG surface density, the PEG chains would be closer to the surface of the nanoparticle with a "mushroom" configuration, while as the density increases, most of the chains are extended away from the surface in a "brush" configuration, which decides the thickness of the PEG shell on the nanoparticle corona [69]. It has been postulated that the brush configuration would create more effective blocking or repulsion of opsonins than the mushroom one [80]. In addition to PEG, some other promising hydrophilic polymers are under investigation for the same purpose, including natural polymers (*e.g.*, heparin, dextran, and chitosan) and synthetic polymers (*e.g.*, poly(amino acids), poly(glycerols), poly(2-oxazolines), and some vinyl polymers) [79, 82].

2.3.4 Ligand Functionalization

The conjugation of targeting ligands to the surface of PEGylated nanoparticles has also been shown to affect their biodistribution [83]. Although targeting ligands could improve the cell- or tissue-specific delivery of nanoparticles, they may compromise the particle surface properties by masking the PEG layer and adversely affecting the nanoparticles' antibiofouling properties in vivo. Our recent study on the effect of ligand density has also revealed a relatively narrow window of ligand density that could result in favorable tumor targeting, while minimizing nanoparticle accumulation in the liver and spleen [50]. Thus, the successful development of targeted nanoparticle technology for efficient drug delivery strongly depends on striking a balance between cellular targeting and immune evasion.

2.4 Targeting Ligands

Despite their enormous potential for drug delivery, the translation of targeted nanoparticle systems has faced considerable challenges, and only a handful of candidates have made it to clinical trials (Table 2.2). The reason targeted nanoparticles have demonstrated limited success in clinical development is complex and could be multifaceted [3]. Among others, an essential aspect for the successful development of targeted nanoparticles relies on the choice of targeting ligands. Several variables that could be considered include ligand biocompatibility, cell specificity, binding affinity, and purity of the ligand [84]. Other important factors that have to be taken into account are the size and charge of the ligand molecule, and their ease of modification and conjugation to the nanoparticles. The choice of ligand, from a practical perspective, is also dependent on production cost, scalability, and stability (*e.g.*, organic solvent and high temperature stability) in mass production. In this section, we discuss five different classes of targeting ligands, including antibodies and antibody fragments, aptamers, peptides, sugars, and small molecules.

2.4.1 Antibodies and Antibody Fragments

Antibodies and antibody fragments form an important class of targeting ligands with a high degree of specificity for cellular receptors and a wide range of binding affinities and have been extensively investigated in targeted drug delivery [85]. Over the past 2 decades, the feasibility of antibody-based tissue targeting has been clinically demonstrated with several different monoclonal antibodies (mAbs) approved by the FDA [86]. The recent advances in hybridoma technology have led to the development of chimeric, humanized, and fully human mAbs to reduce their immunogenicity. The ability of engineered mAbs to target disease processes has been demonstrated by the success of several monoclonal antibody therapeutics, including cetuximab rituximab, trastuzumab, and bevacizumab [19]. mAbs have been used to direct the nanoparticle carriers in a site-specific manner. For example, mAb-conjugated PLA nanoparticles exhibited a sixfold increase in the rate of particle uptake compared with nontargeted particles [87, 88]. Additionally, J591, a mAb against PSMA, was conjugated to G5-PAMAM dendrimers and showed enhanced binding affinity for LNCaP cells, as compared to nontarget PC3 cells [89]. Nevertheless, mAb-conjugated nanoparticles encounter considerable challenges and limitations for drug delivery, since mAb are complex and large (~150 kDa) molecules and require significant engineering at the molecular level to be effective [90, 91].

Compared to mAbs, antibody fragments have demonstrated higher potential for the engineering of targeted nanoparticles as they are smaller in size and lack the complement activation region of mAbs, while retaining the antigen binding specificity [92]. Recent advances in protein engineering have led to the development of antibody fragments such as scFv (single-chain variable fragments), Fab (fragments of antigen binding), their dimers ($F(ab')_2$ and diabody), and recombinant products [93]. Some pioneering examples of antibody fragment-targeted liposomes (immunoliposomes) in clinical trials include MCC-465 that uses $F(ab')_2$ for the targeted delivery of doxorubicin [28, 94] and SGT-53 that uses scFv to deliver tumor suppressor gene, p53 [95].

2.4.2 Aptamers

Nucleic acid aptamers are single-stranded DNA or RNA oligonucleotides with well-defined, three-dimensional structures. Selected by systematic evolution of ligands by exponential enrichment (SELEX), aptamers can recognize a wide variety of molecules (*e.g.*, proteins, phospholipids, sugars, and nucleic acids) with high affinity and specificity [96–98]. This SELEX process uses the concepts of evolution, diversification, selection, and replication, where a library of ~10^{15} random oligonucleotides is enriched to identify specific aptamers that can specifically recognize the target [97]. Aptamers identified through the SELEX process can be chemically synthesized with minimal batch-to-batch variation in a fast and cost-effective manner. When compared with antibodies, aptamers exhibit lower immunogenicity and a relatively smaller size compared with ~150 kD for antibodies, which enables better tissue penetration [99–101]. To further improve on their low serum stability, aptamers can be modified by incorporating 2′-amino, 2′-fluoro, or 2′-*O*-alkyl nucleotides in their backbone [102].

To date, more than 200 aptamers against a variety of biological targets have been isolated, such as cell surface antigens, therapeutic targets, and various growth factors like VEGF [103, 104]. Most notably, the FDA approved an aptamer against $VEGF_{165}$, known as Pegaptanib, for the treatment of age-related macular degeneration (AMD) [105]. The PSMA-specific aptamers have been widely used for the targeted delivery of quantum dots [106], gold nanoparticles [107], and polymeric nanoparticles [14, 46, 108]. We have recently tested A10 aptamer-conjugated PLGA–PEG nanoparticles for targeted drug delivery using prostate cancer model. These PLGA–PEG-Apt nanoparticles can enhance the therapeutic effect of anticancer drugs and reduce systematic toxicity when compared to nontargeted nanoparticles [109]. More recently DNA aptamers, generated through cell-SELEX, have been conjugated with different types of nanoparticles (*e.g.*, magnetic and gold nanoparticles) for cancer detection and treatment [110–112].

2.4.3 Peptides

Peptide ligands have shown significant targeting potential because of their small size, high stability, and relative ease of large-scale synthesis with excellent quality control. The development of phage display techniques [113, 114] and other screening methods has enabled the discovery of new peptide-targeting domains and the isolation of new cell-specific peptide ligands [115, 116]. Peptide-conjugated nanoparticles have been widely used for targeting cancer cells and tumor vasculature [117, 118]. For example, the peptide SP5-52 can recognize tumor neovasculature, while avoiding normal blood vessels in severe, combined immunodeficiency mice bearing human tumors. The SP5-52 peptide-linked liposome has shown to greatly enhance the therapeutic effect of doxorubicin, decrease the growth of tumor blood vessels, and enable high survival rates among human lung and oral cancer-bearing xenograft mice [119]. Recently, this system has been used to target non-small-cell lung cancer (NSCLC) cells and demonstrated increased drug accumulation in tumor tissues by 5.7-fold compared with free drugs [120].

In the case of targeting integrin receptors (e.g., $\alpha_v\beta_3$ and $\alpha_v\beta_5$), short peptide antagonists have been developed based on a 2-benzazepine Gly–Asp mimetic or screened from an Arg–Gly–Asp-based (RGD) peptidomimetic library [121, 122]. For example, the cyclic version of the RGD motif has demonstrated effective binding toward integrins and has been extensively investigated in targeting nanoparticles for disrupting tumor angiogenesis [123]. For intra-articular targeting and retention in cartilage, the peptide ligand WYRGRL against collagen, type II, α1 (COL2A1) was used in the targeted delivery of polymeric nanoparticles [15]. We have recently screened specific targeting peptides against collagen, type IV of the basement membrane and conjugated to lipid-coated polymeric nanoparticles for vascular wall targeting [13].

2.4.4 Sugars

Specific sugar molecules (*e.g.*, lactose, galactose, and mannose) can recognize lectins that are overexpressed on the surface of numerous cancer cells [124, 125]. Thus, sugar molecules represent another interesting approach to specifically target nanoparticle systems to cancer cells. For example, galactose could recognize the asialoglycoprotein receptor which is expressed on hepatocytes, and its high expression is retained on primary liver cancer cells [125]. The galactosamine-conjugated *N*-(2-hydroxypropyl) methacrylamide copolymers (HPMA) (PK2) is currently under clinical evaluation for the treatment of primary liver cancer [8]. In another study, lectin-mediated endocytosis of sugar-conjugated HPMA copolymer conjugates in three different human colon cancer cell lines suggested their potential use for targeted delivery of chemotherapeutics to colon adenocarcinoma [126]. However, to compensate for the weak binding affinity of carbohydrates, multiple or multivalent molecules should be conjugated to the surface of nanoparticles to achieve multivalent interactions. In the case of galactosylated liposomal carriers, it was shown that the targeting efficacy depended on the galactose ligand density [127].

2.4.5 *Small Molecules*

Small molecules have also attracted considerable attention as potential targeting ligands due to their low molecular weights, low production costs, and easy conjugation with nanoparticles. The small size of this kind of targeting ligand allows the functionalization of multiple ligand molecules on single nanoparticles. Folic acid, which is essential in many metabolic processes for cell survival, has shown high specificity in recognizing folate receptors that are overexpressed in many types of tumor cells [128]. There are several examples of folate-conjugated nanoparticles in drug delivery [129, 130], including liposomes, polymeric nanoparticles, and dendrimers [67, 131–134]. These nanoparticles have demonstrated to be effective in treating ovarian, breast, lung, renal, and colon cancers [135, 136]. However, immunochemistry studies have shown overexpression of folate receptors in normal tissues such as the placenta and kidneys as well, raising some concerns for the translation of folate-targeted nanoparticles from bench to bedside.

The development of small molecule-targeting ligands that demonstrate a high affinity and specificity toward cellular receptors has proven to be a challenging task. One strategy to improve the targeting of small molecule-conjugated nanoparticles is through multivalent binding effects, by conjugating multiple ligands on the nanoparticle surface. Another strategy is to select small molecules with high affinity and specificity by using high-throughput screening methods. For example, using fluorescent magnetic nanoparticles, Weissleder *et al.* have recently screened several small molecular ligands from a library of 146 small molecules (≤500 Da), which can specifically bind to endothelial cells, activated human macrophages, and pancreatic cancer cells, respectively [137].

2.5 Conclusions

The application of nanoparticle technologies to drug delivery has demonstrated significant impact on many areas of medicine. The approval of more than two-dozen therapeutic nanoparticle products for clinical use has generated great enthusiasm in both academia and industry, although these first-generation nanoparticle therapeutics are relatively simple and only provide clinical benefits across a narrow range of clinically validated drugs. Toward the development of next-generation nanoparticles, the introduction of controlled release properties and targeting ligands is expected to enable the development of safer and more effective therapeutic nanoparticles. With continuous advances in identifying new biomarkers and associated targeting ligands, and in engineering nanoparticle delivery systems with optimal biophysicochemical properties, it will be increasingly feasible to develop targeted and controlled release nanoparticle products as promising candidates for clinical translation.

References

1. Allen TM, Cullis PR (2004) Drug delivery systems: entering the mainstream. Science 303:1818–1822
2. Wagner V, Dullaart A, Bock A-K, Zweck A (2006) The emerging nanomedicine landscape. Nat Biotechnol 24(10):1211–1217
3. Farokhzad OC, Langer R (2009) Impact of nanotechnology on drug delivery. ACS Nano 3(1):16–20
4. Petros RA, DeSimone JM (2010) Strategies in the design of nanoparticles for therapeutic applications. Nat Rev Drug Discov 9(8):615–627
5. Shi J, Votruba AR, Farokhzad OC, Langer R (2010) Nanotechnology in drug delivery and tissue engineering: from discovery to applications. Nano Lett 10(9):3223–3230
6. Cai W, Chen X (2007) Nanoplatforms for targeted molecular imaging in living subjects. Small 3(11):1840–1854
7. Gao X et al (2005) In vivo molecular and cellular imaging with quantum dots. Curr Opin Biotechnol 16(1):63–72
8. Davis ME, Chen Z, Shin DM (2008) Nanoparticle therapeutics: an emerging treatment modality for cancer. Nat Rev Drug Discov 7(9):771–782
9. Zhang L et al (2007) Nanoparticles in medicine: therapeutic applications and developments. Clin Pharmacol Ther 83(5):761–769
10. Heath T, Fraley R, Papahdjopoulos D (1980) Antibody targeting of liposomes: cell specificity obtained by conjugation of F(ab′)2 to vesicle surface. Science 210:539–541
11. Leserman LD, Barbet J, Kourilsky F, Weinstein JN (1980) Targeting to cells of fluorescent liposomes covalently coupled with monoclonal antibody or protein A. Nature 288:602–604
12. Maeda H (2001) The enhanced permeability and retention (EPR) effect in tumor vasculature: the key role of tumor-selective macromolecular drug targeting. Adv Enzyme Regul 41:189–207
13. Chan JM et al (2010) Spatiotemporal controlled delivery of nanoparticles to injured vasculature. Proc Natl Acad Sci USA 107(5):2213–2218
14. Dhar S, Kolishetti N, Lippard SJ, Farokhzad OC (2011) Targeted delivery of a cisplatin prodrug for safer and more effective prostate cancer therapy in vivo. Proc Natl Acad Sci USA 108(5):1850–1855
15. Rothenfluh DA, Bermudez H, O'Neil CP, Hubbell JA (2008) Biofunctional polymer nanoparticles for intra-articular targeting and retention in cartilage. Nat Mater 7(3):248–254
16. Georgieva JV et al (2011) Surface characteristics of nanoparticles determine their intracellular fate in and processing by human blood–brain barrier endothelial cells in vitro. Mol Ther 19(2):318–325
17. Farokhzad OC, Langer R (2006) Nanomedicine: developing smarter therapeutic and diagnostic modalities. Adv Drug Deliv Rev 58(14):1456–1459
18. Brigger I, Dubernet C, Couvreur P (2002) Nanoparticles in cancer therapy and diagnosis. Adv Drug Deliv Rev 54(5):631–651
19. Wang AZ et al (2008) Biofunctionalized targeted nanoparticles for therapeutic applications. Expert Opin Biol Ther 8(8):1063–1070
20. Ganta S, Devalapally H, Shahiwala A, Amiji M (2008) A review of stimuli-responsive nanocarriers for drug and gene delivery. J Control Release 126(3):187–204
21. Kale AA, Torchilin VP (2010) Environment-responsive multifunctional liposomes. Methods Mol Biol 605:213–242
22. Oh KT, Yin H, Lee ES, Bae YH (2007) Polymeric nanovehicles for anticancer drugs with triggering release mechanisms. J Mater Chem 17(38):3987–4001
23. Moghimi SM, Hunter AC, Murray JC (2005) Nanomedicine: current status and future prospects. FASEB J 19(3):311–330
24. Antimisiaris SG, Kallinteri P, Fatouros DG (2007) Liposomes and drug delivery. Wiley, New York, pp 443–533

25. Moghimi SM, Szebeni J (2003) Stealth liposomes and long circulating nanoparticles: critical issues in pharmacokinetics, opsonization and protein-binding properties. Prog Lipid Res 42(6):463–478
26. Chowdhery R, Gonzalez R (2011) Immunologic therapy targeting metastatic melanoma: Allovectin-7. Immunotherapy 3(1):17–21
27. Matsumura Y et al (2004) Phase I and pharmacokinetic study of MCC-465, a doxorubicin (DXR) encapsulated in PEG immunoliposome, in patients with metastatic stomach cancer. Ann Oncol 15(3):517–525
28. Sankhala KK, Mita AC, Adinin R, Wood L, Beeram M, Bullock S, Yamagata N, Matsuno K, Fujisawa T, Phan AT (2009) A phase I pharmacokinetic (PK) study of MBP-426, a novel liposome encapsulated oxaliplatin. J Clin Oncol 27(15S):2535
29. Lukyanov AN, Elbayoumi TA, Chakilam AR, Torchilin VP (2004) Tumor-targeted liposomes: doxorubicin-loaded long-circulating liposomes modified with anti-cancer antibody. J Control Release 100(1):135–144
30. Schiffelers RM et al (2003) Anti-tumor efficacy of tumor vasculature-targeted liposomal doxorubicin. J Control Release 91(1–2):115–122
31. Pan XQ et al (2002) Strategy for the treatment of acute myelogenous leukemia based on folate receptor beta-targeted liposomal doxorubicin combined with receptor induction using all-trans retinoic acid. Blood 100(2):594–602
32. Torchilin VP (2005) Recent advances with liposomes as pharmaceutical carriers. Nat Rev Drug Discov 4(2):145–160
33. Greco F, Vicent MJ (2009) Combination therapy: opportunities and challenges for polymer-drug conjugates as anticancer nanomedicines. Adv Drug Deliv Rev 61(13):1203–1213
34. Bae Y et al (2004) Preparation and biological characterization of polymeric micelle drug carriers with intracellular pH-triggered drug release property: tumor permeability, controlled subcellular drug distribution, and enhanced in vivo antitumor efficacy. Bioconjug Chem 16(1):122–130
35. Chan JM, Valencia PM, Zhang L, Langer R, Farokhzad OC (2010) Polymeric nanoparticles for drug delivery. Methods Mol Biol 624:163–175
36. Napier ME, DeSimone JM (2007) Nanoparticle drug delivery platform. Polym Rev 47(3):321–327
37. Matsumura Y, Kataoka K (2009) Preclinical and clinical studies of anticancer agent-incorporating polymer micelles. Cancer Sci 100(4):572–579
38. Sutton D, Nasongkla N, Blanco E, Gao J (2007) Functionalized micellar systems for cancer targeted drug delivery. Pharm Res 24(6):1029–1046
39. Matsumura Y, Phase I (2004) Clinical trial and pharmacokinetic evaluation of NK911, a micelle-encapsulated doxorubicin. Br J Cancer 91:1775–1781
40. Hamaguchi T et al (2005) NK105, a paclitaxel-incorporating micellar nanoparticle formulation, can extend in vivo antitumour activity and reduce the neurotoxicity of paclitaxel. Br J Cancer 92(7):1240–1246
41. Wilson RHP, Adam R, Eatock J, Boddy MM, Griffin AV, Miller MR, Matsumura Y, Shimizu T, Calvert V (2008) Phase I and pharmacokinetic study of NC-6004, a new platinum entity of cisplatin-conjugated polymer forming micelles. J Clin Oncol (Meeting Abstracts) 26:2573
42. Hamaguchi T et al (2010) Phase I study of NK012, a novel SN-38-incorporating micellar nanoparticle, in adult patients with solid tumors. Clin Cancer Res 16(20):5058–5066
43. Kim T-Y et al (2004) Phase I and pharmacokinetic study of Genexol-PM, a cremophor-free, polymeric micelle-formulated paclitaxel, in patients with advanced malignancies. Clin Cancer Res 10(11):3708–3716
44. Lee KS et al (2006) Multicenter phase II study of a cremophor-free polymeric micelle-formulated paclitaxel in patients with metastatic breast cancer (MBC). J Clin Oncol (Meeting Abstracts) 24(18_suppl):10520
45. Alexis F et al (2008) HER-2-targeted nanoparticle-affibody bioconjugates for cancer therapy. ChemMedChem 3(12):1839–1843
46. Farokhzad OC et al (2006) Targeted nanoparticle-aptamer bioconjugates for cancer chemotherapy in vivo. Proc Natl Acad Sci USA 103(16):6315–6320

47. Gao W, Chan JM, Farokhzad OC (2010) pH-responsive nanoparticles for drug delivery. Mol Pharm 7(6):1913–1920
48. Zhang L et al (2007) Co-delivery of hydrophobic and hydrophilic drugs from nanoparticle-aptamer bioconjugates. ChemMedChem 2(9):1268–1271
49. Farokhzad OC et al (2004) Nanoparticle-aptamer bioconjugates: a new approach for targeting prostate cancer cells. Cancer Res 64(21):7668–7672
50. Gu F et al (2008) Precise engineering of targeted nanoparticles by using self-assembled biointegrated block copolymers. Proc Natl Acad Sci USA 105(7):2586–2591
51. Service RF (2010) Nanoparticle Trojan horses gallop from the lab into the clinic. Science 330:314–315
52. Chan JM et al (2009) PLGA-lecithin-PEG core-shell nanoparticles for controlled drug delivery. Biomaterials 30(8):1627–1634
53. Salvador-Morales C, Zhang L, Langer R, Farokhzad OC (2009) Immunocompatibility properties of lipid-polymer hybrid nanoparticles with heterogeneous surface functional groups. Biomaterials 30(12):2231–2240
54. Valencia PM et al (2010) Single-step assembly of homogenous lipid-polymeric and lipid-quantum dot nanoparticles enabled by microfluidic rapid mixing. ACS Nano 4(3):1671–1679
55. Wang AZ et al (2010) ChemoRad nanoparticles: a novel multifunctional nanoparticle platform for targeted delivery of concurrent chemoradiation. Nanomedicine 5(3):361–368
56. Zhang L et al (2008) Self-assembled lipid-polymer hybrid nanoparticles: a robust drug delivery platform. ACS Nano 2(8):1696–1702
57. Sengupta S et al (2005) Temporal targeting of tumour cells and neovasculature with a nanoscale delivery system. Nature 436:568–572
58. Paleos CM, Tsiourvas D, Sideratou Z, Tziveleka LA (2010) Drug delivery using multifunctional dendrimers and hyperbranched polymers. Expert Opin Drug Deliv 7(12):1387–1398
59. Lee CC, MacKay JA, Frechet JMJ, Szoka FC (2005) Designing dendrimers for biological applications. Nat Biotech 23(12):1517–1526
60. Liu M, Kono K, Frechet JMJ (2000) Water-soluble dendritic unimolecular micelles: their potential as drug delivery agents. J Control Rel 65(1–2):121–131
61. Xu Q, Wang CH, Pack DW (2010) Polymeric carriers for gene delivery: chitosan and poly(amidoamine) dendrimers. Curr Pharm Des 16(21):2350–2368
62. Gillies ER, Jonsson TB, Frechet JMJ (2004) Stimuli-responsive supramolecular assemblies of linear-dendritic copolymers. J Am Chem Soc 126(38):11936–11943
63. McCarthy TD et al (2005) Dendrimers as drugs: discovery and preclinical and clinical development of dendrimer-based microbicides for HIV and STI prevention. Mol Pharm 2(4):312–318
64. Padilla De Jesus OL, Ihre HR, Gagne L, Frechet JMJ, Szoka FC Jr (2002) Polyester dendritic systems for drug delivery applications: in vitro and in vivo evaluation. Bioconjug Chem 13(3):453–461
65. Patri AK, Kukowska-Latallo JF, Baker JR (2005) Targeted drug delivery with dendrimers: comparison of the release kinetics of covalently conjugated drug and non-covalent drug inclusion complex. Adv Drug Deliv Rev 57(15):2203–2214
66. Yellepeddi VK, Kumar A, Palakurthi S (2009) Surface modified poly(amido)amine dendrimers as diverse nanomolecules for biomedical applications. Expert Opin Drug Deliv 6(8):835–850
67. Kukowska-Latallo JF et al (2005) Nanoparticle targeting of anticancer drug improves therapeutic response in animal model of human epithelial cancer. Cancer Res 65(12):5317–5324
68. Alexis F, Pridgen E, Molnar LK, Farokhzad OC (2008) Factors affecting the clearance and biodistribution of polymeric nanoparticles. Mol Pharm 5(4):505–515
69. Owens DE III, Peppas NA (2006) Opsonization, biodistribution, and pharmacokinetics of polymeric nanoparticles. Int J Pharm 307(1):93–102
70. Eisenstein M (2006) Protein arrays: growing pains. Nature 444:959–962
71. Ostuni E, Chapman RG, Holmlin RE, Takayama S, Whitesides GM (2001) A survey of structure–property relationships of surfaces that resist the adsorption of protein. Langmuir 17(18):5605–5620

72. Nagayama S, Ogawara K-I, Fukuoka Y, Higaki K, Kimura T (2007) Time-dependent changes in opsonin amount associated on nanoparticles alter their hepatic uptake characteristics. Int J Pharm 342(1–2):215–221
73. Fang C et al (2006) In vivo tumor targeting of tumor necrosis factor-[alpha]-loaded stealth nanoparticles: effect of mPEG molecular weight and particle size. Eur J Pharm Sci 27(1):27–36
74. Yuan F et al (1995) Vascular permeability in a human tumor xenograft: molecular size dependence and cutoff size. Cancer Res 55(17):3752–3756
75. Kong G, Braun RD, Dewhirst MW (2000) Hyperthermia enables tumor-specific nanoparticle delivery: effect of particle size. Cancer Res 60(16):4440–4445
76. Roser M, Fischer D, Kissel T (1998) Surface-modified biodegradable albumin nano- and microspheres. II: effect of surface charges on in vitro phagocytosis and biodistribution in rats. Eur J Pharm Biopharm 46(3):255–263
77. Schwendener RA, Lagocki PA, Rahman YE (1984) The effects of charge and size on the interaction of unilamellar liposomes with macrophages. Biochim Biophys Acta (BBA) - Biomembranes 772(1):93–101
78. Davis ME (2009) The first targeted delivery of siRNA in humans via a self-assembling, cyclodextrin polymer-based nanoparticle: from concept to clinic. Mol Pharm 6(3):659–668
79. Knop K, Hoogenboom R, Fischer D, Schubert US (2010) Poly(ethylene glycol) in drug delivery: pros and cons as well as potential alternatives. Angew Chem Int Ed 49(36):6288–6308
80. Vonarbourg A, Passirani C, Saulnier P, Benoit J-P (2006) Parameters influencing the stealthiness of colloidal drug delivery systems. Biomaterials 27(24):4356–4373
81. Gref R et al (2000) 'Stealth' corona-core nanoparticles surface modified by polyethylene glycol (PEG): influences of the corona (PEG chain length and surface density) and of the core composition on phagocytic uptake and plasma protein adsorption. Colloids Surf B Biointerfaces 18(3–4):301–313
82. Moghimi SM, Porter CJH, Illum L, Davis SS (1991) The effect of Poloxamer-407 on liposome stability and targeting to bone marrow: comparison with polystyrene microspheres. Int J Pharm 68(1–3):121–126
83. Takae S et al (2005) Ligand density effect on biorecognition by PEGylated gold nanoparticles: regulated interaction of RCA120 lectin with lactose installed to the distal end of tethered PEG strands on gold surface. Biomacromolecules 6(2):818–824
84. Allen TM (2002) Ligand-targeted therapeutics in anticancer therapy. Nat Rev Cancer 2(10):750–763
85. Torchilin VP (2008) Antibody-modified liposomes for cancer chemotherapy. Expert Opin Drug Deliv 5:1003–1025
86. Gabizon AA (2001) Pegylated liposomal doxorubicin: metamorphosis of an old drug into a new form of chemotherapy. Cancer Invest 19(4):424–436
87. Nellis DF et al (2005) Preclinical manufacture of an anti-HER2 scFv-PEG-DSPE, liposome-inserting conjugate. 1. Gram-scale production and purification. Biotechnol Prog 21(1):205–220
88. Nobs L, Buchegger F, Gurny R, Allemann E (2004) Poly(lactic acid) nanoparticles labeled with biologically active neutravidin for active targeting. Eur J Pharm Biopharm 58(3):483–490
89. Patri AK et al (2004) Synthesis and in vitro testing of J591 antibody-dendrimer conjugates for targeted prostate cancer therapy. Bioconjug Chem 15(6):1174–1181
90. Brennan FR, Shaw L, Wing MG, Robinson C (2004) Preclinical safety testing of biotechnology-derived pharmaceuticals: understanding the issues and addressing the challenges. Mol Biotechnol 27(1):59–74
91. Weinberg WC et al (2005) Development and regulation of monoclonal antibody products: challenges and opportunities. Cancer Metastasis Rev 24(4):569–584
92. Carter P (2001) Improving the efficacy of antibody-based cancer therapies. Nat Rev Cancer 1(2):118–129
93. Pavlinkova G et al (2001) Effects of humanization and gene shuffling on immunogenicity and antigen binding of anti-TAG-72 single-chain Fvs. Int J Cancer 94(5):717–726

94. Mebiopharm Co., Ltd (2009) Safety study of MBP-426 (liposomal oxaliplatin suspension for injection) to treat advanced or metastatic solid tumors. http://clinicaltrials.gov/ct2/show/NCT00355888. Accessed on May 8, 2011
95. SynerGene Therapeutics, Inc. (2010) Safety study of infusion of SGT-53 to treat solid tumors. http://clinicaltrials.gov/ct2/show/NCT00470613. Accessed on May 8, 2011
96. Ellington AD, Szostak JW (1990) In vitro selection of RNA molecules that bind specific ligands. Nature 346:818–822
97. Fang X, Tan W (2010) Aptamers generated from cell-SELEX for molecular medicine: a chemical biology approach. Acc Chem Res 43(1):48–57
98. Tuerk C, Gold L (1990) Systematic evolution of ligands by exponential enrichment: RNA ligands to bacteriophage T4 DNA polymerase. Science 249:505–510
99. Farokhzad OC, Karp JM, Langer R (2006) Nanoparticle-aptamer bioconjugates for cancer targeting. Expert Opin Drug Deliv 3(3):311–324
100. Levy-Nissenbaum E, Radovic-Moreno AF, Wang AZ, Langer R, Farokhzad OC (2008) Nanotechnology and aptamers: applications in drug delivery. Trends Biotechnol 26(8):442–449
101. Nimjee SM, Rusconi CP, Sullenger BA (2005) Aptamers: an emerging class of therapeutics. Annu Rev Med 56:555–583
102. Potti A, Rusconi CP, Sullenger BA, Ortel TL (2004) Regulatable aptamers in medicine: focus on antithrombotic strategies. Expert Opin Biol Ther 4(10):1641–1647
103. Shangguan D et al (2006) Aptamers evolved from live cells as effective molecular probes for cancer study. Proc Natl Acad Sci USA 103(32):11838–11843
104. Daniels DA, Chen H, Hicke BJ, Swiderek KM, Gold L (2003) A tenascin-C aptamer identified by tumor cell SELEX: systematic evolution of ligands by exponential enrichment. Proc Natl Acad Sci USA 100(26):15416–15421
105. Ng EW et al (2006) Pegaptanib, a targeted anti-VEGF aptamer for ocular vascular disease. Nat Rev Drug Discov 5(2):123–132
106. Bagalkot V et al (2007) Quantum dot-aptamer conjugates for synchronous cancer imaging, therapy, and sensing of drug delivery based on bi-fluorescence resonance energy transfer. Nano Lett 7(10):3065–3070
107. Kim D, Jeong YY, Jon S (2010) A drug-loaded aptamer—gold nanoparticle bioconjugate for combined CT imaging and therapy of prostate cancer. ACS Nano 4(7):3689–3696
108. Kolishetti N et al (2010) Engineering of self-assembled nanoparticle platform for precisely controlled combination drug therapy. Proc Natl Acad Sci USA 107(42):17939–17944
109. Cheng JJ et al (2007) Formulation of functionalized PLGA-PEG nanoparticles for in vivo targeted drug delivery. Biomaterials 28(5):869–876
110. Chiu T-C, Huang C-C (2009) Aptamer-functionalized nano-biosensors. Sensors 9(12):10356–10388
111. Delehanty JB, Boeneman K, Bradburne CE, Robertson K, Medintz IL (2009) Quantum dots: a powerful tool for understanding the intricacies of nanoparticle-mediated drug delivery. Expert Opin Drug Deliv 6(10):1091–1112
112. Huang Y-F, Sefah K, Bamrungsap S, Chang H-T, Tan W (2008) Selective photothermal therapy for mixed cancer cells using aptamer-conjugated nanorods. Langmuir 24(20):11860–11865
113. Lam KS et al (1991) A new type of synthetic peptide library for identifying ligand-binding activity. Nature 354:82–84
114. Needels MC et al (1993) Generation and screening of an oligonucleotide-encoded synthetic peptide library. Proc Natl Acad Sci USA 90(22):10700–10704
115. McGuire MJ, Li S, Brown KC (2009) Biopanning of phage displayed peptide libraries for the isolation of cell-specific ligands. Methods Mol Biol 504:291–321
116. Pasqualini R, Ruoslahti E (1996) Organ targeting in vivo using phage display peptide libraries. Nature 380:364–366
117. Arap W et al (2002) Steps toward mapping the human vasculature by phage display. Nat Med 8(2):121–127
118. Lam KS, Zhao ZG (1997) Targeted therapy for lymphoma with peptides. Hematol Oncol Clin North Am 11(5):1007–1019

119. Lee TY, Lin CT, Kuo SY, Chang DK, Wu HC (2007) Peptide-mediated targeting to tumor blood vessels of lung cancer for drug delivery. Cancer Res 67(22):10958–10965
120. Chang DK, Lin CT, Wu CH, Wu HC (2009) A novel peptide enhances therapeutic efficacy of liposomal anti-cancer drugs in mice models of human lung cancer. PLoS ONE 4(1):e4171
121. Li J et al (2004) Fusion protein from RGD peptide and Fc fragment of mouse immunoglobulin G inhibits angiogenesis in tumor. Cancer Gene Ther 11(5):363–370
122. Ruoslahti E, Pierschbacher M (1987) New perspectives in cell adhesion: RGD and integrins. Science 238:491–497
123. Danhier F et al (2009) Targeting of tumor endothelium by RGD-grafted PLGA-nanoparticles loaded with Paclitaxel. J Control Release 140(2):166–173
124. Ohannesian DW et al (1995) Carcinoembryonic antigen and other glycoconjugates act as ligands for galectin-3 in human colon carcinoma cells. Cancer Res 55(10):2191–2199
125. Zubieta MR et al (2006) Galectin-3 expression correlates with apoptosis of tumor-associated lymphocytes in human melanoma biopsies. Am J Pathol 168(5):1666–1675
126. David A, Kopeckova P, Kopecek J, Rubinstein A (2002) The role of galactose, lactose, and galactose valency in the biorecognition of N-(2-hydroxypropyl)methacrylamide copolymers by human colon adenocarcinoma cells. Pharm Res 19(8):1114–1122
127. Managit C, Kawakami S, Nishikawa M, Yamashita F, Hashida M (2003) Targeted and sustained drug delivery using PEGylated galactosylated liposomes. Int J Pharm 266(1–2):77–84
128. Ross JF, Chaudhuri PK, Ratnam M (1994) Differential regulation of folate receptor isoforms in normal and malignant tissues in vivo and in established cell lines. Physiologic and clinical implications. Cancer 73(9):2432–2443
129. Stella B et al (2000) Design of folic acid-conjugated nanoparticles for drug targeting. J Pharm Sci 89(11):1452–1464
130. Park EK, Lee SB, Lee YM (2005) Preparation and characterization of methoxy poly(ethylene glycol)/poly([epsilon]-caprolactone) amphiphilic block copolymeric nanospheres for tumor-specific folate-mediated targeting of anticancer drugs. Biomaterials 26(9):1053–1061
131. Liu Y, Li K, Pan J, Liu B, Feng S-S (2010) Folic acid conjugated nanoparticles of mixed lipid monolayer shell and biodegradable polymer core for targeted delivery of Docetaxel. Biomaterials 31(2):330–338
132. Ni S, Stephenson SM, Lee RJ (2002) Folate receptor targeted delivery of liposomal daunorubicin into tumor cells. Anticancer Res 22(4):2131–2135
133. Pan XQ, Wang H, Lee RJ (2003) Antitumor activity of folate receptor-targeted liposomal doxorubicin in a KB oral carcinoma murine xenograft model. Pharm Res 20(3):417–422
134. Stephenson SM et al (2003) Folate receptor-targeted liposomes as possible delivery vehicles for boron neutron capture therapy. Anticancer Res 23(4):3341–3345
135. Low PS, Henne WA, Doorneweerd DD (2007) Discovery and development of folic-acid-based receptor targeting for imaging and therapy of cancer and inflammatory diseases. Acc Chem Res 41(1):120–129
136. Zhao X, Li H, Lee RJ (2008) Targeted drug delivery via folate receptors. Expert Opin Drug Deliv 5(3):309–319
137. Weissleder R, Kelly K, Sun EY, Shtatland T, Josephson L (2005) Cell-specific targeting of nanoparticles by multivalent attachment of small molecules. Nat Biotechnol 23(11):1418–1423

Chapter 3
Development and Application of Anticancer Nanomedicine

Rong Tong, Li Tang, and Jianjun Cheng

3.1 Introduction: Development of Nanomedicine

There is growing interest in integrating nanotechnology with medicine, creating the so-called nanomedicine for disease diagnosis and treatment with unprecedented precision and efficacy [1]. Nanomedicines are drug- or imaging agent-containing carriers or devices with size ranging from a few to several hundred nanometers [2]. Although the term nanomedicine emerged only recently [1, 3], nanotechnology has been employed in drug delivery for decades [4]. In principle, nanomedicines are designed to enable the delivery of small molecules or macromolecular therapeutics to achieve improved disease treatment by circumventing various physiological barriers. The physiological barriers may prohibit the efficient permeation of nanomedicines with undesired sizes and surface properties. Therefore, there have been significant efforts on controlled formulation of nanomedicines. The majority of current nanotechnology platforms for chemotherapy have involved repackaging of traditional anticancer agents into various forms of nanometer-sized delivery vehicles, such as monomeric polymer–drug conjugates with sizes typically 10 nm or less [2], polymeric nanoparticles [5] or self-assembled amphiphilic block-copolymer micelles [6] in a size range of 20–100 nm, or lipid [7] and polymeric vesicles [8] (also known as liposomes and polymersomes, respectively) with sizes between sub-100 nm to submicrometers.

Liposomes are by far the most successful nanomedicine platform, accounting for 30–40% of nanomedicines that have been approved by the U.S. Food and Drug Administration (FDA) for their usage in the clinic [9]. The report of the first liposomal drug delivery system dates back to the 1960s [5]. Long-circulating liposomes using so-called stealth technique appeared in the literature in the 1980s [10].

R. Tong • L. Tang • J. Cheng (✉)
Department of Materials Science and Engineering, University of Illinois
at Urbana – Champaign, Urbana, IL 61801, USA
e-mail: jianjunc@illinois.edu

S. Svenson and R.K. Prud'homme (eds.), *Multifunctional Nanoparticles for Drug Delivery Applications: Imaging, Targeting, and Delivery*, Nanostructure Science and Technology,
DOI 10.1007/978-1-4614-2305-8_3,

Since then, in particular after the improvement of circulation profiles of liposome through the use of polyethylene glycol (PEG) [11], numerous liposomal nanomedicines have been developed and tested in the clinic, with a handful of them being approved by the FDA. For instance, Doxil®, a PEGylated liposomal doxorubicin, was approved by the FDA in 1995 for treating AIDS-associated Kaposi sarcoma, among other liposomes including Abelcet®, DaunoXome®, DepotDur®, and Ambisome® for treating cancer or other diseases [12]. Polymeric nanomedicine, a subfield of nanomedicine that involves the use of polymeric nanostructures as drug carriers, was first reported in the 1970s [13]. Since then, polymer-based nanomedicines have undergone many preclinical and clinical investigations. Abraxane®, a 130-nm paclitaxel/albumin polymeric nanoparticle, is one such example that has been approved by FDA as a second line treatment of breast cancer [14]. Currently, there are over three dozen nanomedicines approved for clinical use, and more are expected in the coming years [9, 15]. More than 50 companies are developing nanomedicine-based therapeutics or diagnostics for cancer therapy, 34 of which were established since 2006 [16].

The development of the abovementioned therapeutic nanomedicines has been mainly focused on targeting the primary tumors through the so-called enhanced permeation and retention (EPR) effect, a passive targeting mechanism that refers to the accumulation of nanomedicines in tumor tissue facilitated by the highly permeable nature of the tumor vasculature and poor lymphatic drainage of the interstitial fluid in the tumor [17]. The newer generation nanomedicines, however, place greater emphasis on novel strategies to bypass biological barriers at the systemic, tissue, and cellular levels and to locate and target metastatic lesions. New chemistries and fabrication technologies allow precise control of nanomedicine formulation, making it possible to evaluate nanomedicine with the variation of one parameter at a time (e.g., size, surface property, and shape), which provides insight into the fundamental understanding of the interplay of these parameters and the in vivo performance of the nanomedicines. Conjugation chemistry plays a vital role in controlling the incorporation of therapeutics or targeting ligands to nanomedicine. For instance, "click chemistry," a powerful conjugation approach conceived by Barry Sharpless, has become a highly recognized method in the field of nanomedicine.

3.2 In Vitro and In Vivo Studies of Nanomedicines

To achieve the accumulation of nanomedicines in tumor tissue, they must first overcome various systemic barriers, especially the clearance from the circulation system via phagocytic uptake and hepatic filtration. Nanomedicines are then expected to extravasate the tumor vasculature, penetrate the tumor microenvironment, and get internalized into the targeted cancer cells to allow cancer cells—even those situated distal to the tumor vessels—to be exposed to the anticancer agent with sufficiently high concentrations. The nanomedicines' size, shape, and surface property all have a significant impact on the efficiency of bypassing these physiological barriers.

Although the optimal size of nanomedicines for prolonged circulation half-life is still unclear, there is general consensus that their size should be controlled below 200 nm [3] because particles with size over 200 nm tend to induce undesired responses by the reticuloendothelial system (RES) and are quickly cleared from the circulation. Particles 150 nm or smaller may escape through the fenestration of the vascular endothelium and get cleared from the blood circulation. Particles with size less than 20 and 10 nm may be cleared through the lymph nodes and renal systems, respectively [3, 18]. Penetration of intravascularly administered nanomedicines into the tumor mass has been proven difficult because of the high interstitial fluid pressure and complex extracellular matrix of the tumor tissue [19]. Chilkoti and coworkers evaluated and demonstrated the molecular weight (size) dependency on the tumor penetration using dextran-based delivery vehicles [20]. They found that dextrans with low molecular weights (3.3–10 kDa) can efficiently penetrate and homogeneously distribute in the tumor tissue, but dextrans with higher molecular weights (40–70 kDa) were observed only ~15 μm away from the vessel wall, indicating their low penetration/permeation efficiency in the tumor tissue. Using a three-dimensional, multicellular spheroid of human cervical carcinoma cells that simulate a solid tumor, Pun et al. observed similar size dependency of nanoparticles on tumor penetration. Polystyrene nanoparticles with 20 or 40 nm sizes readily penetrated the simulated tumor and distributed homogeneously, whereas 100 and 200 nm particles showed restricted penetration. Interestingly, when nanomedicines were coated with extracellular matrix-disrupting collagenase, tumor penetration of the 20 and 40 nm particles was enhanced by roughly tenfold [21].

Geng et al. recently reported that the shape of delivery vehicles also has a significant effect on biodistribution [22]. They evaluated cylinder-shaped filomicelles (20–60 nm in cross-sectional diameter and a few micrometers in length) in rodents and found that the filomicelles could persist in the circulation up to 1 week after intravenous injection. The circulation half-life is about ten times longer than the half-life of their spherical counterparts, which is presumably caused by the fact that these cylinder-shaped delivery vehicles are more readily extended by flow forces and therefore are less likely to interact with and get taken up by the phagocytic cells. This interesting finding may shed light on the design of a new generation of drug delivery vehicles for enhanced circulation time and improved in vivo performance. A separate study using polymeric nanostructures with various shapes (e.g., cylinder and cube) also demonstrated the high impact of shape on the biological response of nanomedicine [23]. Cylindrical nanostructures with an aspect ratio (height/width) of 3, for example, can be internalized into cells four times faster than those with an aspect ratio of 2. It has yet to be determined whether these uniquely designed nanostructures could outperform the traditional, spherical nanoparticles in terms of biodistribution and antitumor efficacy.

Besides size and shape, surface characteristics and physical properties of nanomedicines can significantly influence the nanoparticle biodistribution. Positively charged particles are typically cleared much more quickly from the circulation than neutral or negatively charged particles [24]. The use of PEG to modify the surface of nanoparticles is critical to improve their circulation half-life and reduce the plasma protein absorption to nanoparticles that could otherwise lead to opsonization, a process that involves surface deposition of blood opsonic factors (such as fibronectin) for enhanced recognition by macrophages [24]. There has been some progress made developing PEG-like, protein-resistant materials, exemplified by zwitterionic polymers [25], which exhibit high resistance to nonspecific protein absorption due in part to their neutral surface charge and hydrophilicity [26]. However, it is unclear at this time whether these materials could be viable, biocompatible alternatives to PEG. A recent study by Verma et al. showed that the surface pattern of nanomedicines can have a dramatic effect on their biological responses [27]. Gold nanoparticles coated with subnanometer striations of alternating anionic (sulfonate) and hydrophobic (methyl) groups can successfully penetrate plasma membrane without disrupting the membrane bilayer. This approach can be particularly useful for direct delivery of cargos to the cytoplasm. The surface-modified nanoparticles also showed improved resistance to protein absorption, providing another potential strategy for surface modification of nanomedicines.

3.3 Preparation of Nanomedicine with Controlled Properties

To develop nanomedicines with consistent in vitro and in vivo performance that can be utilized in targeted or personalized disease treatment, it is crucial to formulate these nanomedicines in a highly controlled manner. Conventional formulation strategies usually give rise to nanomedicines with heterogeneous sizes and predominantly spherical shapes. Particle Replication In Nonwetting Template (PRINT), a top-down nanofabrication technique developed by DeSimone and coworkers, addresses these limitations and allows the formulation of polymeric nanoparticles with precisely controlled sizes in various shapes other than spherical (e.g., cylindrical, cubic, discoid) using soft lithographic molding technology [28, 29]. The DeSimone group utilizes photocurable perfluoropolyether (PFPE) molds to emboss liquid precursor compounds, using highly fluorinated surfaces that are nonwetting to organic materials, which enables the fabrication of isolated objects with excellent control over shape and composition [29]. Another promising device-assisted nanomedicine formulation strategy was developed by Tseng [30, 31], and Karnik and Farokhzad [32–34], utilizing microfluidics to control rapid mixing of polymer and drug and to control droplet size to yield particles with uniform size.

Polymeric nanoparticles are usually prepared by coprecipitation of hydrophobic therapeutics with hydrophobic polymers, such as polylactide (PLA) or polylactide-*co*-glycolide (PLGA). The resulting nanoparticles typically have poorly controlled physicochemical properties such as low drug loading, undesired drug release kinetics, heterogeneous nanoparticle composition, and broad particle size distributions [5]. To address these challenges, a new drug-loading and formulation method was reported by Cheng and coworkers, using drug-initiated lactide polymerization followed by nanoprecipitation [35, 36]. In the presence of a metal catalyst (e.g., (BDI)Zn(II)N(TMS)$_2$ with BDI = 2-((2,6-diisopropylphenyl)amido)-4-((2,6-bisalkyl) imino)-2-pentene), hydroxyl-containing drugs (e.g., paclitaxel, doxorubicin, or camptothecin) can quantitatively form metal-alkoxide complexes, which can subsequently initiate living, ring-opening polymerizations of lactide rings to form drug–PLA conjugates. Nanoprecipitation of the resulting drug–PLA conjugates gives rise to drug–PLA nanoparticles with controllable sizes between 50 and 150 nm and low polydispersity. These nanoparticles have high drug loading (as high as 40 wt%), high loading efficiency (97–100%), and controlled drug release kinetics without burst release effect. The bulky BDI chelating ligand on the metal catalyst also regulates the coordination of the metal catalyst only with the least sterically hindered hydroxyl group of the drug, providing additional control over the polymerization as well as the structure and composition of the polymer–drug conjugates. In a separate study to improve the formulation of nanomedicines via controlled chemistry, Shen and coworkers demonstrated a new concept by using drug molecules (e.g., camptothecin) to control the self-assembly of nanomedicines with minimal amount of carrier materials and therefore substantially enhanced drug loading [37]. Specifically, camptothecin was conjugated to an oligomer ethylene glycol (OEG), and the resulting camptothecin–OEG conjugate self-assembled into liposome-like nanocapsules via the hydrophobic interaction between camptothecin molecules.

Controlled conjugation chemistry is another tool playing a potentially vital role in controlling the incorporation of therapeutics or targeting ligands into nanomedicines. "Click chemistry," a powerful conjugation technique conceived by Barry Sharpless, has become a highly recognized method in the field of nanomedicine, allowing conjugation of therapeutics or targeting ligands to nanomedicines with unprecedented site-specificity [38–40]. The click process involves 1,3-dipolar cycloaddition of an azide to an alkyne to form 1,2,3-triazole rings, a reaction known for its high efficiency and high specificity. Click chemistry proceeds well in aqueous solution [41] or even in live organisms [42, 43], and is independent of other functional groups [38], demonstrating excellent solvent and functionality tolerability. Click chemistry has been widely used lately in the synthesis of polymeric therapeutics, surface modification of nanomedicine, and bioconjugation for in vitro and in vivo applications [44, 45]. In one study, Wooley and coworkers developed a new methodology for the preparation of well-defined core-shell nanoparticles using click chemistry. An amphiphilic diblock copolymer (poly(acrylic acid)-*b*-poly(styrene)), partially functionalized throughout the corona with alkynyl groups, self-assembled in water into micelles and formed nanoparticles after click reaction between the alkynyl shell of the micelles and azide-terminated dendrimers as the cross-linking agent. The remaining azide termini of the dendrimer cross-linker were further utilized for a secondary click reaction to conjugate either fluorescence dye or therapeutics onto the nanoparticles' surface [46]. Murphy et al. have recently demonstrated conjugation between azide-functionalized gold nanorods and an acetylene-functionalized enzyme (trypsin) through click chemistry. The click-conjugated enzyme showed substantially improved specificity and activity compared to the same enzyme linked to the gold nanorods by conventional bioconjugation chemistries [47]. Another innovative utilization of click chemistry was demonstrated by Bertozzi et al. in noninvasive in vivo imaging in developing zebrafish [48]. They first treated zebrafish embryos with azide-containing, unnatural sugars to metabolically label their cell-surface glycans with azides. Subsequently, the embryos were treated with a difluorinated, cyclooctyne-containing fluorophore by means of copper-free click chemistry, enabling the visualization of glycans in vivo at subcellular resolution during the development of the zebrafish embryos.

3.4 Nanomedicine-Mediated Cancer Targeting

There have been enormous efforts of designing nanomedicines aiming for targeted delivery of therapeutics for improved treatment of cancer, cardiovascular diseases, and immunological diseases [49–51]. One of the key challenges is the design and formulation of clinically relevant, targeted nanomedicines [51]. Many nanomedicine platforms have been developed and used in targeted drug delivery applications, including dendrimers, liposomes, polymeric nanoparticles, micelles, protein nanoparticles, ceramic nanoparticles, viral nanoparticles, metallic nanoparticles, and carbon nanotubes [50]. To facilitate the clinical application of targeted nanomedicines, their formulation should involve the use of biocompatible materials and should be completed via simple, robust processes for the assembly of nanomedicine, incorporation of drug and targeting ligand, and purification, postformulation processing, large-scale preparation, sterilization, and storage. The formulation process should also allow facile optimization of physicochemical parameters of the targeted nanomedicines that can be critical to their PK/PD properties, cellular uptake behavior, and in vivo efficacy.

The FDA-approved nanomedicines for cancer therapy function mainly through the accumulation of nanomedicine in tumor tissues via the EPR effect in the leaky tumor vasculature [52] and the subsequent release of the payload to kill the cancer cells. This passive targeting process usually requires long-circulating delivery systems in order to achieve time-dependent accumulation in tumor tissue to substantially improve the biodistribution and pharmacokinetic profile of the therapeutic modality, compared to the conventional administration of unmodified drugs [53]. The efficiency of this passive targeting mechanism is largely determined by the physicochemical properties of the delivery system. Many liposomal or polymeric drug/protein nanomedicines were designed and developed mainly to address issues related to the pharmacological drawbacks of small molecule or protein therapeutics [2, 12, 54]. Without active targeting ligands, certain drug delivery systems with optimized biophysical and chemical properties can still exhibit tissue-specific accumulation [23, 27]. However, to further improve disease targeting, it is inevitable to integrate various active targeting strategies in nanomedicines through the incorporation of targeting ligands.

Targeted ligands can be either incorporated to formulated nanomedicines via surface conjugation or incorporated to prefunctionalized biomaterials prior to the nanomedicine formulation. The latter approach can simplify optimization and potential scale-up of the targeted nanomedicine but can be very difficult to implement, especially in case of macromolecular targeting ligands (e.g., antibodies or aptamers) [35, 36, 55–58]. The majority of targeting ligand incorporation approaches still follow the former strategy. The conjugation of targeting ligands is one of the most critical steps in targeted nanomedicine formulation, which may result in decreased targeting efficiency due to poorly controlled ligand conjugation. Reasons are nonspecific binding prior to reaching the targeted disease tissue or anchoring on the targeted tissue surface too strongly, thus preventing homogeneous diffusion of the nanomedicine throughout the targeted tissue [59]. Therefore, optimization of the ligand density on the nanomedicine surface is a critical step to keep the subtle balance between anchoring affinity and tissue penetration, a key requirement for optimal therapeutic efficacy [60].

The proliferation of tumor cells requires sufficient nutrient supplies from blood. By stopping tumors from making new blood vessels, a process known as antiangiogenesis [61], not only the growth of solid tumors but also the tendency of tumor metastasis may be prohibited [62]. Over the last several decades, a handful of angiogenic targets have been explored in anticancer nanomedicine, which include the vascular endothelial growth factor receptors (VEGFRs), $\alpha_v\beta_3$ integrins, matrix metalloproteinase receptors (MMPs), and vascular cell adhesion molecule-1 (VCAM-1). Cell proliferation markers are another set of targets for cancer therapeutics, as many of these markers are significantly overexpressed on certain tumor cells. Actively targeting nanoparticles have followed the schemes of monoclonal antibodies to target cell proliferation receptors such as human epidermal receptors (HER) [63], transferrin receptors [64–66], and folate receptors [67].

Antibodies (Abs) are the most well-known targeting ligands used in targeted drug delivery. Over a dozen of monoclonal Abs have been approved by the FDA since 1997 [68], including Herceptin® (anti-HER2/neu) for breast cancer and Avastin (anti-VEGF-A) for metastatic colorectal cancer treatment. Hundreds of delivery systems based on Abs or their fragments are in preclinical and clinical investigations [69, 70]. As Abs are derived either from animals [71] or through phage display techniques [72], immunogenicity has always been a concern. The conjugation of Abs to nanomedicine is usually accomplished via coupling chemistry (e.g., carboxylate-to-amine or maleimide-to-thiol couplings). The drawback of this approach is the lack of conjugation site-specificity, which leads to substantially reduced targeting specificity and efficiency [73, 74]. Single-chain variable fragment (scFV) with high affinity to the targeted tumor tissue may restrict the localization and tumor penetration [75]. Another potential issue with the use of antibody-nanomedicine is the nonspecific binding to circulating free antigen or irrelevant receptors, which leads to reduced targeting efficiency [70]. Several strategies can be applied to address these concerns; one such strategy is to use affibody, the fragments of Abs, as the substituent of the high molecular weight Abs. Affibodies have comparable binding affinities and targeting efficiencies as Abs but have substantially reduced sizes (molecular weights of affibodies ~ 6kDa versus those of Abs ~150kDa) [76]; the latter is particularly important when they are used as the targeting ligands in nanomedicine. Engineered methods to increase the circulation time of antibodies have also been reported [77, 78].

Recent development in protein engineering may also facilitate the applications of Abs as targeting agents in nanomedicines [79, 80].

Aptamers (Apts) are single-stranded DNA or RNA that can fold into unique conformations. They can bind to specific targets, either small molecules or macromolecules, with very high affinity. Recently aptamers have been used as a new class of targeting ligands in nanomedicine-mediated cancer targeting and demonstrated great promise [81–84]. Aptamers are usually nonimmunogenic as they are developed via a combinational chemistry approach called systematic evolution of ligands by exponential enrichment (SELEX). As the synthesis of aptamers is achieved via an entirely chemical process, batch-to-batch variability can be substantially reduced. It is also possible to chemically modify aptamers by attaching fluorophores or functional groups for orthogonal bioconjugation; the latter approach holds significant advantage over Abs with respect to site-specific, controlled conjugation to nanomedicines. Aptamers exhibit remarkable stability over a wide range of pH, temperature, and organic solvents without loss of activity, and they can be modified to have improved stability against enzyme degradation, which is critical for their in vivo application. An additional advantage of using aptamers instead of antibodies as targeting ligand in nanomedicine is the potential of controlling the dosage of nanomedicines through the use of complementary DNA as the antidote [85, 86]. This option is particularly important in the case of an accidental overdose of a therapeutic nanomedicine that may otherwise cause significant, acute toxicity. The generalized manufacturing of antidotes to aptamers has recently been described [87]. One issue for using aptamers as targeting ligands for nanomedicines is that the number of available targets is still limited compared to the targets for antibodies. Identification of aptamers via an in vivo selection process has recently been reported, which may address this issue and open a new avenue to a large variety of potentially clinically relevant, tumor-specific aptamers [88].

Oligopeptides as targeting ligands can be selected through phage display. Oligopeptides are usually easy to synthesize and handle as compared to Abs or aptamers. Targeting mediated by oligopeptides, however, can be nonspecific. For instance, RGD (arginine–glycine–aspartic acid), one of the most well-known ligands with strong affinity to the cell adhesion integrin $\alpha_v\beta_3$ that is overexpressed in cancer cells, can target cancer and increase intracellular drug delivery in various preclinical tumor models [89, 90], but it also binds to other integrins such as $\alpha_5\beta_1$ and $\alpha_4\beta_1$. The nonspecific targeting and binding of RGD to other receptors might limit its potential in cancer-specific targeting [91, 92]. *Carbohydrates* in extracellular matrices (ECM) overexpressed in tumors, such as chondroitin sulfate [93] and hyaluronan (HA) receptor [94], allow them to serve as effective targets for cancer targeting. For example, HA coating of liposomes improved their circulation half-life and enhanced their targeting efficiency to HA receptor overexpressing tumors [94]. *Small organic molecule*-based cancer targeting ligands are much easier to prepare in large scale and to incorporate into nanomedicine as Abs, aptamers, or oligopeptides [95, 96]. A few examples such as folate and near-inferred fluorescent dye IR783 show interesting cancer targeting properties and may be promising ligands in nanomedicine-mediated cancer targeting [97, 98].

Tremendous effort has been undertaken to explore whether the incorporation of targeting ligands into nanomedicines can improve their in vivo biodistribution [99]. Early investigations using liposomes containing surface-conjugated, tumor-specific antibody showed that cancer targeting liposomes accumulated in the targeted tumor tissues twice as much as control liposomes [100]. Later, the work by Park and coworkers demonstrated that cancer targeting mediated by antibody–liposome conjugates had enhanced antitumor efficiency compared to control liposomes [101]. However, these antibody–liposome conjugates did not show improved accumulation in tumor tissues, rather the presence of antibody on liposomes improved their localization inside the target cancer cells [102]. Similar results showing improved accumulation inside the targeted cells rather than enhanced total tissue concentration were also reported by Davis and coworkers during their studies on transferrin-polymeric nanoparticle-mediated siRNA delivery [103]. Wittrup and coworker developed a mechanistic model to understand and predict the complex interplay between particle size, affinity, and tumor uptake [104]. Their model showed that particles with diameter of 50 nm or larger should have insignificant tumor uptake for both targeted and nontargeted groups, which is consistent with the observations by Park [102] and Davis [103]. Despite this size limitation, it is generally accepted that cellular uptake and efficacy of nanomedicines can be improved by the incorporation of targeting ligands [102, 103, 105]. Once nanoparticles extravasate into tumor tissue, their retention in the tissue and their uptake by cancer cells are facilitated by active targeting, followed by receptor-mediated endocytosis, both together resulting in higher intracellular drug concentration and increased efficacy [102, 105–107]. One additional aspect of vascular endothelial targeting for oncology or cardiovascular diseases using ligand-mediated active targeting is that the tissue accumulation of targeted nanomedicines is independent from the EPR effect [108]. Similar EPR independence was observed for immunological tissue targeting, utilizing targeted delivery systems as vaccines for active transportation from the lymphatic vessels to the draining lymph nodes, targeting the lymph node-residing dendritic cells [109].

Applying the optimal combination of drug delivery vehicles and suitable targeting ligands for specific disease, targeting may become clinically important. One example supporting this statement is the phase I clinical study of CALAA-01 using Calando Pharmaceutical's RONDEL nanoparticle delivery technology, which demonstrated an RNAi mechanism of action in cancer patients [110]. RONDEL nanoparticle delivery technology, developed by Davis and coworkers [110], is a transferrin-targeting, polymeric system for siRNA delivery for solid tumor therapy. Using multimodal in vivo imaging techniques, Davis and his team showed that non-targeting and transferrin-targeting polymeric nanoparticles have the identical distribution and tendency of accumulation in solid tumors, but the targeted particles led to more pronounced gene inhibition within cancer cells [103, 111]. The transferrin-targeting ligand is used to enhance the cellular uptake of the nanoparticles, rather than concentrating the nanoparticles in the tumor. Davis and coworkers further demonstrated that the presence of intracellularly localized nanoparticles is quantitatively correlated to the dose of the nanoparticles administered [107].

3.5 Current Status and Future Perspective

As the field of cancer nanotechnology further matures with an increasing number of nanotechnologies moving closer to clinical applications, there is plenty of room for continued efforts in developing new, nanometer-sized carriers for the prevention of disease progression and dissemination. To achieve personalized anticancer nanomedicine, there are still many obstacles to overcome. Formulations of nanomedicines with precisely controlled parameters (i.e., drug loading, size, and release kinetics) in large quantity are still challenging. Techniques that can be broadly utilized for the incorporation of therapeutics into a variety of polymers with all translational issues fully addressed are significantly lacking. Much information has been accumulated for the correlation of various physiochemical properties of nanomedicines (e.g., size, surface functional groups, and shape) with the systemic biodistribution, and long-circulating nanomedicines can be prepared for some specific systems. However, long-circulating nanomedicines may not exhibit maximized anticancer effects if these nanomedicines cannot homogeneously distribute in solid tumor tissues and internalize into the target cancer cells. In fact, drug delivery nanomedicines that can successfully penetrate the ECM of tumor tissues are rare. Developing polymeric nanomedicines that can penetrate certain biological barriers (e.g., the blood–brain barrier) is still a formidable task for drug delivery scientists and engineers. Cancer targeting by incorporating homing ligands to the surface of nanomedicines has been attempted for many years. However, formulation of nanomedicines containing protein-based targeting ligands (e.g., antibodies) is extremely difficult to control and may only be made on small scales. Incorporation of antibodies or aptamers into nanomedicines may result in improved in vivo efficacy, but meanwhile may also result in increased accumulation of nanomedicines in undesired organs such as liver or spleen that contain a large number of macrophages cells. Solid formulation of polymeric nanoparticles often resulted in aggregation during postformulation processing (e.g., lyophilization), which substantially reduced their clinical applicability. Although these challenges are difficult to address, synergistic integration of the efforts of chemists, materials scientists, chemical and biomedical engineers, and physicians may facilitate the development of anticancer nanomedicine at an unprecedented pace and may eventually make it possible to develop chemotherapy in time-, tissue-, and patient-specific manner.

References

1. Farokhzad OC, Langer R (2006) Nanomedicine: developing smarter therapeutic and diagnostic modalities. Adv Drug Deliv Rev 58(14):1456–1459
2. Duncan R (2006) Polymer conjugates as anticancer nanomedicines. Nat Rev Cancer 6(9):688–701
3. Moghimi SM, Hunter AC, Murray JC (2005) Nanomedicine: current status and future prospects. FASEB J 19(3):311–330

4. Bangham AD, Standish MM, Watkins JC (1965) Diffusion of univalent ions across lamellae of swollen phospholipids. J Mol Biol 13(1):238–252
5. Soppimath KS, Aminabhavi TM, Kulkarni AR, Rudzinski WE (2001) Biodegradable polymeric nanoparticles as drug delivery devices. J Control Release 70(1–2):1–20
6. Nishiyama N, Kataoka K (2006) Current state, achievements, and future prospects of polymeric micelles as nanocarriers for drug and gene delivery. Pharmacol Ther 112(3):630–648
7. Park JW, Benz CC, Martin FJ (2004) Future directions of liposome- and immunoliposome-based cancer therapeutics. Sem Oncol 31(6):196–205
8. Discher DE, Ahmed F (2006) Polymersomes. Ann Rev Biomed Eng 8:323–341
9. Wagner V, Dullaart A, Bock AK, Zweck A (2006) The emerging nanomedicine landscape. Nat Biotechnol 24(10):1211–1217
10. Allen TM, Chonn A (1987) Large unilamellar liposomes with low uptake into the reticuloendothelial system. FEBS Lett 223(1):42–46
11. Klibanov AL, Maruyama K, Torchilin VP, Huang L (1990) Amphipathic polyethyleneglycols effectively prolong the circulation time of liposomes. FEBS Lett 268(1):235–237
12. Langer R (1998) Drug delivery and targeting. Nature 392(6679):5–10
13. Langer R, Folkman J (1976) Polymers for sustained-release of proteins and other macromolecules. Nature 263(5580):797–800
14. Harries M, Ellis P, Harper P (2005) Nanoparticle albumin-bound paclitaxel for metastatic breast cancer. J Clin Oncol 23(31):7768–7771
15. Allen TM, Cullis PR (2004) Drug delivery systems: entering the mainstream. Science 303(5665):1818–1822
16. Service RF (2010) Nanaotechnology - Nanoparticle Trojan horses gallop from the lab into the clinic. Science 330(6002):314–315
17. Maeda H, Wu J, Sawa T, Matsumura Y, Hori K (2000) Tumor vascular permeability and the EPR effect in macromolecular therapeutics: a review. J Control Release 65(1–2):271–284
18. Gaumet M, Vargas A, Gurny R, Delie F (2008) Nanoparticles for drug delivery: the need for precision in reporting particle size parameters. Eur J Pharm Biopharm 69(1):1–9
19. Minchinton AI, Tannock IF (2006) Drug penetration in solid tumours. Nat Rev Cancer 6(8):583–592
20. Dreher MR, Liu WG, Michelich CR, Dewhirst MW, Yuan F, Chilkoti A (2006) Tumor vascular permeability, accumulation, and penetration of macromolecular drug carriers. J Natl Cancer Inst 98(5):335–344
21. Goodman TT, Olive PL, Pun SH (2007) Increased nanoparticle penetration in collagenase-treated multicellular spheroids. Int J Nanomedicine 2(2):265–274
22. Geng Y, Dalhaimer P, Cai SS, Tsai R, Tewari M, Minko T, Discher DE (2007) Shape effects of filaments versus spherical particles in flow and drug delivery. Nat Nanotechnol 2(4):249–255
23. Gratton SEA, Ropp PA, Pohlhaus PD, Luft JC, Madden VJ, Napier ME, DeSimone JM (2008) The effect of particle design on cellular internalization pathways. Proc Natl Acad Sci USA 105(33):11613–11618
24. Alexis F, Pridgen E, Molnar LK, Farokhzad OC (2008) Factors affecting the clearance and biodistribution of polymeric nanoparticles. Mol Pharm 5(4):505–515
25. Ladd J, Zhang Z, Chen S, Hower JC, Jiang S (2008) Zwitterionic polymers exhibiting high resistance to nonspecific protein adsorption from human serum and plasma. Biomacromolecules 9(5):1357–1361
26. Haag R, Kratz F (2006) Polymer therapeutics: concepts and applications. Angew Chem Int Ed 45(8):1198–1215
27. Verma A, Uzun O, Hu YH, Hu Y, Han HS, Watson N, Chen SL, Irvine DJ, Stellacci F (2008) Surface-structure-regulated cell-membrane penetration by monolayer-protected nanoparticles. Nat Mater 7(7):588–595
28. Kelly JY, DeSimone JM (2008) Shape-specific, monodisperse nano-molding of protein particles. J Am Chem Soc 130(16):5438–5439

29. Rolland JP, Maynor BW, Euliss LE, Exner AE, Denison GM, DeSimone JM (2005) Direct fabrication and harvesting of monodisperse, shape-specific nanobiomaterials. J Am Chem Soc 127(28):10096–10100
30. Wang H, Wang ST, Su H, Chen KJ, Armijo AL, Lin WY, Wang YJ, Sun J, Kamei K, Czernin J, Radu CG, Tseng HR (2009) A supramolecular approach for preparation of size-controlled nanoparticles. Angew Chem Int Ed 48(24):4344–4348
31. Wang H, Liu K, Chen K-J, Lu Y, Wang S, Lin W-Y, Guo F, Kamei K, Chen Y-C, Ohashi M, Wang M, Garcia MA, Zhao X-Z, Shen CKF, Tseng H-R (2010) A rapid pathway toward a superb gene delivery system: programming structural and functional diversity into a supramolecular nanoparticle library. ACS Nano 4(10):6235–6243
32. Karnik R, Gu F, Basto P, Cannizzaro C, Dean L, Kyei-Manu W, Langer R, Farokhzad OC (2008) Microfluidic platform for controlled synthesis of polymeric nanoparticles. Nano Lett 8(9):2906–2912
33. Valencia PM, Basto PA, Zhang LF, Rhee M, Langer R, Farokhzad OC, Karnik R (2010) Single-step assembly of homogenous lipid-polymeric and lipid-quantum dot nanoparticles enabled by microfluidic rapid mixing. ACS Nano 4(3):1671–1679
34. Rhee M, Valencia PM, Rodriguez MI, Langer R, Farokhzad OC, Karnik R (2011) Synthesis of size-tunable polymeric nanoparticles enabled by 3D hydrodynamic flow focusing in single-layer microchannels. Adv Mater 23(12):H79–H83
35. Tong R, Cheng JJ (2009) Ring-opening polymerization-mediated controlled formulation of polylactide-drug nanoparticles. J Am Chem Soc 131(13):4744–4754
36. Tong R, Cheng JJ (2008) Paclitaxel-initiated, controlled polymerization of lactide for the formulation of polymeric nanoparticulate delivery vehicles. Angew Chem Int Ed 47(26):4830–4834
37. Shen Y, Jin E, Zhang B, Murphy CJ, Sui M, Zhao J, Wang J, Tang J, Fan M, Van Kirk E, Murdoch WJ (2010) Prodrugs forming high drug loading multifunctional nanocapsules for intracellular cancer drug delivery. J Am Chem Soc 132(12):4259–4265
38. Kolb HC, Finn MG, Sharpless KB (2001) Click chemistry: diverse chemical function from a few good reactions. Angew Chem Int Ed 40(11):2004–2021
39. Lutz JF, Zarafshani Z (2008) Efficient construction of therapeutics, bioconjugates, biomaterials and bioactive surfaces using azide-alkyne "click" chemistry. Adv Drug Deliv Rev 60(9):958–970
40. Parrish B, Breitenkamp RB, Emrick T (2005) PEG- and peptide-grafted aliphatic polyesters by click chemistry. J Am Chem Soc 127(20):7404–7410
41. Gopin A, Ebner S, Attali B, Shabat D (2006) Enzymatic activation of second-generation dendritic prodrugs: conjugation of self-immolative dendrimers with poly(ethylene glycol) via click chemistry. Bioconjug Chem 17(6):1432–1440
42. Baskin JM, Bertozzi CR (2007) Bioorthogonal click chemistry: covalent labeling in living systems. QSAR Comb Sci 26(11–12):1211–1219
43. Sawa M, Hsu TL, Itoh T, Sugiyama M, Hanson SR, Vogt PK, Wong CH (2006) Glycoproteomic probes for fluorescent imaging of fucosylated glycans in vivo. Proc Natl Acad Sci USA 103(33):12371–12376
44. Algar WR, Prasuhn DE, Stewart MH, Jennings TL, Blanco-Canosa JB, Dawson PE, Medintz IL (2011) The controlled display of biomolecules on nanoparticles: a challenge suited to bioorthogonal chemistry. Bioconjug Chem 22(5):825–858
45. Hein CD, Liu XM, Wang D (2008) Click chemistry, a powerful tool for pharmaceutical sciences. Pharm Res 25(10):2216–2230
46. Joralemon MJ, O'Reilly RK, Hawker CJ, Wooley KL (2005) Shell click-crosslinked (SCC) nanoparticles: a new methodology for synthesis and orthogonal functionalization. J Am Chem Soc 127(48):16892–16899
47. Gole A, Murphy CJ (2007) Azide-derivatized gold nanorods: functional materials for "click" chemistry. Langmuir 24(1):266–272
48. Laughlin ST, Baskin JM, Amacher SL, Bertozzi CR (2008) In vivo imaging of membrane-associated glycans in developing zebrafish. Science 320(5876):664–667

49. Sengupta S, Eavarone D, Capila I, Zhao G, Watson N, Kiziltepe T, Sasisekharan R (2005) Temporal targeting of tumour cells and neovasculature with a nanoscale delivery system. Nature 436(7050):568–572
50. Ferrari M (2005) Cancer nanotechnology: opportunities and challenges. Nat Rev Cancer 5(3):161–171
51. Farokhzad OC, Langer R (2009) Impact of nanotechnology on drug delivery. ACS Nano 3(1):16–20
52. Matsumura Y, Maeda H (1986) A new concept for macromolecular therapeutics in cancer-chemotherapy - Mechanism of tumoritropic accumulation of proteins and the antitumor agent Smancs. Cancer Res 46(12):6387–6392
53. Jain RK (2001) Delivery of molecular and cellular medicine to solid tumors. Adv Drug Deliv Rev 46(1–3):149–168
54. Duncan R (2003) The dawning era of polymer therapeutics. Nat Rev Drug Discov 2(5):347–360
55. Tong R, Yala L, Fan TM, Cheng JJ (2010) The formulation of aptamer-coated paclitaxel-polylactide nanoconjugates and their targeting to cancer cells. Biomaterials 31(11):3043–3053
56. Decuzzi P, Pasqualini R, Arap W, Ferrari M (2009) Intravascular delivery of particulate systems: does geometry really matter? Pharm Res 26(1):235–243
57. Kim B-S, Park SW, Hammond PT (2008) Hydrogen-bonding layer-by-layer assembled biodegradable polymeric micelles as drug delivery vehicles from surfaces. ACS Nano 2(2):386–392
58. Gu F, Zhang L, Teply BA, Mann N, Wang A, Radovic-Moreno AF, Langer R, Farokhzad OC (2008) Precise engineering of targeted nanoparticles by using self-assembled biointegrated block copolymers. Proc Natl Acad Sci USA 105(7):2586–2591
59. Byrne JD, Betancourt T, Brannon-Peppas L (2008) Active targeting schemes for nanoparticle systems in cancer therapeutics. Adv Drug Deliv Rev 60(15):1615–1626
60. Peer D, Karp JM, Hong S, Farokhzad OC, Margalit R, Langer R (2007) Nanocarriers as an emerging platform for cancer therapy. Nat Nanotechnol 2(12):751–760
61. Folkman J (2006) Angiogenesis. Ann Rev Med 57(1):1–18
62. Folkman J (1995) Angiogenesis in cancer, vascular, rheumatoid and other disease. Nat Med 1(1):27–31
63. Bianca SV, Sabrina SA-S, Thea MV, Thomas B, Gert R (1998) Overexpression of EGFR and c-erbB2 causes enhanced cell migration in human breast cancer cells and NIH3T3 fibroblasts. FEBS Lett 425(1):145–150
64. Prost AC, Menegaux F, Langlois P, Vidal JM, Koulibaly M, Jost JL, Duron JJ, Chigot JP, Vayre P, Aurengo A, Legrand JC, Rosselin G, Gespach C (1998) Differential transferrin receptor density in human colorectal cancer: a potential probe for diagnosis and therapy. Int J Oncol 13(4):871–875
65. Singh M (1999) Transferrin as a targeting ligand for liposomes and anticancer drugs. Curr Pharm Des 5(6):443–451
66. Li HY, Qian ZM (2002) Transferrin/transferrin receptor-mediated drug delivery. Med Res Rev 22(3):225–250
67. Leamon CP, Low PS (1991) Delivery of macromolecules into living cells: a method that exploits folate receptor endocytosis. Proc Natl Acad Sci USA 88(13):5572–5576
68. Mehren MV, Adams GP, Weiner LM (2003) Monoclonal antibody therapy for cancer. Ann Rev Med 54(1):343–369
69. Allen TM (2002) Ligand-targeted therapeutics in anticancer therapy. Nat Rev Cancer 2(10):750–763
70. Carter P (2001) Improving the efficacy of antibody-based cancer therapies. Nat Rev Cancer 1(2):118–129
71. van Dijk MA, van de Winkel JGJ (2001) Human antibodies as next generation therapeutics. Curr Opin Chem Biol 5(4):368–374
72. Goletz S, Christensen PA, Kristensen P, Blohm D, Tomlinson I, Winter G, Karsten U (2002) Selection of large diversities of antiidiotypic antibody fragments by phage display. J Mol Biol 315(5):1087–1097

73. Kitamura K, Takahashi T, Yamaguchi T, Noguchi A, Noguchi A, Takashina K-i, Tsurumi H, Inagake M, Toyokuni T, Hakomori S-I (1991) Chemical engineering of the monoclonal antibody A7 by polyethylene glycol for targeting cancer chemotherapy. Cancer Res 51(16): 4310–4315
74. Lee LS, Conover C, Shi C, Whitlow M, Filpula D (1999) Prolonged circulating lives of single-chain Fv proteins conjugated with polyethylene glycol: a comparison of conjugation chemistries and compounds. Bioconjug Chem 10(6):973–981
75. Adams GP, Schier R, McCall AM, Simmons HH, Horak EM, Alpaugh RK, Marks JD, Weiner LM (2001) High affinity restricts the localization and tumor penetration of single-chain Fv antibody molecules. Cancer Res 61(12):4750–4755
76. Nord K, Gunneriusson E, Ringdahl J, Stahl S, Uhlen M, Nygren PA (1997) Binding proteins selected from combinatorial libraries of an alpha-helical bacterial receptor domain. Nat Biotechnol 15(8):772–777
77. Zalevsky J, Chamberlain AK, Horton HM, Karki S, Leung IWL, Sproule TJ, Lazar GA, Roopenian DC, Desjarlais JR (2010) Enhanced antibody half-life improves in vivo activity. Nat Biotechnol 28(2):157–159
78. Schellenberger V, Wang C-w, Geething NC, Spink BJ, Campbell A, To W, Scholle MD, Yin Y, Yao Y, Bogin O, Cleland JL, Silverman J, Stemmer WPC (2009) A recombinant polypeptide extends the in vivo half-life of peptides and proteins in a tunable manner. Nat Biotechnol 27(12):1186–1190
79. Link AJ, Vink MKS, Agard NJ, Prescher JA, Bertozzi CR, Tirrell DA (2006) Discovery of aminoacyl-tRNA synthetase activity through cell-surface display of noncanonical amino acids. Proc Natl Acad Sci USA 103(27):10180–10185
80. Xie JM, Schultz PG (2006) Innovation: a chemical toolkit for proteins—an expanded genetic code. Nat Rev Mol Cell Biol 7(10):775–782
81. Nimjee SM, Rusconi CP, Sullenger BA (2005) Aptamers: an emerging class of therapeutics. Ann Rev Med 56:555–583
82. Keefe AD, Pai S, Ellington A (2010) Aptamers as therapeutics. Nat Rev Drug Discov 9(7):537–550
83. Tuerk C, Gold L (1990) Systematic evolution of ligands by exponential enrichment—RNA ligands to bacteriophage-T4 DNA-polymerase. Science 249(4968):505–510
84. Ellington AD, Szostak JW (1990) In vitro selection of RNA molecules that bind specific ligands. Nature 346(6287):818–822
85. Cao ZH, Tong R, Mishra A, Xu WC, Wong GCL, Cheng JJ, Lu Y (2009) Reversible cell-specific drug delivery with aptamer-functionalized liposomes. Angew Chem Int Ed 48(35):6494–6498
86. Rusconi CP, Roberts JD, Pitoc GA, Nimjee SM, White RR, Quick G, Scardino E, Fay WP, Sullenger BA (2004) Antidote-mediated control of an anticoagulant aptamer in vivo. Nat Biotechnol 22(11):1423–1428
87. Oney S, Lam RTS, Bompiani KM, Blake CM, Quick G, Heidel JD, Liu JYC, Mack BC, Davis ME, Leong KW, Sullenger BA (2009) Development of universal antidotes to control aptamer activity. Nat Med 15(10):1224–1228
88. Mi J, Liu YM, Rabbani ZN, Yang ZG, Urban JH, Sullenger BA, Clary BM (2010) In vivo selection of tumor-targeting RNA motifs. Nat Chem Biol 6(1):22–24
89. Li JJ, Ji JF, Holmes LM, Burgin KE, Barton LB, Yu XZ, Wagner TE, Wei YZ (2004) Fusion protein from RGD peptide and Fc fragment of mouse immunoglobulin G inhibits angiogenesis in tumor. Cancer Gene Ther 11(5):363–370
90. Almutairi A, Rossin R, Shokeen M, Hagooly A, Ananth A, Capoccia B, Guillaudeu S, Abendschein D, Anderson CJ, Welch MJ, Frechet JMJ (2009) Biodegradable dendritic positron-emitting nanoprobes for the noninvasive imaging of angiogenesis. Proc Natl Acad Sci USA 106(3):685–690
91. Peters D, Kastantin M, Kotamraju VR, Karmali PP, Gujraty K, Tirrell M, Ruoslahti E (2009) Targeting atherosclerosis by using modular, multifunctional micelles. Proc Natl Acad Sci USA 106(24):9815–9819

92. Simberg D, Duza T, Park JH, Essler M, Pilch J, Zhang L, Derfus AM, Yang M, Hoffman RM, Bhatia S, Sailor MJ, Ruoslahti E (2007) Biomimetic amplification of nanoparticle homing to tumors. Proc Natl Acad Sci USA 104(3):932–936
93. Bergemann C, Muller-Schulte D, Oster J, a Brassard L, Lubbe AS (1999) Magnetic ion-exchange nano- and microparticles for medical, biochemical and molecular biological applications. J Mag Mag Mater 194(1–3):45–52
94. Eliaz RE, Szoka FC (2001) Liposome-encapsulated doxorubicin targeted to CD44: a strategy to kill CD44-overexpressing tumor cells. Cancer Res 61(6):2592–2601
95. Basu S, Harfouche R, Soni S, Chimote G, Mashelkar RA, Sengupta S (2009) Nanoparticle-mediated targeting of MAPK signaling predisposes tumor to chemotherapy. Proc Natl Acad Sci USA 106(19):7957–7961
96. Kano MR, Bae Y, Iwata C, Morishita Y, Yashiro M, Oka M, Fujii T, Komuro A, Kiyono K, Kaminishi M, Hirakawa K, Ouchi Y, Nishiyama N, Kataoka K, Miyazono K (2007) Improvement of cancer-targeting therapy, using nanocarriers for intractable solid tumors by inhibition of TGF-β signaling. Proc Natl Acad Sci USA 104(9):3460–3465
97. Esmaeili F, Ghahremani MH, Ostad SN, Atyabi F, Seyedabadi M, Malekshahi MR, Amini M, Dinarvand R (2008) Folate-receptor-targeted delivery of docetaxel nanoparticles prepared by PLGA-PEG-folate conjugate. J Drug Target 16(5):415–423
98. Yang X, Shi C, Tong R, Qian W, Zhau HE, Wang R, Zhu G, Cheng J, Yang VW, Cheng T, Henary M, Strekowski L, Chung LW (2010) Near IR heptamethine cyanine dye-mediated cancer imaging. Clin Cancer Res 16(10):2833–2844
99. Pirollo KF, Chang EH (2008) Does a targeting ligand influence nanoparticle tumor localization or uptake? Trends Biotechnol 26(10):552–558
100. de Menezes DEL, Pilarski LM, Allen TM (1998) In vitro and in vivo targeting of immunoliposomal doxorubicin to human B-cell lymphoma. Cancer Res 58(15):3320–3330
101. Park JW, Hong KL, Kirpotin DB, Colbern G, Shalaby R, Baselga J, Shao Y, Nielsen UB, Marks JD, Moore D, Papahadjopoulos D, Benz CC (2002) Anti-HER2 immunoliposomes: enhanced efficacy attributable to targeted delivery. Clin Cancer Res 8(4): 1172–1181
102. Kirpotin DB, Drummond DC, Shao Y, Shalaby MR, Hong KL, Nielsen UB, Marks JD, Benz CC, Park JW (2006) Antibody targeting of long-circulating lipidic nanoparticles does not increase tumor localization but does increase internalization in animal models. Cancer Res 66(13):6732–6740
103. Bartlett DW, Su H, Hildebrandt IJ, Weber WA, Davis ME (2007) Impact of tumor-specific targeting on the biodistribution and efficacy of siRNA nanoparticles measured by multimodality in vivo imaging. Proc Natl Acad Sci USA 104(39):15549–15554
104. Schmidt MM, Wittrup KD (2009) A modeling analysis of the effects of molecular size and binding affinity on tumor targeting. Mol Cancer Ther 8(10):2861–2871
105. Farokhzad OC, Cheng JJ, Teply BA, Sherifi I, Jon S, Kantoff PW, Richie JP, Langer R (2006) Targeted nanoparticle-aptamer bioconjugates for cancer chemotherapy in vivo. Proc Natl Acad Sci USA 103(16):6315–6320
106. Pun SH, Tack F, Bellocq NC, Cheng JJ, Grubbs BH, Jensen GS, Davis ME, Brewster M, Janicot M, Janssens B, Floren W, Bakker A (2004) Targeted delivery of RNA-cleaving DNA enzyme (DNAzyme) to tumor tissue by transferrin-modified, cyclodextrin-based particles. Cancer Biol Ther 3(7):641–650
107. Davis ME, Zuckerman JE, Choi CHJ, Seligson D, Tolcher A, Alabi CA, Yen Y, Heidel JD, Ribas A (2010) Evidence of RNAi in humans from systemically administered siRNA via targeted nanoparticles. Nature 464(7291):1067–1070
108. Zhang N, Chittasupho C, Duangrat C, Siahaan TJ, Berkland C (2007) PLGA nanoparticle-peptide conjugate effectively targets intercellular cell-adhesion molecule. Bioconjug Chem 19(1):145–152
109. Reddy ST, van der Vlies AJ, Simeoni E, Angeli V, Randolph GJ, O'Neil CP, Lee LK, Swartz MA, Hubbell JA (2007) Exploiting lymphatic transport and complement activation in nanoparticle vaccines. Nat Biotechnol 25(10):1159–1164

110. Davis ME (2009) The first targeted delivery of siRNA in humans via a self-assembling, cyclodextrin polymer-based nanoparticle: from concept to clinic. Mol Pharm 6(3): 659–668
111. Choi CHJ, Alabi CA, Webster P, Davis ME (2010) Mechanism of active targeting in solid tumors with transferrin-containing gold nanoparticles. Proc Natl Acad Sci USA 107(3):1235–1240

Chapter 4
Macrophage-Targeted Nanoparticle Delivery Systems

Shardool Jain and Mansoor Amiji

4.1 Introduction

4.1.1 Monocytes and Macrophages

Macrophages originate from a mononuclear phagocyte system (MPS) in the bone marrow. The MPS system comprises of monocytes and tissue macrophages in their various forms. The primary function of these monocytes is phagocytosis [1]. Upon exiting the bone marrow, monocytes circulate in the blood and become activated in various tissues. Once the monocyte reaches the extravascular tissue, it matures into a larger phagocytic cell or *the macrophage*. The main functions of macrophages are phagocytosis, destruction and clearance of microorganisms and apoptotic cells, chemotaxis, antigen processing and presentation, secretion of enzymes and other biologically active substances such as cytokines, and destruction of tumor cells [2, 3]. Macrophages mainly reside in the liver (Kupffer cells), lungs (alveolar macrophages), spleen, lymph nodes, thymus, gut, marrow, brain, connective tissue, and serous cavities [1]. Table 4.1 presents important functions of macrophages along with the relevant substances secreted/produced by these cells.

In addition to this differentiated phenotypic pattern, macrophages can also acquire a distinct functional pattern upon encountering stimuli in tissue microenvironment, and appropriately, this stimuli-induced distinct macrophage state has been coined as *activation* [4]. The activated cells have an enhanced capability to attack and kill microbes and tumor cells. The activation state of macrophages has been broadly classified into classical (M1a and M1b) and alternate (M2a, b, and c) types [5]. The detailed information about the role and features of the above-mentioned subtype activation states have been reviewed elsewhere [5–7], but for the purpose of

S. Jain • M. Amiji (✉)
Department of Pharmaceutical Sciences, School of Pharmacy, Northeastern University, 110 Mugar Life Sciences Building, Boston, MA 02115, USA
e-mail: m.amiji@neu.edu

S. Svenson and R.K. Prud'homme (eds.), *Multifunctional Nanoparticles for Drug Delivery Applications: Imaging, Targeting, and Delivery*, Nanostructure Science and Technology, DOI 10.1007/978-1-4614-2305-8_4,

Table 4.1 Selected biomolecules secreted by activated macrophages

Microbicidal activity
Tumoricidal activity
Chemotaxis
Phagocytosis/pinocytosis
Glucose transport and metabolism
Generation of gaseous mediators
Reactive nitrogen intermediates
Reactive oxygen intermediates
Enzymes
Neutral proteases, elastase, lysozyme, acid hydrolases, collagenases, plasminogen activator, arginase, lipases, phosphatases
1α-hydroxylase
Plasma proteins
Complement components (C1-C5, properdin)
Coagulation factors (factors V, VIII, tissue factor)
Fibronectin
Cytokines and chemokines
IL-1, IL-6, IL-10, IL-12, IL-15, IL-18, TNF-α, TGF-β, GM-CSF, M-CSF, G-CSF
IL-8, MCP-1, MIP-1 α/ β regulated on activation normal T expressed and secreted
Growth factors
Platelet-derived growth factor, endothelial growth factor, fibroblast growth factor
Lipid mediators
Eicosanoids

Adapted from [1]

this chapter, we only have outlined the major differences between the classical and alternative activation pathway. The classical activation state is characterized by killing of intracellular pathogens and tumor resistance and can be induced by interferon-γ (IFN-γ) alone or in conjunction with microbial products such as lipopolysaccharide (LPS) or cytokine such as tumor necrosis factor alpha (TNF-α). The alternative state can be induced by cytokines such as IL-4 and IL-13 and mainly results in anti-inflammatory responses and resolution of injury. Activation of macrophage via the classical pathway is marked by high antigen presentation capacity, high IL-12, IL-23, nitric oxide (NO), and reactive oxygen species production. On the other hand, alternate activation stage is characterized by an increase in the IL-10 and IL-1ra cytokines, mannose and scavenger receptors, arginase production, and decrease in the production of inducible nitric oxide synthase enzyme [7, 8]. Figure 4.1 highlights the central role played by macrophages in mediating innate as well as adaptive immune response.

4.1.2 Role of Macrophages in Inflammation

Inflammation is a defense mechanism adopted by the body in response to the variety of stimuli including pathogens, injury, and autoimmune responses [1, 10, 11].

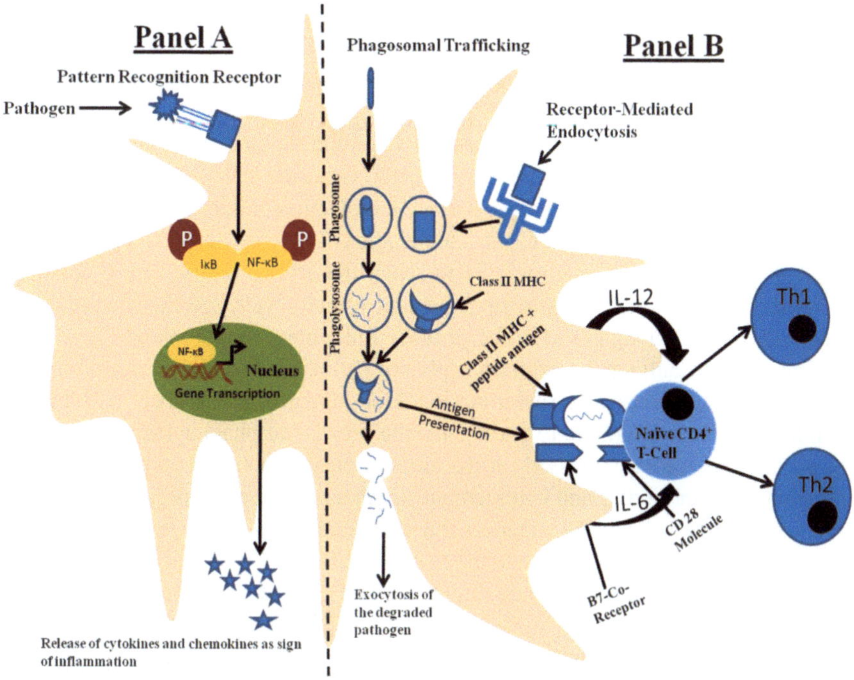

Fig. 4.1 The diagram above depicts the key functions of a macrophage cell. Panel (**a**) shows the ability of macrophages to detect the presence of bacteria (pathogen) via specific receptors (pathogen recognition receptors or toll-like receptors (TLR)) on its surface, which leads to a cascade of the downstream signals that ultimately lead to release of cytokines and chemokines to signal other cells about bacterial invasion. Panel (**b**) highlights the ability of these cells to phagocytose, either via non-specific or receptor-mediated endocytosis, and degrades the bacteria as seen in the phagolysosomal compartment. The degraded bacterial fragments can be excreted out of the cell via process called exocytosis. Also, the degraded antigenic peptide products can be displayed at the cell surface with the help of class II MHC molecule. This process helps in communicating with the T-helper cells. However, naïve T cell requires a secondary signal for activation, which is provided by interactions between B7 receptor on macrophages and CD28 ligand molecule on T cell. Upon activation, these cells can further lead to generation of Th1 and Th2 immune response. In addition, macrophages are also capable of releasing cytokines such as IL-6, and IL-12 that can influence adaptive immune response. NF-κB nuclear factor-κB; IκB inhibitor of NF-κB; IL-6 interleukine-6; IL-12 interleukin-12. Adapted from [9]

The vascularized connective tissues including plasma, circulating cells, blood vessels, and cellular and extracellular components are capable of exhibiting an inflammatory response [1]. The inflammatory response is marked by redness, heat, swelling, and pain. The process is also marked by enhanced vascular permeability, recruitment of leukocytes, and release of inflammatory mediators. The process of inflammation can be broadly divided into acute and chronic inflammation. In acute inflammation, the response lasts for only few hours to few days. The main characteristics of acute inflammation are exudation of fluid and plasma proteins due to altered permeability leading to edema and the emigration of leukocytes, primarily neutrophils [1]. On the other hand, chronic inflammation lasts for longer duration of time and is typically a more intense response, which includes recruitment of lymphocytes and macrophages, tissue necrosis, proliferation of blood vessels, and fibrosis [1].

In an event of injury, there is release of chemokines and soluble mediators from local cells such as vascular endothelial cells, dendritic cells, macrophages, and interstitial fibroblasts. This event acts as a chemotactic gradient to attract the circulating monocytes in the blood to the site of injury. The polymorphonuclear neutrophils are the first inflammatory cells to reach the site of injury. The specialized mononuclear cells, such as macrophages and lymphocytes, are then recruited by further downstream signals [1]. The primary functions of macrophages in inflammation include antigen presentation, phagocytosis, and modulation of immune response through production of various cytokines and growth factors [1, 12]. In case of inflammation caused due to pathogens, the process of phagocytosis is mediated by specific receptors expressed on the surface of macrophages. Additionally, the attachment of antibodies and complement fragments, by a process called opsonization, to the microbes greatly enhances the phagocytotic ability of macrophages [1]. The attachment of the phagocyte to the foreign material also results in the release of cytokines that further exert a wide range of effects in order to efficiently deal with infection.

4.1.3 Role of Macrophages in Infectious Diseases

Macrophages are a very important part of the defense mechanism of the body. As mentioned above, they have a definite role to play in the second line of defense that constitutes inflammation, secretion of cytokines, as well as activation of natural killer (NK) cells, and the complement system [1]. However, there are instances when the bacteria, bacterial fragments, or foreign molecules gain access to bloodstream and have the potential to reach spleen, lymph nodes, or other lymphoid tissues. Thus, macrophages are also important for their ability to stimulate immune response upon engulfing the microbes. Macrophages and dendritic cells (DCs) constitute the so-called antigen-presenting cells (APC) and belong to class II major histocompatibility complex (MHC) type of molecules [1]. After engulfing and degrading antigens early in the infectious process, the activated macrophages translocate the fragments of these antigens to cell surface where they are recognized by the T cell receptors (TCR) on $CD4^+$ T cells. These cells are then able to initiate cellular and humoral immunity by secretion of various cytokines. The $CD4^+$ cells are further divided into two types depending on the type of cytokine secreted. These are Th1 (T-helper-1) cells and Th2 (T-helper-2) cells. The Th1 cells secrete interleukin-2 (IL-2) and IFN-γ and antibody responses of IgG2a isotype in mice. On the other hand, Th2 cells secrete IL-4 and IL-5 and induce antibody responses of IgG1 and IgE type in mice [13, 14]. Table 4.2 summarizes the function of selected cytokines secreted by activated helper T cells and their target cells.

Table 4.2 Target cells and functions of select cytokines

Cytokines	Target cells	Effects upon target cells
Interleukin-2 (IL-2)	Helper T cells and cytotoxic T cells B cells Natural killer (NK) cells	Stimulates proliferation Stimulates proliferation and plasma cell development Enhances activity
Interleukin-4 (IL-4)	B cells Helper cells Macrophages Mast cells	Stimulates proliferation and plasma cell development; induces plasma cells to secrete IgE and IgG; increases number of surface class II MHC molecules Stimulates proliferation Increases number of surface class II MHC molecules; enhances phagocytosis Stimulates proliferation
Interleukin-5 (IL-5)	B cells Hematopoietic stem cells	Stimulates proliferation; induces plasma cells to secrete IgA Induces proliferation and development of eosinophils
Interleukin-10 (IL-10)	Macrophages	Inhibits cytokine production (helps downregulate immune response)
Interferon-γ	Multiple cell types Macrophages B cells Cancer cells Cytotoxic T cells and NK cells	Confers resistance to viruses Enhances phagocytosis Enhances antibody production Inhibits proliferation Enhances killing capacity of cytotoxic T cells and NK cells

Adapted from [11]

4.1.4 *Role of Macrophages in Cancer*

Cancer is defined as the rapid and uncontrolled progression of the cells. Tumor cells achieve this goal by undergoing an array of processes such as angiogenesis, metastases, and immunosuppression. There has been a common belief now that tumor-associated macrophages (TAMs) seem to play a prominent role in tumor survival by helping the tumor to carry out above-mentioned processes [15, 16]. The proof of close association between an abundance of TAMs and poor prognosis in breast, prostate, colon, and cervical cancer has already been established. In a typical scenario, tumor secretes various chemoattractants such as colony-stimulating factor-1 (CSF-1), chemokines (CCL2, 3, 4, 5, and 8), and vascular endothelial growth factor (VEGF). As a result, monocytes in the blood stream get attracted to the tumor tissue where they differentiate and take up the role of the resident macrophages [17]. Additionally, in comparison to macrophages derived from healthy tissues, these cells seem to have a distinct phenotype. TAMs are characterized by low expression of the differentiation-associated macrophage antigens, carboxypeptidase M and CD51, high constitutive expression of interleukin (IL-1) and IL-6, and low

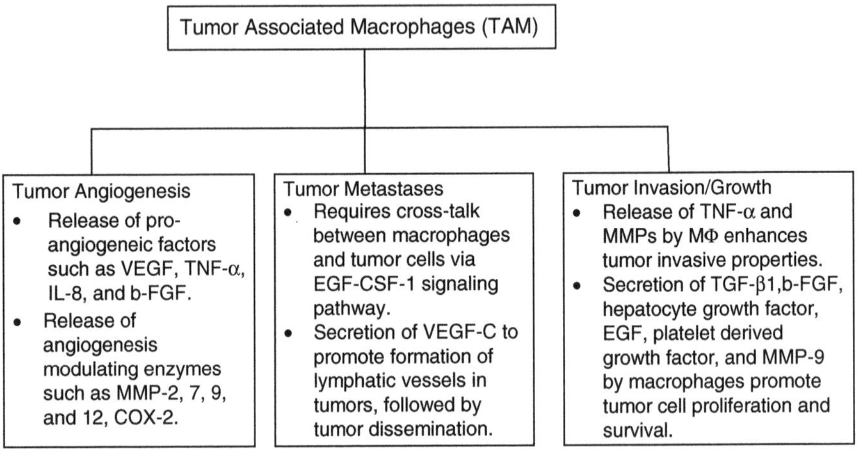

Fig. 4.2 Different functions of tumor-associated macrophages in promoting tumor growth and metastases

expression of TNF-α. It is believed that tumor-derived cytokines, chemokines, and proteases such as IL-4, 6, 10, TGF-β1, and PGE_2 act on macrophages so that they develop into polarized type II or M2 macrophages [18, 19]. As a result, unlike in healthy tissue, the anti-tumor activity of the macrophages is compromised in tumor microenvironment. Additionally, it has also been reported that hypoxia in the tumor microenvironment can also contribute to reduce the cytotoxic activity of TAM toward tumor cells by promoting the secretion of PGE_2 and IL-10 [17]. Thus, it appears that upon differentiation, resident macrophages are trained or molded by tumor cells to perform specific functions. The role of TAMs in the above-mentioned processes is summarized in Fig. 4.2 [17].

4.2 Macrophage-Targeted Delivery: The *Trojan Horse* Concept

The idea of employing macrophages as Trojan horse vector comes from the knowledge whereby microorganisms, such as bacteria and viruses, use monocytes and macrophages to proliferate and hide from immune-surveillance system [20–22]. Especially, in the case of HIV-1 progression, there is mounting evidence depicting the role of macrophages in disease progression, pathogenesis, and latent reservoir formation. Like other retroviruses and lentiviruses, HIV-1 has the ability to infect and replicate in nondividing cells such as cells of monocyte/macrophage lineage, although $CD4^+$ T lymphocytes appear to be the primary target of HIV virus [20, 23]. The virus may affect the macrophages by influencing the cytokine production profile, and this may result in chronic inflammation and extensive tissue damage. Interestingly, the infected macrophages appear to be resistant to the toxic effects of the virus and thus serve as a *Trojan horse* to the HIV virus [23]. In a recent model, proposed by Herbein and colleagues [24], it was mentioned that during the early stage of the disease, the viral proteins are responsible for establishing the infection and activation of macrophages to M1 (classical) state. The classical activation pathway leads to Th1-type response where cytokines and chemokines such as IFN-γ, TNF-α, IL-1β, MIP-1α, MIP-1β, and RANTES are produced and this leads to formation of viral reservoirs by increasing the viral transcription and also act to inhibit the viral HIV-1 entry. This stage is also marked by increasing rate of T-cell apoptosis. Further, during the later stage, characteristic features of alternative macrophage activation pathway are present indicating a shift from M1 to M2 state. Additionally, during these changes, intermediate levels of T-cell apoptosis were reported. Also, the presence of IL-4/13 and proinflammatory cytokines was related to reduction in the expansion of macrophage HIV-1 reservoirs. Last stage of the disease is marked by a dramatic increase in the depletion of $CD4^+$ and $CD8^+$ T-cells. Also, IL-10-deactivated macrophages are prominent at this stage and are involved in clearance of the apoptotic cells, inhibition of proinflammatory cytokines and chemokines, and reducing the expression of class II MHC molecules on plasma membrane of macrophages. In addition, high expression of type I IFN during this stage was reported to strongly inhibit the HIV-1 replication. According to the authors, another layer of complexity surrounds this model. The M1/M2/Md model can also be reversible, and this phenomenon may be dependent on the tissues affected and the local microenvironment. The validity and therapeutic interventions based on this model may still need to be established, but for our purposes, it describes to us the role of macrophages as a Trojan horse that protects and propagates the viral load.

4.2.1 Nanoparticle Systems for Macrophage-Targeted Delivery

Nanotechnology can be defined as the understanding and control of matter in 1–100 nm dimension range [25]. The precise use of these engineered materials within this range has been explored in various fields, including drug delivery and diagnostics. It is very well known that nanoplatform systems offer many advantages over conventional delivery systems in terms of solubility, diffusivity, blood circulation, drug release, and immunogenicity [25]. Nanoparticle systems can be made from a variety of different materials including polymers, metals, ceramics, and lipids. Based on their method of preparation, they can be engineered into various shapes and sizes, thus providing adaptable platforms for therapeutic and diagnostic approaches. Readers interested in exploring the various available polymer and lipid-based nanosystems are directed to excellent reviews by Yih et al. [26], Rawat et al. [27], and Ulrich et al. [28].

With the advancement in the applications of nanotechnology for medical diagnosis and therapy over the past two decades, the ability to modify the surface of the nanoparticles to enhance the specificity of the delivery system for a particular cell/tissue/organ has been explored in great detail. In the same vein, the idea of having multicompartment nanoparticle-based delivery systems for combination delivery of drugs/genes or combination therapy and diagnostic (so-called theranostic systems) approaches has also been explored. The readers interested in multifunctional nanocarriers beyond the scope of the other chapters in this book are directed to a recent review by Torchilin [29].

4.2.2 Approaches for Targeted Delivery to Macrophages

As described above, the role of macrophages as phagocytic and antigen-presenting cells is very well established. However, over or underactivity of these cells can result in the onset and/or progression of pathologic conditions such as growth and spread of malignant tumors, sepsis, chronic inflammation in rheumatoid arthritis, lysosomal storage disease, atherosclerosis, and major infections including HIV/AIDS and tuberculosis. Thus, therapies targeting macrophages appears to be a very promising strategy. Since, the focus is on nanoparticle-based therapeutics, the two approaches that can be employed are either passive or active targeting of the particles to these cells.

Passive targeting refers to the accumulation of the drug or drug-carrier nanosystem by exploiting the pathophysiological condition and anatomical route [30, 31]. For example, nanoparticle systems that are >100 nm are readily opsonized and removed from the circulation by the reticuloendothelial system (RES) and degraded by the macrophages in the liver and spleen [30]. Thus, the tendency of unmodified nanoparticles to localize in the RES has been exploited as a way to passively deliver the payload to macrophages. Gupta et al. [32] showed that amphotericin B (AmB) emulsions were rapidly taken up by the macrophages of the RES in comparison to the free drug (AmB-deoxycholate or AmB-Doc). Additionally, the nanoparticles have also been observed to passively accumulate at inflamed sites due to the enhanced permeation and retention (EPR) effect. The presence of macrophages at these sites presents an excellent opportunity to passively target these cells. Corvo et al. [33] showed that in a mouse model of rheumatoid arthritis, intravenous administration of liposomes grafted with poly(ethylene glycol) (PEG) of molecular weight 1,900 Da resulted in their passive accumulation at the inflamed arthritic sites. Furthermore, Boerman et al. [34] exploited this phenomenon to show that PEGylated liposomes labeled with technetium-99 can be used as a vector for scintigraphic imaging of arthritis. Lobenberg et al. [35] performed an in vivo study with the objective of investigating the possibility of specific drug targeting of the antiretroviral drugs to the reticuloendothelial cells (such as macrophages) by the oral route. Poly(hexylcyanoacrylate) (PHCA) nanoparticles were loaded with ^{14}C-labeled azidothymidine (AZT). The area under the curve (AUC) of the radiolabeled AZT in the blood stream was found to be 30% higher when it was bound to the nanoparticles as opposed to the solution form of the drug. Similarly, many other studies have indicated the potential of nanoparticles for targeting macrophages from drug delivery perspective. In an attempt to further enhance the delivery to a specific cell line

Table 4.3 Examples of receptor systems utilized for targeting macrophages

Macrophage receptor targeted	Brief description	References
Mannose receptor	C-type lectin	[2, 37]
	Involved in pathogen recognition	
	Ligand binding to the receptor is dependent on Ca^{2+} concentration	
	Facilitates receptor-mediated endocytosis	
Scavenger receptor	Receptors bind a variety of poly-anionic macro-molecules and modified (acetylated) LDL	[2, 37]
	Three classes of receptor are known: SR-A, B, and C	
	Scavenger receptors are implicated in pathological deposition of cholesterol during atherogenesis	
Dectin-1 receptor	C-type lectin	[38, 39]
	Unlike, other lectins, does not require Ca^{2+} coordination for calcium binding	
	Involved in pattern recognition and phagocytosis of unopsonized β-glucan particles and soluble β-glucan (component of yeast cell wall)	
Tuftsin peptide	Tetra-peptide sequence L-Thr-L-Lys-L-Pro-L-Arg (TKPR)	[40, 41]
	Synthesized by the enzymatic processing of CH2 domain of the Fc fragment of the heavy chain of IgG molecule	
	Activates Mφ and enhances their phagocytic ability	
Hyaluronate	Ligand for CD44 receptor	[42, 43]
	The receptor is a glycoprotein and is known to be involved in phagocytosis of large particles	

(macrophages in this case), newer drug delivery systems are being designed that are exploiting the presence of certain receptors on the surface of macrophages.

Active targeting approaches in the context of delivery systems are based on the surface modification of the systems with an agent (e.g., ligand, antibody and peptide) that has the selective affinity for recognizing and interacting with the particular cell type, tissue, or organ in the body [36]. The phenomenon of active targeting can be achieved by utilizing the presence of various receptors and lipid components on the plasma membrane of the cells (macrophages in this case). These receptors may be uniquely expressed on specific cells or may show a differentially higher expression on diseased cells in comparison to the normal cells. Table 4.3 briefly describes some of the key properties of the different receptors that are present on the surface of macrophages and have been utilized for actively targeting these cells.

Review of the literature shows several examples where the presence of various receptors such as Fc and complement receptors; lectin-based receptors specific for mannose, galactose, and sialic acid; lipoprotein receptors; and scavenger receptors have been exploited for the delivery of the payload [2, 44]. Kaur et al. [45] looked at mannose-modified gelatin nanoparticles for macrophage targeting of didanosine. The formulation was characterized in both in vitro and in vivo settings. The particle size range was reported to be 140 ± 19 nm, and the drug loading was $79.5 \pm 4.7\%$. The cell-based studies indicated a 5-fold higher uptake of the drug encapsulated in the mannose-modified formulation as compared to the free drug in phosphate-buffered saline (PBS) solution. Furthermore, biodistribution studies revealed a higher localization of the drug in the nanoparticle formulation in the spleen, lymph node, and brain (a key target for HIV infected macrophages) as compared to free drug solution. Jain et al. [46] took a step further and also compared the unmodified nanoparticles with mannose-modified, gelatin-based formulation. The cellular uptake by surface-modified nanoparticles was reported to be 2.7 times higher than the plain gelatin nanoparticles. More importantly, in vivo studies indicated that in comparison to the gelatin nanoparticles, modification with mannose significantly enhanced the uptake of the drug in liver, lymph nodes, and lung. The above-mentioned study exploited the presence of surface receptors to enhance the drug delivery to the macrophage cells. Additionally, encapsulating the drug in a nanoparticle-based formulation increases the possibility of delivering a higher payload to the cells.

Schmitt et al. [47] developed chitosan-based nanogels decorated with hyaluronate for selective delivery to macrophages. The group mentioned the modification was advantageous as hyaluronate is known to interact with CD44 cell surface receptor of macrophages, which is involved in phagocytosis. Additionally, the modification resulted in reversal of the surface charge from positive (due to chitosan polymer) to negative (due to hyaluronate), hence improving the biocompatibility. The results indicated that local injection of the modified nanogel encapsulating the photosensitizers was retained in the inflamed joint over a longer time period as compared to the free photosensitizer. Additionally, photodynamic therapy resulted in the reduction of inflammation as compared to the standard corticoid treatment.

In addition to the above-mentioned approaches, there are some additional considerations that should be taken into account when designing the delivery system to target macrophages. For example, it has been reported that particle size, surface charge, and hydrophobicity can influence the uptake by macrophages [44, 48]. Roser et al. [49] reported that particles with a neutral surface charge are less phagocytable in comparison to the charged particles. Additionally, it was reported that phagocytosis rate did not differ for anionic and cationic particles having the same absolute charge value. However, there have been conflicting reports on the effect of particle size on the cell uptake, especially when comparing liposomal and polymer based formulations. Allen et al. [50] performed a study on the effect of liposomal composition, size, concentration, and incubation time on the uptake by murine bone marrow macrophages. The results suggested that higher uptake was observed with smaller and negatively charged liposomes. The liposomal uptake increased linearly with the incubation time and concentration. On the other hand, Schäfer et al. [51] looked at poly(methylmethacrylate), poly(alkylcyanoacrylate), and human serum albumin (HSA) particle uptake by human macrophages. The results indicated that nanoparticles made from the same material but of larger diameter were phagocytosed to a larger extent. For example, phagocytosis of the nanoparticles made from HSA of 1.5 μm in diameter was higher in comparison to 200 nm particles made from the same material. Lastly, studies conducted by Tabata et al. [52] have shown that hydrophobic nanoparticles or particles coated with lipophilic material are preferred by macrophages in comparison to their hydrophilic counterparts. This phenomenon could be due to the hydrophobic interactions with the cell surface.

4.2.3 *Nanosystems for Delivery of Vaccines*

Vaccination is a proven strategy in the prevention of the infectious diseases and cancer. They can be developed from various sources to generate an immune response and at the same time potentiate the harmful effects associated with an actual infection. Conventional vaccines include live-attenuated or inactivated pathogens, antigenic peptides, proteins, and polysaccharides, while novel approaches are based on generation of vaccines from genetic material [53]. Some of the vaccine strategies mentioned here have suffered from poor bioavailability, primarily due to the delivery issues. For example, subunit vaccines, such as antigenic proteins, peptides, and polysaccharides, are not ideal candidates for oral and rectal administration as these therapeutics are prone to enzymatic degradation in the gastrointestinal (GI) tract. DNA vaccines also suffer from similar problems of degradation by harsh pH environment and enzymes [54]. The studies conducted with the naked plasmid DNA have shown that intravenous or intramuscular administration of the DNA elicited a weak immune response due to the restrictive movement, degradation by macrophages, and negligible uptake by myocytes [54, 55]. Besides all the inherent problems, the advantages associated with these modalities in comparison to the live-attenuated or inactivated pathogen-based vaccines have prompted researchers to exploit the nanocarrier-based approach to enhance the efficacy of such vaccines. In addition, the nanocarriers may also act as adjuvants to further enhance the immune response by protecting the antigen, modulating cytokine release, activating $CD8^+$ CTL responses, or delivering the antigen to target tissue [56].

Toward this end, the nanocarrier systems can be emulsions, liposomes, solid lipid nanoparticles and polymeric nanoparticles, and immunostimulatory complexes (ISCOMS). Since the theme of the chapter is on macrophage-targeted delivery systems, Table 4.4 lists the nanoplatforms that have been used to deliver the vaccine specifically to the macrophages.

4.2.4 *Macrophage-Targeted Nanosystems for Imaging*

The arrangement and spacing of atoms at scale of nanometer imparts unique physical and chemical properties to the material [36]. Particles such as colloidal gold, iron oxide crystals, and quantum dots are some of the examples of inorganic nanoparticles that are in the size range of 1–30 nm [70]. These particles have been explored for their ability to act as imaging/contrast enhancing agents for multimodality, noninvasive imaging techniques such as magnetic resonance imaging (MRI), positron emission tomography (PET), and computed tomography (CT). In addition, long circulation half-lives and ease of surface modification can further result in improvement of signal intensity and specificity. Overall, these features makes such particles as attractive diagnostic or contrast enhancing agents.

Macrophages have been identified as suitable imaging targets in diseases such as inflammation, atherosclerosis, and rheumatoid arthritis as these cells are present in

Table 4.4 Select examples of vaccines in nano-delivery systems for targeting macrophages

Vaccine type	Nano-delivery system	Size (nm)	Incorporated antigen	References
Protein/peptide vaccines	PCL, PLGA, PCL-PLGA blend, and PCL-PLGA copolymer	~250 nm	Diphtheria toxoid	[57]
	Poly(ethylene glycol)-poly(lactic acid)	150–170 nm	Tetanus toxoid	[58]
	RGD peptide-modified PLGA containing PCL-PEG	~211 nm	Ovalbumin	[59]
	N-trimethyl chitosan (TMC, polycationic) and mono-*N*-carboxy-methyl chitosan (MCC, polyampholytic)	~285 nm	Tetanus toxoid	[60]
	Galactosylated LDL nanoparticles	~23 nm	Ovalbumin	[61]
	Sulfobutylated poly(vinyl-alcohol)-grafted PLGA	100 nm, 200 nm, >1,000 nm	Tetanus toxoid	[62]
	Liposomes	0.15–3.9 μm	Cryptococcal culture filtrate antigen (CneF)	[63]
	Soy-bean oil nanoemulsion	20±5 nm	MAGE-HSP70 fusion protein and SEA (MHS)	[64]
	Calcium phosphate nanoparticles	<1.2 μm	HSV-2	[65]
DNA vaccines	Chitosan	337±27 nm 150–300 nm	HBsAg, pCMVArah2	[66, 67]
	Mannosylated liposomes	98±2.4 nm	pUb-M	[68]
	PLL	Not determined	psOVA-C1	[69]

PCL poly(epsilon-caprolactone); *PLGA* poly(D,L-lactide-*co*-glycolide); *LDL* low density lipoproteins; *PLL* poly(L-lactic acid)

abundance and also are involved in multiple functions leading to inflammation and angiogenesis. Therefore, monitoring their role in such conditions can provide valuable insight into the progression of the disease. Lipinski and colleagues [71] have described the formation of a scavenger receptor-targeted nanoparticles system for imaging macrophages in atherosclerosis. The micelle-based nanoparticles contained gadolinium, an MRI contrast agent, in the core, while the surface of the micelles was decorated with antibodies specific for scavenger receptor-B (CD36). The nanoparticles were reported to have a mean diameter of 125 nm, and on average, each nanoparticle was reported to contain 14,900 Gd atoms. Additionally, for the sake of comparison, another set of formulation containing the antibody against the Fc receptors on macrophages was also included, along with the untargeted nanoparticles. The *ex vivo* studies were conducted on the excised human aorta, harvested at the time of autopsy, with moderate to severe atherosclerosis. Pre-contrast images of the aorta sections were compared with the postnanoparticle containing contrast agent treatment using a 1.5-T MR system. The results indicated that the targeted nanoparticles increased the contrast-to-noise ratio (CNR) by 52.5% as compared to the Fc-NP (CNR increased by 17.2%) and untargeted nanoparticles (CNR increased 18.7%) ($p=0.001$). Additionally, confocal fluorescent microscopy revealed that targeted nanoparticles were localized into the macrophages, whereas the nontargeted and Fc-NPs were reported to distribute throughout the plaque region. Thus, this study demonstrated that nanoparticles targeting macrophages have the potential to improve the detection and characterization of the human aortic atherosclerosis. Specific examples demonstrating the efficacy of nanoparticles as macrophage-targeted contrast agents in an *in-vivo* model will be discussed in detail in later sections. Table 4.5 considers select examples of nanoplatform systems, containing suitable imaging agents that have been used to target macrophages and demonstrated the advantage of such systems in the treatment of atherosclerosis and rheumatoid arthritis disease models.

4.2.5 *Targeted Systems for Delivery of Small Molecule and Biological Therapeutics*

The diversity of macrophage function in various disorders such as HIV/AIDS, inflammation-related disorders, cancer, and infectious diseases makes it a valid pharmaceutical target. For example, macrophages are known to phagocytose and kill various microorganisms, but with the advancement in the field of molecular biology, it has been established that many of these pathogens have developed subtle means for residing in the macrophages by mainly avoiding phagocytosis or developing lysosomal tolerance [79]. Additionally, the drugs available for such diseases suffer from the problem of side effects and low solubility. Therefore, nanoparticle-based drug delivery systems can improve the therapeutic index of such drugs by lowering the toxicity and enhancing the targeting ability to macrophages. For example, Khan et al. [80] have developed 50–100 nm tuftsin-bearing liposomes to specifically deliver amphotericin B to the phagolysosomal compartment of macrophages in a mouse model of fungal infection, *Candida albicans*. The pharmacokinetic studies

Table 4.5 Nanoparticle systems targeting macrophages in atherosclerosis and rheumatoid arthritis

Nanoparticle system	Particle size	Disease state	References
Crystalline ethyl-3,5bis(acetylamino)-2,4,6-triiodobenzoate iodinated particles dispersed in surfactant for stabilization	259 nm	Atherosclerosis	[72]
Dextranated and DTPA modified magnetofluorescent labeled with ^{64}Cu	20 nm	Atherosclerosis	[73]
POPC, DPPE-NBD, DPPE-Biotin, Gd-DOTA-BSA	125 nm, 89±13 nm and 107.3±0.21 nm	Atherosclerosis	[71, 74, 75]
Self-assembled gold coated iron oxide NPs stabilized with dextran	~30 nm	Atherosclerosis	[76]
Silica NPs containing a luminescent $[Ru(bpy)_3]Cl_2$ core and a paramagnetic monolayer coating of a silylated Gd complex	37 nm	Rheumatoid arthritis	[77]
Ultrasmall iron oxide NP	~30 nm	Antigen-induced arthritis	[78]

POPC palmitoyl-oleoyl-phosphatidylcholine; *DPPE-NBD* 1,2-dipalmitoyl-*sn*-glycero-3-phosphoethanolamine-N-7-nitro-2-1,3-benzoxadiazol-4-yl; *DPPE-Biotin* 1,2-dipalmitoyl-*sn*-glycero-3-phosphoethanolamine-N-Biotinyl; *bpy* 2,2′-bypyridine

revealed that 32 mg/L of amphotericin B was available in the systemic circulation of mice treated with tuftsin-bearing amphotericin B liposome, while it was 25 mg/L for amphotericin B liposomes, 4 h post-drug administration. In vivo toxicity studies demonstrated that the amphotericin B deoxycholate formulation induced elevations in serum creatinine (approximately 300% of control) and blood urea (approximately 380% of control) values, while a significant decrease (blood urea approximately 150% of control and serum creatinine approximately 210% of control) was observed in the animals treated with the tuftsin-loaded amphotericin B liposomal formulation. Hence, overall, the tuftsin-bearing liposomal formulation resulted in minimizing the toxicity and side effects of the drug and improved the safety and efficacy profile of the amphotericin B.

Gagne et al. [81] prepared sterically stabilized (i.e., PEG-modified) immunoliposomes to deliver high concentrations of indinavir (protease inhibitor). Additionally, the immunoliposomes were coupled to Fab' fragment of the anti-HLA-DR antibody as the HLA-DR (class II MHC molecules) are highly expressed on the surface of macrophages. The particle size of the liposomes was reported to be 100–120 nm. The tissue and plasma distribution studies were conducted in female C3H mice that were given a single subcutaneous injection of free or encapsulated drug. Indinavir incorporated into sterically stabilized anti-HLA-DR immunoliposomes was shown to be highly efficient in delivering the drug to lymphoid tissues leading to 21–126-fold increased accumulation when compared to the free drug. Thus, based on these lines, several examples have been listed in Table 4.6, which summarizes the nanoplatform systems containing small molecule drugs for targeting macrophages.

Table 4.6 Select examples of small molecule therapeutics encapsulated in macrophage-targeted nano-delivery systems

Formulation type	Composition	Particle size	Active molecule	Disease condition	References
Mannosylated solid lipid nanoparticles		Not available	Rifabutin	*Mycobacterium tuberculosis*	[82]
Mannosylated liposomes	DSPC/Chol/Man-4-chol	110±6.9 nm	Dexamethasone palmitate	Endotoxin-induced lung inflammation	[83]
Mannan-coated gelatin nanoparticles		120±12 nm	Didanosine	HIV/AIDS	[45]
Mannosylated liposomes		1,000 nm	Ciprofloxacin	Parasitic infection	[84]
O-Palmitoyl mannan-coated emulsions (nanosized lipid particles)	Trilaurin/ phosphatidylcholine	0.32±0.02 μm	Amphotericin B (AmB)	Visceral Leishmaniasis	[32]
Hexylcyanoacrylate nanoparticles		230±20 nm	Azidothymidine	HIV/AIDS	[85]
Tuftsin-bearing liposomes	Egg phosphatidylcholine liposomes	Not determined	Rifampin	*Mycobacterium tuberculosis*	[86]
Tuftsin-bearing liposomes	Egg phosphatidylcholine liposomes	50–100 nm	Amphotericin B	*Candida albicans*	[80]
Folate-PEG-PAMAM dendrimers		Not available	Indomethacin	Rheumatoid arthritis	[87]
Acetylated LDL microemulsion	DPPE/DPPC/seal oil	27–32 nm	Azidothymidine and pro-drug forms	HIV/AIDS	[88]
Liposomes	Soy PC/cholesterol/D, L-α-tocopherol	135±55 nm	Clodronate	Anti-angiogenic therapy	[89]
Poly(isohexylcyanoacrylate) nanoparticles		250±20 nm	3′-Azido 3′-deoxy-thymidine (AZT)	HIV/AIDS	[90]
Immunoliposomes	DPPC/DPPG/DSPE-PEG-MAL	100–120 nm	Indinavir	HIV/AIDS	[81]

Alternatively, biological therapies such as therapeutic gene delivery or gene silencing have been implied to target macrophages. Additionally, in some cases, attempts have been made to transfer the therapeutic gene to these cells and use them as "guided" missiles [91]. The two general approaches used for gene delivery include viral and non-viral vector systems. The viral methods generally give higher transfection efficiencies and a longer transgene expression, but immunogenicity associated with such vectors and the cost of producing such vectors on large scale have led to exploration of efficient and safe nonviral vector platforms such as polymer- and lipid-based nanoparticles. However, the main disadvantage with non-viral vector systems is lower transfection efficiency as compared to viral counterparts. This mainly occurs because nonviral vectors are endocytosed, and it is very likely that DNA in such vectors can get degraded by nucleases, which are especially abundant in lysosomes of macrophages, following endosome–lysosome fusion.

4.3 Illustrative Examples for Delivery of Vaccines

4.3.1 Mucosal Vaccination

The mucosal delivery most commonly involves gastrointestinal, urogenital, and respiratory tracts [54]. The delivery of vaccines via the mucosal route is preferred as it can not only generate systemic immune response but can also provide local immune protection [92, 93]. Additionally, mucosal surfaces are considered to be the most common route for pathogen entry into the body, and hence, targeting such sites can prevent the invasion by the foreign antigen [92]. The delivery systems for mucosal vaccines have been primarily designed to target mucosal-associated lymphoid tissues (MALT) of the Peyer's patches in the gut and respiratory tract. The tissue is separated from the lumen by the follicle-associated epithelium (FAE), which is composed of enterocytes and specialized microfold (M) cells. These cells are capable of transcytosis of foreign matter from the apical to basal side of the membrane. M cell basolateral membrane contains a pocket that is deeply invaginated with lymphocytes and APCs such as macrophages [94]. Therefore, the strategic placement of macrophages at this site allows the sampling and processing of the foreign antigens, which can ultimately lead to activation of T cells and B cells, and hence the generation of cellular and humoral immune responses. Therefore, targeting the M cells for mucosal vaccine delivery can be considered as a passive and effective way to deliver the antigen to APCs such as macrophages.

Garinot et al. [59] developed PEG-modified poly(D,L-lactide-*co*-glycolide) (PLGA)-based nanoparticles displaying integrin-binding, arginine–glycine–aspartic acid (RGD) peptides at the particle surface. The surface modification was done to target the β-integrin receptors on M cells. Besides, the PLGA, the formulation also included poly(ε-caprolactone-*co*-ethylene glycol) (PCL-PEG), an amphiphilic copolymer. Photografting (i.e., light-induced chemical reaction) method was employed to covalently link the peptide to PEG moiety of the PCL–PEG, included in the formulation. The particle size and the zeta-potential of the peptide-modified formulation were 211 ± 2.9 nm and −13.8 ± 4.3 mV, respectively. Ovalbumin was used as a model antigen for the in vivo study, and the antigen encapsulation efficiency in the modified formulation was determined to be 40%. The in vitro studies were designed to study the particle uptake and transport in monoculture (Caco-2) and coculture models (Caco-2 and Raji cells). Basolateral solutions were then sampled, and the particles were counted using flow cytometry. Results obtained from these studies indicated that peptide-modified formulation enhanced the particle uptake by 3.5-folds as compared to the unmodified formulation in coculture cell model system. Next, in order to evaluate the ability of the formulation as an oral vaccine delivery system, in vivo studies were conducted in the specific antigen-free female NMRI mice with 5 μg of the ovalbumin antigen. The IgG titers in serum and IFNγ production level were measured after 10 and 13 weeks of first immunization, respectively. Additionally, in vivo localization of nanoparticles in mouse Peyer's patches was evaluated by performing confocal microscopy on FITC-labeled nanoparticles. The oral immunization studies revealed that number of mice producing IgG was slightly higher in the case of targeted nanoparticle formulation as compared to the nontargeted formulation. The group also reported that significant IFNγ production was induced in some of the groups immunized orally. Lastly, the confocal microscopy studies showed that targeted nanoparticles were found to be colocalized with M cells; however, the nontargeted nanoparticles were less numerous in Peyer's patches and more scattered.

With a similar objective, Wei et al. [64] encapsulated MAGE1-HSP 70 and SEA (MHS) complex protein in nanoemulsion as an antitumor vaccine delivery strategy via oral route. The average diameter of the nanoemulsion was 20 ± 5 nm, and the MHS complex protein encapsulation efficiency was reported to be 87%. The group then compared the efficacy of the delivery system in inducing immune response by delivering the nanoemulsion via peroral and subcutaneous routes. The results indicated that nanoemulsion containing the MHS protein complex could fiercely elicit cellular immune response as compared to MHS or nanoemulsion alone. Furthermore, encapsulating MHS in the emulsion system delayed the tumor growth and deferred tumor occurrence of mice challenged with B16-MAGE-1 tumor cells. The group also suspected that transcytosis of the emulsion by M cell and subsequent uptake by local macrophages and antigen presentation was the main pathway for the immune response observed via per oral delivery.

Roy et al. [67] formulated chitosan–DNA nanoparticles for oral allergen gene immunization. This strategy resulted in the modulation of peanut antigen-induced murine anaphylactic responses. The plasmid employed for this study was pCMVArah2. The particle size and surface charge of the nanoparticles were reported to be between 150 and 300 nm and +10 mV, respectively. The studies indicated that mice immunized with nanoparticles showed a significant reduction in allergen-induced anaphylaxis associated with alleviated levels of IgE, plasma histamine, and vascular leakage. Additionally, mice receiving the plasmid-containing nanoparticles produced secretory IgA and serum IgG2a. Lastly, the gene expression studies were performed by evaluating β-galactosidase expression after the chitosan–p43LacZ plasmid delivery. The studies revealed that due to the mucoadhesive nature of the chitosan polymer, the nanoparticles might adhere to the gastrointestinal epithelia and transported across the mucosal boundary by M cells and transfect epithelial and/or immune cells.

The other popular route for mucosal vaccination has been intranasal administration of nanoparticles. Khatri et al. [66] also developed plasmid/chitosan–DNA nanoparticles-based vaccine formulation against hepatitis B. The plasmid DNA construct used in this study was pRC/CMV-HBs(S). The nanoparticles were reported to be spherical in shape with a mean diameter of 337 ± 27 nm. Moreover, as part of formulation characterization, DNA stability in the nanoparticles was also confirmed and DNA encapsulation efficiency was reported to be $96.2 \pm 1.8\%$. Animal studies were conducted in female Balb/c mice (6–8 weeks) via intranasal administration of chitosan–DNA (CH) NP, with total 100 μg of the plasmid DNA. The serum anti-HBsAg titer in the CH-NP group was reported to be less in comparison to that elicited by naked plasmid DNA and alum-adsorbed HBsAg. However, the group claimed that mice were seroprotective within 2 weeks, and the IgG level was above the clinical protective level (>10 mIU/ml), indicating successful generation of systemic immunity. Also, as depicted by the level of secretory IgA antibody, chitosan–DNA nanoparticles were reported to be more efficient in eliciting mucosal immune response as compared to intramuscular (i.m.) administration of naked plasmid and alum-adsorbed recombinant protein vaccine. Lastly, the group suspected that chitosan nanoparticles may be able to pass the mucosal membrane and directly transfect the antigen-presenting cells, or they might be first taken up by the M cell-like cells in nasal-associated lymphoid tissue (NALT) and then presented to the underlying APCs.

4.3.2 Systemic Vaccination

In addition to the mucosal vaccination routes (oral and intranasal), nanoparticle-based vaccine administration routes such as subcutaneous, intravenous, intraperitoneal, and intramuscular have also been explored for targeting macrophages. For example, Hattori et al. [95] have investigated the potency of mannosylated, cationic (Man-C4-Chol) liposomes as a DNA vaccine carrier. Ovalbumin antigen-expressing plasmid (pCMV-OVA) was constructed to evaluate DNA vaccination. Furthermore, the potency of the construct was compared with naked plasmid vector and 3β-[N-(N′-N′-dimethylaminoethane)-carbamoyl] cholesterol (DC-Chol) liposomes, upon intravenous administration. The mean particle size of the Man-liposome and DC-Chol liposomes complexed with plasmid DNA were 182.3 ± 7.4 nm and 175.8 ± 5.2 nm, respectively. In an attempt to evaluate antigen presentation by macrophages in vitro, the authors adopted a coculture model of macrophages, cultivated with CD8OVA1.3 T cell hybridomas. The group stated that after antigen is expressed in APCs and subsequently processed and presented as peptide epitopes on MHC class I molecules, it would result in stimulation of antigen epitope-specific CD8$^+$ T cells. Indeed, a higher secreted level of IL-12 from CD8OVA1.3 T cell hybridoma was reported in case of Man-liposomes then in naked plasmid or DC-Chol treatment group. The in vivo studies looked at OVA mRNA expression and MHC class I-restricted antigen presentation on CD11c$^+$ cells and also levels of inflammatory cytokines TNF-α, IL-12, and IFN-γ. Additionally, OVA-specific cytokine release after intravenous administration of different nanoparticles versions was also evaluated. Overall, these results indicated that targeted delivery by mannose-liposomes was superior in their ability to target APCs and enhance the Th1 response and, hence, was a potent method for DNA vaccine therapy.

Other groups such as Tang et al. [69, 96] and Lu et al. [68] have also utilized mannan-coated polylysine and liposomal delivery systems for APC-targeted DNA vaccination via intradermal and intraperitoneal routes of administration, respectively. Ribeiro et al. [97] synthesized dendriplexes (dendrons/plasmid DNA complex) and encapsulated them in PLGA particles using the double emulsion method. The plasmid DNA employed for the study was protective antigen (PA) of *Bacillus anthracis*. The particle size of the nanoparticles was reported to be approximately 500 nm. The group tested different versions of the nanoparticle constructs by incorporating 14 μg of the plasmid DNA per dose, administered at weekly intervals via intramuscular route. Antibody titers were measured after three immunization injections. The authors concluded the PLGA construct with a C-18 hydrocarbon chain was superior to the rest of the formulations in producing the anti-PA IgG antibodies in the immunized mice. More importantly, the authors hinted that sustained release of the DNA from the PLGA particles, and hence, transfection of muscle cells were aiding antigen presentation by recruited mononuclear cells and subsequent generation of humoral and cell-mediated immune response.

4.4 Illustrative Examples for Imaging

Fayad et al. [98] have developed contrast agents that can be used in CT imaging of atherosclerosis by targeting the high-density localized macrophages. Studies from this group mentioned that such an approach may be beneficial over conventional CT scan of atherosclerotic areas. In a conventional CT imaging process, the atherosclerotic plaques are characterized on the basis of their densities as hypodense, dense, or calcified. However, plaque density strongly depends on intensity of luminal enhancement, which may ultimately depend on factors such as cardiac blood flow and time delay between injected contrast agent and CT scan acquisition [99]. Thus, when measuring plaque density, high variability in luminal enhancement intensity is a clear limitation in CT angiographies. Thus, in order to circumvent the problem, a more plausible route would be to target macrophages that are present in abundance in the plaque region [100].

In collaborations with NanoScan Imaging (Lansdale, PA), the Fayad group has extensively studied an iodinated nanoparticle-based agent, N1177. It was reported that the agent was composed of crystalline iodinated particles dispersed in surfactant for stabilization. The nanoparticles were made from ethyl-3,5-bis (acetylamino)-2,4,6-triiodobenzoate. Nanoparticle suspension was made by milling the iodinated particles with the surfactant in the presence of inert beads. The mean particle size of the particle was 259 nm, and on-shelf stability was reported to be 8 months. The concentration of the iodine in the particles was measured to be 67 mg/ml. Next, the uptake of N1177 by macrophages was evaluated both qualitatively and quantitatively by optical microscopy and inductively coupled plasma mass spectroscopy (ICP-MS), respectively. The results obtained via microscopy revealed the presence of dark granules, indicative of nanoparticles, only in the cytoplasm of the cells. In order to quantify the uptake of N1177 by macrophages, the cells were incubated for 1 h with both nanoparticle agent and conventional CT agent, adjusted to same iodine concentration. The group showed via ICP–MS that the uptake of N1177 was $4{,}920 \pm 1{,}019$ μg iodine/g wet weight versus 56 ± 11 μg iodine/g wet weight of contrast agent.

Additionally, the in vivo biodistribution of N117 and the detection of macrophages in atherosclerotic plaques were also performed in the New Zealand white rabbit model. The aim of the distribution study was to evaluate the effectiveness of N1177 in enhancing the circulation time in blood and accumulation in macrophage-rich region upon intravenous injection. The control used in the study was a conventional CT contrast agent, iopamidol. The dose of both the contrast agents was maintained at 250 mg iodine/kg of body weight. The results showed that the X-ray absorption value or Hounsefield unit (HU) of the aortic lumen was higher for N1177 (125 ± 14.1 HU) as compared to iopamidol (61.8 ± 13.9 HU). The group reported that absorption value obtained via N1177 was sufficient to achieve a clear delineation of all the major arterial and venous contours by CT. However, 2 h postinjection, the absorption value, and density measurement in aortic lumen obtained for both contrast agents were not statistically significant from precontrast values. Moreover, at this time point, it was shown that N1177 had distributed to organs containing macrophages such as liver and spleen, whereas no enhancement was detected in these organs 2 h after the injection of the conventional contrast agent.

Lastly, in order to detect the macrophages in atherosclerotic plaques, eight rabbits were given atherosclerotic lesions by combination of double balloon injury in the aortas (1 month apart) and a hypercholesterolemic diet (for 4 months). The lesions generated via this procedure were reported to contain high levels of macrophage infiltration. Four additional, non-injured rabbits fed with chow diet were used as controls. All the animals were first imaged with traditional contrast agent (iopamidol) and 1 week later with N1177. Previous results obtained from kinetic study revealed that the optimal imaging time was 2 h post-N1177 administration. The same time frame was used for the detection of macrophages in this part of the study, and the results showed that a significant elevation in the detection of the macrophage-rich plaque region was obtained with N1177 as compared to iopamidol ($p < 0.001$). Densities measured in plaque region increased from 29.7 ± 6.0 HU (before injection) to 43.0 ± 7.3 HU (2 h postinjection) in the case of N1177. In comparison, density values obtained with iopamidol were 31.5 ± 6.9 HU (before injection) and 35.6±8.1 HU (2 h postinjection). No substantial enhancement in the arterial wall of control nonatherosclerotic rabbits 2 h after the injection of the N1177 (0.9±1.2 HU) or conventional (0.3±1.3 HU) CT contrast agent was detected. The next step was to correlate the data generated via CT imaging with intensity of the macrophage infiltration in the corresponding lipid-rich core of the plaque via histological analysis. Thus, after reconstruction of the CT axial slices of the aorta, the corresponding sections were excised and sectioned for immuno-histochemistry analysis. Fibrous cap and lipid-rich core of the plaques were stained and analyzed for the distribution of macrophages by using a specific monoclonal antibody to RAM-11 (a marker for macrophage cytoplasm). The results confirmed that the N1177-enhanced CT showed a macrophage area extending to more than 20% of the intimal area in 90% of the corresponding histological sections. No macrophages were detected in the aortic wall of control rabbits. Additionally, the data was also supported by TEM images and energy dispersion spectrometric data, showing the presence of a large number of electron-dense granules in the lysosomes of macrophages in the atherosclerotic plaques of rabbits, killed 2 h after injection of N1177. Thus, overall, the study successfully showed that the in vivo imaging of macrophage infiltration in atherosclerotic plaques of rabbits can be improved via development of contrast agents such as N1177.

Lutz et al. [78] conducted a preliminary study to evaluate the ability of ultrasmall, superparamagnetic iron oxide (USPIO) nanoparticles to detect synovial macrophages in an experimental rabbit model of antigen-induced arthritis, using molecular resonance imaging (MRI). The iron oxide nanoparticles were reported to be 18–30 nm in size. Thirteen female rabbits (NZW) were divided into two groups: control ($n=3$) and antigen-injected ($n=10$). Intra-articular injections of methylated bovine serum albumin were performed to induce unilateral arthritis. In addition to the three control animals, the contralateral knee joint of the antigen-injected animal group also served as controls in this study. It was reported that after the onset of arthritis, all the knees were imaged prior to and 24 h post-USPIO particle injection. The nanoparticles were administered via i.v. route at a dose of 150 μmol of iron per kilogram of body weight. Two MR imaging sessions comprising of baseline readings and 24 h post-nanoparticle administration were performed using a 1.5-T MRI system. Various MR imaging protocols including T1-weighted spin echo, T2-weighted fast spin echo, T2*-weighted gradient echo, and short inversion time inversion-recovery sequences were adopted for the study. The images were obtained in the transverse, sagittal, and coronal planes for all the sequences. It was also mentioned that qualitative and quantitative image analysis was performed with regards to signal characteristics and pattern. In addition, histopathological studies were conducted to locate the nanoparticles in the synovial tissue. As a first step in the study, the difference between the arthritic knee and control or contralateral knee was established in terms of synovial thickening and joint fluid. It was mentioned that the mean thickness of the arthritic synovium as measured by T2-weighted images was 2.8±1 mm as compared to nonarthritic knee, which was reported to be below 1 mm in thickness. Images obtained with T2 and T2* weighted and short inversion time inversion-recovery sequences showed moderate-to-large joint effusion in arthritic knees, whereas little or no joint fluid was reported in nonarthritic knees. The same pattern was confirmed with histopathologic studies, where extensive synovial hyperplasia was reported in all the arthritic knees and normal synovial lining was observed in nonarthritic knees.

More importantly, the images obtained 24 h after the intravenous administration of the USPIO nanoparticles revealed a significant increase in the signal-to-noise ratio (SNR) in case of T1 ($p=0.03$). However, more predominant T2 ($p=0.01$) and T2* ($p=0.02$) effects were observed in the synovium of the all the arthritic knees. It was reported that only T2* effects were present in the joint effusion of the arthritic knees ($p=0.01$). The authors reported that no significant changes were observed in the contralateral joints of the antigen-injected and the joints of the animals in the control group ($p=0.6$–0.91). Lastly, results of the histologic examination revealed numerous blue iron-positive cells within the hyperplastic synovial tissue. The cells were identified to be macrophages. It was mentioned that the areas of the focal signal loss on T2-weighted fast SE or susceptibility effects on T2-weighted images correlated to foci of iron staining upon histologic analysis. Histologic examination of the contralateral knee of the antigen-injected did not reveal accumulation of the iron nanoparticles-containing macrophages. The authors suspected that iron particles may permeate through the capillary walls of the blood vessel into the interstitial space at sites of inflammation and are ultimately phagocytosed by the macrophages in the inflamed tissue. The group also acknowledged that kinetics of the particle uptake and quantification of the USPIO particles into the resident macrophages is critical and needs to be explored in greater detail. In addition, there is a need to construct the biodistribution profile of such small nanoparticles and determine the toxicity of these particles in an in vivo setting. In summary, this study showed that nanoparticle-based contrast agents provide an opportunity to selectively visualize arthritic joint in an experimental model of antigen-induced arthritis and that these agents have the potential to serve as an imaging modality in monitoring the therapeutic effects of drugs that affect macrophage activity.

4.5 Targeted Therapeutic Systems in Inflammatory Diseases

4.5.1 Illustrative Examples of Small Molecule Therapeutics

The rationale for employing macrophage-targeted therapy in the treatment of inflammation-related diseases is based on their direct or indirect involvement in such disease areas such as atherosclerosis/restenosis, inflammatory bowel disease (IBD), neuroinflammation, and rheumatoid arthritis. As a general strategy, bisphosphonates have been explored for their properties to act as anti-inflammatory drugs [89]. This class of drugs has been known to chemists since the early nineteenth century. Bisphosphonates are pyrophosphate analogues where oxygen bridges have been replaced by a carbon with various side chains [101]. It has also been known that bisphosphonates (e.g., clodronate) in liposomes are selectively taken up by macrophages and induce apoptotic cell death after delivery to the cytoplasm. The elimination of subsets of macrophages in various tissues is dependent on the route of administration and dose of the drug. Based on this observation, these drugs have been explored as anti-inflammatory drug candidates to treat diseases such as restenosis.

Restenosis can be described as formation of neo-intima, leading to re-obstruction of vessel wall after balloon injury. Upon injury, macrophages accumulate in the vessel wall triggering smooth muscle cell proliferation and extracellular matrix formation via secretion of numerous growth factors, cytokines, and enzymes leading to vessel obstruction [102]. Danenberg et al. [103] were the first group to report that intravenous administration of clodronate-containing liposomes (LC) resulted in macrophage depletion and reduced neo-intimal formation after balloon injury in rat (Sabra male) and New Zealand white rabbit models. The liposomal formulation comprised of 50 μmol/L distearoylphosphatidylglycerol (DSPG), 100 μmol/L cholesterol, and 150 μmol/L of 1,2-distearoyl-*sn*-glycero-3-phosphocholine (DSPC) by reverse-phase evaporation technique. The concentration of the drug in the liposomes was 24.5 mmol/L, and the average particle size of the clodronate-containing liposomes was 190 ± 18 nm. The rabbits were kept on hypercholesterolemic diet for 30 days prior to angioplasty. Upon confirming hypercholesterolemia (plasma cholesterol >1,200 mg/dl), animals were anesthetized and balloon injury was performed on the left common carotid artery with a 3-mm angioplasty balloon catheter. However, to ascertain the effect of macrophage depletion in a nonhypercholesterolemic animal model, the rat carotid injury model was used. In both animal models, the dose of the free drug or encapsulated drug was maintained at 15 mg/kg. Other controls included empty liposomes and buffer. The liposomal formulation and free drug were administered intravenously at days −1 and +6 in both animal models. Morphometric analysis was performed to estimate the degree of neointimal thickening and the degree of remodeling. The degree of neointimal thickening was expressed as the ratio between the area of the neointima and the original lumen (i.e., percent stenosis) and as the ratio between the neointimal area to the area of the media (N/M). The degree of remodeling, constrictive (negative) and expansive (positive), and remodeling ratio was measured by comparing the ratio of the total arterial area of the balloon-injured segment with that of an adjacent, noninjured segment. Overall, the results indicated that marked reduction in the neointima formation was observed upon treatment with clodronate-containing liposomes (LC). It was also mentioned that the treatment with the LC was the major contributing factor for the observed increased in the luminal area in comparison to the mild increase observed via expansive remodeling. Additionally, no systemic adverse effects were reported.

Additional studies were performed to unveil the underlying mechanism responsible for marked neointima reduction. The anti-coagulated blood of rabbits was analyzed to count the population of circulating monocytes via flow cytometry. Immunohistochemistry analysis was performed to evaluate the macrophage prevalence in the arterial sections of the hypercholesterolemic rabbits and other organs such as liver and spleen. Lastly, IL-1β concentration and matrix metalloproteinase-2 (MMP-2) activity in the arterial tissue of the rabbits were also measured

6 days after treatment with clodronate-containing liposomes. Both IL-1β and MMP-2 are secreted by activated macrophages in response to injury and, hence, contribute to the process of neointimal proliferation. Thus, measurement of these parameters before and after the treatment can act as a yardstick to further authenticate the presence of macrophages in the injured arterial tissue. The combined results of all these studies indicated that the population of circulating blood monocytes and infiltrated macrophages in the injured tissue were transiently reduced, along with IL-1β concentration and MMP-2 activity, upon treatment with clodronate liposomes. Thus, it was concluded that LC administration reduced the neointimal hyperplasia after balloon injury in the rat and hypocholesterolemic rabbit models. The suggested mechanism was systemic selective, transient modulation of monocyte/macrophage activity.

In a similar study, Cohen-Sela et al. [104] investigated the effect of polymeric nanoparticle-based formulation containing bisphosphonates alendronate in alleviating restenosis. The drug was encapsulated into PLGA nanoparticles prepared via double emulsion-solvent evaporation technique. The ALN nanoparticle had an average diameter of 223 ± 64 nm, with a negative surface charge of -4.4 ± 0.9 mV. The drug encapsulation efficiency was determined to be $55.1 \pm 7.4\%$. In vivo studies indicated that ALN nanoparticles resulted in significant reduction in neointima-to-media ratio and stenosis after balloon injury. Moreover, the nanoparticle treatment led to reduction in the levels of both IL-1β and matrix metalloproteinases [2, 9]. The authors concluded that ALN nanoparticles have the capability to reduce the neointimal formation in vivo by systemic depletion of monocytes.

In the case of rheumatoid arthritis, an increase in the expression of Fc receptors on the surface of macrophages has been reported. Chandrasekar et al. [87] exploited this observation to deliver indomethacin, a non-steroidal anti-inflammatory drug (NSAID), encapsulated in folate-targeted PEG conjugates of an anionic dendrimer (G3.5 PAMAM), to the inflamed arthritic site. The pharmacokinetic studies revealed an increased AUC, circulatory half-life, and mean residence time in the case of the folate-PEG-PAMAM conjugates. Additionally, the time-averaged relative drug exposure r(e) was reported to be between 1.81 and 2.37, and the overall drug targeting efficiency T(e) was reported to be 3.44 as compared to the native dendrimer (1.72). Thus, it was concluded that such a construct is an ideal choice for targeted drug delivery for anti-arthritic therapy.

4.5.2 *Illustrative Examples for Delivery of Biological Therapeutics*

Howard et al. [105] explored the possibility of down-regulating/silencing the expression of tumor necrosis factor (TNF-α) using the siRNA (small interfering RNA) approach. Towards this end, the investigators explored the possibility of using a chitosan–siRNA polyplex system in downregulating the TNF-α expression in peritoneal macrophages for anti-inflammatory treatment in a murine arthritis model.

The selection of systemic macrophage-derived TNF-α is based on its predominant role in modulating local proinflammatory and regulatory cytokine effects. Additionally, recruitment of systemic macrophages to local sites and their role during local inflammation makes them an ideal target for such therapies. The investigators chose the intraperitoneal route over intravenous administration as the intraperitoneal administration allows delivery into a blood-free, macrophage-rich environment, and serum protein-induced polyplex aggregation associated with systemic delivery can also be avoided.

Self-assembled nanoparticles comprising of 27-mer dicer-substrate siRNA (DsiRNA) or control siRNA and chitosan (84% deacetylated; N:P ratio of 63) were formed. The particle size was reported to be ranging from 350 to 450 nm. A significant TNF-α knockdown was reported in in vitro experiments conducted with primary murine peritoneal macrophages, after 24 and 48 h of particle administration. The in vivo studies looked at both the therapeutic and prophylactic treatment strategies. The therapeutic potential of silencing TNF-α production in systemic macrophages with chitosan-based nanoparticles was investigated in a collagen type II DBA/I arthritic model. The treatment groups for animal studies were as follows: unmodified anti-TNF-α DsiRNA (5 μg), 2′-O-Me modified anti-TNF-α DsiRNA (2.5 μg), control DsiRNA (5 μg), sodium acetate buffer (0.2 mol/L), and dexamethasone positive control group (400 μg/kg; daily (day 1–14)), subcutaneously. Except for the dexamethasone-treated group, the rest of the groups were given i.p. treatment on days 1, 3, 5, 7, and 9 with 200 μL of nanoparticles. Five animals were treated per group. The starting arthritic score was determined to be ~3 (day 1). The end points of this study were based on the arthritic score (before and after treatment) and survival (%). Results indicated that there was a significant difference between the modified chit/DsiRNA nanoparticles and control siRNA nanoparticles ($p<0.028$). However, no significant difference between the control and buffer-treated group was reported. The animal survival data indicated a marked difference between anti-TNF-α DsiRNA (modified and unmodified formulations) and dexamethasone (100% survival), or control siRNA (60% survival) and buffer (40% survival). Furthermore, histological analysis of the paws, extracted on day 5, revealed that joint and cartilage integrity was maintained in animals treated with unmodified anti-TNF-α DsiRNA and dexamethasone control groups. On the other hand, control siRNA nanoparticle and buffer-treated group exhibited extreme cartilage and bone destruction. Interestingly, the authors reported that some level of cartilage damage and cellular infiltration was observed in the mice treated with modified 2′-O-Me anti-TNF-α DsiRNA; however, the damage was to a lesser extent as compared to the control siRNA and buffer-treated groups. The inflammation observed in the case of modified formulation was speculated to result from a combined suppression of TNF-α and type I IFN effect (innate response observed due to the non-specific interaction of the double-stranded RNA with toll-like receptors). However, it was mentioned that these abnormalities can be improved if the administered dose of the modified formulation is increased.

Finally, as part of prophylactic studies, the investigators were interested in evaluating the efficiency of the formulations (modified and unmodified) to silence TNF-α production during the inductive phase (day 0–28 after collagen immunization) of the disease so that the onset of inflammation can be delayed. A similar protocol to therapeutic studies was applied in this study. The modified and unmodified formulations were administered 2 days before immunization and after immunization on days 1, 5, 9, 13, 17, and 21. The control siRNA nanoparticles (score = 1.3) and buffer-treated group (score = 1.5) showed signs of joint inflammation as compared to modified formulation (score = 0.3) from day 26 onwards. Interestingly, rapid progression of the joint inflammation was noticeable by the end of second day of withdrawing the treatment with modified formulation. The arthritic score of 0.3 on day 26 rose to ~5.3 by day 31. The score for control siRNA nanoparticles and untreated group, during the same period, was reported to be 1.5–2.5 and 1.5–3.2, respectively. The investigators attributed the rapid onset of inflammation to the immunoregulatory effect of TNF-α suppression. The rapid onset of inflammation upon termination of the treatment, however, points to the transient nature of RNAi-based therapeutics. While there is a need to modify the design of the current formulation to compensate for the transient siRNA expression, the studies of Howard et al. provide an insight into the potential of siRNA based formulation in the treatment of inflammation-related diseases.

Similarly, Zuo et al. [106] investigated the effect of alveolar macrophage-targeted NFκB decoy by mannosylated (Man) cationic liposomes in a LPS-induced lung inflammation model, upon intratracheal administration. The rationale for using antisense ODN is attributed to its ability to specifically modify cellular gene expression by either disabling or degrading the target RNA [107]. The nanocomplex was reported to be around 100 nm in size and was aerosolized by Microsprayer® for intratracheal administration. The particles were reported to be physically stable during the spraying process. Confocal microscopy was performed on the lung tissue sections excised after intratracheal administration of the FAM-labeled Man-cationic liposome, loaded with 50 μg of NF-κB decoy. As evident from the cell uptake ratios, the Man-liposomes (uptake ratio 4.0) were highly selective towards alveolar macrophages as compared to the naked NFκB decoy (uptake ratio 3.0) ($p < 0.01$) and its complex with cationic liposome (uptake ratio 2.3) ($p < 0.01$). Additionally, it was reported that enhanced uptake of NF-κB decoy by Man-cationic liposomes was significantly inhibited upon co-administration of mannan ($p < 0.05$), further confirming mannose receptor-mediated endocytosis. The therapeutic potential of the nanoplatform was tested in the LPS-induced lung inflammation model. The nanoparticles were first administered and lung inflammation was induced 30 min post-nanoparticle treatment by intratracheal instillation of LPS at 0.5 mg/kg. The BAL samples and lung tissue were collected 3 h post-LPS challenge. The cytokine levels of TNF-α, IL-1β, and chemokine CINC-1 along with MPO activity were measured. The electrophoretic mobility shift assay of the nuclear extracts from lung tissue was also assessed to further examine the inhibitory mechanism of NFκB decoy. The results indicated that Man-cationic liposomes were superior in their ability to inhibit TNF-α, IL-1β, and CINC-1 in BAL fluid as well as in the lung tissue ($p < 0.05$). Moreover, the levels of cytokine and chemokine achieved with the mannosylated formulation were significantly lower when compared to that of LPS, naked NFκB

Table 4.7 Select examples of macrophage-targeted nanosystems for biological therapy

Therapeutic moiety	Nano-delivery system	Particle size	Disease condition	References
NF-κB decoy	Mannosylated liposome	100 nm	LPS-induced lung inflammation	[110]
Anti-TNF-α oligonucleotides	Galactosylated low molecular weight chitosan	Not available	Experimental colitis	[106]
TNF-α siRNA	Solid polyketal PK3 and choloroquine	800–900 nm	Acute liver failure	[111]
Plasmid CMV-luciferase	Mannosylated cationic liposomes/DNA complex	~200 nm	Reporter gene expression	[112]
Plasmid CMV-luciferase	Mannosylated poly (L-lysine)/DNA complex	10–20 nm	Reporter gene expression	[113]
Plasmid CMV-luciferase	Histidine-conjugated mannosylated cationic liposomes	116±15.8 nm	Reporter gene expression	[114]
IL-1,6, and IL-8 siRNA	Lipoplex	~700 nm	Experimental rheumatoid arthritis	[115]
Oligonucleotides	Galactosylated low molecular weight chitosan	Not available	Kupffer cells targeted anti-inflammatory therapy	[116]

decoy, unmodified cationic liposome/NFκB decoy complex, and mannosylated cationic liposomes containing the scrambled decoy complex. A similar trend was reported for the neutrophilic MPO enzyme activity levels. Lastly, the EMSA analysis revealed stronger inhibition of the activated NF-κB in lung tissue after treatment with mannosylated liposomes/ NF-κB κ = kappa decoy complex as opposed to other controls. No inhibitory effects were reported with mannosylated cationic liposomes containing the scrambled decoy complex. Overall, this study indicated that active targeting to macrophage population can be achieved by incorporating mannose sugar residues into the formulation. Such modification proved to be an effective anti-inflammatory strategy for lung inflammation.

The illustrative examples mentioned in this section have been focused on the suppression of anti-inflammatory cytokines such as TNF-α. However, it will also be interesting to look at examples that have taken the alternate route for treating inflammation by delivering plasmid gene encoding for anti-inflammatory cytokine protein such as IL-10 to macrophages, and comparing it to the silencing effect of TNF-α. As far as the therapeutic gene delivery is concerned, the studies mentioned in Table 4.7 have been conducted with a reporter plasmid DNA system and have not clearly demonstrated the efficacy of the formulations to deliver the therapeutic gene to macrophages in an in vivo setting. Towards this end, the examples that we found were focused on delivery systems that are in the micrometer range, and hence, were excluded as the focus of the chapter is mainly on nanometer range technologies.

However, interested readers are encouraged to look at studies conducted by Bhavsar et al. [108] and Nakase et al. [109] that evaluated the efficacy of the therapeutic IL-10 plasmid DNA in a microsphere-based formulation for the treatment of inflammatory bowel disease.

4.6 Targeted Therapeutic Systems in Cancer

4.6.1 Illustrative Examples of Small Molecule Therapeutics

The role of tumor-associated macrophages (TAM) has already been reviewed in the previous sections. The general approach in this field has focused on either eliminating the macrophages from the tumor environment or use of macrophages as a Trojan horse to carry therapeutics to solid tumor cells [117]. There are also few relevant examples that have looked at nanoparticle-based systems for targeting tumor stromal cells from immunotherapy perspective.

Miselis et al. [118] examined the contribution of TAM to tumor growth and metastasis using an orthotopic immunocompetent mouse model of diffuse malignant peritoneal mesothelioma. Clodronate-containing liposomes (CLIP) were used to target and eliminate the macrophages from the tumor environment. Additionally, the CLIP liposomes were labeled with a fluorescent dye, chloromethylbenzamido-octadecyl (C_{18}) indocarbocyanine (CM-DiI). The 40 L mesothelioma cell line and C57BI/6 wild type (WT) and C57BI/6 TgN (bACT-eGFP) 10sb mice were used for the studies. As part of the ex vivo studies, the group evaluated the macrophage targeting ability of the formulation. Toward this end, WT or eGFP mice were injected i.p. with 2×10^6 mesothelioma cells and tumor spheroids were harvested. The tumor spheroids extracted from eGFP mice were then treated with CM-DiI labeled CLIP, CM-DiI labeled liposomes, or mock CM-DiI labeled PBS. Immunofluorescence labeling was done with CD68, CD11c, and GFP to identify macrophages, dendritic cells, and host cells. The authors reported that the majority of the cells targeted with CM-DiI liposomes were macrophages in the explanted spheroids, which was evident from the colocalized fluorescence images of labeled liposomes and F4/80 (red fluorescence)-labeled macrophages.

As part of the in vivo studies the group looked at the distribution of the CLIP, induction of apoptosis by CLIP, and overall efficacy of the treatment with CLIP. The biodistribution studies were done with CM-DiI labeled liposomes, injected into tumor-bearing mice. The tissues were collected posttreatment and counterstained with 4,6-diamidino-2-phenylindole, and organs including liver, spleen, intestine with mesentery, kidney, and lungs were examined. The biodistribution profile indicated extensive labeling in the marginal zone of spleen, with scattered labeling of the individual cells in the white pulp. Histopathological studies indicated extensive apoptosis and focal congestion in the red pulp of the spleen. Additionally, liver uptake by Kupffer cells and focal injury of liver cells was also revealed. More importantly, CM-DiI labeled CLIP and mock CM-DiI labeled PBS showed extensive labeling of the solid tumor located at the periphery. Additionally, it was reported that CLIP treatment resulted in a decrease in the tumor cell density and enhanced apoptosis within mesenteric tumors. Based on these results and additional studies, the group concluded that clodronate-containing liposomes induced apoptosis in both tumor spheroids and in established tumors in vivo.

Lastly, three separate experiments were conducted to determine the efficacy of the CLIP treatment in animal tumor model. In summarization, these studies indicated that tumor-bearing mice injected with CLIP resulted in 4-fold reduction in tumor frequency as compared to mock CM-DiI labeled PBS. In addition, a 17-fold reduction in tumor burden, and a 5-fold reduction in the tumor invasion and metastasis was observed in comparison to the mock CM-DiI labeled PBS group. They also reported that following transplantation of tumor spheroids and treatment with CLIP, a 4-fold decrease in tumor number and 15-fold decrease in tumor burden was observed as compared to the control. A similar trend was observed in mice bearing established tumors, where treatment with CLIP resulted in a 2-fold reduction in tumor number and relative tumor burden. Thus, based on the overall studies the authors concluded that targeting peritoneal macrophages in the orthotopic model of malignant mesothelioma can suppress the tumor growth, invasion, and metastasis.

4.6.2 *Illustrative Examples for Biological Therapeutics*

Opanasopit et al. [119] studied the inhibitory effect of a nanosized, mannosylated liposomal formulation containing an immunomodulator, muramyl dipeptide (MDP), in an experimental liver metastases model. MDP is a component of bacterial cell wall and has been used as an adjuvant to stimulate macrophages. Upon activation by MDP, macrophages can produce prostaglandins and collagenase, super-oxide anions, and more importantly, cytolytic activity against tumor cells [120]. The composition of the formulation consisted of Man-4-chol, distearoylphosphatidylcholine (DSPC), and cholesterol in 5:1:2.5 molar ratio. Liposomal mixtures were hydrated with MDP solution at a ratio of 20 μg/mg of lipid. The particle size of the formulation was reported to be approximately 95 nm. The murine colon carcinoma cell line, CT-26 was used for tumor development in male ddY and CDF1 mice. Particle distribution studies were carried out with [^{3}H]-labeled liposomes in male ddY mice. The particle distribution was analyzed in terms of organ uptake, with particular emphasis on the distribution between parenchymal cells (PC) and non-parenchymal (Kupffer) cells (NPC) of the liver. It was reported that 2.5 and 25 mg/kg doses of the injected formulation resulted in 75% of radioactivity recovery from the liver. Further dose increase led to a decrease in radioactivity recovery, an observation the authors attributed to saturation of the mannose receptor-mediated particle uptake in the liver. Moreover, the particle accumulation ratio in the PC versus NPC cells was reported to be 0.39 in case of Man-liposomes. In case of unmodified, ^{3}[H]-free liposomes, equal amounts of particles were found in both types of liver cells (PC and NPC).

Next, experimental liver metastasis was induced in the CDF1 mice by injecting 10^5 CT-26 tumor cells into the portal vein. Free MDP, a mixture of empty liposomes with free MDP, or liposomes containing MDP were administered on days −3, 0, 3, 7, and 10 of tumor injection. Mice were sacrificed on day 14, and the number of tumor modules on the liver surface and liver weight were measured. The results revealed that the number of tumor modules in the control (treated with saline) and free MDP groups was approximately 50. A slightly better effect was seen with bare liposomes containing MDP ($p<0.01$); however, mannosylated liposomes containing MDP were most promising in their ability to inhibit liver metastasis ($p<0.005$). No improvement in liver metastasis inhibition was seen with bare or Man-liposomes plus free MDP groups. The survival time of the tumor-bearing mice was significantly increased upon treatment with the Man-liposomes ($p<0.05$). Overall, the study successfully demonstrated the ability of nanoparticle formulation to actively target non-parenchymal (Kupffer) cells of the liver and further exploited the tumoricidal activity of macrophages to inhibit tumor metastasis.

4.7 Concluding Remarks and Future Directions

The examples discussed above have demonstrated the therapeutic efficacy of nanoparticles and the benefits of targeting macrophages with such platforms in treating various diseases for both small molecule drugs as well as biological therapeutics that include antibodies or nucleic acid constructs. Additionally, examples of both passive and active targeted nanoparticles towards these cells have been provided. However, another approach—complementing active targeting—is transcriptional targeting. This approach is particularly applicable to gene delivery systems for macrophages and takes advantage of the tissue-specific/inducible promoters or enhancers that control gene expression. The concept has been very well established and particularly explored for cancer gene therapy [121, 122]. In our case, the promoters could be induced by a physiological state such as inflammation or secreted proteins such as cytokines. Moreover, these elements could be incorporated into plasmid DNA encoding for the therapeutic protein. Such DNA system could be either complexed with a nanoparticle vector or physically encapsulated into the vector. The nanoparticles could be surface-modified to target the macrophages and, since the therapeutic gene will be under the control of macrophage-specific promoter, the protein will only be expressed in macrophages and would minimize the off-target cells or tissue effects.

A relatively new field has started to emerge where nanoparticles have been utilized to monitor the progression of the disease by targeting macrophages. In this chapter, examples of such systems were discussed in detail; however, the future application in this field will probably focus on the design of multifunctional nanoplatforms that incorporate both diagnostic (and imaging) and therapeutic moieties in a single system. The preliminary results obtained by McCarthy et al. [123] and Ma et al. [76] with multifunctional nanoparticles have been encouraging.

The McCarthy group designed a nanoparticle system containing an iron oxide (λ_{max}<300 nm) core (5 nm size and 8,000 Fe/particle) with a shell of dextran (8–10 kDa). These monocrystalline nanoparticles were reported to be 33 nm in size. The dye Alexa Fluor 750 (AF750; λ_{max}=755 nm) was conjugated to the nanoparticles, and a potent photosensitizer, 5-(4-carboxyphenyl)-10,15,20-triphenyl-2,3-dihydroxychlorin (TPC; λ_{max}=648 nm) was covalently attached to the primary amines of the nanoparticle. Upon illumination, photosensitizers generate cytotoxic reactive singlet oxygen species (ROS). The main advantage of this approach has been envisioned in diseases that may benefit from the removal of macrophages such as atherosclerosis and cancer [124]. The approximately 100-nm difference between the longest wavelength absorption for TPC and AF750 resulted in minimal energy transfer upon excitation at 650 nm (therapeutic wavelength). Cell uptake studies in RAW 264.7 macrophage cells indicated that uptake and localization of the particles was time-dependent, and an average of 10^6–10^7 particles per macrophage cell were reported at saturation. More importantly, when the cells were incubated for 1 h with 0.1 mg Fe per ml, 35% of the cells remained viable. It was also reported that longer incubation times and higher dose resulted in complete cell killing. Based on these experiments, the authors have concluded that such system has the capability to target macrophages in an in vivo setting and can lead to efficient cell killing. In addition, such system and can be utilized as an agent for both MRI and near-infrared fluorescence imaging.

In summary, research in the field of nanotechnology or nanomedicine has redefined the traditional ways of delivering therapeutic agents, and moreover, has imparted selectivity to limit off-target side effects. There is also growing interest in developing multifunctional nanoprobes that are compatible with non-invasive imaging techniques such as MRI, PET, and SPECT. The combination of such agents has opened new doors to reach out to potential targets such as macrophages that are responsible for orchestrating some of the key events in the pathogenesis of various diseases.

References

1. Fujiwara N, Kobayashi K (2005) Macrophages in inflammation. Curr Drug Targets Inflamm Allergy 5:281–286(3)
2. Ross JA, Auger MJ (2002) The biology of the macrophage. In: Burke B, Lewis CE (eds) The macrophage, 2nd edn. Oxford University Press, New York, pp 16–23
3. Lopes MF, Freire-De-Lima CG, Dosreis GA (2000) The macrophage haunted by cell ghosts: a pathogen grows. Immunol Today 21(10):489–494
4. Hamilton TA (2002) Molecular basis of macrophage activation: from gene expression to phenotypic diversity. In: Burke B, Lewis CE (eds) The macrophage, 2nd edn. Oxford University Press, New York, pp 74–75
5. Mantovani A, Sica A, Sozzani S, Allavena P, Vecchi A, Locati M (2004) The chemokine system in diverse forms of macrophage activation and polarization. Trends Immunol 25(12):677–686
6. Gordon S (2003) Alternative activation of macrophages. Nat Rev Immunol 3(1):23–35

7. Wilson HM, Barker RN, Erwig LP (2009) Macrophages: promising targets for the treatment of atherosclerosis. Curr Vasc Pharmacol 7(2):234–243
8. Mosser DM (2003) The many faces of macrophage activation. J Leukoc Biol 73(2):209–212
9. Rosenberger CM, Finlay BB (2003) Phagocyte sabotage: disruption of macrophage signaling by bacterial pathogens. Nat Rev Mol Cell Biol 4(5):385–396
10. Owais M, Gupta CM (2005) Targeted drug delivery to macrophages in parasitic infections. Curr Drug Deliv 2(4):311–318
11. Germann WJ, Stanfield CL, Cannon JG, Niles MJ (2002) The immune system. In: Brassert C (ed) Principles of human physiology. Benjamin Cummings Publishing Co, San Francisco, pp 708–740
12. Tanner AR, Arthur MJ, Wright R (1984) Macrophage activation, chronic inflammation and gastrointestinal disease. Gut 25(7):760–783
13. Figarella-Branger D, Civatte M, Bartoli C, Pellissier JF (2003) Cytokines, chemokines, and cell adhesion molecules in inflammatory myopathies. Muscle Nerve 28(6):659–682
14. Godillot AP, Madaio M, Weiner DB, William VW (2000) DNA vaccination as anti-inflammatory strategy. In: Evans CH, Robbins PD (eds) Gene therapy in inflammatory diseases. Birhauser Verlag, Basel, pp 205–230
15. Lewis C, Murdoch C (2005) Macrophage responses to hypoxia: implications for tumor progression and anti-cancer therapies. Am J Pathol 167(3):627–635
16. Ono M (2008) Molecular links between tumor angiogenesis and inflammation: inflammatory stimuli of macrophages and cancer cells as targets for therapeutic strategy. Cancer Sci 99(8):1501–1506
17. Lewis CE, Pollard JW (2006) Distinct role of macrophages in different tumor microenvironments. Cancer Res 66(2):605–612
18. Leek RD, Harris AL (2002) Tumor-associated macrophages in breast cancer. J Mammary Gland Biol Neoplasia 7(2):177–189
19. Elgert KD, Alleva DG, Mullins DW (1998) Tumor-induced immune dysfunction: the macrophage connection. J Leukoc Biol 64(3):275–290
20. Herbein G, Coaquette A, Perez-Bercoff D, Pancino G (2002) Macrophage activation and HIV infection: can the Trojan horse turn into a fortress? Curr Mol Med 2(8):723–738
21. Guidi-Rontani C (2002) The alveolar macrophage: the Trojan horse of bacillus anthracis. Trends Microbiol 10(9):405–409
22. Nguyen L, Pieters J (2005) The Trojan horse: survival tactics of pathogenic mycobacteria in macrophages. Trends Cell Biol 15(5):269–276
23. Verani A, Gras G, Pancino G (2005) Macrophages and HIV-1: dangerous liaisons. Mol Immunol 42(2):195–212
24. Herbein G, Varin A (2010) The macrophage in HIV-1 infection: from activation to deactivation? Retrovirology 7(33):1–15
25. Zhang L, Gu FX, Chan JM, Wang AZ, Langer RS, Farokhzad OC (2008) Nanoparticles in medicine: therapeutic applications and developments. Clin Pharmacol Ther 83(5):761–769
26. Yih TC, Al-Fandi M (2006) Engineered nanoparticles as precise drug delivery systems. J Cell Biochem 97(6):1184–1190
27. Rawat M, Singh D, Saraf S, Saraf S (2006) Nanocarriers: promising vehicle for bioactive drugs. Biol Pharm Bull 29(9):1790–1798
28. Ulrich AS (2002) Biophysical aspects of using liposomes as delivery vehicles. Biosci Rep 22(2):129–150
29. Torchilin VP (2006) Multifunctional nanocarriers. Adv Drug Deliv Rev 58(14):1532–1555
30. Moghimi SM, Hunter AC, Murray JC (2001) Long-circulating and target-specific nanoparticles: theory to practice. Pharmacol Rev 53(2):283–318
31. Vasir JK, Reddy MK, Labhasetwar VD (2005) Nanosystems in drug targeting: opportunities and challenges. Curr Nanosci 1(1):47–64
32. Gupta S, Dube A, Vyas SP (2007) Antileishmanial efficacy of amphotericin B bearing emulsomes against experimental visceral leishmaniasis. J Drug Target 15(6):437–444

33. Corvo ML, Boerman OC, Oyen WJ, Van Bloois L, Cruz ME, Crommelin DJ, Storm G (1999) Intravenous administration of superoxide dismutase entrapped in long circulating liposomes. II. In vivo fate in a rat model of adjuvant arthritis. Biochim Biophys Acta 1419(2):325–334
34. Boerman OC, Oyen WJ, Storm G, Corvo ML, Van Bloois L, Van Der Meer JW, Corstens FH (1997) Technetium-99m labelled liposomes to image experimental arthritis. Ann Rheum Dis 56(6):369–373
35. Löbenberg R, Araujo L, Kreuter J (1997) Body distribution of azidothymidine bound to nanoparticles after oral administration. Eur J Pharm Biopharm 44(2):127–132
36. Nie S, Xing Y, Kim GJ, Simons JW (2007) Nanotechnology applications in cancer. Annu Rev Biomed Eng 9:257–288
37. Mukhopadhyay A, Basu SK (2003) Intracellular delivery of drugs to macrophages. Adv Biochem Eng Biotechnol 84:183–209
38. Brown GD, Taylor PR, Reid DM, Willment JA, Williams DL, Martinez-Pomares L, Wong SY, Gordon S (2002) Dectin-1 is a major beta-glucan receptor on macrophages. J Exp Med 196(3):407–412
39. Herre J, Gordon S, Brown GD (2004) Dectin-1 and its role in the recognition of beta-glucans by macrophages. Mol Immunol 40(12):869–876
40. Najjar VA (1983) Tuftsin, a natural activator of phagocyte cells: an overview. Ann N Y Acad Sci 419:1–11
41. Fridkin M, Najjar VA (1989) Tuftsin: its chemistry, biology, and clinical potential. Crit Rev Biochem Mol Biol 24(1):1–40
42. Naor D, Nedvetzki S (2003) CD44 in rheumatoid arthritis. Arthritis Res Ther 5(3):105–115
43. Vachon E, Martin R, Plumb J, Kwok V, Vandivier RW, Glogauer M, Kapus A, Wang X, Chow CW, Grinstein S, Downey GP (2006) CD44 is a phagocytic receptor. Blood 107(10):4149–4158
44. Ahsan F, Rivas IP, Khan MA, Torres Suarez AI (2002) Targeting to macrophages: role of physicochemical properties of particulate carriers-liposomes and microspheres-on the phagocytosis by macrophages. J Control Rel 79(1–3):29–40
45. Kaur A, Jain S, Tiwary AK (2008) Mannan-coated gelatin nanoparticles for sustained and targeted delivery of didanosine: in vitro and in vivo evaluation. Acta Pharm 58(1):61–74
46. Jain SK, Gupta Y, Jain A, Saxena AR, Khare P, Jain A (2008) Mannosylated gelatin nanoparticles bearing an anti-HIV drug didanosine for site-specific delivery. Nanomedicine 4(1):41–48
47. Schmitt F, Lagopoulos L, Kauper P, Rossi N, Busso N, Barge J, Wagnieres G, Laue C, Wandrey C, Juillerat-Jeanneret L (2010) Chitosan-based nanogels for selective delivery of photosensitizers to macrophages and improved retention in and therapy of articular joints. J Control Rel 144(2):242–250
48. Chellat F, Merhi Y, Moreau A, Yahia L (2005) Therapeutic potential of nanoparticulate systems for macrophage targeting. Biomaterials 26(35):7260–7275
49. Roser M, Fischer D, Kissel T (1998) Surface-modified biodegradable albumin nano- and microspheres. II: effect of surface charges on in vitro phagocytosis and biodistribution in rats. Eur J Pharm Biopharm 46(3):255–263
50. Allen TM, Austin GA, Chonn A, Lin L, Lee KC (1991) Uptake of liposomes by cultured mouse bone marrow macrophages: influence of liposome composition and size. Biochim Biophys Acta 1061(1):56–64
51. Schäfer V, von Briesen H, Andreesen R, Steffan AM, Royer C, Troster S, Kreuter J, Rübsamen-Waigmann H (1992) Phagocytosis of nanoparticles by human immunodeficiency virus (HIV)-infected macrophages: a possibility for antiviral drug targeting. Pharm Res 9(4):541–546
52. Tabata Y, Ikada Y (1988) Effect of the size and surface charge of polymer microspheres on their phagocytosis by macrophage. Biomaterials 9(4):356–362
53. Azad N, Rojanasakul Y (2006) Vaccine delivery-current trends and future. Curr Drug Deliv 3(2):137–146
54. Chadwick S, Kriegel C, Amiji M (2009) Delivery strategies to enhance mucosal vaccination. Expert Opin Biol Ther 9(4):427–440

55. Srivastava IK, Singh M (2005) DNA vaccines: focus on increasing potency and efficacy. Int J Pharm Med 19(1):15–28
56. Singh M, O'Hagan DT (2002) Recent advances in vaccine adjuvants. Pharm Res 19(6):715–728
57. Singh J, Pandit S, Bramwell VW, Alpar HO (2006) Diphtheria toxoid loaded poly-(epsilon-caprolactone) nanoparticles as mucosal vaccine delivery systems. Methods 38(2):96–105
58. Tobio M, Sanchez A, Vila A, Soriano II, Evora C, Vila-Jato JL, Alonso MJ (2000) The role of PEG on the stability in digestive fluids and in vivo fate of PEG-PLA nanoparticles following oral administration. Colloids Surf B Biointerfaces 18(3–4):315–323
59. Garinot M, Fievez V, Pourcelle V, Stoffelbach F, Des Rieux A, Plapied L, Theate I, Freichels H, Jerome C, Marchand-Brynaert J, Schneider YJ, Preat V (2007) PEGylated PLGA-based nanoparticles targeting M-cells for oral vaccination. J Control Rel 120(3):195–204
60. Sayin B, Somavarapu S, Li XW, Sesardic D, Senel S, Alpar OH (2009) TMC-MCC (N-trimethyl chitosan-mono-N-carboxymethyl chitosan) nanocomplexes for mucosal delivery of vaccines. Eur J Pharm Sci 38(4):362–369
61. Wu F, Wuensch SA, Azadniv M, Ebrahimkhani MR, Crispe IN (2009) Galactosylated LDL nanoparticles: a novel targeting delivery system to deliver antigen to macrophages and enhance antigen specific T cell responses. Mol Pharmaceutics 6(5):1506–1517
62. Jung T, Kamm W, Breitenbach A, Hungerer KD, Hundt E, Kissel T (2001) Tetanus toxoid loaded nanoparticles from sulfobutylated poly(vinyl alcohol)-graft-poly(lactide-co-glycolide): evaluation of antibody response after oral and nasal application in mice. Pharm Res 18(3):352–360
63. Lambros MP, Schafer F, Blackstock R, Murphy JW (1998) Liposomes, a potential immunoadjuvant and carrier for a cryptococcal vaccine. J Pharm Sci 87(9):1144–1148
64. Ge W, Li Y, Li ZS, Zhang SH, Sun YJ, Hu PZ, Wang XM, Huang Y, Si SY, Zhang XM, Sui YF (2009) The antitumor immune responses induced by nanoemulsion-encapsulated MAGE1-HSP70/SEA complex protein vaccine following peroral administration route. Cancer Immunol Immunother 58(2):201–208
65. He Q, Mitchell A, Morcol T, Bell SJ (2002) Calcium phosphate nanoparticles induce mucosal immunity and protection against herpes simplex virus type 2. Clin Diagn Lab Immunol 9(5):1021–1024
66. Khatri K, Goyal AK, Gupta PN, Mishra N, Vyas SP (2008) Plasmid DNA loaded chitosan nanoparticles for nasal mucosal immunization against hepatitis B. Int J Pharm 354(1–2): 235–241
67. Roy K, Mao HQ, Huang SK, Leong KW (1999) Oral gene delivery with chitosan-DNA nanoparticles generates immunologic protection in a murine model of peanut allergy. Nat Med 5(4):387–391
68. Lu Y, Kawakami S, Yamashita F, Hashida M (2007) Development of an antigen-presenting cell-targeted DNA vaccine against melanoma by mannosylated liposomes. Biomaterials 28(21):3255–3262
69. Tang CK, Lodding J, Minigo G, Pouniotis DS, Plebanski M, Scholzen A, Mckenzie IF, Pietersz GA, Apostolopoulos V (2007) Mannan-mediated gene delivery for cancer immunotherapy. Immunology 120(3):325–335
70. Moghimi SM, Hunter AC, Murray JC (2005) Nanomedicine: current status and future prospects. FASEB J 19(3):311–330
71. Lipinski MJ, Frias JC, Amirbekian V, Briley-Saebo KC, Mani V, Samber D, Abbate A, Aguinaldo JG, Massey D, Fuster V, Vetrovec GW, Fayad ZA (2009) Macrophage-specific lipid-based nanoparticles improve cardiac magnetic resonance detection and characterization of human atherosclerosis. JACC Cardiovasc Imaging 2(5):637–647
72. Fabien H, Jean-Christophe C, Jonathan FE, Gordon R, Vucic E, Amirbekian V, Fisher EA, Fuster V, Feldman LJ, Fayad ZA (2007) Noninvasive detection of macrophages using a nanoparticulate contrast agent for computed tomography. Nat Med 13(5):636–641

73. Nahrendorf M, Zhang H, Hembrador S, Panizzi P, Sosnovik DE, Aikawa E, Libby P, Swirski FP, Weissleder R (2007) Nanoparticle PET-CT imaging of macrophages in inflammatory atherosclerosis. Circulation 117:379–387
74. Lipinski MJ, Amirbekian V, Frias JC, Aguinaldo JG, Mani V, Briley-Saebo KC, Fuster V, Fallon JT, Fisher EA, Fayad ZA (2006) MRI to detect atherosclerosis with gadolinium-containing immunomicelles targeting the macrophage scavenger receptor. Magn Reson Med 56(3):601–610
75. Amirbekian V, Lipinski MJ, Briley-Saebo KC, Amirbekian S, Aguinaldo JG, Weinreb DB, Vucic E, Frias JC, Hyafil F, Mani V, Fisher EA, Fayad ZA (2007) Detecting and assessing macrophages in vivo to evaluate atherosclerosis noninvasively using molecular MRI. Proc Natl Acad Sci USA 104(3):961–966
76. Ma LL, Feldman MD, Tam JM, Paranjape AS, Cheruku KK, Larson TA, Tam JO, Ingram DR, Paramita V, Villard JW, Jenkins JT, Wang T, Clarke GD, Asmis R, Sokolov K, Chandrasekar B, Milner TE, Johnston KP (2009) Small multifunctional nanoclusters (nanoroses) for targeted cellular imaging and therapy. ACS Nano 3(9):2686–2696
77. Kim JS, An H, Rieter WJ, Esserman D, Taylor-Pashow KM, Sartor RB, Lin W, Lin W, Tarrant TK (2009) Multimodal optical and Gd-based nanoparticles for imaging in inflammatory arthritis. Clin Exp Rheumatol 27(4):580–586
78. Lutz AM, Seemayer C, Corot C, Gay RE, Goepfert K, Michel BA, Marincek B, Gay S, Weishaupt D (2004) Detection of synovial macrophages in an experimental rabbit model of antigen-induced arthritis: ultrasmall superparamagnetic iron oxide-enhanced MR imaging. Radiology 233(1):149–157
79. Heinzen RA, Scidmore MA, Rockey DD, Hackstadt T (1996) Differential interaction with endocytic and exocytic pathways distinguish parasitophorous vacuoles of *Coxiella burnetii* and *Chlamydia trachomatis*. Infect Immun 64(3):796–809
80. Khan MA, Owais M (2006) Toxicity, stability and pharmacokinetics of amphotericin B in immunomodulator tuftsin-bearing liposomes in a murine model. J Antimicrob Chemother 58(1):125–132
81. Gagne JF, Desormeaux A, Perron S, Tremblay MJ, Bergeron MG (2002) Targeted delivery of indinavir to HIV-1 primary reservoirs with immunoliposomes. Biochim Biophys Acta 1558(2):198–210
82. Nimje N, Agarwal A, Saraogi GK, Lariya N, Rai G, Agrawal H, Agrawal GP (2009) Mannosylated nanoparticulate carriers of rifabutin for alveolar targeting. J Drug Target 17(10):777–787
83. Wijagkanalan W, Higuchi Y, Kawakami S, Teshima M, Sasaki H, Hashida M (2008) Enhanced anti-inflammation of inhaled dexamethasone palmitate using mannosylated liposomes in an endotoxin-induced lung inflammation model. Mol Pharmacol 74(5):1183–1192
84. Chono S, Tanino T, Seki T, Morimoto K (2008) Efficient drug targeting to rat alveolar macrophages by pulmonary administration of ciprofloxacin incorporated into mannosylated liposomes for treatment of respiratory intracellular parasitic infections. J Control Release 127(1):50–58
85. Lobenberg R, Araujo L, von Briesen H, Rodgers E, Kreuter J (1998) Body distribution of azidothymidine bound to hexyl-cyanoacrylate nanoparticles after i.v. injection to rats. J Control Release 50(1–3):21–30
86. Agarwal A, Kandpal H, Gupta HP, Singh NB, Gupta CM (1994) Tuftsin-bearing liposomes as rifampin vehicles in treatment of tuberculosis in mice. Antimicrob Agents Chemother 38(3):588–593
87. Chandrasekar D, Sistla R, Ahmad FJ, Khar RK, Diwan PV (2007) Folate coupled poly(ethyleneglycol) conjugates of anionic poly(amidoamine) dendrimer for inflammatory tissue specific drug delivery. J Biomed Mater Res A 82(1):92–103
88. Hu J, Liu H, Wang L (2000) Enhanced delivery of AZT to macrophages via acetylated LDL. J Control Release 69(3):327–335

89. Zeisberger SM, Odermatt B, Marty C, Zehnder-Fjallman AH, Ballmer-Hofer K, Schwendener RA (2006) Clodronate-liposome-mediated depletion of tumour-associated macrophages: a new and highly effective antiangiogenic therapy approach. Br J Cancer 95(3):272–281
90. Dembri A, Montisci MJ, Gantier JC, Chacun H, Ponchel G (2001) Targeting of 3′-azido 3′-deoxythymidine (AZT)-loaded poly(isohexylcyanoacrylate) nanospheres to the gastrointestinal mucosa and associated lymphoid tissues. Pharm Res 18(4):467–473
91. Burke B, Sumner S, Maitland N, Lewis CE (2002) Macrophages in gene therapy: cellular delivery vehicles and in vivo targets. J Leukoc Biol 72(3):417–428
92. Singh M, Chakrapani A, O'Hagan D (2007) Nanoparticles and microparticles as vaccine-delivery systems. Expert Rev Vaccines 6(5):797–808
93. Shahiwala A, Vyas TK, Amiji MM (2007) Nanocarriers for systemic and mucosal vaccine delivery. Recent Pat Drug Deliv Formul 1(1):1–9
94. Clark MA, Jepson MA, Hirst BH (2001) Exploiting M-cells for drug and vaccine delivery. Adv Drug Deliv Rev 50(1–2):81–106
95. Hattori Y, Kawakami S, Suzuki S, Yamashita F, Hashida M (2004) Enhancement of immune responses by DNA vaccination through targeted gene delivery using mannosylated cationic liposome formulations following intravenous administration in mice. Biochem Biophys Res Commun 317(4):992–999
96. Tang CK, Sheng KC, Pouniotis D, Esparon S, Son HY, Kim CW, Pietersz GA, Apostolopoulos V (2008) Oxidized and reduced mannan mediated MUC1 DNA immunization induce effective anti-tumor responses. Vaccine 26(31):3827–3834
97. Ribeiro S, Rijpkema SG, Durrani Z, Florence AT (2007) PLGA-dendron nanoparticles enhance immunogenicity but not lethal antibody production of a DNA vaccine against anthrax in mice. Int J Pharm 331(2):228–232
98. Fabien Hyafil J-C, Feig JE, Gordon R, Vucic E, Amirbekian V, Fisher EA, Fuster V, Feldman LJ, Fayad ZA (2007) Noninvasive detection of macrophages using a nanoparticulate contrast agent for computed tomography. Nat Med 13(5):636–641
99. Schroeder S, Kopp AF, Ohnesorge B, Flohr T, Baumbach A, Kuettner A, Herdeg C, Karsch KR, Claussen CD (2001) Accuracy and reliability of quantitative measurements in coronary arteries by multi-slice computed tomography: experimental and initial clinical results. Clin Radiol 56(6):466–474
100. Pohle K, Achenbach S, Macneill B, Ropers D, Ferencik M, Moselewski F, Hoffmann U, Brady TJ, Jang IK, Daniel WG (2007) Characterization of non-calcified coronary atherosclerotic plaque by multi-detector row CT: comparison to IVUS. Atherosclerosis 190(1):174–180
101. Rodan GA, Fleisch HA (1996) Bisphosphonates: mechanisms of action. J Clin Invest 97(12):2692–2696
102. Bayes-Genis A, Campbell JH, Carlson PJ, Holmes DR Jr, Schwartz RS (2002) Macrophages, myofibroblasts and neointimal hyperplasia after coronary artery injury and repair. Atherosclerosis 163(1):89–98
103. Danenberg HD, Fishbein I, Gao J, Mönkkönen J, Reich R, Gati I, Moerman E, Golomb G (2002) Macrophage depletion by Clodronate-containing liposomes reduces neointimal formation after balloon injury in rats and rabbits. Circulation 106:599–605
104. Cohen-Sela E, Rosenzweig O, Gao J, Epstein H, Gati I, Reich R, Danenberg HD, Golomb G (2006) Alendronate-loaded nanoparticles deplete monocytes and attenuate restenosis. J Control Release 113(1):23–30
105. Howard KA, Paludan SR, Behlke MA, Besenbacher F, Deleuran B, Kjems J (2009) Chitosan/siRNA nanoparticle-mediated TNF-α knockdown in peritoneal macrophages for anti-inflammatory treatment in a murine arthritis model. Am Soc Gene Ther 17(1):162–168
106. Zuo L, Huang Z, Dong L, Xu L, Zhu Y, Zeng K, Zhang C, Chen J, Zhang J (2010) Targeting delivery of anti-TNFalpha oligonucleotide into activated colonic macrophages protects against experimental colitis. Gut 59(4):470–479
107. Crooke ST (2004) Progress in antisense technology. Annu Rev Med 55:61–95

108. Bhavsar MD, Amiji MM (2008) Oral IL-10 gene delivery in a microsphere-based formulation for local transfection and therapeutic efficacy in inflammatory bowel disease. Gene Ther 15(17):1200–1209
109. Nakase H, Okazaki K, Tabata Y, Ozeki M, Watanabe N, Ohana M, Uose S, Uchida K, Nishi T, Mastuura M, Tamaki H, Itoh T, Kawanami C, Chiba T (2002) New cytokine delivery system using gelatin microspheres containing interleukin-10 for experimental inflammatory bowel disease. J Pharm Exp Ther 301(1):59–65
110. Wijagkanalan W, Kawakami S, Higuchi Y, Yamashita F, Hashida M (2011) Intratracheally instilled mannosylated cationic liposome/NFkappaB decoy complexes for effective prevention of LPS-induced lung inflammation. J Control Release 149(1):42–50
111. Lee S, Yang SC, Kao CY, Pierce RH, Murthy N (2009) Solid polymeric microparticles enhance the delivery of siRNA to macrophages in vivo. Nucleic Acids Res 37(22):e145
112. Kawakami S, Sato A, Nishikawa M, Yamashita F, Hashida M (2000) Mannose receptor-mediated gene transfer into macrophages using novel mannosylated cationic liposomes. Gene Ther 7(4):292–299
113. Ferkol T, Perales JC, Mularo F, Hanson RW (1996) Receptor-mediated gene transfer into macrophages. Proc Natl Acad Sci USA 93(1):101–105
114. Nakamura K, Kuramoto Y, Mukai H, Kawakami S, Higuchi Y, Hashida M (2009) Enhanced gene transfection in macrophages by histidine-conjugated mannosylated cationic liposomes. Biol Pharm Bull 32(9):1628–1631
115. Khoury M, Escriou V, Courties G, Galy A, Yao R, Largeau C, Scherman D, Jorgensen C, Apparailly F (2008) Efficient suppression of murine arthritis by combined anticytokine small interfering RNA lipoplexes. Arthritis Rheum 58(8):2356–2367
116. Dong L, Gao S, Diao H, Chen J, Zhang J (2008) Galactosylated low molecular weight chitosan as a carrier delivering oligonucleotides to Kupffer cells instead of hepatocytes in vivo. J Biomed Mater Res A 84(3):777–784
117. Choi MR, Stanton-Maxey KJ, Stanley JK, Levin CS, Bardhan R, Akin D, Badve S, Sturgis J, Robinson JP, Bashir R, Halas NJ, Clare SE (2007) A cellular Trojan horse for delivery of therapeutic nanoparticles into tumors. Nano Lett 7(12):3759–3765
118. Miselis NR, Wu ZJ, van Rooijen N, Kane AB (2008) Targeting tumor-associated macrophages in an orthotopic murine model of diffuse malignant mesothelioma. Mol Cancer Ther 7(4):788–799
119. Opanasopit P, Sakai M, Nishikawa M, Kawakami S, Yamashita F, Hashida M (2002) Inhibition of liver metastasis by targeting of immunomodulators using mannosylated liposome carriers. J Control Rel 80(1–3):283–294
120. Taniyama T, Holden HT (1979) Direct augmentation of cytolytic activity of tumor-derived macrophages and macrophage cell lines by muramyl dipeptide. Cell Immunol 48(2):369–374
121. Robson T, Hirst DG (2003) Transcriptional targeting in cancer gene therapy. J Biomed Biotechnol 2003(2):110–137
122. Griffiths L, Binley K, Iqball S, Kan O, Maxwell P, Ratcliffe P, Lewis C, Harris A, Kingsman S, Naylor S (2000) The macrophage—a novel system to deliver gene therapy to pathological hypoxia. Gene Ther 7(3):255–262
123. Mccarthy JR, Jaffer FA, Weissleder R (2006) A macrophage-targeted theranostic nanoparticle for biomedical applications. Small 2(8–9):983–987
124. Demidova TN, Hamblin MR (2004) Macrophage-targeted photodynamic therapy. Int J Immunopathol Pharmacol 17(2):117–126

Chapter 5
Paclitaxel-Triazine Dendrimer Constructs: Efficacy, Toxicity, and Characterization

Eric E. Simanek and Jongdoo Lim

5.1 Introduction

For those interested in macromolecular drug delivery [1–3], paclitaxel represents an appropriate drug for study for a number of reasons. First, as a chemotherapeutic, paclitaxel has activity against a broad spectrum of cancers including those of the head and neck, breast, ovaries, lung as well as Kaposi's sarcoma [4, 5]. Second, the drug represents a challenge for formulation due to its sparing solubility in water. The historic formulation Taxol®, wherein the agent is delivered in Cremophor® EL, has been more recently surpassed by Abraxane®, a particle comprising the drug and serum albumin [6, 7]. Third, the drug itself presents a reactive hydroxyl group that can be selectively functionalized, usually through acylation to append the agent to the polymer. Fourth, any new effort benefits from the opportunity for comparison to the aforementioned clinical standards and a host of other reported prodrugs, including those derived from poly(glutamic acid), poly(ethylene glycol) (PEG), poly(*N*-(2-hydroxypropyl)methacrylamide) (HPMA), as well as others [8–19].

The body of this chapter is divided into subchapters. Listed in Table 5.1, these subchapters are matched with the relevant question that can be considered as a metric for success. The chapter ends with conclusions and our perspective on the future.

E.E. Simanek (✉) • J. Lim
Department of Chemistry, Texas Christian University,
Box 298860, Fort Worth, TX 76129, USA
e-mail: e.simanek@tcu.edu

S. Svenson and R.K. Prud'homme (eds.), *Multifunctional Nanoparticles for Drug Delivery Applications: Imaging, Targeting, and Delivery*, Nanostructure Science and Technology,
DOI 10.1007/978-1-4614-2305-8_5, © Springer Science+Business Media, LLC 2012

Table 5.1 The challenges for realizing nanomedicines

Item	Topic	Question of merit
1.	Synthesis	Can it be made?
2.	Scale	Can enough be made?
3.	Drug conjugation and release	Can the drug be attached and released?
4.	Reproducibility	Is it the same each time it is made?
5.	Biodistribution	Does it get to the tumor?
6.	Biocompatibility, safety, and toxicity	How does it elsewise interact with the body?
7.	Efficacy	Does it work?
8.	Therapeutic advantage	Does it work sufficiently better than what is currently available?

5.2 Synthesis

Our strategy for preparing a macromolecular drug construct progresses in three phases: (1) making the macromolecule, (2) appending the drug, and (3) functionalizing the construct (historically with PEG). The macromolecules that we have pursued are triazine dendrimers [20]. By taking advantage of the chemoselective reaction of cyanuric chloride and diamine linkers, these materials can be prepared using convergent and divergent approaches.

Early energies focused on convergent routes to these materials wherein a surface group, oftentimes a monochlorotriazine bearing two protected amines, was reacted iteratively with a diamine linker, and then cyanuric chloride to yield low generation materials. As an advantage, these materials were commonly "pure" as defined by the limits of the conventional techniques of organic chemistry. Inherent to the success of this strategy was judicious choice of the linking diamine [21, 22]. A series of efforts measuring relative reactivities of amines for monochlorotriazines allowed us to identify diamines like 4-aminomethylpiperidine, which showed a 20× difference in reactivity between the reactive, constrained piperidine group and the primary amine (Fig. 5.1). The reactivity of primary amine was advantageously retarded in the subsequent reaction with cyanuric chloride—only adding twice to afford the desired, advanced monochlorotriazine[1].

Recently, divergent routes have been pursued that allow us access to higher generation materials [23, 24]. Initially, more reactive dichlorotriazine monomers were reacted with less reactive polyamine cores presenting primary amines (Scheme 5.1). The use of a dichlorotriazine monomer has an additional consequence. Upon reaction, the resulting dendrimer comprises multiple monochlorotriazines. These groups offer sites for the introduction of chemical diversity or must be capped, usually with piperidine, before iteration. While pure materials can be obtained through generation

[1]Examples of this strategy are pervasive in our early literature up through 2007. As an alternative, using diamines like piperazine often leads to addition at both amines and the production of a side product that can be difficult to separate and sometimes to detect.

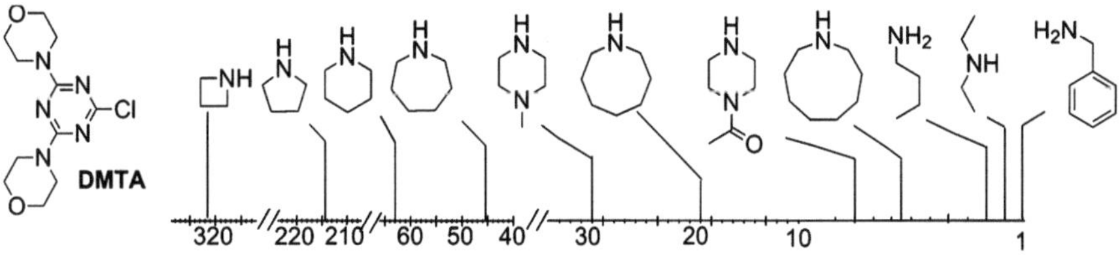

Fig. 5.1 The relative reactivity of amines toward monochlorotriazines. Combining two such amines into a single molecule provides a linking diamine that might be used successfully for chemoselective reaction

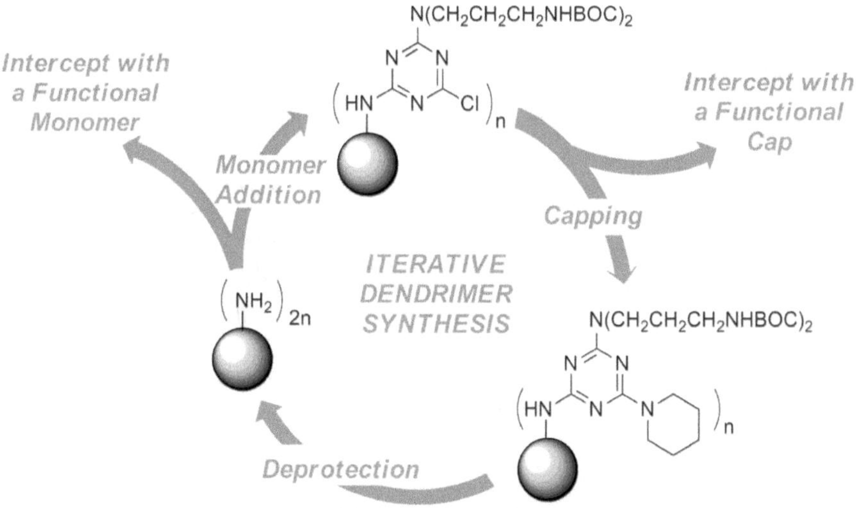

Scheme 5.1 Divergent routes rely on iterative reactions of monomer addition, capping, and deprotection. This 3-step per generation route can be intercepted either with a functional monomer or with a functional cap. To date, paclitaxel-bearing constructs have been realized through the former strategy. The capping step is omitted when reactive, constrained secondary amines are paired with monochlorotriazines in an iterative 2-step per generation method (not shown)

four or five, the predicted heterogeneity arising from incomplete reaction appears in higher generation dendrimers. More recently, however, monochlorotriazines and more reactive and constrained secondary amine linkers have been employed [25]. This route reduces the number of reactions per generation from three to one by removing the capping step and employing a macromonomer.

All routes can afford materials that can be subsequently functionalized. Initially, the polyamine dendrimer derived from the convergent route was stoichiometrically manipulated with electrophilic groups, including Michael reagents and active esters [26]. The divergent route using dichlorotriazine monomers offers two routes for the installation of groups of interest. These groups can be incorporated during capping of the poly(monochlorotriazine) using the so-called functional cap strategy [27].

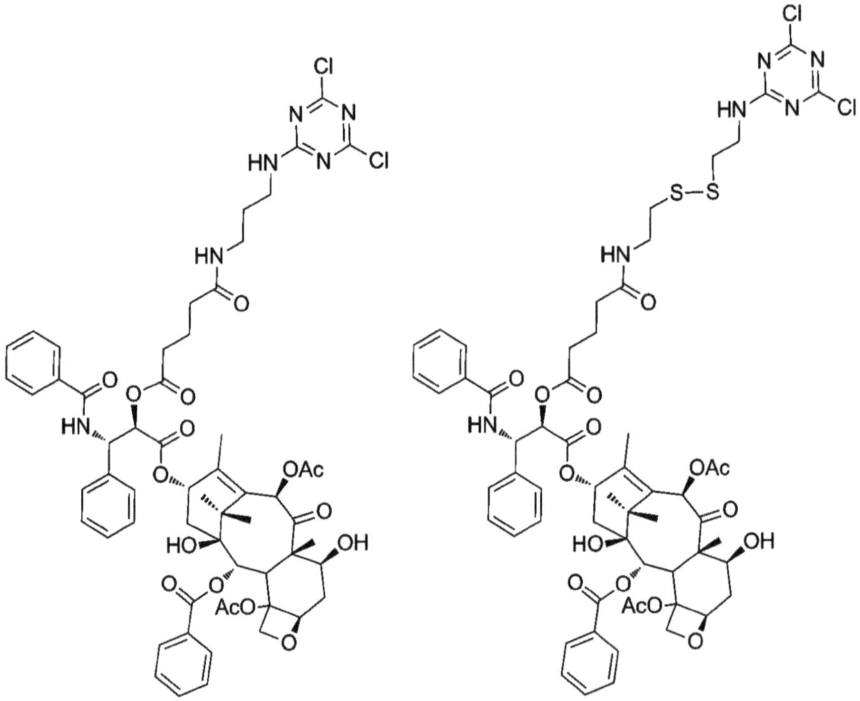

Chart 5.1 Functional monomers used to prepare paclitaxel-containing dendrimers comprise a single, biolabile ester or an ester and disulfide

Alternatively, these groups can be incorporated into a dichlorotriazine using the so-called functional monomer strategy [28]. Both routes have been explored, and agents including camptothecin [27], desferrioxamine [28], and paclitaxel [29, 30] have been incorporated. Paclitaxel is installed as a functional monomer (Chart 5.1), an opportunity afforded by the lack of competing reactive groups.

Incorporating paclitaxel into a functional monomer confers a number of advantages. First, it reduces the number of synthetic steps that paclitaxel (or any other agent) is exposed to. Second, it provides us with an efficient strategy to make functional monomers that differ in linker composition. Third, it provides an opportunity to survey a range of different surface groups by varying the capping group, although to date, studies have been confined to PEG.

5.3 Scale

Criteria used to evaluate success in a synthesis and the opportunity for realizing scale include the number of steps, the yields and waste streams, the cost of reagents, and the efforts required for purification.

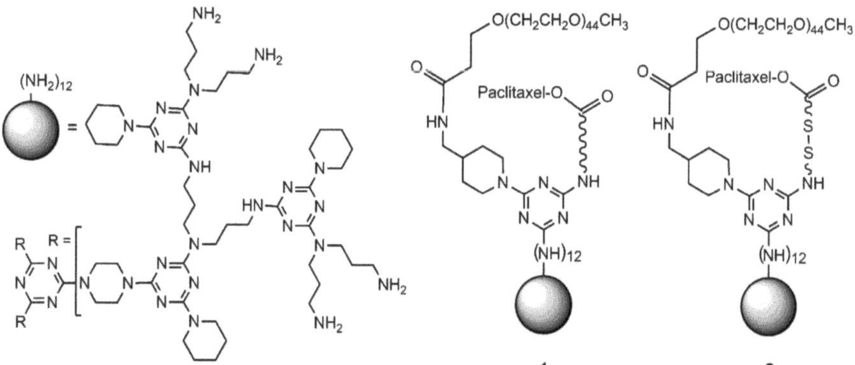

1

2

Chart 5.2 The lead architectures **1** and **2** derive from a common generation two triazine dendrimer (*left*). **1** contains an ester linkage to paclitaxel, and **2** contains the same ester and a disulfide bond

5.3.1 Steps, Yields, and Purity

The paclitaxel constructs of interest, **1** and **2** (shown in Chart 5.2), are prepared in three steps from the generation two (G2) dendrimer and the dichlorotriazines previously identified. Reaction of these reagents, followed by capping with 4-aminomethylpiperidine and PEGylation, provides the desired materials. The yield for this three-step transformation is approximately 80%, although the materials that result are clearly mixtures resulting from both (1) incomplete PEGylation (typically 9 of the desired 12 PEG chains installed) and, to a much lesser extent, (2) incomplete dichlorotriazine addition (trace amounts). Addressing this heterogeneity is central to our ongoing efforts (*vide infra*).

The dichlorotriazines are each prepared in four steps in 60% and 49% yields. Lower yields here, and in unpublished ongoing efforts, appear to correlate well with increasing lability of the tether. The generation two dendrimer is available at kilogram scale in 70% overall yield from the monomers when a single chromatographic step is performed [31]. However, the resulting material in this case is approximately 92% pure. When the generation one materials are produced at more modest scales (50–100 g) and subjected to chromatographic purification at each step, the resulting materials have been externally validated as pure using HPLC methods [32]. This purity is hypothesized to be maintained through generation two materials when chromatographic separations are employed after each step, but this purity has not been rigorously established by HPLC, only through MALDI-TOF.

5.3.2 PEGylation

PEGylation represents a critical step that conveys solubility to the drug-laden hydrophobic dendrimer. Specifically, 2-kDa PEG is required as shorter PEGs failed to convey solubility in water. PEGylation also provides size, which is believed to be critical for targeting [33, 35]. While PEGylation was not required in another PTX construct relying on a G5 PAMAM dendrimer, the average number of PTX groups was three [36]. This compromise comes with costs, as the heterogeneity obtained through PEGylation is replaced with the heterogeneities intrinsic to the synthesis of a larger generation dendrimer. While we believe that triazines will provide a narrower distribution of species than other platforms, additional efforts are necessary to explore this potential.

A less transparent motivation in abandoning PEGylation is our currently held belief that PEGylation promotes aggregation of these materials into what we envision to be micellar structures, with PEG-rich exteriors and hydrophobic (triazine and paclitaxel) interiors. Evidence from dynamic light scattering methods suggests that the 40-kDa monomeric species envisioned likely exists in equilibrium with a larger, 400-kDa aggregate in saline. Additionally, micron-size particles are observed in pure water.

5.3.3 *Cost of Chemicals and Time*

Our efforts now focus on divergent routes to take advantage of common monomers that can be used repeatedly in a synthesis. The generation one dendrimer was prepared at kilogram scale at a cost less than $10/g. For the paclitaxel constructs, we rely on the generation two dendrimers as starting material. Costs can be reduced almost tenfold if alternatives to BOC-protecting groups can be identified. While the intrinsic low costs of most of the reagents are beneficial aspects of these efforts, there is room for considerable improvement. First, reaction times are long—often more than a day is required for complete reaction. Second, purification is nontrivial—column chromatography after each step is required. Third, organic solvents are used – replacement with aqueous acetone and mineral bases would reduce waste streams. Fourth, greater convergency in these routes must be explored. All said, however, we are optimistic that process chemistry can surmount these challenges when we provide compelling motivation in the form of a preclinical candidate.

5.4 Drug Conjugation and Release

Drug conjugation is achieved by introduction of dichlorotriazines with pendant paclitaxel groups. We have reported two constructs to date **1** and **2** [29, 30]. Paclitaxel conjugation relies on acylation of the side chain hydroxyl with glutaric anhydride and subsequent amidation to an amine intermediate that can be installed on cyanuric chloride. These reactions offer versatility in the composition of the linking group, as both simple aliphatic- and disulfide-containing tethers have been incorporated. We continue to explore more labile alternatives.

Drug release from one of these constructs has been extensively studied in collaboration with the Nanotechnology Characterization Laboratory (NCL). NCL developed an HPLC assay using docetaxel as an internal standard for both quantification and extraction efficiency when monitoring stability and release in serum. The conjugate showed less than 1% drug released in saline at neutral and low pH or in organic solvent. In the presence of serum from mice, humans, or rats, the half-lives for release were measured to be 870 h, 190 h, and 64 h, respectively. These half-lives correspond to less than 20%, 10%, and 8% of drug being released over a 48 h period. Balancing release rates with tumor uptake and clearance rates will be critically important in defining a therapeutic window. We currently view the aliphatic linker of **1** to be too stable, and the tether of **2** to be an improvement. More labile linkers that facilitate ester hydrolysis are being sought.

5.5 Reproducibility

To ultimately realize clinical relevance, the agent of interest must be prepared with batch-to-batch reproducibility. The material need not be a single agent, but the diversity of species must be consistently represented. Accomplishing this task is advanced by having knowledge of the identity of the species present and knowing their origins. The inherent stability of the triazine scaffold offers clear advantage here, as spontaneous decomposition is not an issue. Heterogeneities arise from incomplete synthetic steps or failed purifications. Signal-to-noise limitations and the inherent degeneracy in the polymer limit the usefulness of the NMR spectroscopy. High pressure liquid chromatography (HPLC) and mass spectrometry offer the greatest insights into purity. MALDI-TOF mass spectrometry allows the course of a reaction that involves multiple substitutions to be followed unambiguously. Historically, we have relied heavily on this technique. HPLC has provided corroborating evidence, but has been underutilized in our efforts. Given the difficulty in unambiguously fingerprinting a complex mixture of similarly composed molecules, alternative strategies need attention.

Future efforts will benefit if we continue to refine both routes and targets to avoid heterogeneity. While low generation dendrimers are produced successfully and arguably as single-chemical entities, the introduction of the functional monomer and subsequent PEGylation step leads to significant heterogeneity. This heterogeneity is reflected in the broadness of the signal derived from mass spectrometry, and multiple peaks observed in the HPLC chromatogram. The greatest source of heterogeneity appears to be the PEGylation, with group dendrimers displaying between 6 and the desired 12 PEG chains as revealed in the HPLC trace in Fig. 5.2. Finding alternatives to PEGylation is a current focus of our efforts.

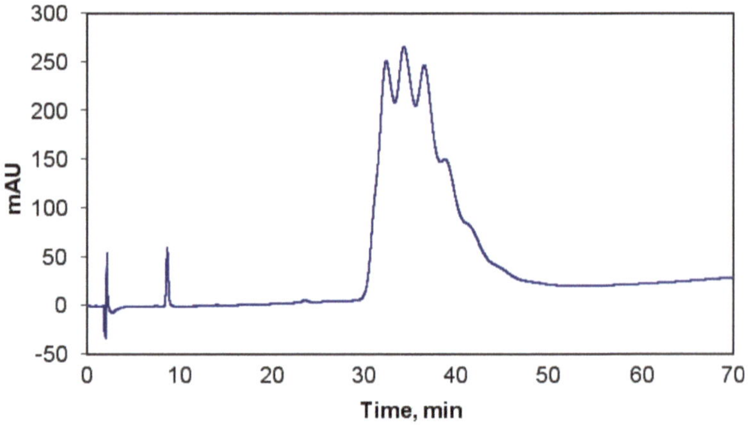

Fig. 5.2 HPLC of dendrimer **1** showing, presumably, differences in the degree of PEGylation. Free paclitaxel elutes at 9 min

5.6 Biodistribution

Adopting a stable, largely bioinert triazine dendrimer for drug delivery conveys an intention of having the material either localize to the site of action or be cleared from the circulation by renal filtration. Accordingly, dendrimers large enough to benefit from the enhanced permeability and retention (EPR) effect [33, 35], yet small enough to be cleared by the kidneys are the target. Practically, this means species with diameters between 5 and 10 nm and/or molecular weights around 50 kDa. These criteria suggest dendrimers varying between generation 3 and 5. Experiment has shown that changes in structure lead to differences in behavior in vivo. Earlier, we have found that drug-free, PEGylated generation three dendrimers show anomalously long retention times in the vasculature when compared with other architectures [34]. These retention times were reflected by long elimination half-lives calculated using a two-compartment open model. We now attribute the origins of these differences to the aggregation state of these dendrimers in solution. Interestingly, **1** and **2**, which differ from these earlier species by the addition of paclitaxel, appear to behave much more like monomers, although not exclusively.

Figure 5.3 and Table 5.2 provide summaries of the distribution data of **1** and **2**. Half-lives observed are consistent with expectations derived from the literature. Tumor uptake is also consistent with many other constructs: approximately 3% of the injected dose localizes. Increasing over time, the tumor-to-blood and tumor-to-muscle ratios provide a source of enthusiasm. The observed renal clearance route is also desirable. Over the course of the experiment, tenfold higher levels of construct

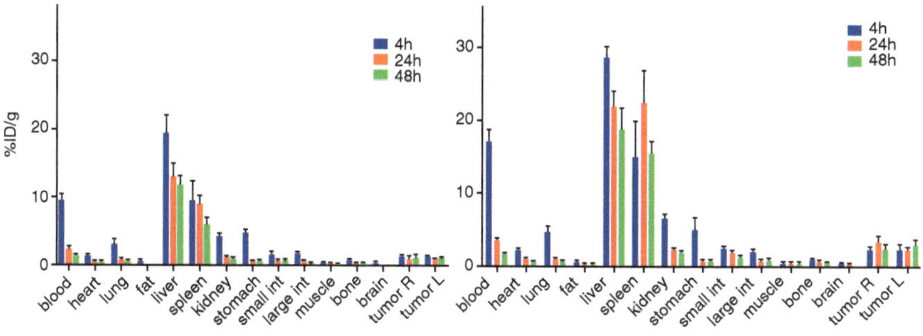

Fig. 5.3 Biodistribution of **1** (*left*) and **2** (*right*)

Table 5.2 Biodistribution data for **1** and **2**

	% in urine			% in feces			Half-lives			
	24 h	48 h	72 h	24 h	48 h	72 h	$T_{1/2\alpha}$	$T_{1/2\beta}$	Tumor-to-blood	Tumor-to-muscle
1	26.6	31.5	34.8	1.6	2.1	2.5	0.4 h	15.3 h	0.8	14
2	42.7	49.7	54.9	2.7	3.3	3.8	0.4 h	19.3 h	1.8	23

were observed in the urine over the feces. However, the amount of liver and spleen uptake remains higher than desired.

5.7 Biocompatibility, Safety, and Toxicity

Based exclusively on acute dosing, paclitaxel-containing dendrimer **1** shows no adverse effects on dosing in amounts up to 400 mg/kg construct (or equivalently, 100 mg/kg paclitaxel) in healthy mice. NCL performed a thorough hematological and clinic chemistry panel with these materials including assessments of cell populations; clotting potential; electrolyte, serum constituent, and carbohydrate balance; and organ function including liver, pancreas, and kidney. No difference from saline was observed. The original data are reported in Table 5.3. No data are available concerning chronic exposure.

5.8 Efficacy

In cell culture, toxicity is observed at low millimolar concentrations for **1** and **2**. Cytotoxicity is enhanced by the addition of reducing agents when a disulfide group is incorporated into the linker. Similar toxicities between both **1** and **2**, however, lead us to hypothesize that ester hydrolysis is a critical, rate-limiting step.

Table 5.3 Hematology and clinical chemistry of **1**[a]

		Prodrug **1** dose (mg PTX/kg body weight)			
	Vehicle control (PBS)	10	25	50	100
	Number of animals				
White blood cells ($10^3/\mu L$)	3	2	2	2	3
Total leukocytes	8.86 ± 3.82	9.13 ± 1.09	9.34 ± 2.74	7.35 ± 4.79	8.75 ± 3.00
Neutrophils	2.01 ± 0.67	2.30 ± 0.21	2.95 ± 0.72	2.09 ± 0.98	2.20 ± 0.91
Lymphocytes	6.41 ± 2.98	6.29 ± 0.83	5.94 ± 1.92	4.57 ± 3.75	6.14 ± 2.02
Macrophage/monocytes	0.41 ± 0.15	0.52 ± 0.04	0.40 ± 0.08	0.50 ± 0.22	0.28 ± 0.15
Eosinophils	0.020 ± 0.026	0.025 ± 0.007	0.040 ± 0.014	0.13 ± 0.11	0.09 ± 0.06
Basophils	0.007 ± 0.006	0.010 ± 0.00	0.0100 ± 0.0001	0.06 ± 0.07	0.037 ± 0.012
Clotting potential					
Platelet count (10^3/mL)	811 ± 66	750 ± 43	664 ± 90	564 ± 73[a]	578 ± 54[a]
Mean platelet volume (fL)	5.1 ± 0.2	5.1 ± 0.2	5.2 ± 0.1	5.3 ± 0.2	4.9 ± 0.2
	Number of animals				
Red blood cells	3	2	2	2	2
Total erythrocyte count (M/μL)	9.35 ± 0.41	9.53 ± 0.14	8.63 ± 0.37	8.78 ± 0.74	8.63 ± 1.62
Hemoglobin (hb) conc. (g/dL)	14.8 ± 0.5	15.1 ± 0.1	13.9 ± 0.1	13.8 ± 1.8	14.3 ± 1.7
Hematocrit (%)	51.6 ± 0.7	51.8 ± 1.4	46.9 ± 0.1	49.8 ± 6.0	48.9 ± 8.9
Mean corp. volume (fL)	54.9 ± 3.6	54.8 ± 0.1	54.3 ± 2.3	56.6 ± 2.1	56.7 ± 0.4
Mean corp. hb. (Pg)	15.8 ± 0.9	15.8 ± 0.1	16.1 ± 0.8	15.7 ± 0.8	16.7 ± 1.4
Mean corp. hb. conc. (g/dL)	28.6 ± 0.8	29.1 ± 0.6	29. 6 ± 0.2	27.8 ± 0.4	29.5 ± 2.2
Red cell distr. width (%)	17.2 ± 1.2	16.5 ± 1.1	16.4 ± 0.4	15.6 ± 0.2	16.0 ± 0.4

	Number of animals				
Electrolyte balance (mmol/L)	3	3	3	3	3
Calcium	10.1±0.1	9.9±0.1	9.8±0.2	9.8±0.3	9.7±0.1
Phosphate	8.2±0.9	7.3±1.0	8.0±0.6	8.3±1.0	9.1±0.2
Potassium	7.6±0.5	7.3±0.2	7.3±0.2	17.6±17.2	7.4±0.2
Sodium	155±24	154±3	155±3	157±1	154±2
Carbohydrate metabolism					
Glucose (mmol/L)	172±24	177±15	166±16	150±14	159±10
Pancreatic function					
Amylase (U/L)	963±81	182±108	928±63	807 ±171	893±110
Liver function: hepatobiliary					
Total bilirubin (mg/dL)	0.4±0.1	0.4±0.0	0.3±0.1	0.3±0.0	0.4±0.1
Liver function: hepatocellular					
Alanine aminotransferase (U/L)	79±16	86±5	74±20	82±10	81±5
Kidney function					
Creatinine (mmol/L)	0.2±0.0	0.2±0.0	0.2±0.0	0.2±0.0	0.2±0.0
Urea nitrogen (mg/dL)	19±1	19±4	20±4	24±6	21±3
Others					
Albumin (g/L)	4.1±0.1	3.7±0.3	3.9±0.3	4.1±0.1	3.0±0.1
Alkaline phosphatase (U/L)	79±16	86±5	74±20	82±10	81±5
Globulin (calculated, g/L)	1.4±0.1	1.4±0.3	1.4±0.4	1.2±0.3	1.1±0.3
A/G ratio	3.0±0.2	2.8±0.8	2.9±0.8	3.7±0.9	3.6±0.8
Total protein (g/L)	5.5±0.2	5.1±0.2	5.3±0.3	5.3±0.2	5.1±0.3

[a]Deviation of platelet count is currently attributed to sample preparation

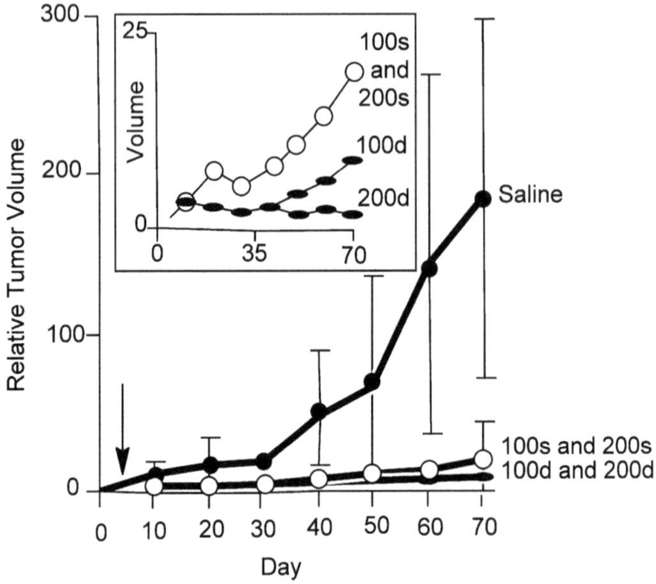

Fig. 5.4 Efficacy studies in a prostate cancer model

With collaborators at the University of Texas Southwestern Medical Center, efficacy has been observed in vivo. Using a human prostate tumor expressing luciferase, tumor volumes could be measured using both calipers and bioluminescent imaging over the course of 70-day trials using four treatment groups and a control group. Treatment groups received either 100 or 200 mg/kg paclitaxel in a single dose or twice. A heterogeneous tumor population was used, and accordingly, data are reported as relative tumor volume, which compares the measured size to the size on day one of the study. Shown in Fig. 5.4 are the results of the experiment for **1**, data which have undergone peer review. Treatment groups showed tumor growth retardation at single doses of 100 mg/kg (100s) or 200 mg/kg (200s). This effect was more pronounced in the twice-dosed (100d, 200d) animals.

Preliminary, and unreviewed, evaluation of **2** confirms the hypothesis that the introduction of a more labile linker in the form of a disulfide will increase toxicity. In these studies, growth inhibition/regression was recorded in the group receiving the highest treatments (200d). Bioluminescence imaging revealed similar trends, with a lack of metabolic activity observed in animals receiving the highest dose. Unfortunately, mortality was also an issue in this treatment group, with two of the animals lost within the first 48 h. Currently, the dosing regime is being optimized with the goal of minimizing dose size, increasing survival, and achieving "cures" as defined by tumor reduction and absence of metabolic activities as measured by bioluminescence imaging. We remain optimistic in these pursuits.

5.9 Therapeutic Index

Currently, uncertainties in both dose schedule and size preclude any conclusions to be drawn. Certainly, 1 is not competitive using the Abraxane dosing regimen, as the level of efficacy is not matched even at higher doses of paclitaxel. The second paclitaxel prodrug has proven to be more toxic and allows for reduced dose sizes in animal models of prostate cancer. The results of these studies will be reported in due course.

5.10 Future Perspective

We remain committed to the long-term goal of realizing an efficacious nanomedicine for the treatment of solid cancerous tumors. Our efforts with paclitaxel have been successful as measured by a number of iterations/refinements applied to all stages of these efforts, from chemical synthesis to characterization and biological assessment. Indeed, more iterations are on the horizon with the goal of reducing dose size and optimizing dosing regimen, with the realization that tumor saturation appears to happen rapidly. To capture future efforts, we return to the first outline (Table 5.4) and articulate the goals of the individual activities in greater detail.

Table 5.4 The challenges for realizing our nanomedicine

Item	Topic	Goal for future efforts
1.	Synthesis	Can the synthesis be optimized to reduce the number of steps required?
		Can the number of species that result be reduced through alternatives to PEG?
2.	Scale	Can we extend scale to these new routes? Can we reduce costs, reduce waste, and increase atom economy?
3.	Drug conjugation and release	Can dose be further reduced by modifying the linker to facilitate cooperative release?
4.	Reproducibility	Can we make multiple, identical batches?
5.	Biodistribution	If PEG is abandoned, is biodistribution affected adversely?
6.	Biocompatibility, safety, and toxicity	Similarly, do the new generations maintain high safety?
7.	Efficacy	Can we reduce doses and optimize regimes to address therapeutic advantage?

Throughout the course of these efforts, however, we are compelled to continually ask the question, "Are we fooling ourselves?" That is, do the data and climate suggest that while academically interesting, our attention might be better focused on other problems given the challenges we will face and the terrain we have already covered? Indeed, the ability for momentum to obfuscate the most meritorious path should not be underestimated. At present, we think the efforts are justified. The weight of each biological experiment profoundly affects this course, and in truth, we may be slowing. Our commitment to these efforts, even in the absence of current funding support, is still driven by a handful of observations that we have made including:

1. The materials proposed are tractable synthetically. While not optimized, triazine dendrimers hold great promise.
2. The materials are as safe as Abraxane (serum albumin) in acute dosing in animal models.
3. The materials are efficacious. We are seeing tumor regression under some circumstances.
4. These materials might be competitive with clinically relevant Abraxane. The cytotoxicity in equivalent paclitaxel units for Abraxane and **1** are identical in LS174T cells (0.1 μM at 48 h and 72 h). Based on literature reports, the dosing used for Abraxane and **1** is within a factor of two, although direct comparisons have not been executed.
5. These materials are still subject to optimization and enhanced function such as targeting or reporting.

5.11 Conclusions

While we have been pursuing triazine dendrimers for almost 10 years now, our recent turn toward paclitaxel-laden constructs has provided us with compelling motivation for the continued pursuit of these materials. These efforts have led us to explore other chemotherapeutics. To this end, our interest in camptothecin has been communicated, and agents currently precluded from the clinic for reasons of toxicity and/or bioavailability are also being studied, notably Brefeldin A. In addition to chemotherapy, these efforts have inspired partnerships with other groups that are continuing to be explored. The applications of these scaffolds to other areas of medical interest, including gene and RNAi delivery, infectious disease, and diagnostic medical imaging, are yielding interesting insights. Computation is playing an increasingly important role. We remain optimistic that time and energy will provide a rich set of lessons learned and perhaps a clinically relevant candidate in due course.

Acknowledgments The authors thank Dr. Sonke Svenson for the opportunity to convey the contents of our presentation at the National ACS Meeting in Boston (2010) in this forum. In addition to the long-standing efforts of coworkers identified in the relevant citations, the authors wish to thank Dr. Xiankai Sun and Su-Tang Lo of the University of Texas Southwestern Medical Center for ongoing collaboration in this area. Additionally, individuals of the Nanotechnology Characterization Laboratory who contributed to the collection of data on these dendrimers include Stephan Stern, Jeffrey D. Clogston, Jiwen Zheng, Pavan P. Adiseshaiah, Marina Dobrovolskaia, as led by Anil Patri. This work is supported with funds from Texas Christian University. The work is impacted by partnerships with others in related areas, including Drs. Thomas Kissel and Olivia Merkel in Marburg, Dr. Sunil Shaunak in London, Drs. Pete Choyke and Hisataka Kobayashi in Bethesda, and Giovanni Pavan at SUPSI.

References

1. Duncan R (2003) The dawning era of polymer therapeutics. Nat Rev Drug Discov 2:347–360
2. Putnam D (2008) Drug delivery: the heart of the matter. Nat Mater 7:836–837
3. Menjoge AR, Kannan RM, Tomalia DA (2010) Dendrimer-based drug and imaging conjugates: design considerations for nanomedical applications. Drug Discov Today 15:171–185
4. Scripture CD, Figg WD, Sparreboom A (2005) Paclitaxel chemotherapy: from empiricism to a mechanism-based formulation strategy. Ther Clin Risk Manag 1:107–114
5. Crown J, O'Leary M, Ooi WS (2004) Docetaxel and paclitaxel in the treatment of breast cancer: a review of clinical experience. Oncologist 9:24–32
6. Blum JL, Savin MA, Edelman G, Pippen JE, Robert NJ, Geister BV, Kirby RL, Clawson A, O'Shaughnessy JA (2007) Phase II study of weekly albumin-bound paclitaxel for patients with metastatic breast cancer heavily pretreated with taxanes. Clin Breast Cancer 7:850–856
7. Hawkins MJ, Soon-Shiong P, Desai N (2008) Protein nanoparticles as drug carriers in clinical medicine. Adv Drug Deliv Rev 60:876–885
8. Bonomi P (2007) Paclitaxel poliglumex (PPX, CT-2103): macromolecular medicine for advanced non-small-cell lung cancer. Expert Rev Anticancer Ther 7:415–422
9. Albain KS, Belani CP, Bonomi P, O'Byrne KJ, Schiller JH, Socinski M (2006) PIONEER: a phase III randomized trial of paclitaxel poliglumex versus paclitaxel in chemotherapy-naive women with advanced-stage non-small-cell lung cancer and performance status of 2. Clin Lung Cancer 7:417–419
10. Beeram M, Rowinsky EK, Hammond LA, Patnaik A, Schwartz GH, de Bono JS, Forero L, Forouzesh B, Berg KE, Rubin EH, Beers S, Killian A, Kwiatek J, McGuire J, Spivey L, Takimoto CH (2002) A phase I pharmacokinetic (PK) study of PEG-paclitaxel in patients with advanced solid tumors (Abstract). Proc Am Soc Clin Oncol 21:405
11. Meerum Terwogt JM, ten Bokkel Huinink WW, Schellens JH, Schot M, Mandjes IA, Zurlo MG, Rocchetti M, Rosing H, Koopman FJ, Beijnen JH (2001) Phase I clinical and pharmacokinetic study of PNU166945, a novel water-soluble polymer-conjugated prodrug of paclitaxel. Anticancer Drugs 12:315–323
12. Khandare JJ, Jayant S, Singh A, Chandna P, Wang Y, Vorsa N, Minko T (2006) Dendrimer versus linear conjugate: influence of polymeric architecture on the delivery and anticancer effect of paclitaxel. Bioconjug Chem 17:1464–1472
13. Gao Y, Chen L, Gu W, Xi Y, Lin L, Li Y (2008) Targeted nanoassembly loaded with docetaxel improves intracellular drug delivery and efficacy in murine breast cancer model. Mol Pharm 5:1044–1054
14. Vrudhula VM, MacMaster JF, Li Z, Kerr DE, Senter PD (2002) Reductively activated disulfide prodrugs of paclitaxel. Bioorg Med Chem Lett 12:3591–3594
15. Majoros IJ, Myc A, Thomas T, Mehta CB, Baker JR (2006) PAMAM dendrimer-based multifunctional conjugate for cancer therapy: synthesis, characterization, and functionality. Biomacromolecules 7:572–579
16. Papas S, Akoumianaki T, Kalogios C, Hadjiarapolglou L, Theodoropoulous PA, Tsikaris V (2007) Synthesis and antitumor activity of peptide-paclitaxel conjugates. J Pept Sci 13:662–671

17. El Alaoui A, Saha N, Schmidt F, Monneret C, Florent J-C (2006) New Taxol (paclitaxel) prodrugs designed for ADEPT and PMT strategies in cancer chemotherapy. Bioorg Med Chem Lett 14:5012–5019
18. Zou Y, Fu H, Ghosh S, Farquhar D, Klostergaard J (2004) Antitumor activity of hydrophilic paclitaxel copolymer prodrug using locoregional delivery in human orthotopic non-small cell lung cancer xenograft models. Clin Cancer Res 10:7382–7391
19. Guillemard V, Saragovi HU (2001) Taxane-antibody conjugates afford potent cytotoxicity, enhanced solubility, and tumor target selectivity. Cancer Res 61:694–699
20. Simanek EE, Hanan A, Lalwani S, Lim J, Mintzer M, Venditto VJ, Vittur B (2010) The eight year thicket of triazine dendrimers: strategies, targets, and applications. Proc R Soc A 466: 1445–1468
21. Steffensen MB, Simanek EE (2003) Chemoselective building blocks for dendrimers from relative reactivity data. Org Lett 5:2359–2361
22. Moreno KX, Simanek EE (2008) Identification of diamine linkers with differing reactivity and their application in the synthesis of a melamine dendrimers. Tetrahedron Lett 49:1152–1154
23. Hollink E, Simanek EE (2006) A divergent route to diversity in macromolecules. Org Lett 8:2293–2295
24. Crampton H, Hollink E, Perez LM, Simanek EE (2007) A divergent route towards single-chemical entity triazine dendrimers with opportunities for structural diversity. New J Chem 31:1283–1290
25. Lim J, Mintzer MA, Perez LM, Simanek EE (2010) Synthesis of odd generation triazine dendrimers using a divergent, hypermonomer approach. Org Lett 12:1148–1151
26. Chen HT, Neerman MF, Parrish AR, Simanek EE (2004) Cytotoxicity, hemolysis, and acute in vivo toxicity of dendrimers based on melamine, candidate vehicles for drug delivery. J Am Chem Soc 126:10044–10048
27. Venditto VJ, Allred K, Allred CD, Simanek EE (2009) Intercepting triazine dendrimer synthesis with nucleophilic pharmacophores as a general strategy toward drug delivery vehicles. Chem Commun 5541–5542
28. Lim J, Venditto VJ, Simanek EE (2010) Synthesis and characterization of a triazine dendrimer that sequesters iron(III) using 12 desferrioxamine B groups. Bioorg Med Chem 18: 5749–5753
29. Lim J, Simanek EE (2008) Synthesis of water-soluble dendrimers based on melamine bearing 16 paclitaxel groups. Org Lett 10:201–204
30. Lim J, Chouai A, Lo S-T, Liu W, Sun X, Simanek EE (2009) Design, synthesis, characterization, and biological evaluation of triazine dendrimers bearing paclitaxel using ester and ester/disulfide linkages. Biconjug Chem 20:2154–2161
31. Chouai A, Simanek EE (2008) Kilogram-scale synthesis of a second-generation dendrimer based on 1,3,5-triazine using green and industrially compatible methods with a single chromatographic step. J Org Chem 73:2357–2366
32. Chouai A, Venditto VJ, Simanek EE, Vanderplas BC, Ragan JA (2009) Large scale green synthesis of a generation-1 melamine (triazine) dendrimer. Organic Synth 86:151
33. Greish K, Fang J, Inutsuka T, Nagamitsu A, Maeda H (2003) Macromolecular therapeutics: advantages and prospects with special emphasis on solid tumour targeting. Clin Pharmacokinet 42:1089–1105
34. Lim J, Guo Y, Rostollan CL, Stanfield J, Hsieh J-T, Sun X, Simanek EE (2008) The role of the size and number of polyethylene glycol chains in the biodistribution and tumor localization of triazine dendrimers. Mol Pharm 5:540–547
35. Matsumura Y, Maeda H (1986) A new concept for macromolecular therapeutics in cancer chemotherapy: mechanism of tumoritropic accumulation of proteins and the antitumor agent smancs. Cancer Res 6:6387–6392
36. Majoros IJ, Myc C, Thomas T, Mehta CB, Baker JR, Jr. (2006) PAMAM dendrimer-based multifunctional conjugate for cancer therapy: Synthesis, characterization, and functionality. Biomacromolecules 7: 572

Part II
Dual or Multiple Drug Delivery (Combination Therapy)

Chapter 6
Versatile Fixed-Ratio Drug Combination Delivery Using Hydrophobic Prodrug Nanoparticles

Barry D. Liboiron, Paul G. Tardi, and Lawrence D. Mayer

6.1 Introduction

For over 30 years, drug combination therapy has been the standard of care for the treatment of most forms of cancer. The use of drug combinations was pioneered by the works of Frei and Freireich who demonstrated that the treatment of childhood leukemia could be dramatically enhanced through the use of multiple agents [1, 2]. When methotrexate was used as a single agent therapy, response rates of only 40% were observed but could be improved to >95% through the addition of 6-mercaptopurine, prednisone, and vincristine. This dramatic improvement in cancer therapy resulted in the widespread use of combination treatments for all forms of cancer. The development of new drug combinations was based on selecting therapeutic agents with nonoverlapping toxicities and then dosing each drug to its maximum tolerated dose. The rationale underlying this approach was based on the concept that maximum dose intensity will result in maximum therapeutic activity. From the days of these early clinical studies to present day, our knowledge of the molecular pathways leading to the generation of cancer has grown dramatically. Scientists and clinicians have used this information to try to identify the most therapeutically active anticancer drug combinations. Since the rapid evaluation of drug combinations is not practical in a clinical setting, research scientists have performed the majority of the drug combination screening. Based on overall activity, each drug combination can be classified as either additive, antagonistic, or synergistic. By definition, an additive drug combination is the sum of the effects contributed by each drug while synergistic and antagonistic combinations are defined as more or less than the additive effect, respectively.

B.D. Liboiron • P.G. Tardi • L.D. Mayer (✉)
Celator Pharmaceuticals Corp., 1779 West 75th Avenue,
Vancouver, BC, Canada V6P 6P2
e-mail: lmayer@celatorpharma.com

S. Svenson and R.K. Prud'homme (eds.), *Multifunctional Nanoparticles for Drug Delivery Applications: Imaging, Targeting, and Delivery*, Nanostructure Science and Technology, DOI 10.1007/978-1-4614-2305-8_6,

To determine the classification of a drug combination, the individual drugs as well as the combination are incubated with tumor cells in vitro. Following a fixed time frame of drug exposure, the cell viability is determined at various drug concentrations. The cell viability is then plotted against drug concentration to identify the amount of drug required to cause 50% tumor cell killing (IC_{50} value). There have been numerous mathematical models reported in the literature that utilize cell viability data to determine the classification of a drug combination. These models range in complexity from general techniques requiring simple manual calculations to sophisticated algorithms aided by computers. A comprehensive review of in vitro synergy analysis methods has been recently published [3]. The most commonly used model for drug combination analysis is the median-effect method of Chou and Talalay [4–6]. This method has several advantages over other approaches including the following: (1) the fundamental equations used were derived from basic mass action enzyme kinetic models, (2) the fitting of data uses linear regression which is statistically accepted, (3) the experimental design requires fewer data points than other methods, and (4) the method is available as a software package which facilitates data entry and modeling. Based on these advantages, we chose to evaluate various drug combinations using this model.

In vitro evaluation of drug combinations commonly utilizes a series of concentrations of drug A and drug B to establish a dose response curve for each drug. Based on the effective concentrations for each drug, the combination is subsequently evaluated. Each concentration of drug A is mixed with each concentration of drug B to produce a matrix of drug combinations at various concentrations and drug ratios. Although this approach results in drug combination analysis at a wide range of drug ratios, it is not systematic and often occurs at non-therapeutically relevant amounts of drug. In order to evaluate the role of drug ratios on the classification of a drug combination, we devised a systematic approach to drug ratio screening. A series of fixed drug ratios, each diluted over a range of concentrations, is evaluated for tumor growth inhibition. Using high-throughput screening, tumor growth inhibition curves are generated for the series of drug ratios in a panel of tumor cell lines. The cytotoxicity curves are analyzed by the median-effect model to determine the combination index (CI) for each drug combination [7, 8]. The results of a gemcitabine and cisplatin combination are presented as a synergy heat map in Table 6.1. The CI values reflecting synergy (<0.9), additivity (0.9–1.1), or antagonism (>1.1) are color-coded as green, yellow, and red, respectively. When CI values are examined at high cell kill (essential for effective chemotherapy), we observed all three different classifications. Using the synergy heat map, it is readily apparent that synergistic, additive, and antagonistic effects can be achieved with different drug ratios of gemcitabine and cisplatin. The consistent trend in drug ratio dependency across multiple tumor types indicates the widespread importance of this effect on combination chemotherapy. These results highlight three critical features associated with the use of combination chemotherapy: (1) a drug combination may act synergistically or antagonistically, depending on the ratio of the two drugs; (2) to capture synergy in vivo, the target drug ratio must be successfully delivered to the tumor site; and (3) the intravenous administration of a synergistic drug ratio cannot be maintained due to the different pharmacokinetic properties of anticancer agents.

Table 6.1 In vitro synergy heat map for gemcitabine and cisplatin, displaying the CI values at an ED80 for 16 different drug ratios screened against nine tumor cell lines

Cell lines		CI@ED80															
screened	Tumor	64:1	32:1	16:1	8:1	4:1	2:1	1:1	1:2	1:4	1:8	1:16	1:32	1:64	1:128	1:256	1:512
HT-29	Colon	1.80	1.30	4.50	2.30	0.62	0.85	0.69	0.50	0.33	0.16	0.15	0.14	0.03	0.03	0.01	0.02
H1299	Lung	1.70	1.10	3.60	9.70	0.74	0.95	0.40	0.44	0.58	0.44	0.55	0.64	0.18	0.24	0.27	0.54
A549(3)	Lung	1.10	0.90	0.85	1.10	0.90	0.85	0.88	0.81	0.55	0.71	0.41	0.51	0.67	1.20	1.90	2.60
A2780	Ovarian	1.30	1.10	0.82	0.96	0.73	0.87	0.76	0.76	1.40	1.20	3.20	8.30	2.90	4.30	3.40	2.30
BxPC-3	Pancreatic	1.30	1.10	0.83	0.52	0.54	0.94	0.68	0.52	0.38	0.23	0.35	0.31	0.48	0.58	1.10	1.10
H460	Lung	1.10	0.88	0.85	1.40	1.20	0.65	0.96	0.82	0.91	1.10	0.67	0.71	1.30	0.76	0.92	0.93
H1299	Lung	9.00	1.70	0.62	0.51	0.49	0.86	0.66	0.62	0.34	0.55	0.49	0.27	0.69	2.50	1.80	4.70
IGROV-1	Ovarian	3.00	0.79	2.90	1.40	1.10	0.70	1.30	0.65	0.87	0.41	0.31	0.57	0.64	1.10	1.20	1.30
MCF-7	Breast	0.80	0.50	1.00	1.10	1.10	1.00	1.00	0.84	0.97	0.81	0.42	0.44	0.39	0.38	0.81	1.30

CI values between 0 and 0.89 are synergistic (*green*), values between 0.9 and 1.1 are additive (*yellow*), and values greater than 1.1 are antagonistic (*red*). A high prevalence of drug synergy is observed between gemcitabine to cisplatin ratios of 4:1 and 1:64. Reprinted with permission from [3]

Drug delivery vehicles (e.g., liposomes or polymeric micelles) are particularly well suited for controlling administered drug ratios. Drug carriers ranging in size from 20 to 100 nm do not readily escape the blood vessels of healthy tissue [9], thereby eliminating the diverse early distribution phase associated with various anticancer drugs. They also accumulate at sites of tumor growth due to the enhanced permeability and retention (EPR) effects associated with tumors [10, 11]. Specifically, gaps in the tumor vasculature result in a "sieving" effect, which traps the nanoparticle within the extracellular space of the tumor. Delivery systems carrying synergistic anticancer drug combinations will provide a depot for drug release at the tumor site. Most importantly, nanoparticulate delivery systems can be manipulated so that the release rates of the anticancer agents are controlled following systemic administration.

The delivery system best suited for the coordinated release of both agents is dictated by the physical properties of the drug combination. For water-soluble drugs, liposomal formulations can be readily adapted through changes in membrane composition. In the case of drugs with poor water solubility, the hydrophobic core of a polymer nanoparticle can be modified through the use of different base polymer systems. Although the approaches may be quite different, control of synergistic ratios in vivo can be readily achieved.

6.1.1 The Importance of Drug Ratios: Development of CPX-351

Cytarabine and daunorubicin have been used as the standard of care in the treatment of acute myelogenous leukemia (AML) for over three decades [12, 13]. Attempts to improve the clinical activity of this drug combination through the use of high-dose cytarabine [14, 15] or modifying daunorubicin dose and schedule [16] increased the toxicity of the combination but did not provide an increase in overall survival. Despite the widespread use of this combination in a clinical setting, there has been little research focused on understanding the interactions between these two drugs on a cellular level. Considering the overall activity of a drug combination can be influenced by the drug ratio exposed to tumor cells [7, 17–19], we investigated the role of drug ratio-dependent synergy for cytarabine and daunorubicin. Extensive analysis of ratio dependency for this drug combination was reported elsewhere [20], with the most synergistic ratio identified as a 5:1 molar ratio of cytarabine to daunorubicin. Based on the high solubility of cytarabine and daunorubicin (greater than 300 mg/ml and 25 mg/ml, respectively), a liposomal formulation represented the ideal delivery vehicle for this drug combination. The liposomal membrane composition chosen to coordinate the release of these two agents was based on a manuscript reporting the co-encapsulation of irinotecan and floxuridine [21]. In this manuscript, a phosphatidylcholine-based formulation was stabilized with phosphatidylglycerol and a small amount of cholesterol. This formulation was readily manipulated to alter the drug release rates of each drug independently. When this formulation was utilized for the co-encapsulation of cytarabine and daunorubicin,

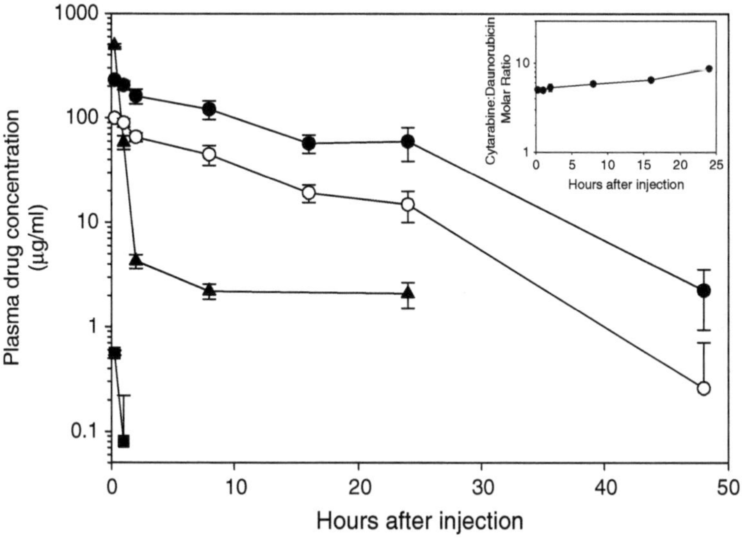

Fig. 6.1 Plasma drug concentrations of cytarabine (*filled circle*) and daunorubicin (*open circle*) following i.v. administration of liposomes co-encapsulated with cytarabine and daunorubicin at a 5:1 molar ratio (12:5.3 mg/kg) to CD-1 nude mice ($n=3$ per time point). Cytarabine (*filled triangle*) and daunorubicin (*filled square*) concentrations were also evaluated in CD-1 nude mice ($n=3$ per time point) following administration of free drug cocktail (600:9 mg/kg) (**a**) *Insert*: Circulating plasma cytarabine-to-daunorubicin molar ratios for the liposomal formulation was calculated from absolute plasma concentrations. Reproduced with permission from [20]

coordinated drug release was observed in vivo with drug half-lives of 11.6 and 8.5 h, respectively (Fig. 6.1). The circulating drug ratio was maintained in the plasma between 5:1 and 9:1 for over 24 h (Fig. 6.1 inset). These drug ratios were within the range of 5:1 and 10:1 shown to be the most synergistic ratios in vitro [20]. This observation is in sharp contrast to the free drug cocktail, where plasma concentrations changed rapidly and independently. At 15 min after injection, the plasma cytarabine-to-daunorubicin ratio had changed 13-fold, and drug levels had decreased by more than 95%. These results indicate that co-encapsulation of cytarabine and daunorubicin using liposome nanoparticles provides an effective means of controlling drug release and maintenance of a synergistic drug ratio in vivo.

In order to confirm that the ratios identified through in vitro cytotoxicity assays translate into the most therapeutically active ratios in vivo, liposomal formulations of cytarabine and daunorubicin were generated at a variety of fixed ratios, ranging from 1:1 to 12:1 [20]. In a P388 leukemia model, these liposomal formulations were administered at the maximum tolerated dose (MTD). At the 12:1 ratio, a high level of activity was observed as reflected in a long-term survival rate of 83%. Although impressive, the most therapeutically active formulation was the synergistic 5:1 ratio which resulted in 100% survival, even at 80% of the MTD (Fig. 6.2). Interestingly at the 3:1 ratio, the daunorubicin dose was almost twice as high as that

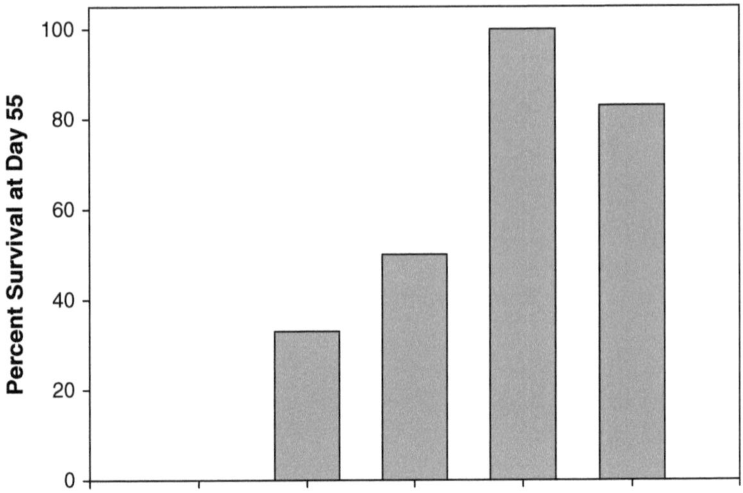

Molar Drug Ratio	Saline	1:1	3:1	5:1	12:1
Cytarabine	N/A	5.4 mg/kg	10 mg/kg	10 mg/kg	15 mg/kg
Daunorubicin	N/A	12.5 mg/kg	7.7 mg/kg	4 mg/kg	3 mg/kg

Fig. 6.2 Survival of BDF-1 mice bearing P388 ascites tumors at day 55 following i.v. administration treatment on days 1, 4, and 7 with saline or liposome co-encapsulated with cytarabine and daunorubicin at different drug molar ratios ($n=6$ mice per group). Formulations were dosed at their MTD with the exception of CPX-351 (5:1 molar drug ratio), which was dosed at 0.8 of its MTD. Reproduced with permission from [20]

of the synergistic 5:1 ratio with an equivalent dose of cytarabine, yet only 50% survival was observed. The observation that higher drug doses can result in inferior therapeutic activity highlights the importance of formulating and delivering synergistic drug ratios to the tumor site. Decreasing the cytarabine-to-daunorubicin ratio even further to 1:1 resulted in the lowest long-term survival, at 33%. These results provide a strong correlation between in vitro synergy results and therapeutic activity in vivo. The ability to generate such a correlation could result in the rapid identification and formulation of superior anticancer drug combinations. This liposomal formulation of a 5:1 molar ratio of cytarabine and daunorubicin, referred to as CPX-351, has recently completed Phase II clinical trials in AML, where impressive clinical results were observed [22].

6.2 Dual Drug Nanoparticle Systems

Liposomal carriers are well suited for the stable encapsulation of hydrophilic drugs as the interior aqueous core provides an ideal medium for the accumulation and retention of water-soluble drugs within the carrier. Hydrophilic agents are not likely to pass through a lipid bilayer, and the interior buffer can be manipulated in many ways to promote intraliposomal precipitation, complexation [23], or oligomerization [24] that serves to prevent egress of the entrapped drug(s) from the carrier [25, 26]. Hydrophobic drugs, however, tend to reside not in the aqueous internal buffer of liposomal carriers but intercalate within the lipid membrane itself. Drug residence within this lipid compartment is typically short-lived, and therefore, retention of hydrophobic drugs in liposomes is typically poor. For example, more than 90% of the injected drug of a liposomal formulation of paclitaxel was eliminated from the plasma within 1 h postinjection [27]. Polymer nanoparticles are far better suited to the delivery of poorly water-soluble agents, and many reports and reviews of successful single formulations of hydrophobic drugs in polymer nanoparticles have been presented. Interested readers are referred to several recent reviews in addition to other chapters in this book [28–30].

To achieve the augmented efficacy of a dual drug formulation such as CPX-351 where at least one of the drugs is poorly water-soluble, the use of a polymeric nanoparticle delivery system is indicated. Stable formulation of *one* agent into a polymeric nanoparticle, while at least maintaining in vivo efficacy, can be challenging; these difficulties are only amplified when one attempts to introduce a second agent into the carrier system. Potential problems include increased particle instability, decreased overall drug loading (expressed as a mass drugs/mass polymer ratio), deleterious drug–drug interactions within the particle, and/or increased particle size.

Encapsulation of more than one drug in a single particle system (co-formulation) has been of increasing interest as researchers attempt to circumvent problems associated with poor tumor delivery of multiple agents. Decades of experience have shown that chemotherapy agents are vastly more effective in combination, and it may not be possible to deliver efficacious amounts of the second drug when administered as a free agent. Indeed, CPX-351 shows that delivery of the two agents to the tumor site should occur at the optimal synergistic drug ratio to achieve maximized efficacy. Many reports describing the development, in vitro properties, and in vivo performance of dual drug nanoparticle formulations have appeared in the recent literature (Tables 6.2–6.4).

These dual drug nanoparticle systems can be classified by the manner in which the drugs are encapsulated within the nanoparticle. The simplest approach involves mixing of the drugs with the polymeric carrier, followed by either solvent exchange and/or evaporation, a method that we refer to as passive encapsulation. The hydrophobic drugs preferentially associate with the hydrophobic portions of the polymer as the latter self-associates into micelles or nanoparticles with increasing water-to-solvent ratio. This method has the advantages of being relatively simple to develop, and it requires no chemical modifications of the drugs or polymer. Introduction of the drug(s) to the polymer can take place either before or after nanoparticle formation, although encapsulation efficiencies are typically much higher when the drug(s) is present during nanoparticle formation. The second strategy uses direct conjugation of the drug to the polymer through a biodegradable linkage, typically an ester or peptide bond, to produce a polymer-conjugated nanoparticle system. Chemical modification of the polymer chain is usually required to prepare multiple sites for

Table 6.2 Passively encapsulated dual drug nanoparticle systems

Drug A	Drug B	Particle composition	Drug release properties	References
Paclitaxel	Ceramide	PEG-PCL	Not reported	[32]
Vincristine	Quercetin	PLGA nanoparticles	IVR: quercetin release slower than vincristine	[40]
Vincristine	Verapamil	PLGA	IVR: drug release well coordinated over 25 h at pH 6.5. At pH 7.4, verapamil release was slower than vincristine	[41]
Paclitaxel	Tariquidar	PLGA	IVR: slow but coordinated release. Presence of second drug has marginal effect on release profile of first	[34]
Etoposide, docetaxel, paclitaxel	17-AAG	PEG-PLA	Relatively rapid ($t_{1/2}$<2 h) but well-coordinated drug release for either etoposide or docetaxel with 17-AAG over 24 h. Paclitaxel release much slower ($t_{1/2}$=5 h)	[45]
Paclitaxel	Doxorubicin	GEG and LE polymer micelles	IVR: singly formulated particles had similar release properties	[31]
Paclitaxel	Curcumin	Flaxseed oil nanoemulsion±DSPE-PEG	Not reported	[33]
Paclitaxel	Rapamycin	Glycerol monooleate-iron oxide	IVR: burst release (~15%) over 24 h followed by uncoordinated release over 20 days	[44]
Paclitaxel	siRNA (P-gp target)	PLGA-PEI	IVR: siRNA release rate 2–3 times that of paclitaxel. Particle functionalization slows release of both drugs	[35]
Paclitaxel	Combretastatin	PLGA	Not reported	[36]
Doxorubicin	Methylene blue	Sodium alginate, Aerosol-OT™	Not reported	[38]
Doxorubicin	Indocyanine green	PLGA	Doxorubicin release reported and found to correlate with earlier studies. Indocyanine green release not studied	[39]

PEG poly(ethylene glycol); *PCL* poly(ε-caprolactone); *PLGA* poly(D,L-lactide-*co*-glycolide); *IVR* in vitro release; *17-AAG* 17-allylamino-17-demethoxygeldanamycin; *PLA* poly(lactide); *GEG* poly(γ-benzyl-L-glutamate) and poly(ethylene glycol) block copolymer; *LE* poly(L-lactide)/poly(ethylene glycol) block copolymer; *DSPE* distearoylphosphatidylethanolamine; *PEI* poly(ethylene imine)

Table 6.3 Polymer-conjugated dual drug nanoparticle formulations

Drug A	Drug B	Particle composition	Drug release properties	Reference
Doxorubicin	Combretastatin	PLGA, PEG-DSPE, PC, chol	In vitro release: Combretastatin release much faster, doxorubicin releases as DOX-PLGA fragments	[55]
Doxorubicin	AGM	Single chain HPMA polymer micelle	First report of polymer-drug conjugate for combination therapy. Both drugs conjugated to HPMA polymer backbone via peptide linkages	[47, 49]
Doxorubicin	Wortmannin	PEG-poly(Asp)	Not reported	[50]
Doxorubicin	Gemcitabine	HMPA-based polymer-drug conjugate, dual formulation	Enzymatic release assay: uncoordinated release from polymer backbone	[53]
Paclitaxel	Combretastatin	PEG-PLA	In vitro release: Combretastatin release much faster	[56]
Cytarabine	Floxuridine	Random copolymer of vinylsebacoyl-cytarabine and vinyladipoyl-floxuridine	Release of 2 drugs widely different over 50 h at pH 7.4. Very small release of drugs (<10%) under strong acidic conditions	[54]
Pt(IV)	Docetaxel	PEG-PLGA	Pt(IV) conjugated to succinic acid and conjugated to PLA backbone (10% wt Pt). Passive docetaxel loading. Docetaxel release much faster than Pt(IV)	[57]

PLGA poly(D,L-lactide-co-glycolide); *PEG* poly(ethylene glycol); *DSPE* distearoylphosphatidylethanolamine; *PC* phosphatidylcholine; *Chol* cholesterol; *Asp* aspartate; *PLA* poly(lactide)

Table 6.4 Dual drug nanoparticle formulations containing drug conjugates

Drug A	Drug B	Particle composition	Drug release properties	References
Paclitaxel	Gemcitabine	PEG-PS, POPC	Both drugs conjugated to succinate/diglycolate linkers and nonpolar anchors. Release of both drugs dictated by hydrophobicity of anchor	[74, 75]
Paclitaxel	Gemcitabine	Paclitaxel-gemcitabine drug conjugate, self-assembled	Hydrolysis assay confirms slow hydrolysis of parent prodrugs	[60]

PEG poly(ethylene glycol); *PS* poly(styrene); *POPC* palmitoyl oleoyl phosphatidylcholine

drug conjugation, while some polymer systems already possess chemically active sites either on each monomer or the terminal end that can be used as linkage sites. In either case, this method generally results in very stable drug encapsulation and retention. The third encapsulation method involves reversible conjugation of the drug to another chemical species. The goal of this conjugation is to alter the physicochemical properties of the drug in order to induce significant changes in its stability, biodistribution, and/or release from a nanoparticle. In this method, a prodrug is synthesized and then encapsulated in the nanoparticle, typically through the aforementioned passive method. We refer to these systems as drug-conjugated nanoparticles.

Many drug combinations show synergistic and/or antagonistic interactions in vitro and in vivo. We have shown that these interactions can be highly dependent on the ratio of the two drugs at the target site. To achieve ratio-dependent synergistic activity in vivo from a drug combination delivered in a nanoparticle formulation, the following criteria must be met: (1) the drugs should circulate entrapped in the particle for extended periods (typically $t_{1/2}$>4 h) to allow for passive accumulation at the tumor site via the EPR effect and (2) the drugs should be released from the particle at equivalent rates, such that the drug-to-drug ratio established during formulation of the particle is also delivered to the tumor. In the next section, we highlight recent reports of dual drug nanoparticle systems and consider the degree to which the systems satisfy the above criteria. Dual drug nanoparticle systems that can successfully deliver their drug payload to the tumor site while maintaining a synergistic drug ratio within the particle are likely to have superior antitumor efficacy over both the free drug cocktail and other delivery systems in which the importance of drug ratios is neglected.

6.2.1 Passive Dual Drug Nanoparticle Systems

Formulation of multiple drugs into polymeric nanoparticles has received increased attention recently, and passive encapsulation is the most commonly used method to produce a multidrug nanoparticle. The vast majority of systems reported in Table 6.2

were developed for co-delivery of paclitaxel with another agent, highlighting both the importance of the taxane in chemotherapy and its suitability for stable, passive formulation in a polymeric nanoparticle. Paclitaxel was co-formulated with a diverse group of second agents: a cytotoxic [31], an apoptosis inhibitor [32], multidrug resistance inhibitors [33, 34] including siRNA [35], and an antiangiogenic drug [36]. Other examples were combinations of established cytotoxic drugs such as doxorubicin with 5-fluorouracil [37], but also with the photosensitizers methylene blue [38] and indocyanine green [39].

The limitations of passive encapsulation in nanoparticles become apparent when we consider the considerable evidence for drug ratio-dependent synergy/antagonism. Since the drugs are loaded by what is largely an equilibrium process (drug solubility in the polymeric particle vs. aqueous solution), there is only limited control of each drug's release rate from the particle, particularly in vivo. Without coordinated drug release rates for both agents, the ratio established during formulation will be quickly disrupted upon administration in vivo, as one drug likely leaves the particle faster than the other. In some cases, drug release rates were not even considered. Devalapally et al. prepared singly formulated PEG-PCL nanoparticles of paclitaxel and ceramide but did not study drug stability or retention [32]. Similarly, a nanoemulsion of flax-seed oil, stabilized with DSPE-PEG, containing paclitaxel and curcumin showed excellent drug stability over 3 months, but drug level measurements were made on the bulk solution only, which does not consider whether slow release of both agents from the hydrophobic phase of the emulsion was taking place [33]. The novel combinations of doxorubicin with the photosensitizing agents methylene blue [38] and indocyanine green [39] were also incompletely characterized with respect to drug release. In both cases, however, photodynamic therapy in combination with the co-formulated nanoparticle resulted in increased cytotoxicity [39] or survival of mice with xenograft tumors [38] over administration of the dual drug nanoparticle alone. Other examples emphasize the difficulty in coordinating drug release rates to maintain a desired drug-to-drug ratio. For vincristine and quercetin, PLGA nanoparticles successfully encapsulated both drugs; however, quercetin release was considerably slower than the release of more hydrophilic vincristine [40]. Drug release of verapamil was pH sensitive when co-formulated with vincristine, also in PLGA, with drug release correlated over 25 h at pH 6.5, but verapamil release was inhibited and not equivalent to vincristine at pH 7.4. Vincristine release was unaffected by the difference in pH [41]. Iron oxide nanoparticles coated with glycerol monooleate (GMO, a substitute for the more commonly used oleic acid) are capable of co-encapsulating paclitaxel and rapamycin within the GMO corona [42, 43]. The drugs are quite stable within the coated nanoparticle; after an initial burst release of approximately 15% of both drugs in an in vitro assay, both agents are slowly released over a span of 20 days. The release, however, is not well coordinated: at 10 days, rapamycin release is ~80% compared to ~40% for paclitaxel [44].

Some systems are designed to exploit large differences in drug release rate by first releasing either an antiangiogenic agent to collapse tumor vasculature [36] or an inhibitor of multidrug resistance (MDR) proteins such as permeability glycoprotein (P-gp) [35]. In the latter case, P-gp silencing siRNA was co-encapsulated with paclitaxel in a mixture of PLGA and poly(ethylene imine) (PEI), a cationic polymer.

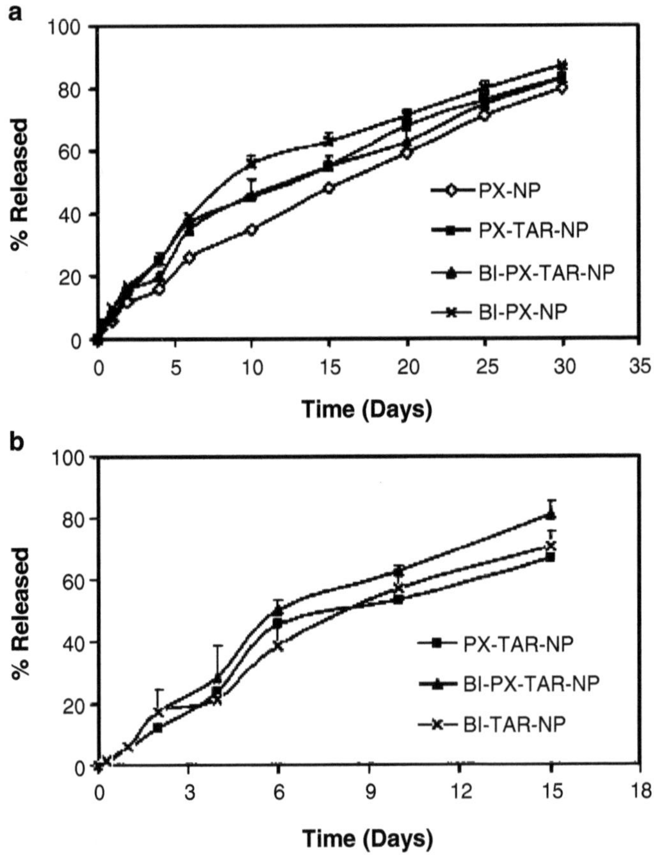

Fig. 6.3 In vitro release of passively loaded (**a**) paclitaxel and (**b**) tariquidar from PLGA dual drug nanoparticles in 0.5% w/v Tween 80 in PBS buffer at 37°C. Data is presented as the mean of three experiments ± standard deviation. *PX-NP* paclitaxel nanoparticles, *PX-TAR-NP* paclitaxel-tariquidar nanoparticles, *BI-PX-NP* paclitaxel nanoparticles functionalized with biotin, *BI-TAR-NP* tariquidar nanoparticles functionalized with biotin, *BI-PX-TAR-NP* paclitaxel-tariquidar nanoparticles functionalized with biotin. Reproduced with permission from [34]

Release of the RNA oligomer was approximately two or three times the rate of paclitaxel release for bare and biotin-functionalized nanoparticles. Functionalization with biotin significantly decreased the rate of paclitaxel release from the nanoparticles. It is not currently known, however, whether cytotoxic drugs such as paclitaxel possess any ratio-dependent synergy with RNA silencing agents [36]. While no study reported in Table 6.2 considered the drug-to-drug ratio encapsulated within the nanoparticle system, some systems were capable of normalizing drug release rates over long periods. Patil et al. loaded paclitaxel and tariquidar into PLGA nanoparticles at a 5:1 molar ratio and studied the in vitro release of the drugs in phosphate-buffered 0.5% w/v Tween 80 at 37°C (Fig. 6.3). Reasonably coordinated release of both drugs was maintained over 2 weeks, with approximately 65% of tariquidar compared to 55% of paclitaxel released over 15 days [34]. The authors also studied the release of each drug in the absence of the second component. Such studies can be helpful in iterative formulation campaigns to coordinate drug release. The close correlation of release rates of these two agents is likely a product of the similar chemical structures and molecular weights of the encapsulated agents.

Kwon and coworkers successfully encapsulated etoposide, docetaxel, and/or paclitaxel with the preclinical stage compound 17-AAG (17-allylamino-17-demethoxygeldanamycin) in PEG-PLA polymeric micelles. Both docetaxel and etoposide release were very well coordinated with 17-AAG release, albeit the release half-lives of these compounds were all less than 2 h, which may not be sufficient time to exploit the EPR effect in vivo. Paclitaxel release was considerably slower ($t_{1/2}=5$ h) and thus not well coordinated with any of the other agents [45]. Further detail on these systems appears in a following chapter of this book.

Preparation of two distinct nanoparticles, followed by mixing of the two systems at the desired ratio, can be used to overcome differences in physicochemical properties that may lead to uncoordinated drug release in a co-formulation. Na et al. formulated doxorubicin in poly(γ-benzyl L-glutamate)/PEG (GEG) polymeric micelles and paclitaxel in PEG-PLA block copolymeric micelles (LE) separately [31]. Doxorubicin and paclitaxel release from the disparate particles was well coordinated over 3 days. A rudimentary assay, similar to an isobologram analysis, suggested synergistic in vitro cytotoxicity between these two agents, which serves to further emphasize the importance of normalizing the drug release rates. The use of two separate nanoparticles would also allow selecting of the drug-to-drug ratio by mixing the two particles in the desired proportion. However, using two different polymeric systems might affect the in vivo biodistribution of each system differently, which may result in delivery of non-synergistic or even antagonistic drug ratios to the tumor site. Further in vivo studies would be required to ensure that the assigned ratio is truly delivered to the target site [17, 46].

6.2.2 Polymer-Conjugated Dual Drug Nanoparticle Systems

Passive encapsulation of anticancer drugs can provide marked improvement in the drug's plasma elimination half-life, plasma solubility, and tumor delivery; however, the magnitude of these improvements is largely dictated by the properties of the desired drug and polymeric system selected. As we have seen, encapsulation of multiple drugs can affect the retention of both agents, as both drugs can affect particle formation and stability. The list of biocompatible polymers is finite, and so for some agents, it may not be possible to produce a nanoparticle platform, which encapsulates more than one agent and still achieves improvements in drug pharmacokinetics. Further, if a dual drug polymer-based nanoparticle system is to exploit drug ratio-dependent synergy between the two agents (e.g., as in CPX-351), the system must provide a means to coordinate the retention and release of each drug

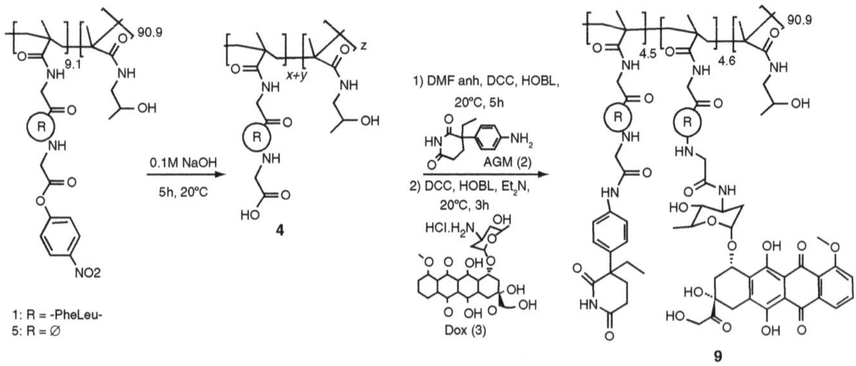

Scheme 6.1 Synthesis of dual drug polymer-conjugated formulation of doxorubicin and aminoglutethimide (AGM) using a peptide-functionalized HPMA backbone (*DCC* 1,3-dicyclohexylcarbodiimide; *DMF* *N,N*-dimethylformamide; *HOBt* 1-hydroxybenzotriazole). Reproduced with permission from [47]

such that the encapsulated ratio is maintained and a synergistic drug-to-drug ratio is delivered to the target site. While many successful formulations of single agents have been reported, passive formulation of two agents into a single particle may not exert enough control on the release of both agents to achieve this aim.

Direct covalent attachment of anticancer agents to a polymer backbone has several potential advantages over passive encapsulation methods. Firstly, drug circulation times are typically increased as the retention of the drug is dependent on the degradation rate of the drug–polymer linkage and the accessibility of the linkage site to enzymes and/or water that achieve the liberation of the active agent. Second, drug payloads of polymer-conjugated drug delivery systems can be much higher than what could be achieved with an equivalent passive system. The release of drugs from the polymer can also be controlled through careful selection of an appropriate linker (e.g., an amino acid or peptide sequence) between the polymer and drug. Incorporation of two drugs into a single polymer-conjugated system presents additional challenges. In this section, we review a selection of polymer-conjugated dual drug nanoparticle systems to highlight the means by which encapsulation and delivery to tumor sites of two active agents have been achieved using this method, as summarized in Table 6.3.

The first example of a dual drug-conjugated nanoparticle was reported by Duncan and coworkers in 2005 [47]. Encouraged by clinical activity observed with an *N*-(2-hydroxypropyl)methacrylamide (HPMA)-based copolymer functionalized with doxorubicin (HPMA-Dox) [48], a polymer chain with peptide-conjugated doxorubicin and the aromatase inhibitor aminoglutethimide (AGM) was synthesized (HPMA-Dox-AGM), shown in Scheme 6.1. Small angle neutron scattering (SANS) supported the hypothesis that each single polymer chain formed a single micelle in solution and that conjugation to either or both drugs had little effect on the particle size determined by SANS [47]. While it was not determined whether doxorubicin and AGM possess any drug-to-drug synergy, the co-formulated particle did provide superior in vitro cytotoxicity against MCF-7 breast cancer cells when compared to HPMA-Dox. Enzyme-mediated release of the two agents from the polymer showed that AGM was initially released much more rapidly than doxorubicin, with peak release of ~17% at 3 h compared to approximately 7% for doxorubicin. No further AGM release was observed; however, doxorubicin release was in fact greater than AGM release at 5 h [47]. Subsequent in vitro cell studies determined that the improved efficacy of the HPMA-Dox-AGM polymer over HPMA-Dox was due to the drug release properties of the polymer rather than any preferential lysosomal uptake of the dual drug polymer over that of HPMA-Dox [49]. Further improvements in efficacy may be realized by investigating the potential synergy between these two drugs to direct future optimization of drug release rates from this system.

Kwon and coworkers reported a dual drug-loaded polymeric micelle based on PEG-poly(aspartate hydrazide), in which doxorubicin and wortmannin (a kinase inhibitor) were conjugated to the polymer via hydrazone bonds [50]. This combination was selected based on earlier evidence that inhibition of phosphatidylinositol-3 kinase signaling pathways sensitizes cells to doxorubicin [51], suggesting possible synergistic interactions between the pair of drugs. Drug ratio control could be achieved simply by varying the amount of each drug present during conjugation to the pendant hydrazide groups of each monomer in the polymer chain. Reaction with the carbonyl group of each drug produced an acid-sensitive hydrazone linkage. Analysis by nuclear magnetic resonance (NMR) spectroscopy confirmed that the mixing ratio of the resultant polymer corresponded well to the ratio present in the reaction solution, possibly due to the chemical similarity of the two drugs [50]. To apply this technology to other combinations, fixing of the drug ratio would likely require a pre-assessment of the reactivity of both drugs to the hydrazide group of the polymer to roughly determine reaction rates. Despite the claim that these systems deliver drugs to tumors at fixed ratios, no in vitro or in vivo study is reported to confirm that both doxorubicin and wortmannin are retained equally well within the polymer. In vitro cytotoxicity studies indicated that the polymer chain with 50% doxorubicin and 50% wortmannin had similar activity compared to free doxorubicin, but the lack of in vivo PK data or biodistribution makes it difficult to assess whether this 1:1 drug ratio would in fact reach the target site. However, the platform eases encapsulation of a desired drug ratio, and the acid-sensitive hydrazone linkage provides selectivity as to the eventual site of linker hydrolysis (putatively lysosomes in tumor cells). Kwon and coworkers later demonstrated the flexibility of the platform through a report on the conjugation of doxorubicin and the preclinical compound 17-AAG, using identical hydrazone linkages to a PEG-poly(aspartate hydrazide) polymer backbone [52].

6.2.2.1 Polymer-Conjugated Systems of Hydrophilic Drugs

An advantage of the polymer conjugation method for dual drug nanoparticles is that drugs with widely different physical properties can be formulated into a single particle. As most nanoparticles are self-assembled, typically through a solvent emulsion or rapid mixing method, hydrophilic drugs will typically associate with

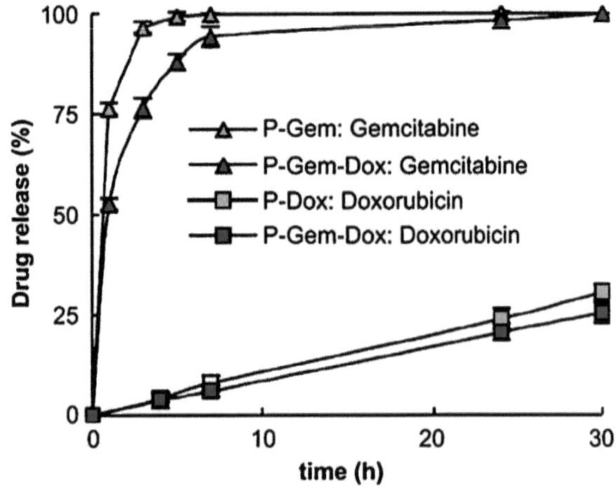

Fig. 6.4 In vitro drug release of HPMA polymer-conjugated gemcitabine and doxorubicin, mediated by cathepsin B at pH 6.0 (*P-Gem* poly(HPMA-peptide-gemcitabine), *P-Dox* poly(HPMA-peptide-doxorubicin), *P-Gem-Dox* poly(HPMA-(peptide-gemcitabine)-(peptide-doxorubicin))). Reproduced with permission from [53]

the aqueous phase during formulation and not the hydrophobic core of the forming particle. Conjugated to the polymer backbone, hydrophilic drugs such as gemcitabine [53], cytarabine, and floxuridine [54] can be stably formulated into a nanoparticle. Lammers et al. conjugated gemcitabine and doxorubicin to an HPMA-based polymer system in a stepwise manner. Drug conjugation sites were produced by conducting the radical polymerization with a mixture of free and peptide-functionalized monomers ($M_w = 23.5$ kDa, polydispersity index PDI = 1.6), followed by reaction of the polymer chains with a sub-stoichiometric amount of gemcitabine, and finally reaction with doxorubicin to conjugate to the remaining linker sites on the polymer chains. This stepwise method, once optimized for the reactivity of the two drugs (i.e., conjugation "yield"), would allow for accurate control of the drug-to-drug ratio. The extent of drug loading can also be exquisitely controlled by varying the proportion of free and functionalized monomers prior to polymerization [53]. The in vitro drug release was investigated at pH 6 in the presence of the lysosomal protease cathepsin B to cleave the peptide linker (Fig. 6.4). Gemcitabine release from the conjugated polymer was considerably faster than doxorubicin release (~100% for gemcitabine compared to <25% for doxorubicin at 24 h), attributed to the smaller size of gemcitabine, imparting less steric hindrance to the enzyme-mediated hydrolysis reaction. Encouragingly, release of either drug was largely unaffected by the presence of the second drug; release of gemcitabine or doxorubicin was very similar for either the single or dual drug-conjugated polymer nanoparticle. Further work would be required to tune the reactivity of the amide linkage to either drug before reasonable maintenance of the established drug-to-drug ratio could be achieved.

Yin et al. used a different approach in which monomers of divinyl dicarboxylates were enzymatically conjugated with cytarabine and floxuridine via ester bond formation on the 5′ and 3′ positions of the sugar moieties of the two drugs, respectively [54]. Polymerization of the prepared monomers produced polydisperse

($M_w/M_n = 1.42$) polymers with low yield (33%). In vitro studies revealed that the release of cytarabine was approximately five times greater than that of floxuridine at 50 h at pH 7.4 (~35% for cytarabine compared to ~7% for floxuridine). Coordinated but very slow drug release (<10% over 50 h) could be achieved under highly acidic conditions (pH 1.2). The authors attributed the substantial difference at pH 7.4 to the presence of the amino group of cytarabine, although the difference in the starting monomers (divinyl sebacate for cytarabine versus divinyl adipate for floxuridine) may also contribute to the observed difference in hydrolysis rate.

6.2.2.2 Hybrid Polymer-Conjugated/Passive Nanoparticle Systems

Conjugation of one or more drugs to the polymer can increase the retention of the drug within the particle when compared to the same drug passively encapsulated. Recent studies have exploited this difference to achieve sequential release of two drugs from a single particle. Sengupta first reported a "nanocell" formulation of PLGA-conjugated doxorubicin, enveloped in a mixture of DSPE-PEG, phosphatidylcholine, and cholesterol, into which the antiangiogenesis agent combretastatin was passively incorporated [55]. This work was followed by Wang and Ho who utilized polymer-conjugated paclitaxel in place of doxorubicin, conjugated onto the terminal end of a carboxyl acid-functionalized PEG-PLA copolymer. They reported a simpler nanoparticle system that did not require subsequent lipid coating of the nanoparticles, and they conjugated paclitaxel via an ester bond to the PLA polymer rather than an amide linkage to the amino sugar of doxorubicin [56]. In both cases, combretastatin release was much faster than release of the polymer-bound cytotoxic agent, which had observable effects on endothelial cells used in antiangiogenesis assays conducted in both studies. This sequential release may cause a collapse of a target tumor vasculature, which both studies claim may trap the nanoparticles within the tumor site. It is difficult to gauge the extent of doxorubicin release in the earlier study; however, maximal doxorubicin release was observed at approximately 100 h during an in vitro assay, followed by a steady decline in the total amount of drug released over the next 225 h (14 days total) [55]. Doxorubicin-PLGA fragments were also detected during the release assay. Combretastatin release was largely complete by 100 h. In the study by Wang et al., approximately 20% of the paclitaxel released from the polymer in less than 24 h, compared to over 40% of the combretastatin. Total paclitaxel release, however, plateaued at approximately 60% over 11 days. Combretastatin release was largely complete (~70% of the total combretastatin) by 6 days [56].

Langer, Farokhzad, Lippard, and coworkers recently reported an elegant system that combined coordination chemistry of platinum(IV) complexes with nanoparticle technology to produce a polymer-conjugated dual drug delivery system [57]. A platinum(IV) complex was first coordinated with succinic acid at the apical position, the resultant complex then conjugated to pendant hydroxyl groups of a derivatized PLA polymer chain to approximately 10% Pt by weight (Scheme 6.2a). The PLA-Pt(IV) prodrug was then nanoprecipitated with docetaxel and PLGA-PEG by

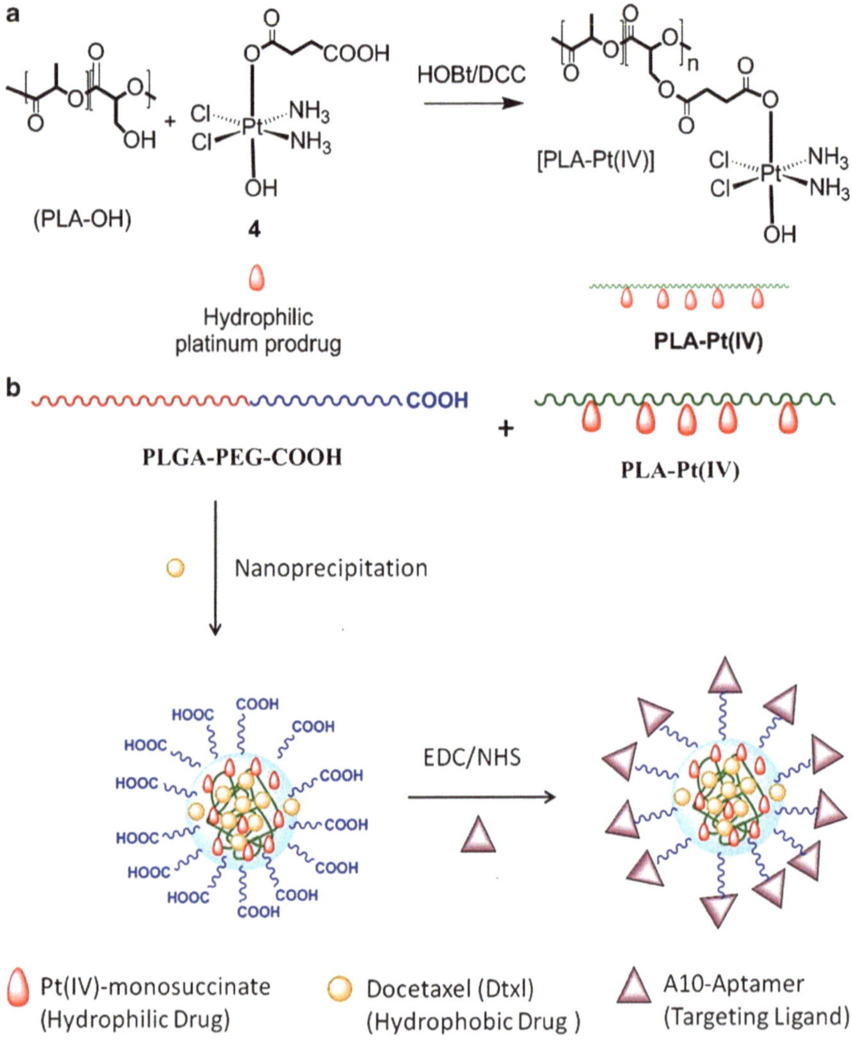

Scheme 6.2 (**a**) Synthesis of PLA-conjugated Pt(IV) prodrug through conjugation of succinic acid Pt(IV) with PLA-OH. (**b**) Design and formulation of Pt(IV)-docetaxel dual drug polymer nanoparticles. Reproduced with permission from [57]

hydrodynamic flow focusing, a microfluidic rapid mixing method [58], producing particles of approximately 100 nm in diameter (Scheme 6.2b). Encapsulation efficiency of PLA-Pt(IV) was 95%, starting with 50% of the PLA prodrug and 50% PLGA-PEG. This efficiency corresponds to approximately 5% Pt(IV) content by weight within the nanoparticle. Docetaxel was loaded up to 1% by weight at 80% efficiency. Neither free cisplatin nor Pt(IV) prodrug (no polymer) could be adequately loaded into the nanoparticle [57]. The release properties of this system highlight the benefit of polymer conjugation of hydrophilic drugs such as a platinum drug to fundamentally change the water solubility of the drug. Release of the Pt(IV) from the PLA-Pt(IV) prodrug in the nanoparticle was considerably slower than that of hydrophobic docetaxel, with approximately 40% Pt(IV) released at 24 h compared to over 75% of docetaxel. Release of Pt(IV) was likely mediated by ester hydrolysis and not reduction to Pt(II) [57]. The authors note that the release rate of the platinum prodrugs can likely be tuned to that of docetaxel by varying the molecular weight of the polymer, as previously shown for PLGA-PEG conjugates of cisplatin [59].

6.2.3 Conjugated Drug Nanoparticle Systems

An alternative strategy to direct covalent linking of drugs to the polymer backbone is to chemically alter the physicochemical properties of the drug(s). Conjugation of the drug(s) to some moiety via a hydrolysable linkage (producing a prodrug) can dramatically alter the water solubility, charge state, or other physical property of the drug, leading to improved pharmacokinetic properties or making the drug more amenable to encapsulation in a drug carrier. This approach is particularly attractive for hydrophilic drugs, such as nucleoside analogs. Two examples of conjugated drug nanoparticle systems are reported in Table 6.4.

A simple example of drug conjugation was recently reported by Aryal et al. in which paclitaxel was functionalized at the 2′-OH with glutaric anhydride as a linker, forming an ester bond between drug and linker and leaving a pendant carboxylate on the linker for subsequent conjugation to gemcitabine via a second ester bond (Fig. 6.5a) [60]. The dual drug conjugates were then formulated in PLGA nanoparticles, which were further coated in lipid and lipid-PEG conjugates. Hydrolysis rates, however, were studied only for the unformulated conjugate and found to be pH dependent with significantly more hydrolysis in pH 6.0 buffer solution than at pH 7.4 (90% vs. ~35% at 24 h, respectively; Fig. 6.5b). The dual drug particles did possess superior cytotoxicity compared to that of the unformulated conjugate, suggesting that prodrug hydrolysis took place intracellularly [60]. It is likely that this increased in vitro activity is related to the relative solubilities of the prodrug in the nanoparticle and in aqueous solution. No comparison was made with a 1:1 mixture of free paclitaxel and gemcitabine. Obviously, this system suffers severe limitations if the synergistic drug-to-drug ratio for paclitaxel and gemcitabine is anything other than 1:1. By using different linker chemistries, it is possible that the system might be able to accommodate slightly larger ratios; small excesses of one drug over the other (e.g., 2:1 or 3:1) might be adaptable but would require further development of the linker chemistry. Moreover, each addition to the conjugate increases the overall size of the conjugate, likely decreasing the absolute drug loading in a nanoparticle, or decreasing the affinity of the conjugate for the forming nanoparticle.

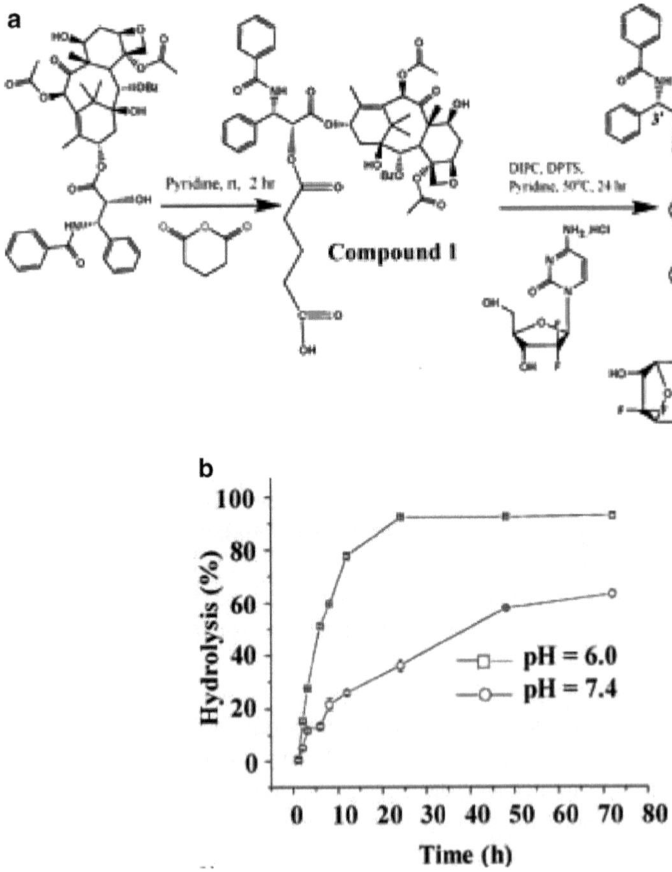

Fig. 6.5 (**a**) Synthesis of a dual drug conjugate of paclitaxel and gemcitabine via a diester linker (compound 2) (*DIPC* 1,3-diisopropylcarbodiimide; *DPTS* 4-(*N*,*N*-dimethylamino)pyridinium-4-toluenesulfonate). (**b**) Hydrolysis kinetics of paclitaxel-gemcitabine conjugate at pH 6.0 and 7.4. Reproduced with permission from [60]

What is required is a system in which the drug ratio can be accurately controlled and that provides sufficient flexibility that virtually any synergistic drug ratio can be accommodated in the particle. Secondly, the retention and release of the two drugs should be coordinated, such that the ratio of the two agents is maintained and the synergistic drug ratio is delivered to the tumor site. Failure to control drug release from the nanoparticle could lead to delivery of an antagonistic drug ratio, resulting in compromised in vivo efficacy. Lastly, the system should accommodate drugs with different physicochemical properties, not just hydrophobic drugs.

6.3 The CombiPlex® Platform: Application to Nanoparticle Delivery

We have reported several formulations of drug combinations in liposomes using the CombiPlex approach, all of which stably encapsulate two agents and maintain the synergistic drug ratio for extended periods of time in vivo [7, 20, 21, 61, 62]. Maintaining the synergistic ratio of the drugs has been shown to lead to increased in vivo efficacy. Several disparate classes of drugs have been successfully formulated using the CombiPlex technology platform, including topoisomerase I inhibitors (irinotecan) [7, 21, 61], antimetabolites (floxuridine, cytarabine), and anthracycline antibiotics (daunorubicin) [20]. By formulating the reactive agent cisplatin in a separate liposome, we can combine liposomal cisplatin with liposomal irinotecan at a 7:1 ratio and achieve greater in vivo efficacy against solid tumors than the free drug cocktail at its respective MTD [62]. However, these formulations involved liposomal encapsulation of hydrophilic drugs, for which liposomes are ideal. We have recently demonstrated that the CombiPlex approach can also be used with hydrophobic drugs such as the taxanes, using the conjugated drug approach described above. Here, we will discuss the use of this approach to develop a dual drug-loaded nanoparticle system that can stably encapsulate two drugs with wildly different physicochemical properties (i.e., paclitaxel and gemcitabine) at a fixed drug ratio. Through careful selection of linker and conjugate moieties, the release rate of the two drugs can be coordinated such that the ratio of these disparate agents can be maintained within a nanoparticle for extended periods of time in vivo.

Many drug conjugates for the taxanes and camptothecins [63, 64] have focused on increasing the *hydrophilicity* (i.e., aqueous solubility) of these agents in an attempt to increase delivery to tumor sites. Some of the conjugates used included water-soluble polymers such as PEG [65–67], polypeptides [68–70], and polylactides [71]. Our approach is to greatly increase the *hydrophobicity* of each agent in a drug combination through conjugation to highly nonpolar fatty acids, cholesterol, or diacyl glycerol, producing prodrugs composed of the parent drug, a hydrolysable linker, and the nonpolar anchor. Rapid mixing of the hydrophobic prodrugs with appropriate amounts of amphiphilic, PEGylated polymer leads to the spontaneous formation of nanoscale (20–100 nm), solid-core nanoparticles [72, 73]. Thus, the plasma circulation lifetime of both prodrugs is dictated by three factors: (1) the distribution properties of the polymer nanoparticle, (2) the hydrolysis kinetics of the ester bond between the parent drug and the linker, and (3) the hydrophobicity of the anchor. The goal is to render both drugs sufficiently hydrophobic so that the choices of anchor and linker chemistry are the critical factors that determine the plasma elimination rate, not the physicochemical properties of the parent drugs themselves. By generating a library of different anchor and linker combinations for each drug, the retention time of each prodrug within the particle can be normalized to each other, thus maintaining the synergistic drug ratio of the retained drugs. Scheme 6.3 provides an overview of the library of linkers and hydrophobic anchors we have used to generate a variety of hydrophobic prodrugs of paclitaxel [74, 75].

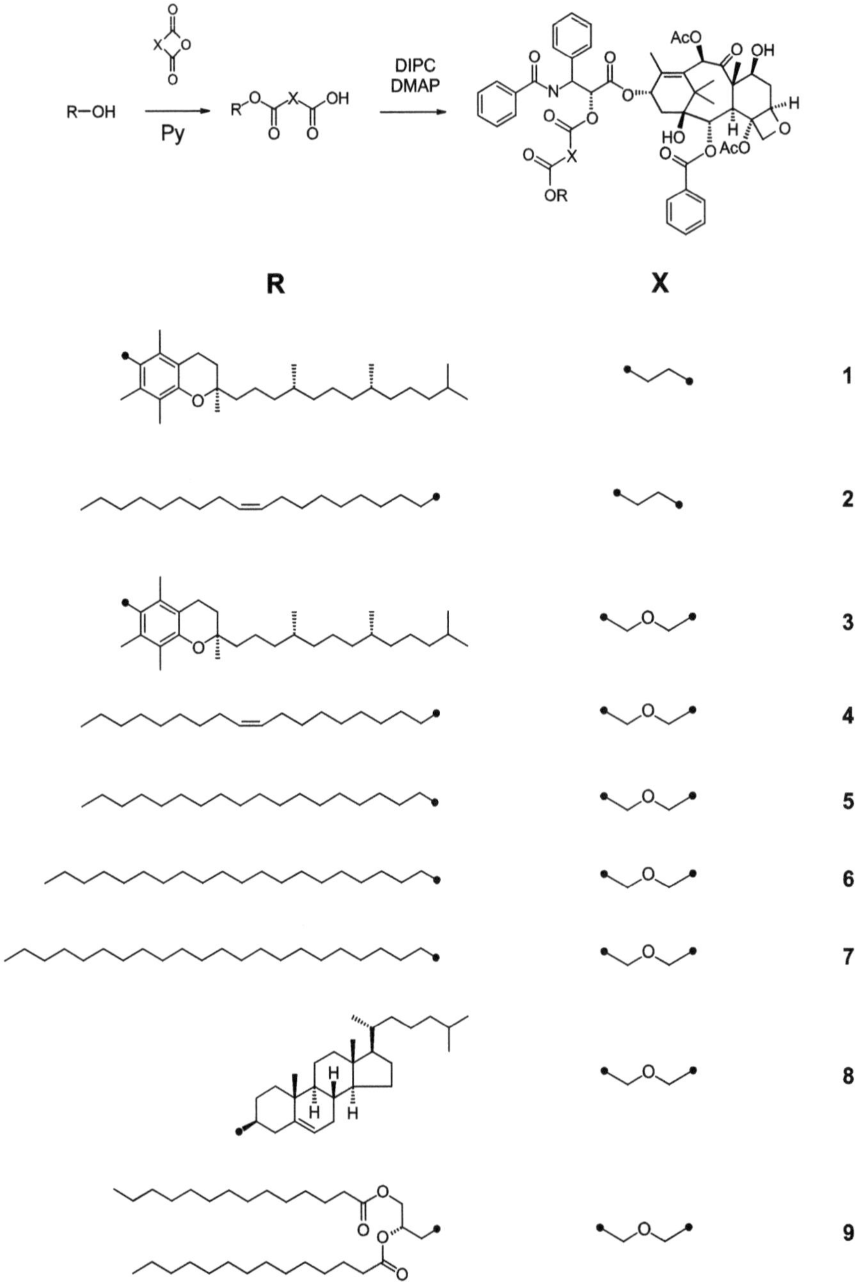

Scheme 6.3 Synthesis of hydrophobic paclitaxel prodrugs. The numbers indicate the designation of the prodrug, that is, ProPac#, where # is the indicated combination of linker X (succinate for ProPac1 and 2, diglycolate for ProPac3–9) and anchor R (*py* pyridine; *DIPC* diisopropylcarbodiimide; *DMAP N,N*-4-dimethylaminopyridine). Reproduced with permission from [75]

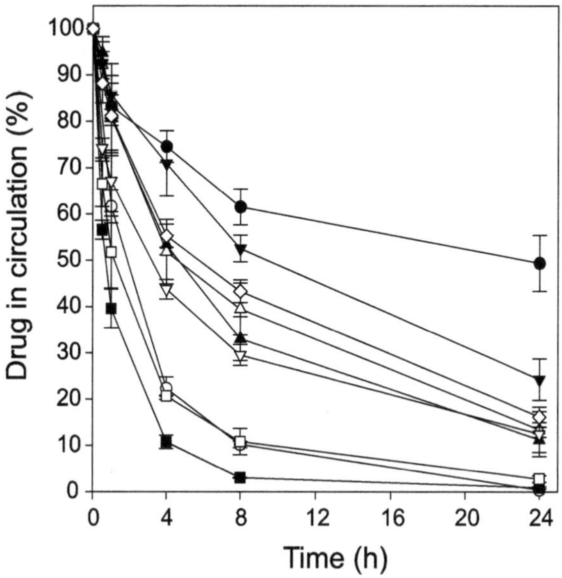

Fig. 6.6 Elimination of paclitaxel prodrugs formulated as prodrug/POPC/PEG$_{2000}$-PS$_{2500}$ (1:1:2; w/w) formulations and administered intravenously to athymic nude *Foxn*1^{nu} mice at a dose of 7 mg/kg (n=3/time point). The prodrugs used were (1) succinate-α-tocopherol (*filled circle*), (2) succinate-oleyl alcohol (*open circle*), (3) diglycolate-α-tocopherol (*open diamond*), (4) diglycolate-oleyl alcohol (*filled square*), (5) diglycolate-octadecanol (*open square*), (6) diglycolate-cosanol (*filled triangle*), (7) diglycolate-docosanol (*open triangle*), (8) diglycolate-cholesterol (*filled inverted triangle*), and (9) diglycolate-1,2-dimiristoyl-*sn*-glycerol (*open inverted triangle*). Error bars represent standard deviation (n=3). Reproduced with permission from [75]

Paclitaxel was first selected as a model compound. Paclitaxel prodrugs (named ProPac#, where # represents a combination of a particular linker and anchor; see Scheme 6.3) were formulated into 20 to 30 nm nanoparticles composed of PEG-polystyrene and POPC as a lipid stabilizer. Through iterative combination of paclitaxel with two linkers (succinate and diglycolate) and a variety of nonpolar anchors, the relationships between anchor hydrophobicity and linker lability with pharmacokinetic profile and in vitro and in vivo efficacy were established.

In vitro release assays can provide some insight into the relative release rates of drugs from particulate carriers; however, in our experience, an in vivo pharmacokinetic assessment of plasma elimination rates of drug combinations provides greater assurance that the synergistic drug ratio established in the administered particle is in fact delivered to the site of action. Therefore, we evaluated plasma drug elimination kinetics of formulated ProPac nanoparticles by intravenous (i.v.) administration in mice (Fig. 6.6). Plasma elimination of the prodrug was found to be dictated by two main processes: elimination of the intact nanoparticle from the circulation and release of the prodrug from the nanoparticle, which is then expected to be rapidly eliminated from the circulation as free paclitaxel. ProPac1, for example, composed of a succinate linker and tocopherol as the nonpolar anchor, showed very little release of the prodrug from the particle. This was determined by labeling the nanoparticle with tritiated cholesteryl hexadecyl ether (^{3}H-CHE), a non-metabolizable radioactive marker, which tracked clearance of the polymer. In the case of ProPac1, the slow plasma clearance of the polymer and the prodrug was virtually identical, indicating negligible prodrug release over 24 h. Consequently, this formulation also exhibited poor in vivo efficacy against HT-29 solid tumor xenografts; sufficient bioavailability of the prodrug from the particle is critical to antitumor efficacy. To increase the release rate of the prodrug from the particle, the succinate linker was replaced with the more labile diglycolate linker. Pharmacokinetic analysis revealed that this change increased the release rate of the prodrug from the particle (Fig. 6.6, ProPac3). Prodrug clearance rate of ProPac3 in plasma was significantly faster than ProPac1. We prepared and tested a large set of paclitaxel-diglycolate-anchor prodrugs, in which the hydrophobicity of the anchor was systematically varied. While paclitaxel conjugated to oleyl C18:1 alcohol (Fig. 6.6, ProPac4) resulted in only 10% of the injected dose remaining after 4 h, saturated docosanoic (behenic) C22:0 acid and cholesterol anchors increased circulation times dramatically, with 20–30% of the administered prodrug remaining in the plasma 24 h after injection [75]. By subtracting the clearance rate of the particle itself (as determined by radiolabeled CHE content), the half-life of ProPac release from the particle could be calculated. This method revealed that parent drug-release half-lives increased from 1 to 2 h for unsaturated C18:1 and saturated C18:0 lipid anchors to 12–24 h for the more hydrophobic C22:0 and cholesterol anchors, respectively [75].

Linker chemistry, prodrug hydrophobicity, and prodrug pharmacokinetic behavior were correlated with efficacy against HT-29 human colorectal cancer solid tumor xenografts in mice. Tumor growth inhibition of the various systems was compared for identical paclitaxel molar equivalents and at lower dose levels to differentiate therapeutic activities. Diglycolate-linked prodrugs showed activity against the xenograft, while succinate-based prodrugs (ProPac1, as described above) were largely inactive. Within the diglycolate series, increased tumor growth inhibition was observed with increasing hydrophobicity of the lipid anchor (from C18:1 up to C22:0). The most active formulations (Propac7, C22:0 anchor and Propac8, cholesterol anchor) were compared to conventional Taxol® at each formulation's MTD. Propac8 nanoparticles were found to induce greater tumor growth inhibition than Taxol, with complete tumor regression observed and regrowth delayed approximately 6 weeks beyond that observed with Taxol [75].

With the relationship between linker/anchor construct and therapeutic efficacy defined with the ProPac series, co-formulation with a second drug was attempted with the goal of achieving similar retention times for both prodrugs within the particle to ensure delivery of the established drug ratio. Gemcitabine was selected as the second agent due to its current clinical use in combination with paclitaxel [76] and its high hydrophilicity, which would demonstrate the flexibility of the CombiPlex prodrug nanoparticle platform. Single aliphatic anchors were insufficiently hydrophobic to retain the gemcitabine prodrugs in nanoparticles for extended times. Conjugation of gemcitabine to diacyl glycerol (C18:0, ProGem12) enabled the prodrug to be stably incorporated into nanoparticles, co-formulated with ProPac7 conjugates (C22:0 anchor). The two prodrugs exhibited coordinated PK and maintenance of the formulated 4:1 paclitaxel-to-gemcitabine ratio for 24 h after i.v.

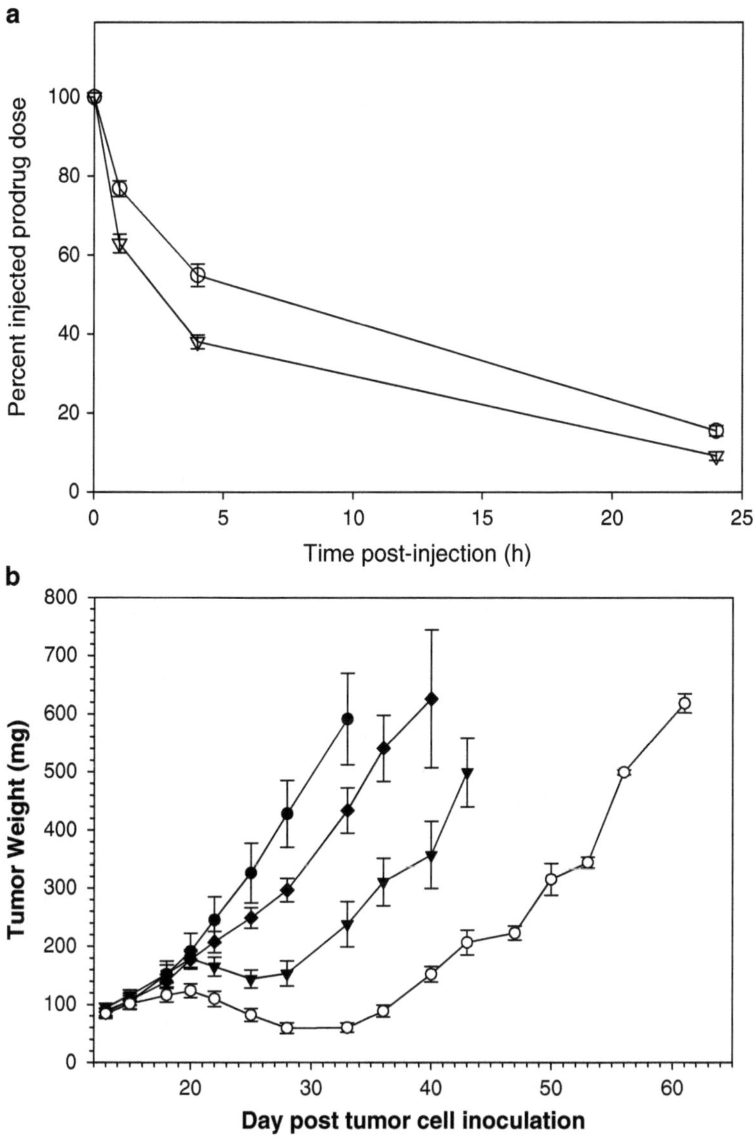

Fig. 6.7 (**a**) Plasma elimination of ProPac7 (*open circle*) and ProGem12 (*open inverted triangle*) prodrugs co-formulated as ProPac7:ProGem12:POPC:PEG$_{2000}$-PS$_{2500}$ (4:1:2:8 w/w). Error bars represent standard deviation. Figure adapted from data presented in [74]. (**b**) Efficacy of ProPac7 (*filled inverted triangle*, 30 mg/kg), ProGem12 (*filled diamond*, 12 mg/kg), and co-formulated ProPac7:ProGem12 prodrug nanoparticles (*open circle*, 30:12 mg/kg) against HT-29 human colon carcinoma xenograft (Q2Dx5, *filled circle*, saline). Reproduced with permission from [25]

administration (Fig. 6.7a). Co-formulated ProGem12-ProPac7 nanoparticles exhibited superior efficacy against HT-29 xenografts than the individual nanoparticle agents, as shown in Fig. 6.7b [74]. The data presented indicates that the CombiPlex nanoparticle platform can be successfully applied to drug combinations with disparate physicochemical properties through the generation of hydrophobic prodrugs and formulation into nanoparticles. Careful selection of linker and anchor properties allows for careful tuning of the release rates of both drugs and coordination of plasma elimination rates, ensuring that the drug-to-drug ratio within the particle is maintained for delivery to the target site. This technology therefore expands the scope of drug combinations that can be evaluated for the benefits of drug delivery vehicle-based ratiometric dosing.

6.4 Conclusions

We have presented a rationale for delivery of chemotherapeutic drug combinations at carefully defined synergistic ratios to ensure maximal efficacy. A liposomal formulation of cytarabine and daunorubicin, CPX-351, demonstrated that drug interactions, either synergistic, antagonistic, or additive, can be highly ratio dependent, and only by locking in this ratio with a drug carrier can we expect to deliver efficacious drug combinations.

Development of dual drug formulations has now expanded into polymer-based nanoparticles. These systems use a variety of encapsulation techniques such as passive incorporation and polymer or drug conjugation to generate a nanoparticle formulation containing two active agents. While many examples of these two-drug nanoparticle systems can now be found in the literature, few provide adequate control over the drug retention properties of each component. In most cases, drug release from the particle is not coordinated, which could lead to rapid loss of the synergistic drug ratio within the particle and potential exposure of the target site to an antagonistic ratio, further compromising in vivo efficacy.

Our approach of generating a library of prodrugs of each agent, examining their PK properties in vivo and finally co-formulating the two prodrugs in a single system, represents an extension of the successful CombiPlex technology previously used to formulate several clinical combinations of hydrophilic drugs in liposomal carriers. We have now extended this technique to allow for formulation of hydrophilic and hydrophobic drugs within a single particle. Correlation of prodrug composition to in vivo PK and efficacy properties permits the selection of the optimal prodrug for each active agent, such that when the prodrugs are co-formulated, the drug-to-drug ratio can be maintained for extended periods of time in vivo, and superior efficacy is observed.

References

1. Frei E III, Freireich EJ (1964) Leukemia. Sci Am 210:88–96
2. Freireich EJ, Frei E III (1964) Recent advances in acute leukemia. Prog Hematol 27:187–202
3. Harasym TO, Liboiron BD, Mayer LD (2010) Drug ratio-dependent antagonism: a new category of multidrug resistance and strategies for its circumvention. In: Zhou J (ed) Multi-drug resistance in cancer. Humana, New York, pp 291–323

4. Chou TC (1991) The median-effect principle and the combination index for quantitation of synergism and antagonism. In: Chou TC, Rideout DC (eds) Synergy and antagonism in chemotherapy. Academic, New York, pp 61–102
5. Chou TC (2006) Theoretical basis, experimental design, and computerized simulation of synergism and antagonism in drug combination studies. Pharmacol Rev 58:621–681
6. Chou TC, Talalay P (1984) Quantitative analysis of dose-effect relationships: the combined effects of multiple drugs or enzyme inhibitors. Adv Enzyme Regul 22:27–55
7. Mayer LD, Harasym TO, Tardi PG, Harasym NL, Shew CR, Johnstone SA, Ramsay EC, Bally MB, Janoff AS (2006) Ratiometric dosing of anticancer drug combinations: controlling drug ratios after systemic administration regulates therapeutic activity in tumor-bearing mice. Mol Cancer Ther 5:1854–1863
8. Mayer LD, Janoff AS (2007) Optimizing combination chemotherapy by controlling drug ratios. Mol Interv 7:216–223
9. Allen TM, Cullis PR (2004) Drug delivery systems: entering the mainstream. Science 303:1818–1822
10. Maeda H (2001) The enhanced permeability and retention (EPR) effect in tumor vasculature: the key role of tumor-selective macromolecular drug targeting. Adv Enzyme Regul 41:189–207
11. Yuan F, Dellian M, Fukumura D, Leunig M, Berk DA, Torchilin VP, Jain RK (1995) Vascular permeability in a human tumor xenograft: molecular size dependence and cutoff size. Cancer Res 55:3752–3756
12. Kimby E, Nygren P, Glimelius B (2001) A systematic overview of chemotherapy effects in acute myeloid leukemia. Acta Oncol 40:231–252
13. McCauley DL (1992) Treatment of adult acute leukemia. Clin Pharm 11:767–796
14. Bishop JF, Matthews JP, Young GA, Szer J, Gillett A, Joshua D, Bradstock K, Enno A, Wolf MM, Fox R, Cobcroft R, Herrmann R, Van Der Weyden M, Lowenthal RM, Page F, Garson OM, Juneja S (1996) A randomized study of high-dose cytarabine in induction in acute myeloid leukemia. Blood 87:1710–1717
15. Weick JK, Kopecky KJ, Appelbaum FR, Head DR, Kingsbury LL, Balcerzak SP, Bickers JN, Hynes HE, Welborn JL, Simon SR, Grever M (1996) A randomized investigation of high-dose versus standard-dose cytosine arabinoside with daunorubicin in patients with previously untreated acute myeloid leukemia: a southwest oncology group study. Blood 88:2841–2851
16. Curtis JE, Messner HA, Hasselback R, Elhakim TM, McCulloch EA (1994) Contributions of host and disease-related attributes to the outcome of patients with acute myelogenous leukemia. J Clin Oncol 2(Suppl 4):253–259
17. Harasym TO, Tardi PG, Johnstone SA, Mayer LD, Bally MB, Janoff AS (2007) Fixed drug ratio liposome formulations of combination cancer therapeutics. In: Gregoriadis G (ed) Liposome technology volume III: interactions of liposomes with biological milieu, 3rd edn. Informa Healthcare USA Inc, New York, pp 25–46
18. Pavillard V, Kherfellah D, Richard S, Robert J, Montaudon D (2001) Effects of the combination of camptothecin and doxorubicin or etoposide on rat glioma cells and camptothecin-resistant variants. Br J Cancer 85:1077–1083
19. Swaffar DS, Ang CY, Desai PB, Rosenthal GA, Thomas DA, Crooks PA, John WJ (1995) Combination therapy with 5-fluorouracil and L-canavanine: in vitro and in vivo studies. Anticancer Drugs 6:586–593
20. Tardi PG, Johnstone SA, Harasym NL, Xie S, Harasym TO, Zisman N, Harvie P, Bermudes D, Mayer LD (2009) In vivo maintenance of synergistic cytarabine: daunorubicin ratios greatly enhances therapeutic efficacy. Leuk Res 33:129–139
21. Tardi P, Gallagher R, Johnstone S, Harasym N, Webb M, Bally M, Mayer L (2007) Coencapsulation of irinotecan and floxuridine into low cholesterol-containing liposomes that coordinate drug release in vivo. Biochim Biophys Acta 1768:678–687
22. Lancet JE, Cortes JE, Hogge DE, Tallman M, Kovacsovics T, Damon LE, Ritchie E, Komrokji RS, Louie AC, Feldman EJ. (2010) Phase 2B randomized study of CPX-351 vs. cytarabine (CYT)+daunorubicin (DNR) (7+3 regimen) in newly diagnosed AML patients aged 60–75. In 52nd American Society of Hematology (ASH) Annual Meeting, Orlando, FL

23. Dicko A, Kwak S, Frazier AA, Mayer LD, Liboiron BD (2010) Biophysical characterization of a liposomal formulation of cytarabine and daunorubicin. Int J Pharm 391:248–259
24. Dicko A, Frazier AA, Liboiron BD, Hinderliter A, Ellena JF, Xie X, Cho C, Weber T, Tardi PG, Cabral-Lilly D, Cafiso DS, Mayer LD (2008) Intra and inter-molecular interactions dictate the aggregation state of irinotecan co-encapsulated with floxuridine inside liposomes. Pharm Res 25:1702–1713
25. Dicko A, Mayer LD, Tardi PG (2010) Use of nanoscale delivery systems to maintain synergistic drug ratios in vivo. Expert Opin Drug Deliv 7:1329–1341
26. Drummond DC, Noble CO, Hayes ME, Park JW, Kirpotin DB (2008) Pharmacokinetics and in vivo drug release rates in liposomal nanocarrier development. J Pharm Sci 97:4696–4740
27. Gua W, Johnson JL, Khan S, Ahmad A, Ahmad I (2005) Paclitaxel quantification in mouse plasma and tissues containing liposome-entrapped paclitaxel by liquid chromatography-tandem mass spectrometry: application to a pharmacokinetic study. Anal Biochem 336:213–220
28. Lukyanov AN, Torchilin VP (2004) Micelles from lipid derivatives of water-soluble polymers as delivery systems for poorly soluble drugs. Adv Drug Deliv Rev 56:1273–1289
29. Torchilin VP (2007) Targeted pharmaceutical nanocarriers for cancer therapy. AAPS J 9:E128–E147
30. Vicent MJ, Dieudonne L, Carbajo RJ, Pineda-Lucena A (2008) Polymer conjugates as therapeutics: future trends, challenges and opportunities. Expert Opin Drug Deliv 5:593–614
31. Na HS, Lim YK, Jeong YI, Lee HS, Lim YJ, Kang MS, Cho CS, Lee HC (2010) Combination antitumor effects of micelle-loaded anticancer drugs in a CT-26 murine colorectal carcinoma model. Int J Pharm 383:192–200
32. Devalapally H, Duan Z, Seiden MV, Amiji MM (2007) Paclitaxel and ceramide co-administration in biodegradable polymeric nanoparticulate delivery system to overcome drug resistance in ovarian cancer. Int J Cancer 121:1830–1838
33. Ganta S, Amiji M (2009) Coadministration of paclitaxel and curcumin in nanoemulsion formulations to overcome multidrug resistance in tumor cells. Mol Pharm 6:928–939
34. Patil Y, Sadhukha T, Ma L, Panyam J (2009) Nanoparticle-mediated simultaneous and targeted delivery of paclitaxel and tariquidar overcomes tumor drug resistance. J Control Release 136:21–29
35. Patil Y, Swaminathan SK, Sadhukha T, Ma L, Panyam J (2010) The use of nanoparticle-mediated targeted gene silencing and drug delivery to overcome tumor drug resistance. Biomaterials 31:358–365
36. Wang Z, Chui W-K, Ho PC (2011) Nanoparticulate delivery system targeted to tumor neovasculature for combined anticancer and antiangiogenesis therapy. Pharm Res 28:585–596
37. Zhang H, Zhao C, Cao H, Wang G, Song L, Niu G, Yang H, Ma J, Zhu S (2010) Hyperbranched poly(amine-ester) based hydrogels for controlled multi-drug release in combination chemotherapy. Biomaterials 31:5445–5454
38. Khdair A, Chen D, Patil Y, Ma L, Dou QP, Shekhar MPV, Panyam J (2010) Nanoparticle-mediated combination chemotherapy and photodynamic therapy overcomes tumor drug resistance. J Control Release 141:137–144
39. Tang Y, Lei T, Manchanda R, Nagesetti A, Fernandez-Fernandez A, Srinivasan S, McGoron AJ (2010) Simultaneous delivery of chemotherapeutic and thermal-optical agents to cancer cells by a polymeric (PLGA) nanocarrier: an in vitro study. Pharm Res 27:2242–2253
40. Song X, Zhao Y, Hou S, Xu F, Zhao R, He J, Cai Z, Li Y, Chen Q (2008) Dual agents loaded PLGA nanoparticles: systematic study of particle size and drug entrapment efficiency. Eur J Pharm Biopharm 69:445–453
41. Song XR, Cai Z, Zheng Y, He G, Chui FY, Gong DQ, Hou SX, Xiong SJ, Lei XJ, Wei YQ (2009) Reversion of multidrug resistance by co-encapsulation of vincristine and verapamil in PLGA nanoparticles. Eur J Pharm Sci 37:300–305
42. Gupta AK, Berry C, Gupta M, Curtis A (2003) Receptor-mediated targeting of magnetic nanoparticles using insulin as surface ligand to prevent endocytosis. IEEE Trans Nanobiosci 2:256–261

43. Jain TK, Morales MA, Sahoo SK, Leslie-Pelecky DL, Labhasetwar V (2005) Iron oxide nanoparticles for sustained delivery of anticancer agents. Mol Pharm 2:194–205
44. Dilnawaz F, Singh A, Mohanty C, Sahoo SK (2010) Dual drug loaded superparamagnetic iron oxide nanoparticles for targeted cancer therapy. Biomaterials 31:3694–3706
45. Shin H-C, Alani AWG, Rao DA, Rockich NC, Kwon GS (2009) Multi-drug loaded polymeric micelles for simultaneous delivery of poorly soluble anticancer drugs. J Control Release 140:294–300
46. Liboiron BD, Tardi PG, Harasym TO, Mayer LD (2011) Nanoscale delivery systems for combination chemotherapy. In: Kratz F (ed) Cancer drug delivery. Wiley VCH, Weinheim, Germany
47. Vicent MJ, Greco F, Nicholson RI, Paul A, Griffiths PC, Duncan R (2005) Polymer therapeutics designed for a combination therapy of hormore-dependent cancer. Angew Chem Int Ed 44:4061–4066
48. Vasey PA, Kaye SB, Morrison R, Twelves C, Wilson P, Duncan R, Thomson AH, Murray LS, Hilditch TE, Murray T, Burtles S, Fraier D, Frigerio E, Cassidy J (1999) Phase 1 clinical and pharmacokinetic study of PK1 [N-(2-hydroxypropyl)methacrylamide copolymer doxorubicin]: first member of a new class of chemotherapeutic agents - drug-polymer conjugates. Clin Cancer Res 5:83–94
49. Greco F, Vicent MJ, Gee S, Jones AT, Gee J, Nicholson RI, Duncan R (2007) Investigating the mechanism of enhanced cytotoxicity of HPMA copolymer-Dox-AGM in breast cancer cells. J Control Release 117:28–39
50. Bae Y, Diezi TA, Zhao A, Kwon GS (2007) Mixed polymeric micelles for combination cancer chemotherapy through the concurrent delivery of multiple chemotherapeutic agents. J Control Release 122:324–330
51. Fujiwara Y, Kawada K, Takano D, Tanimura S, Ozaki K, Kohno M (2006) Inhibition of the PI3 kinase/Akt pathway enhances doxorubicin-induced apoptotic cell death in tumor cells in a p53-dependent manner. Biochem Biophys Res Commun 340:560–566
52. Bae Y, Alani AW, Rockich NC, Lai TS, Kwon GS (2010) Mixed pH-sensitive polymeric micelles for combination drug delivery. Pharm Res 27:2421–2432
53. Lammers T, Subr V, Ulbrich K, Peschke P, Huber PE, Hennink WE, Storm G (2009) Simultaneous delivery of doxorubicin and gemcitabine to tumors in vivo using prototypic polymeric drug carriers. Biomaterials 30:3466–3475
54. Yin C, Li X, Wu Q, Wang J-L, Lin X-F (2010) Multidrug nanoparticles based on novel random copolymer containing cytarabine and fluorodeoxyuridine. J Colloid Interface Sci 349:153–158
55. Sengupta S, Eavarone D, Capila I, Zhao G, Watson N, Kiziltepe T, Sasisekharan R (2005) Temporal targeting of tumour cells and neovasculature with a nanoscale delivery system. Nature 436:568–572
56. Wang Z, Ho PC (2010) A nanocapsular combinatorial sequential drug delivery system for antiangiogenesis and anticancer activities. Biomaterials 31:7115–7123
57. Kolishetti N, Dhar S, Valencia PM, Lin LQ, Karnik R, Lippard SJ, Langer R, Farokhzad OM (2010) Engineering of self-assembled nanoparticle platform for precisely controlled combination drug therapy. Proc Natl Acad Sci USA 107:17939–17944
58. Karnic R, Gu F, Basto P, Cannizzaro C, Dean L, Kyei-Manu W, Langer R, Farokhzad OC (2008) Microfluidic platform for controlled synthesis of polymeric nanoparticles. Nano Lett 8:2906–2912
59. Avgoustakis K, Beletsi A, Panagi Z, Klepetsanis P, Karydas AG, Ithakissios DS (2002) PLGA-mPEG nanoparticles of cisplatin: In vitro nanoparticle degradation, in vitro drug release and in vivo drug residence in blood properties. J Control Release 79:123–135
60. Aryal S, Hu C-MJ, Zhang L (2010) Combinatorial drug conjugation enables nanoparticle dual-drug delivery. Small 6:1442–1448
61. Harasym T, Tardi P, Harasym N, Harvie P, Johnstone S, Mayer L (2007) Increased preclinical efficacy of irinotecan and floxuridine coencapsulated inside liposomes is associated with tumor delivery of synergistic drug ratios. Oncol Res 16:361–374

62. Tardi PG, Dos Santos N, Harasym TO, Johnstone SA, Zisman N, Tsang AW, Bermudes DG, Mayer LD (2009) Drug ratio-dependent antitumor activity of irinotecan and cisplatin combinations in vitro and in vivo. Mol Cancer Ther 8:2266–2275
63. Hu X, Jing X (2009) Biodegradable amphiphilic polymer-drug conjugate micelles. Expert Opin Drug Deliv 6:1079–1090
64. Kratz F, Abu Ajaj K, Warnecke A (2007) Anticancer carrier-linked prodrugs in clinical trials. Expert Opin Investig Drugs 16:1037–1058
65. Fleming AB, Haverstick K, Saltzman WM (2004) In vitro cytotoxicity and in vivo distribution after direct delivery of PEG-camptothecin conjugates to the rat brain. Bioconjug Chem 15:1364–1375
66. Gopin A, Ebner S, Attali B, Shabat D (2006) Enzymatic activation of second-generation dendritic prodrugs: conjugation of self-immolative dendrimers with poly(ethylene glycol) via click chemistry. Bioconjug Chem 17:1432–1440
67. Yu D, Peng P, Dharap SS, Wang Y, Mehlig M, Chandna P, Zhao H, Filpula D, Yang K, Borowski V, Borchard G, Zhang Z, Minko T (2005) Antitumor activity of poly(ethylene glycol)-camptothecin conjugate: the inhibition of tumor growth in vivo. J Control Release 110:90–102
68. Cavallaro G, Licciardi M, Caliceti P, Salmaso S, Giammona G (2004) Synthesis, physico-chemical and biological characterization of a paclitaxel macromolecular prodrug. Eur J Pharm Biopharm 58:151–159
69. Singer JW, Bhatt R, Tulinsky J, Buhler KR, Heasley E, Klein P, de Vries P (2001) Water-soluble poly-(L-glutamic acid)-gly-camptothecin conjugates enhance camptothecin stability and efficacy in vivo. J Control Release 74:243–247
70. Veronese ML, Flaherty K, Kramer A, Konkle BA, Morgan M, Stevenson JP, O'Dwyer PJ (2005) Phase I study of the novel taxane CT-2103 in patients with advanced solid tumors. Cancer Chemother Pharmacol 55:497–501
71. Tong R, Yala L, Fan TM, Cheng J (2010) The formulation of aptamer-coated paclitaxel-polylactide nanoconjugates and their targeting to cancer cells. Biomaterials 31:3043–3053
72. Johnson BK, Prud'homme RK (2003) Flash nanoprecipitation of organic actives and block copolymers using a confined impinging jets mixer. Aust J Chem 56:1021–1024
73. Johnson BK, Prud'homme RK (2003) Mechanism for rapid self-assembly of block copolymer nanoparticles. Phys Rev Lett 91:118302
74. Ansell SM, Johnstone SA, Tardi PG, Lo L, Beck J, Xie S, Bermudes D, Prud'homme RK, Mayer LD (2008) Development of highly efficacious hydrophobic paclitaxel prodrugs delivered in nanoparticles for fixed-ratio drug combination applications. AACR Meeting Abstracts, Apr: 5734
75. Ansell SM, Johnstone SA, Tardi PG, Lo L, Xie S, Shu Y, Harasym TO, Harasym NL, Williams L, Bermudes D, Liboiron BD, Saad W, Prud'homme RK, Mayer LD (2008) Modulating the therapeutic activity of nanoparticle delivered paclitaxel by manipulating the hydrophobicity of prodrug conjugates. J Med Chem 51:3288–3296
76. Albain KS, Nag SM, Calderillo-Ruiz G, Jordan JP, Llombart AC, Pluzanska A, Rolski J, Melemed AS, Reyes-Vidal JM, Sekhon JS, Simms L, O'Shaughnessy J (2008) Gemcitabine plus paclitaxel versus paclitaxel monotherapy in patients with metastatic breast cancer and prior anthracycline treatment. J Clin Oncol 26:3950–3957

Chapter 7
Polymeric Micelles for Multiple-Drug Delivery

Glen S. Kwon

7.1 Introduction

The supramolecular core/shell architecture of a polymeric micelle is shown in Fig. 7.1. Assembly of amphiphilic block copolymers (ABCs) in water occurs at a critical micelle concentration (CMC), forming nanoscopic micelles composed of several hundred ABCs. For drug delivery, the hydrophilic block is often poly(ethylene glycol) (PEG), a biocompatible, highly hydrated polymer that is approved for use in humans. The hydrophobic block largely dictates the chemical and physical nature of the core region, the usual site for drug solubilization of polymeric micelles. Given the versatility in the chemistry of the core-forming block, it is easy to imagine that a variety of poorly water-soluble drugs can be incorporated in polymeric micelles, resulting in drug solubilization, and there are a growing number of scientific papers and review articles on the capacity of polymeric micelles for drug solubilization [1–4]. However, the hydrophobic block must also be biocompatible for preclinical drug development, limiting chemical diversity. ABCs that have a poly(lactic acid) (PLA) or a poly(L-amino acid) block are prototypes that have been safely injected intravenously in humans besides the better known Pluronics [5]. From a safety perspective, ABCs have molecular weights that permit renal clearance and have a low tendency for cellular membrane disruption relative to low-molecular-weight surfactants. Furthermore, ABC micelles can be sterilized simply by filtration (0.22 μm) as a step during the manufacturing process or prior to intravenous (IV) injection or infusion.

In this chapter, I will discuss recent progress in multiple-drug solubilization via polymeric micelles [6, 7]. While it is evident that drug delivery nanotechnology has exciting potential in drug targeting via the EPR effect, it is easy to surmise that a

G.S. Kwon (✉)
Pharmaceutical Sciences Division, School of Pharmacy, University of Wisconsin,
777 Highland Avenue, Madison, WI 53705, USA
e-mail: gskwon@pharmacy.wisc.edu

S. Svenson and R.K. Prud'homme (eds.), *Multifunctional Nanoparticles for Drug Delivery Applications: Imaging, Targeting, and Delivery*, Nanostructure Science and Technology,
DOI 10.1007/978-1-4614-2305-8_7, © Springer Science+Business Media, LLC 2012

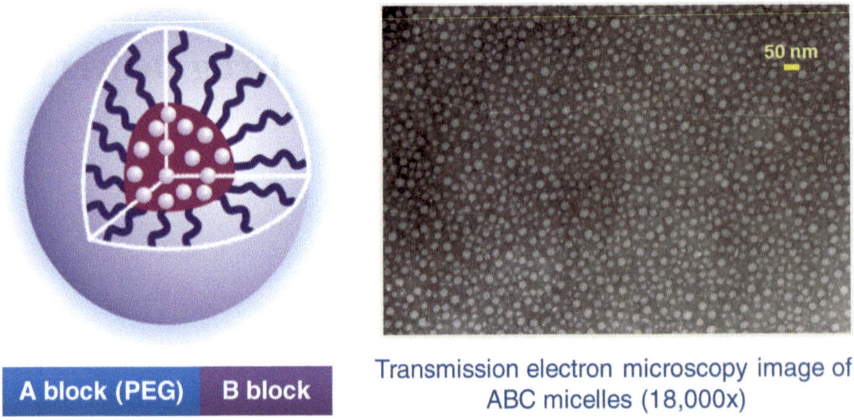

Fig. 7.1 Sketch of polymeric micelles containing a poorly water-soluble drug and a transmission electron microscopy (TEM) image of polymer micelles. Note from the TEM picture that polymeric micelles are spherical in shape and have low polydispersity in terms of size

single anticancer agent delivered by a "nanocarrier" will not be sufficient for cancer therapy given the likelihood of drug resistance. Multiple-drug delivery is common in current cancer therapy, but it is done in an arcane manner: it requires sequential IV drug infusion; it utilizes toxic IV vehicles, e.g., ethanol, Cremophor EL, even DMSO in clinical studies [8], contributing to deleterious toxicity of chemotherapy; and it is incapable of multiple-drug delivery into solid tumors at the same time, a situation that is mirrored by cell culture studies seeking synergistic combinations of drugs [9]. In recent drug delivery research, there is a greater emphasis on the delivery of multiple anticancer agents via water-soluble drug conjugates and liposomes [9–11], aiming for synergistic efficacy. I will describe recent studies on polymeric micelles that suggest that they will play a key part in cancer therapy and in multiple-drug delivery, especially for poorly water-soluble anticancer agents that are common in cancer drug development. In the future, we can imagine that polymeric micelles will be multimodal for cancer therapy, i.e., possess capability in drug delivery and other modalities, such as molecular imaging.

7.2 PEG-*b*-PLA Micelles for Paclitaxel Delivery

PEG-*b*-PLA micelles have been widely studied for drug delivery and have gained approval in Korea as a nanocarrier for paclitaxel in cancer therapy as an alternative to a mixture of Cremophor® EL (PEG–triglycerol triricinoleate) and ethanol in the standard IV formulation [12]. Paclitaxel stabilizes microtubules and thereby acts as a mitotic inhibitor, and it is used to treat patients with lung, breast, and ovarian cancers [13]. However, paclitaxel is poorly water-soluble, ca. 1 mg/L, hampering early clinical development for about 10 years, and it required Cremophor EL and ethanol in its original formulation termed Taxol. One gram of paclitaxel requires about 80 g of Cremophor EL for drug solubilization. While the primary dose-limiting toxicities of paclitaxel are neutropenia and peripheral neuropathy, Cremophor EL, a nonionic surfactant, has been implicated in acute hypersensitivity reactions, peripheral neuropathy, and dyslipidemia after the IV infusion of paclitaxel and several other poorly water-soluble drugs, including cyclosporin A, photosensitizers, and fat-soluble vitamins [14, 15]. The pharmacological and PK effects of Cremophor EL and Tween 80 have been reviewed [12, 14, 15]. Hypersensitivity reactions range from mild pruritus to systemic anaphylaxis, resulting in severe clinical outcomes that include respiratory arrest, cardiac arrest, and death. About 2–4% of patients have severe hypersensitivity reactions despite prophylaxis and have to discontinue therapy. Thus, there has been wide interest in alternative IV vehicles for paclitaxel, culminating in the approval of Abraxane®, an albumin nanoparticle-based vehicle that has a low propensity to cause hypersensitivity reactions [16].

PEG-*b*-PLA micelles readily increase the water solubility of paclitaxel to >1 mg/mL, which is sufficient for cancer therapy [17]. The CMC of PEG-*b*-PLA is 40 μM at 25°C in water, based on the measurement of fluorescence changes of 1,6-diphenyl-1,3,5-hexatriene. The process for the solubilization of paclitaxel by PEG-*b*-PLA micelles is straightforward: both are dissolved in acetonitrile, and the organic solvent removed by heat and reduced pressure. PEG-*b*-PLA and paclitaxel form a solid in which the crystallinity of PEG is low relative to PEG alone, and paclitaxel is in a molecularly dispersed or an amorphous state [17]. The dissolution of the solid is accomplished by heating at 60°C, resulting in a gel, which is dissolved by addition of water and stirring. In this way, PEG-*b*-PLA micelles increase the water solubility of paclitaxel to >1 mg/mL at 25% drug loading. In the approved IV formulation of paclitaxel, Genexol-PM, the molecular weight of PEG and poly(D, L lactic acid) is 2,000 and 1,750 g/mol, respectively, and it is a solid product that is reconstituted with an aqueous vehicle prior to IV infusion.

In preclinical experiments, Genexol-PM was surprisingly less toxic than Taxol® [18–20]. After intraperitoneal (IP) injections on five consecutive days in B6D2F1 mice with IP P388 leukemia, the maximum tolerated dose (MTD) of paclitaxel injected as part of PEG-*b*-PLA micelles was 100 mg/kg, whereas the MTD of Taxol was 20 mg/kg [18]. On a daily schedule for five consecutive days, the MTD of paclitaxel after IV injection was 25 and 20 mg/kg, respectively, in nude mice [19]. After three IV injections on days 0, 4, and 8, the MTD of Genexol-PM and Taxol in nude mice was 60 and 20 mg/kg, respectively [20]. The MTD was defined as <15% change in body weight relative to controls and no deaths or remarkable changes in appearance within 1 week of injections. Clinically, Taxol is administered every week or once every 3 weeks for cancer therapy. All mice that received Taxol showed apathy for 30 s after injection, whereas there were no visible signs of distress for Genexol-PM. Cremophor EL alone also caused mice to show signs of apathy. The median lethal dose (LD_{50}) of Genexol-PM in Sprague–Dawley rats was 205 and 222 mg/kg for males and female rats, respectively, whereas the LD_{50} for Taxol was 8.3 and 8.8 mg/kg for male and female rats, respectively.

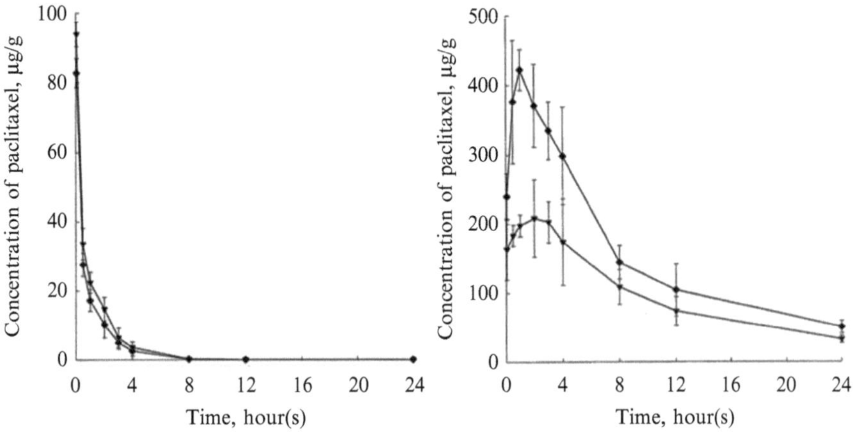

Fig. 7.2 Plasma profile and tumor biodistribution of paclitaxel as Genexol-PM dosed IV at 50 mg/kg (*diamond*) and Taxol dosed IV at 20 mg/kg (*inverted triangle*) in a B16 melanoma tumor model. $n=4$ for each time point. Reproduced with Permission from [20]

The low acute toxicity of Genexol-PM relative to Taxol is likely due to the absence of pharmacological effects caused by PEG-*b*-PLA; however, changes in the pharmacokinetics (PK) of paclitaxel brought about by PEG-*b*-PLA micelles relative to Cremophor EL micelles play a role [15, 20]. In a subcutaneous (SC) B16 melanoma murine tumor model, Genexol-PM dosed IV at 50 mg/kg had a lower maximum plasma concentration (C_{max}) and area under the curve (AUC) than Taxol at 20 mg/kg, as shown in Fig. 7.2 [20]. The elimination half-life of paclitaxel ($t_{1/2\beta}$) was 0.21 and 0.34 h for Genexol-PM and Taxol, respectively. At these doses, the biodistribution of paclitaxel in organs was 2–3 times greater for Genexol-PM over Taxol, including the SC tumor (Fig. 7.2). At an equivalent dose of 20 mg/kg, the biodistribution of paclitaxel was similar; however, the plasma AUC was 10–20 times less for Genexol-PM relative to Taxol. Cremophor EL micelles have a unique ability to affect the PK of paclitaxel: a nonlinear PK profile for paclitaxel, resulting in a dose-proportional increase in various organs, but a disproportionate increase in plasma AUC [21]. It has been postulated that Cremophor EL micelles entrap paclitaxel in blood, resulting in a non-PK profile.

While quite a bit is known on the PK of paclitaxel as Taxol, there is a surprising lack of detailed mechanistic insight into the role(s) of Cremophor EL and its PK in relation to the kinetic stability of its micelles in blood. Any discussion on the PK of paclitaxel must take into account Cremophor EL's effects on serum lipoproteins: a decrease in electrophoretic mobility and production of lipoprotein dissociation products, which bind paclitaxel [15]. Furthermore, Cremophor EL is an inhibitor of P-glycoprotein, a multidrug transporter, associated with drug resistance in cancer [22]. Cremophor EL micelles display nonlinear PK, showing reduced clearance with elevated dose [15, 21]. The bioanalytical assay of Cremophor EL involves saponification that produces ricinoleic acid, which is modified by naphthylamine for reverse-phase HPLC [23]. Other bioanalytical assays for Cremophor EL have been described [15]. The terminal half-life of Cremophor EL is 89 h with values ranging from 10 to 140 h, depending on the method of its bioanalysis. Interestingly, little is known about the ultimate fate of Cremophor EL after infusion, noting <0.1% renal clearance and insignificant accumulation at solid tumors, which is surprising considering its long half-life in plasma. In summary, while it is known that Cremophor EL has effects on the PK of paclitaxel and causes drug interactions for hydrophobic drugs, e.g., etoposide, the relative contribution of entrapment in Cremophor EL micelles versus effects on serum lipoproteins is ill defined.

PK studies on paclitaxel, PEG-*b*-PLA, and Genexol-PM are even fewer in number, but suggest that paclitaxel is released within a few minutes from PEG-*b*-PLA micelles, and it distributes among serum proteins, classified as lipoprotein-deficient plasma fraction, HDL fraction, LDL fraction, and VLDL fraction [24]. After 5 min, the distribution of free paclitaxel and paclitaxel added as part of PEG-*b*-PLA micelles in human plasma was remarkably similar. PEG was 2,000 g/mol, and the weight ratio of PEG to PLA was 40:60. Paclitaxel distributed equally between serum lipoprotein and serum lipoprotein deficient fractions in both cases. In the serum lipoprotein fractions, 70–75% of paclitaxel was associated with HDL. PEG-*b*-PLA at 2.0 mg/mL did not alter the composition of plasma lipoproteins, indicating an absence of intermolecular interaction that has been noted for Cremophor EL.

The release of paclitaxel from PEG-*b*-PLA micelles on a timescale of several minutes is consistent with recent work on PEG-*b*-PLA micelles that showed a loss of Förster resonance energy transfer (FRET) in vivo between a pair of lipophilic dyes ($DiIC_{18}$ and $DiOC_{18}$) within 15 min after IV injection [25]. FRET for $DiIC_{18}$ and $DiOC_{18}$ in PEG-*b*-PLA micelles is readily observed due to their distance proximity in the cores after solubilization at 0.75%. Disruption of PEG-*b*-PLA micelles leads to the leakage of $DiIC_{18}$ and $DiOC_{18}$ and loss of FRET. In this way, it was shown that α- and β-globulins play major roles in the leakage of $DiIC_{18}$ and $DiOC_{18}$ and loss of FRET for PEG-*b*-PLA micelles, whereas γ-globulins, serum albumin, and red blood cells play minor roles in determining the stability of PEG-*b*-PLA micelles in blood.

In summary, PEG-*b*-PLA micelles are unable to increase the circulation time of hydrophobic drugs such as paclitaxel because of their disruption in blood after IV injection and leakage of drug, caused by the action of α- and β-globulins. Instead, PEG-*b*-PLA micelles are safer than Cremophor EL, permitting dose escalation; a relative absence of pharmacological effects, e.g., hypersensitivity reactions; and evidence of a linear PK profile for paclitaxel. At this stage, it is tempting to state that PEG-*b*-PLA micelles cannot carry paclitaxel into solid tumors by the EPR effect. However, there are not enough studies to fully support this conclusion, noting evidence for prolonged circulation in mice for PEG-*b*-PLA micelles (25% in blood after 24 h) when the molecular weights of the PEG and PLA blocks were 5,100 and 5,300 g/mol, respectively [26]. The integrity of PEG-*b*-PLA micelles was confirmed after 24 h by collection of plasma and a gel filtration assay, noting that this result was at odds with the FRET study on PEG-*b*-PLA micelles [25]. Additional PK studies on paclitaxel delivered by PEG-*b*-PLA micelles are required that define effects of the molecular weights of PEG and PLA blocks, paclitaxel content, and drug targeting mediated by targeting ligands attached on the distal ends of PEG.

Owing to an elevated MTD over Taxol and superior PK profile, Genexol-PM had higher antitumor activity in tumor xenograft models [20]. In vitro, Genexol-PM and Taxol showed comparable cytotoxicity against a human ovarian cancer cell line, OVCAR-3, and a human breast cancer cell line, MCF-7. However, Genexol-PM dosed IV at its MTD (60 mg/kg) had significantly greater antitumor efficacy (reduction in tumor volume) than Taxol at its MTD (20 mg/kg), dosed on the same schedule in a SKOV3 human ovarian cancer implanted SC in nude (nu/nu) athymic mice and in a MX-1 human breast cancer implanted SC in Tac:Cr:(NCr)-nu athymic mice. Taxol delayed tumor growth in a SKOV3 human ovarian xenograft model; by contrast, Genexol-PM caused tumor shrinkage and, in some cases, tumor regression. Similar results were obtained in an MX-1 human breast xenograft model, noting undetectable tumors after 18 days from the start of treatment for Genexol-PM.

Clinical studies have largely confirmed the safety of Genexol-PM over Taxol, higher MTD, linear PK profile in humans, and better antitumor response [27–29]. In a phase I clinical trial, 21 patients received IV Genexol-PM over 3 h every 3 weeks at escalating doses (135–390 mg/m^2) and were evaluated for toxicity and response [27]. Peripheral neuropathy and myalgia (muscle pain) were the most common toxicities of Genexol-PM. One patient had a grade 3 myalgia during cycle 1 at 230 and 300 mg/m^2, and two of three patients developed grade 4 neutropenia or grade 3 neuropathy at 390 mg/m^2. There were no acute hypersensitivity reactions despite the lack of premedication with hydrocortisone and histamine blocker. The MTD of Genexol-PM was set at 390 mg/m^2, which is higher than the MTD for Taxol (175 mg/m^2) infused IV in a 3-week regimen. Three patients had partial responses among the 21 patients, and of the three, two were refractory to Taxol. Lastly, the AUC of paclitaxel in humans ($n=13$) increased with dose with the exception of 230 mg/m^2, suggesting that Genexol-PM has a linear PK profile in humans.

In a single-arm Phase II study, the recommended IV dose of Genexol-PM was set at 230–300 mg/m^2 (3 h infusion every 3 weeks) for patients ($n=69$) with advanced non-small cell lung cancer (NSCLC), and therapy was evaluated in combination with cisplatin at 60 mg/m^2. Cisplatin was infused IV after Genexol-PM (sequential drug administration) [28]. The objective response rate was 37.7%; all were partial responders, noting that the objective response rate for Abraxane as a single agent in advanced NSCLC was 16% [30]. The median time to progression was 5.8 months, and the median survival time was 21.7 months for Genexol-PM plus cisplatin. The response rates and survival data compared favorably to results in phase II and III clinical trials on Taxol (175–200 mg/m^2) in combination with cisplatin (75–80 mg/m^2) [31–33]. In spite of a higher dose of paclitaxel, adverse effects of Genexol-PM plus cisplatin were comparable to Taxol plus cisplatin. The major toxic side effects were grade 3/4 neutropenia (29 and 17%, respectively), grade 3 peripheral neuropathy (13%), and grade 3/4 arthralgia, i.e., joint pain (7.3%). Four patients experienced grade 3/4 hypersensitivity reactions. After implementation of prophylaxis, incidence of grade 3/4 hypersensitivity reactions decreased, but still a few patients had hypersensitivity reactions despite premedication. While somewhat surprising based on the earlier phase I clinical trial, the authors suggested that paclitaxel itself may cause hypersensitivity reactions and that patients that receive paclitaxel in a combination with cisplatin are at higher risk for hypersensitivity reactions.

In a single-arm phase II study in women with histologically confirmed metastatic breast cancer ($n=41$), IV infusion of Genexol-PM at 300 mg/m^2 over 3 h every 3 weeks with an average of 8 cycles per patient produced an overall response rate of 58.5% with 5 complete responses and 19 partial responses [29]. Thirty seven patients who received Genexol-PM as a first-line therapy had a 59.5% response rate, and two responses were observed in four patients treated in a second-line setting. This response rate compared favorably with Abraxane, which had a response rate of 47.6% at 300 mg/m^2 in a 3-week dosing regimen [34] and Taxol, which had a response rate of 21–54% as a first-line therapy in patients with metastatic breast cancer [35–38]. The major toxic side effects were grade 3/4 neutropenia (51 and 17.1%, respectively), grade 3 peripheral neuropathy (51%), grade 1 and 2 thrombocytopenia (22%), and grade 3/4 myalgia (2.4%). Eight patients experienced hypersensitivity reactions with 3, 3, and 2 patients at grades 1, 2, and 3, respectively. Only one patient who had a grade 2 hypersensitivity reactions had to prematurely halt therapy. The two patients who had grade 3 hypersensitivity reactions responded well to corticosteroid and/or antihistamine. Subsequent prophylaxis prevented additional hypersensitivity reactions. It appears that Genexol-PM does illicit hypersensitivity reactions in cancer patients, although at a lower level than that of Cremophor EL.

In summary, Genexol-PM has significant antitumor efficacy in advanced NSCLC with cisplatin, and in metastatic breast cancer as a single agent. The high response rates of Genexol-PM are probably due to a higher MTD and higher tumor accumulation of paclitaxel, and these phase II results compare favorably to Abraxane and Taxol, although the results to date are not definitive and require head-to-head clinical trials. The major toxic side effects are the same as those observed for Taxol and are comparable even though a higher dose of paclitaxel was used in the phase II clinical trials on Genexol-PM. The hypersensitivity reactions noted for Genexol-PM were somewhat surprising although seemingly less for Cremophor EL in Taxol. Genexol-PM represents a safe, simple, and soluble nanocarrier for paclitaxel, and this clinical work done in Korea culminated in its premarket approval in this country in 2006. A phase II clinical trial on Genexol-PM plus gemcitabine for pancreatic cancer is ongoing in the United States of America.

7.3 PEG-*b*-PLA Micelles for Multiple-Drug Delivery

PEG-*b*-PLA micelles have a proven ability for single-drug solubilization, and there is a growing body of preclinical and clinical knowledge on PEG-*b*-PLA micelles due to the significance of Genexol-PM. Besides the major advantages of safety and evidence of a superior PK profile, i.e., linear PK, over Cremophor EL, there is solid evidence that PEG-*b*-PLA micelles can serve as a nanocontainer for multiple poorly water-soluble anticancer agents (Fig. 7.3), raising unexplored avenues of combination

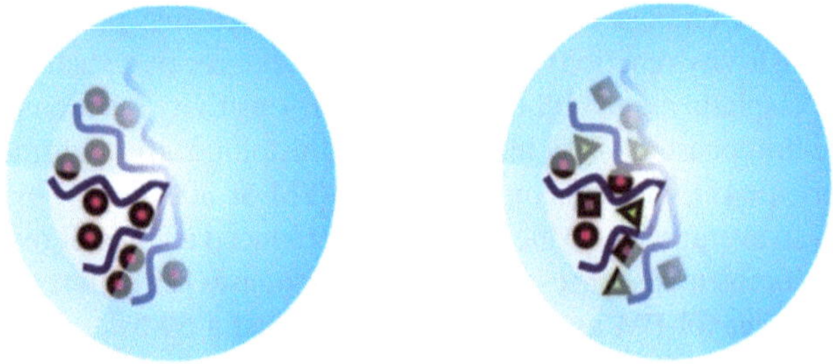

Fig. 7.3 Single-drug- and multiple-drug-loaded polymeric micelles

cancer therapy. Combination drug therapy is commonplace in cancer therapy, noting that paclitaxel as Genexol-PM has been used in combination with one other anticancer agent for NSCLC and pancreatic cancers [28]. There is a lot of interest in combining chemotherapy with signal transduction inhibitors in efforts that seek to disable cancer cell survival pathways and overcome drug resistance [39, 40]. Chemotherapy and signal transduction inhibitors are often poorly water soluble and require Cremophor EL; cosolvent, such as ethanol or DMSO; or other drug solubilization vehicles for IV infusion.

We envision combination cancer strategies that benefit from sequential drug administration can be achieved via single-drug-loaded PEG-*b*-PLA micelles. Sequential administration of anticancer agents may be required to prevent cell cycle arrest and obtain synergistic cancer cell apoptosis [41]. The replacement of two or more toxic IV vehicles by PEG-*b*-PLA micelles is a simple but significant advance for combination drug therapy, especially when it involves chemotherapy, such as paclitaxel. Beyond paclitaxel, PEG-*b*-PLA micelles have been used to solubilize β-lapachone, etoposide, and 17-allylamino-17-demethoxygeldanamycin (17-AAG) in water [42–44]. It is expected that PEG-*b*-PLA micelles will increase the water solubility of other poorly water-soluble anticancer agents in preclinical development.

Multiple-drug-loaded PEG-*b*-PLA micelles are unique in drug delivery and offer a simple and safe solution to concurrent drug delivery in combination cancer strategies. Instead of a separate IV vehicle for every poorly water-soluble anticancer agent in a combination cancer strategy, PEG-*b*-PLA micelles entrain multiple anticancer agents, lowering costs associated with the production of multiple-drug formulations, simplifying IV infusion of drug "cocktails," and raising the possibility that anticancer agents access solid tumor at the same time for synergistic antitumor responses. All anticancer agents in combination cancer strategies are now infused sequentially, lowering the likelihood that they act simultaneously at solid tumors even though proof of synergy was obtained in cell culture after simultaneous drug exposure.

Table 7.1 summarizes results for PEG-*b*-PLA micelles on the solubilization individual drugs and 2- and 3-drug combinations of paclitaxel, docetaxel, 17-AAG, and

Table 7.1 Single agent and multiple-drug solubilization by PEG-*b*-PLA micelles (mean ± SD; *n* = 3)

Anticancer agent	Drug level in water (mg/mL)	% drug loading (wt drug(s)/wt polymer)	PEG-*b*-PLA micelle diameter (nm ± SD)
Paclitaxel	3.54 ± 0.32	11.8 ± 1.1	38.8 ± 0.6
Docetaxel	4.27 ± 0.44	14.2 ± 1.5	37.3 ± 1.7
Etoposide	3.31 ± 0.15	11.0 ± 0.5	32.6 ± 1.0
17-AAG	3.90 ± 0.28	13.0 ± 0.9	39.3 ± 2.9
Paclitaxel+	3.92 ± 0.17	26.0 ± 1.4	38.9 ± 1.1
17-AAG	3.88 ± 0.29		
Docetaxel+	4.62 ± 0.44	28.8 ± 0.2	39.0 ± 0.8
17-AAG	4.01 ± 0.08		
Etoposide+	3.49 ± 0.24	25.6 ± 1.3	35.3 ± 1.2
17-AAG	4.21 ± 0.38		
Etoposide+	3.17 ± 0.04	34.3 ± 1.6	36.5 ± 0.5
Paclitaxel+	3.50 ± 0.20		
17-AAG	3.61 ± 0.33		

etoposide [6]. The molecular weight of PEG and poly(D,L lactic acid) was 4,200 and 1,800 g/mol, respectively. Individually, PEG-*b*-PLA micelles raised the water solubility of paclitaxel, docetaxel, etoposide, and 17-AAG from 0.1 to 10 mg/L to ca. 4.0 mg/mL. This is sufficient water solubility for cancer therapy for each one of these anticancer agents. The % drug loading (wt drug/wt polymer) for PEG-*b*-PLA micelles was 11–13% and consistent with values for polymeric micelles in the literature [45]. In every case, the average hydrodynamic diameter of PEG-*b*-PLA micelles was about 40 nm, irrespective of the anticancer agent. For 2- and 3-drug combinations, solubilization of each anticancer agent was similar to the solubilization achieved for each anticancer agent incorporated individually by PEG-*b*-PLA micelles. Remarkably, % drug loading increased to 26–29% for 2-drug combinations and to 34% for the 3-drug combination without a significant change in hydrodynamic diameter of PEG-*b*-PLA micelles. It is noted that the simultaneous solubilization of multiple poorly water-soluble anticancer agents in water is unprecedented and cannot be achieved with nonionic surfactants, such as Tween 80 [46].

The physical stability of PEG-*b*-PLA micelles with respect to drug precipitation was monitored by reverse-phase HPLC after storage at room temperature for 24 h. Surprisingly, individual anticancer agents with the exception of 17-AAG entrained in PEG-*b*-PLA micelles precipitated from solution over 24 h (data not shown); only 16% of paclitaxel remained in solution, whereas 98.6% of 17-AAG remained in solution. The results suggest that Genexol-PM does not form a thermodynamically stable aqueous solution, and it has to be administered to patients within several hours to avoid the infusion of paclitaxel precipitate, noting that a similar situation is present for Taxol upon dilution with an aqueous vehicle prior to IV infusion. In contrast, 17-AAG and its 2- and 3-drug combinations that have been solubilized by PEG-*b*-PLA micelles are stable with respect to drug precipitation over 24 h, indicating thermodynamic stability. The origin of this unique physical stability of PEG-*b*-PLA micelles containing multiple poorly water-soluble anticancer agents

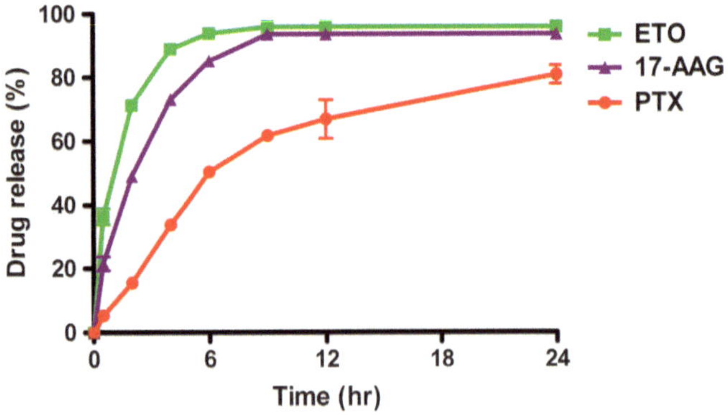

Fig. 7.4 In vitro release kinetics of paclitaxel, etoposide, and 17-AAG from PEG-*b*-PLA micelles (mean ± SD; *n* = 4). In vitro half-lives for drug release are 5.1, 0.88, and 1.8 h, respectively (fairly rapid release). Reproduced with permission from [6]

has not been defined, but may involve intermolecular interaction of 17-AAG and other anticancer agents in the core region of PEG-*b*-PLA micelles. The absence of ^{1}H NMR signals for paclitaxel, etoposide, and 17-AAG in D_2O after their solubilization together by PEG-*b*-PLA micelles suggests that the site of multiple-drug solubilization is the core region of PEG-*b*-PLA micelles [6].

Figure 7.4 shows the release profile of paclitaxel, etoposide, and 17-AAG co-incorporated in PEG-*b*-PLA micelles. Etoposide had the fastest release, followed by 17-AAG and then paclitaxel. The release profiles of multiple-drug-loaded PEG-*b*-PLA micelles were quite similar to the release profiles for the individual anticancer agents released by PEG-*b*-PLA micelles, and they corresponded well to their oil-in-water partition coefficients: log *P* values increased along with half-lives of drug release for PEG-*b*-PLA micelles (data not shown). One limitation of this study was that it was done above the CMC of PEG-*b*-PLA micelles. In vivo, it is expected that the PEG-*b*-PLA micelles will dissociate due to the action of α- and β-globulins and dilution beneath the CMC, resulting in drug release due to micelle dissociation as a possible alternative mechanism of drug release. In this situation, we expect that PEG-*b*-PLA micelles will not have a major impact on the PK of paclitaxel, etoposide, and 17-AAG co-incorporated in PEG-*b*-PLA micelles, but this hypothesis awaits validation by PK experiments in rodent models.

In summary, PEG-*b*-PLA micelles are capable of multiple-drug solubilization for parenteral drug delivery, offering an alternative drug delivery strategy for combination cancer therapy. For combination cancer therapy involving paclitaxel, etoposide, and 17-AAG, PEG-*b*-PLA micelles simply replace Cremophor EL, ethanol, Tween 80, and DMSO [12, 47, 48]. Under current clinical practice, each anticancer agent would require separate sequential IV infusion, and Cremophor EL, ethanol, Tween 80, and DMSO would add to the significant toxicity burden brought about by paclitaxel, etoposide, and 17-AAG. It is noted that Taxol and 17-AAG have been tested in a phase I clinical trial [49]. However, the toxicity of Cremophor

EL and DMSO has been noted in mice, with several deaths observed after the rapid sequential IP injections of Taxol and 17-AAG solubilized by DMSO [50]. Recently, Cremophor EL replaced DMSO as a vehicle for 17-AAG in phase I clinical trials [51], but this added Cremophor EL in a combination with Taxol will certainly contribute to the overall toxicity burden, which was described earlier. In contrast, PEG-*b*-PLA micelles will minimize the toxicity of 2- and 3-drug combinations of paclitaxel and 17-AAG. Given the satisfactory safety profile of Genexol-PM, PEG-*b*-PLA micelles may facilitate entry of 2- and 3-drug combinations of paclitaxel and 17-AAG into cancer clinical trials. At this point, additional experiments must be done to better understand multiple-drug solubilization by PEG-*b*-PLA micelles: Can we expand the range of drugs that can undergo multiple-drug solubilization? Can we explain the unique results obtained with 17-AAG? Are there PK interactions for multiple-drug-loaded PEG-*b*-PLA micelles? Recently, we have found that other poorly water-soluble anticancer agents can undergo multiple-drug solubilization, and other polymeric micelles besides PEG-*b*-PLA micelles are capable of multiple-drug solubilization (data not shown).

7.4 PEG-*b*-Poly(L-Amino Acid) Micelles for Doxorubicin Delivery

PEG-*b*-poly(L-amino acid) micelles have also been widely studied for drug delivery and have entered clinical trials for doxorubicin (DOX), cisplatin, irinotecan, and paclitaxel [3, 52–55]. The clinical advancement of this major class of drug delivery nanotechnology has been reviewed recently by Matsumura and Kataoka [55]. A major advantage of PEG-*b*-poly(L-amino acid) micelles for drug delivery is the variety in L-amino acids that can make up the core-forming block that permits flexibility in composition for drug solubilization simply by hydrophobic interaction and by reversible chemical linkages (i.e., prodrugs) [56]. The best evidence for tumor targeting by polymeric micelles in murine tumor models has been achieved by PEG-*b*-poly(L-amino acid) prodrug micelles because premature drug release during circulation in blood is minimized, and drug release can be triggered at solid tumors in response to external signals such as pH change.

Kataoka and coworkers have developed a particularly attractive PEG-*b*-poly(L-amino acid) micelle for the pH-sensitive delivery of DOX that set the foundation for efforts in multiple-drug delivery via mixed polymeric micelles [57–60]. The chemical structure of PEG-*b*-poly(aspartate-hydrazone) with bound DOX (PEG-*b*-poly(aspartate-hyd-DOX)) is shown in Fig. 7.5; the synthetic procedure has been described in a review article [56]. The molecular weight of PEG for PEG-*b*-poly(aspartate-hyd-DOX) was 12,000 g/mol, and the numbers of aspartic acid and hydrazide groups were 37 and 28, respectively. The degree of DOX substitution for PEG-*b*-poly(aspartate-hyd-DOX) was 67%. PEG-*b*-poly(aspartate-hyd-DOX) assembled into micelles that release DOX in a pH-sensitive manner (Fig. 7.5): the release of DOX at physiological pH is slow due to hydrolysis of the hydrazone linkage, whereas the release rate increased under slightly acidic conditions (pH=4.5–6).

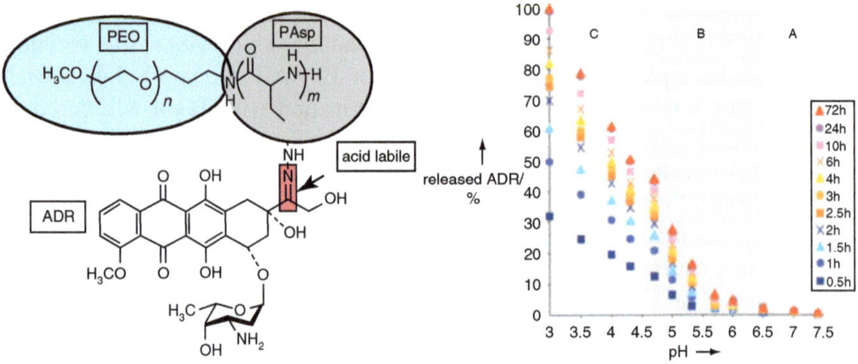

Fig. 7.5 Chemical structure of PEG-*b*-poly(aspartate-hyd-DOX) and ADR (DOX) release from its micelles as a function of time and pH. DOX is the generic name for ADR. Reproduced with permission from [57]

Using multicellular tumor spheroids from C26 colon adenocarcinoma cells, PEG-*b*-poly(aspartate-hyd-DOX) micelles showed little evidence of drug release (quenched fluorescence of DOX) after 1 h, whereas the presence of DOX fluorescence in C26 cells after 3 h signified entry of PEG-*b*-poly(aspartate-hyd-DOX) micelles into many of the C26 cells and drug release. However, most of the nuclei of C26 cells remained blue due to Hoechst 33258 staining and low nuclear entry of DOX. After 24 h, DOX fluorescence in the nuclei was clearly present and overlapped Hoechst 33258 staining, resulting in a pinkish appearance for C26 nuclei. These results suggest that PEG-*b*-poly(aspartate-hyd-DOX) micelles are taken up by C26 cells by endocytosis and trafficked via the endosomal/lysosomal pathway. Once PEG-*b*-poly(aspartate-hyd-DOX) micelles encounter the slightly acidic pH of endosomes and lysosomes, DOX release quickens and proceeds over 24 h, and DOX localizes at its major site of drug action, the nuclei of cancer cells. For a human small cell lung cancer cell line, SBC-3, the IC_{50} of DOX was 0.039 mg/L, whereas the IC_{50} for PEG-*b*-poly(aspartate-hyd-DOX) micelles was 0.27 mg/L after incubation for 24 h. Facile penetration of PEG-*b*-poly(aspartate-hyd-DOX) micelles in multicellular tumor spheroids is encouraging and probably reflects their small size, which is ca. 65 nm.

PEG-*b*-poly(aspartate-hyd-DOX) micelles circulated for over 24 h in mice bearing C26 colon tumors and had higher tumor accumulation than free DOX, probably due to the EPR effect [58]. At 10 mg/kg, the tumor AUC (Area Under the Curve) of PEG-*b*-poly(aspartate-hyd-DOX) micelles was fourfold higher than that of free DOX. As a result, PEG-*b*-poly(aspartate-hyd-DOX) micelles had superior antitumor efficacy. At 20 mg/kg, PEG-*b*-poly(aspartate-hyd-DOX) micelles reduced tumor volumes and resulted in two complete cures (no tumor reoccurrence) after three IV injections on 4-day intervals without a major loss in body weight. In contrast, free DOX only inhibited tumor growth at 10 mg/kg and resulted in a significant loss of body weight in tumor-bearing mice (17% loss in body weight). At 15 mg/kg,

DOX killed 6 out of 6 mice. In a C26 adenocarcinoma murine tumor model, PEG-*b*-poly(aspartate-hyd-DOX) micelles increased the antitumor efficacy of DOX over free DOX without a corresponding increase in toxicity, i.e., increased the therapeutic index of DOX. PEG-*b*-poly(aspartate-hyd-DOX) micelles are in preclinical development in Japan.

7.5 PEG-*b*-Poly(L-Amino Acid) Micelles for Multiple-Drug Delivery

For multiple-drug delivery, PEG-*b*-poly(L-amino acid) prodrugs assemble into mixed polymeric micelles that carry two or more different drugs in their core region (Fig. 7.6). In this strategy, each PEG-*b*-poly(L-amino acid) prodrug is synthesized separately, mixed together in an organic solvent, and assembled into mixed polymeric micelles by the simple replacement of organic solvent by water, using hydrophobic interaction as the driving force for assembly [7]. In this way, it is straightforward to adjust drug ratios in mixed polymeric micelles by varying the degree of drug substitution on PEG-*b*-poly(L-amino acid) prodrugs or varying the ratio of PEG-*b*-poly(L-amino acid) prodrugs in the organic solvent, noting that it is common that the drug ratio will be a key factor in determining synergy in cell culture [9]. If we employ hydrazone chemistry established for PEG-*b*-poly(aspartate-hyd-DOX), mixed polymeric micelles may carry a payload of two or more anticancer agents into solid tumors by the EPR effect with an identical PK profile, undergo endocytosis, and simultaneously release the drug payload within the cancer cells due to a drop in pH, aiming for synergy (Fig. 7.6).

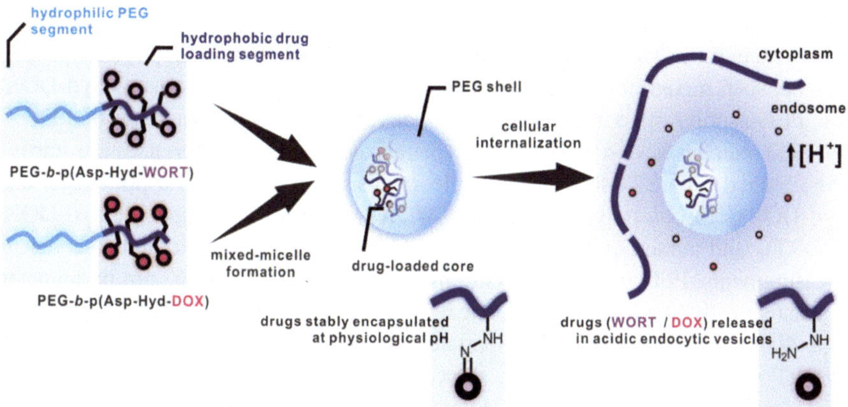

Fig. 7.6 Multiple-drug delivery via mixed polymeric micelles that release drug in response to acidic pH

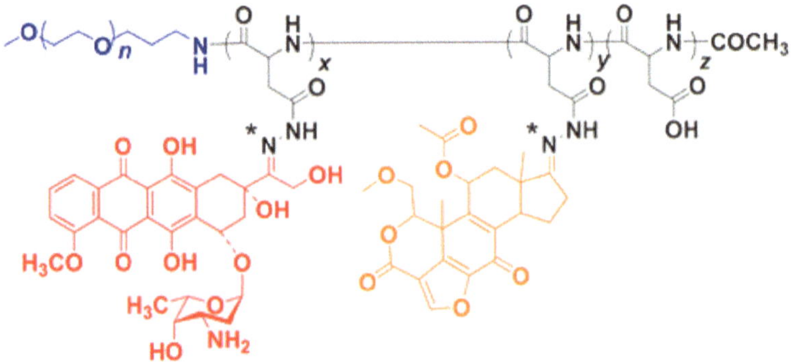

Fig. 7.7 Chemical structure of PEG-*b*-poly(aspartate-hyd-DOX-WORT). Reproduced with permission from [7]

Prodrugs PEG-*b*-poly(aspartate-hyd-DOX) and PEG-*b*-poly(aspartate-hydra zone) with conjugated wortmannin (WORT), a poorly water-soluble, phosphatidylinositol-3 kinase (PI3K) inhibitor, assembled into mixed polymeric micelles [7]. The molecular weight of PEG was 12,000 g/mol, and the numbers of aspartic acid and hydrazide groups were 40 and 31, respectively. The degree of DOX and WORT substitution for PEG-*b*-poly(aspartate-hyd-DOX) and PEG-*b*-poly(aspartate-hyd-WORT) was ca. 30%. DOX and WORT were also attached together on PEG-*b*-poly(aspartate-hydrazide), forming polymeric micelles that contain DOX and WORT in the core region (Fig. 7.7).

At this point, it is likely that mixed polymeric micelles offer the simplest way for multiple-drug delivery. For DOX and WORT, the reactivity of their ketone groups was similar, allowing control of drug ratios on PEG-*b*-poly(aspartate-hydrazide) by adjusting the ratios of DOX and WORT in the reaction mixture: 20, 50, and 75% WORT. However, this may not always be the case, given differences in reactivity between drugs, and it is easier to separately characterize PEG-*b*-poly(aspartate-hyd-DOX) and PEG-*b*-poly(aspartate-hyd-WORT) by ^{1}H NMR spectroscopy because of an absence of overlapping peaks in their spectra. The differences in physical properties between mixed polymeric micelles based on PEG-*b*-poly(aspartate-hyd-DOX) and PEG-*b*-poly(aspartate-hyd-WORT) relative to PEG-*b*-poly(aspartate-hyd-DOX-WORT) micelles have not been clearly delineated; however, differences are starting to be defined. Figure 7.8 shows the particle size distributions for PEG-*b*-poly(aspartate-hyd-DOX), PEG-*b*-poly(aspartate-hyd-WORT), PEG-*b*-poly(aspartate-hyd-DOX-WORT), and mixed polymeric micelles, obtained by dynamic light scattering measurements. PEG-*b*-poly(aspartate- hyd-DOX) micelles were 63 nm in diameter, consistent with the earlier studies done by Kataoka and coworkers [57–60]. The average diameters of PEG-*b*-poly(aspartate-hyd-DOX-WORT) at 25, 50, and 75% WORT were 64, 91, and 73 nm, respectively. PEG-*b*-poly(aspartate-hyd-WORT) micelles were 74 nm in diameter. Interestingly, mixed polymeric micelles at a 1:1 ratio of PEG-*b*-poly(aspartate-hyd-DOX) to PEG-*b*-poly(aspartate-hyd-WORT) were larger, having an average diameter of 220 nm. The larger size of the mixed

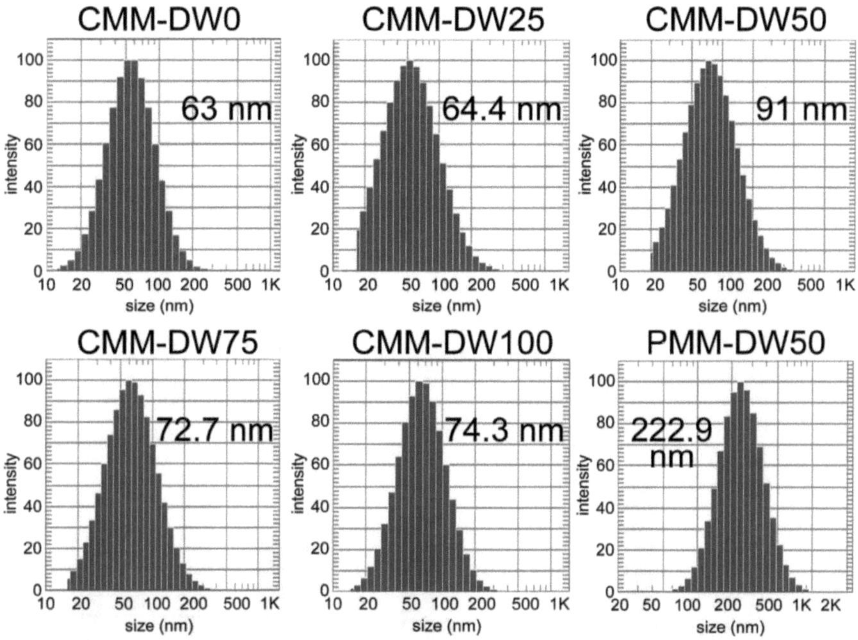

Fig. 7.8 Particle size distributions for PEG-*b*-poly(aspartate-hyd-DOX), PEG-*b*-poly(aspartate-hyd-WORT), PEG-*b*-poly(aspartate-hyd-DOX-WORT), and their mixed polymeric micelles. CMM: chemically mixed micelle, i.e., PEG-*b*-poly(aspartate-hyd-DOX-WORT). CMM varied in WORT content from zero (CMM-DW0) to 100% (CMM-DW100). PMM: physically mixed micelle. Reproduced with permission from [7]

polymeric micelles suggest that assembly of PEG-*b*-poly(aspartate-hyd-DOX) and PEG-*b*-poly(aspartate-hyd-WORT) does occur and that the mixed polymeric micelles have a higher association number than individual polymeric micelles that contain solely DOX, WORT, or both attached onto PEG-*b*-poly(aspartate-hydrazide).

The cytotoxicity of combinations of DOX and WORT as free drug and polymeric micelles against MCF-7 breast cancer cells after incubation for 30 or 72 h is shown in Fig. 7.9. After 30 h, the results for free drugs indicate that DOX and WORT at a 1:1 ratio exerts either additive or synergistic activity against MCF-7 cells. This enhanced potency of the drug combination with half the quantity of DOX likely results from the action of WORT on the PI3K/AKT/mTOR pathway, which is one of the most frequently dysregulated signaling pathways in cancer [40]. A similar trend was noted for mixed polymeric micelles of PEG-*b*-poly(aspartate-hyd-DOX) and PEG-*b*-poly(aspartate-hyd-WORT) and PEG-*b*-poly(aspartate-hyd-DOX-WORT) micelles at a 1:1 ratio. Clearly, polymeric micelles were less potent than free drugs, and this difference was lower at 72 h, presumably due to drug release from polymeric micelles over time. There were no statistical differences in cytotoxicity against MCF-7 cells between mixed polymeric micelles of PEG-*b*-poly(aspartate-hyd-DOX) and PEG-*b*-poly(aspartate-hyd-WORT) and PEG-*b*-poly(aspartate-hyd-DOX-WORT) micelles at a 1:1 ratio.

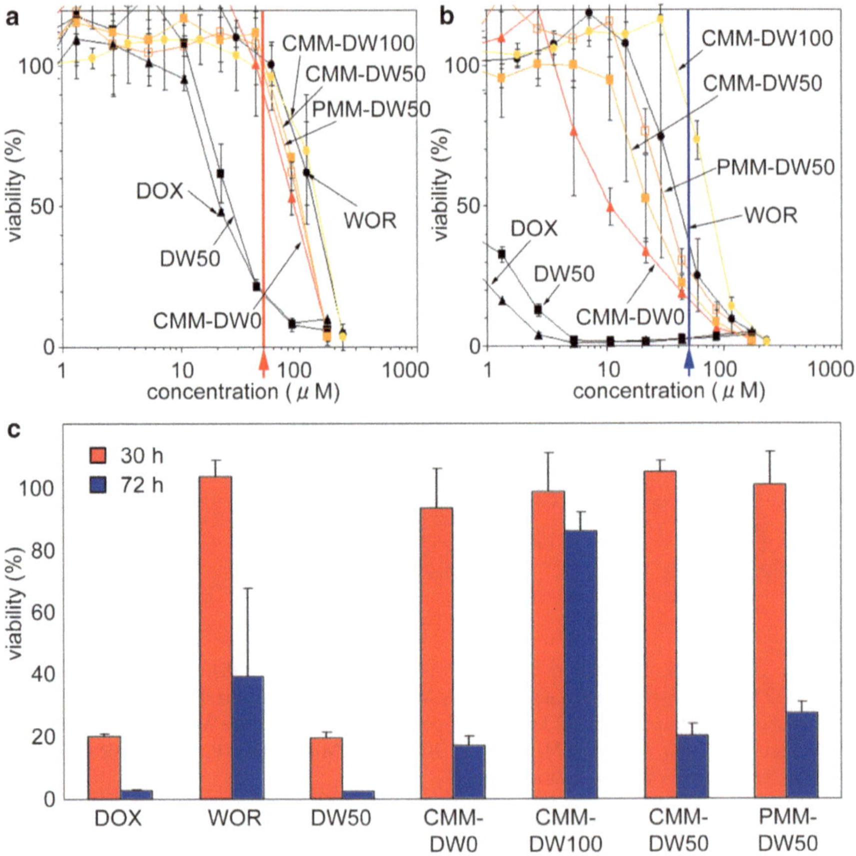

Fig. 7.9 Cytotoxicity of DOX, WORT, and 1:1 combinations as free drugs and polymeric micelles at 50 μM against MCF-7 breast cancer cells (mean ± SD; $n = 4$). Reproduced with permission from [7]

In summary, PEG-*b*-poly(L-amino acid) prodrugs assemble into mixed polymeric micelles that carry two or more different drugs in their core region. Lipophilic drugs that possess an aldehyde or ketone group can be linked to PEG-*b*-poly(aspartate-hydrazide) for assembly into polymeric micelles and mixed polymeric micelles, and facilitate multiple-drug solubilization. When a lipophilic drug does not possess an aldehyde or ketone group, spacer groups can be introduced that utilize an ester group and a pH-sensitive hydrazone linkage [61]. Besides anticancer agents, imaging agents also can be co-incorporated in mixed polymeric micelles in multimodel strategies that have gained interest in the realm of cancer nanotechnology [62].

7.6 Future Perspectives and Conclusions

PEG-*b*-PLA and PEG-*b*-poly(L-amino acid) prodrug micelles are capable of multiple-drug solubilization and multiple-drug delivery of poorly water-soluble anticancer agents. PEG-*b*-PLA micelles physically entrain two to three anticancer agents—paclitaxel, docetaxel, etoposide, and 17-AAG—without the requirement of Cremophor EL, ethanol, or DMSO. Remarkably, the dimensions of PEG-*b*-PLA micelles do not change after multiple-drug solubilization, even though the level of PEG-*b*-PLA remained unchanged. A greater physical understanding of multiple-drug solubilization via PEG-*b*-PLA micelles will undoubtedly provide an impetus toward additional examples of poorly water-soluble drug "cocktails," which might exert synergistic anticancer activity. The attainment of synergistic anticancer activity in a safe and simple drug delivery strategy is satisfying, given the current state of drug delivery via the IV route for poorly water-soluble anticancer agents, e.g., paclitaxel. In this mode of concurrent drug delivery, it will be important to show a lack of PK interactions, e.g., drug metabolism by CYP-450, which may cause unexpected changes in PK and unforeseen toxicity. For drug solubilization in cancer drug development, PEG-*b*-PLA micelles or related ABCs will likely replace Cremophor EL, probably the most commonly used solubilizer despite its known weaknesses. PEG-*b*-poly(L-amino acid) prodrugs assemble into interesting mixed polymeric micelles for multiple-drug delivery. In contrast to PEG-*b*-PLA micelles, there is strong evidence for prolonged circulation in blood and for tumor targeting via the EPR effect for PEG-*b*-poly(aspartate-hyd-DOX) micelles. It is likely that mixed polymeric micelles based upon similar chemistries will enable multiple-drug targeting of solid tumors and triggered drug release due to a drop in intracellular pH, releasing two or more anticancer agents simultaneously in cancer cells for synergistic activity. Both PEG-*b*-PLA and PEG-*b*-poly(L-amino acid) prodrug micelles require additional testing in animal models, but it is expected that they will enter clinical trials for multiple-drug delivery without major safety issues due to their proven track record of safety in humans in single-agent studies.

References

1. Kataoka K, Kwon GS, Yokoyama M, Okano T, Sakurai Y (1993) Block copolymer micelles as vehicles for drug delivery. J Control Release 24:119–132
2. Jones M-C, Leroux J-C (1999) Polymeric micelles—A new generation of colloidal drug carriers. Eur J Pharm Biopharm 48:101–111
3. Lavasanifar A, Samuel J, Kwon GS (2002) Poly(ethylene oxide)-*block*-poly(L-amino acid) micelles for drug delivery. Adv Drug Deliv Rev 54:169–190
4. Gaucher G, Marchessault RH, Leroux JC (2010) Polyester-based micelles and nanoparticles for the parenteral delivery of taxanes. J Control Release 143:2–12
5. Kabanov AV, Batrakova EV, Alakhov VY (2002) Pluronic® block copolymers for overcoming drug resistance in cancer. Adv Drug Deliv Rev 54:759–779

6. Shin H-C, Alani AWG, Rao DA, Rockich NC, Kwon GS (2009) Multi-drug loaded polymeric micelles for simultaneous delivery of poorly soluble anticancer drugs. J Control Release 140:294–300
7. Bae Y, Diezi TA, Zhao A, Kwon GS (2007) Mixed polymeric micelles for combination cancer chemotherapy through the concurrent delivery of multiple chemotherapeutic agents. J Control Release 122:324–330
8. Jonkman-de Vries JD, Flora KP, Bult A, Beijnen JH (1996) Pharmaceutical development of (investigational) anticancer agents for parenteral use—a review. Drug Dev Ind Pharm 22:475–494
9. Mayer LD, Janoff AS (2007) Optimizing combination chemotherapy by controlling drug ratios. Mol Interv 7:216–223
10. Greco F, Vicent MJ (2009) Combination therapy: Opportunities and challenges for polymer-drug conjugates as anticancer nanomedicines. Adv Drug Deliv Rev 61:1203–1213
11. Mayer LD, Harasym TO, Tardi PG, Harasym NL, Shew CR, Johnstone SA, Ramsay EC, Bally MB, Janoff AS (2006) Ratiometric dosing of anticancer drug combinations: Controlling drug ratios after systemic administration regulates therapeutic activity in tumor-bearing mice. Mol Cancer Res 5:1854–1863
12. Hennenfent KL, Govindan R (2006) Novel formulations of taxanes: A review. Old wine in a new bottle? Ann Oncol 17:735–749
13. Jordan MA, Wilson L (2004) Microtubules as a target for anticancer drugs. Nat Rev Cancer 4:253–265
14. Gelderblom H, Verweij J, Nooter K, Sparreboom A, Cremophor EL (2001) The drawbacks and advantages of vehicle selection for drug formulation. Eur J Cancer 37:1590–1598
15. ten Tije AJ, Verweij J, Loos WJ, Sparreboom A (2003) Pharmacological effects of formulation vehicles implications for cancer therapy. Clin Pharmacokinet 42:665–685
16. Stenger M (2005) Abraxane (nanoparticle albumin-bound paclitaxel) in metastatic breast cancer. Commun Oncol 2:214–215
17. Zhang X, Jackson JK, Burt HM (1996) Development of amphiphilic diblock copolymers as micellar carriers of Taxol. Int J Pharm 132:195–206
18. Zhang X, Burt HM, Mangold G, Dexter D, Von Hoff D, Mayer L, Hunter WL (1997) Anti-tumor efficacy and biodistribution of intravenous polymeric micellar paclitaxel. Anti-Cancer Drugs 8:696–701
19. Zhang X, Burt HM, Von Hoff D, Dexter D, Mangold G, Degen D, Oktaba AM, Hunter WL (1997) An investigation of the antitumor activity and biodistribution of polymeric micellar paclitaxel. Cancer Chemother Pharmacol 40:81–86
20. Kim SC, Kim DW, Shim YH, Bang JS, Oh HS, Kim SW, Seo MH (2001) In vivo evaluation of polymeric micellar paclitaxel formulation: Toxicity and efficacy. J Control Release 72:191–202
21. Sparreboom A, van Tellington O, Nooijen WJ, Beijen JH (1996) Tissue distribution, metabolism and excretion of paclitaxel in mice. Anti-Cancer Drugs 7:78–86
22. Woodcock DM, Linsenmeyer ME, Chojnowski G, Kriegler AB, Nink V, Webster LK, Sawyer WH (1992) Reversal of multidrug resistance by surfactants. Br J Cancer 66:62–68
23. Sparreboom A, van Tellington O, Huizing MT, Nooijen WJ, Beijen JH (1996) Determination of polyoxyethyleneglycerol triricinoleate 35 (cremophor el) in plasma by pre-column derivatization and reversed-phase high-performance liquid chromatography. J Chromatogr B 681:355–362
24. Ramaswamy M, Zhang X, Burt HM, Wasan KM (1997) Human plasma distribution of free paclitaxel and paclitaxel associated with diblock copolymers. J Pharm Sci 86:460–464
25. Chen H, Kim S, He W, Wang H, Low PS, Park K, Cheng J-X (2008) Fast release of lipophilic agents for circulating peg-pdlla micelles revealed by in vivo Förster resonance energy transfer imaging. Langmuir 24:5213–5217
26. Yamamoto Y, Nagasaki Y, Kato Y, Sugiyama Y, Kataoka K (2001) Long-circulating poly(ethylene glycol)-poly(D, L-lactide) block copolymer micelles with modulated surface charge. J Control Release 77:27–38
27. Kim T-Y, Kim D-W, Chung J-Y, Shin SG, Kim S-C, Dae SH, Kim NK, Bang Y-J (2004) Phase I and pharmacokinetic study of Genexol-pm, a cremophor-free, polymeric micelle-formulated paclitaxel, in patients with advanced malignancies. Clin Cancer Res 10:3708–3716

28. Kim D-W, Kim S-Y, Kim H-K, Kim S-W, Shin SW, Park K, Lee MY, Heo DS (2007) Multicenter phase II trial of Genexol-pm, a novel cremophor-free, polymeric micelle formulation of paclitaxel, with cisplatin in patients with advanced non-small-cell lung cancer. Ann Oncol 18:2009–2014
29. Lee KS, Chung HC, Im SA, Park YH, Kim CS, Kim S-B, Rha SY, Lee MY, Ro J (2008) Multicenter phase II trial on Genexol-pm, a cremophor-free, polymeric micelle formulation of paclitaxel, in patients with metastatic breast cancer. Breast Cancer Res Treat 108:241–250
30. Green MR, Manikhas GM, Orlov S, Afansyev B, Makhson AM, Hawkins MJ (2006) Abraxane, a novel cremophor-free, albumin-bound particle form of paclitaxel for the treatment of advanced non-small-cell lung cancer. Ann Oncol 17:1263–1268
31. Von Pawel J, Wagner H, Niederle N, Heider A, Koschel G, Hecker D, Hanske M (1996) Phase II study of paclitaxel and cisplatin in patients with non-small lung cancer. Semin Oncol 23:47–50
32. Gatzmeier U, von Pawel J, Gottfried M, ten Velde GP, Mattson K, DeMarins F, Harper P, Salvati F, Robinet G, Lucenti A, Bogaerts J, Gallent GJ (2000) Phase III comparative study of high-dose cisplatin versus a combination of paclitaxel and cisplatin in patients with advanced non-small-cell lung cancer. Clin Oncol 18:3390–3399
33. Rosell R, Gatzmeier U, Betticher DC, Keppler U, Macha HN, Pirker R, Berthet P, Breau JL, Nicholson M, Ardizzoni A, Chmaissani A, Bogaerts J, Gallant G (2002) Phase III randomized trial comparing paclitaxel/carboplatin with paclitaxel/cisplatin in patients with advanced non-small-cell lung cancer: A cooperative multinational trial. Ann Oncol 13:1539–1549
34. Ibrahim NK, Samuels B, Page R, Doval D, Patel KM, Rao SC, Nair MK, Bhar P, Desai N, Hortobagyi GN (2005) Multicenter phase II trial of Abi-007, an albumin-bound paclitaxel, in women with metastatic breast cancer. J Clin Oncol 23:6019–6026
35. Winer EP, Berry DA, Woolf S, Duggan D, Kornblith A, Harris LN, Michaelson RA, Kirshner JA, Fleming GF, Perry MC, Graham ML, Sharp SA, Keresztes R, Henderson IC, Hudis C, Muss H, Norton L (2004) Failure of higher-dose paclitaxel to improve outcome in patients with metastatic breast cancer: Cancer and leukemia group b trial 9342. J Clin Oncol 22:2061–2068
36. Paridaens R, Biganzoli L, Bruning P, Klijn JG, Gamucci T, Huston S, Coleman R, Schachter J, van Vreckem A, Sylvester R, Awada A, Wildiers J, Piccart M (2000) Paclitaxel versus doxorubicin as first-line single-agent chemotherapy for metastatic breast cancer: A European organization for research and treatment of cancer randomized study with cross-over. J Clin Oncol 18:724–733
37. Sledge GW, Neuberg D, Bernado P, Ingle JN, Martino S, Rowinsky EK, Wood WC (2003) Phase III trial of doxorubicin, paclitaxel, and the combination of doxorubicin and paclitaxel as front-line chemotherapy for metastatic breast cancer: An intergroup trial (e1193). J Clin Oncol 21:588–592
38. Bishop JF, Dewar J, Toner GC, Smith J, Tattersall MH, Olver IN, Ackland S, Stephenson J, Canetta R (1999) Initial paclitaxel improves outcome compared with cmfp combination chemotherapy as front-line therapy in untreated metastatic breast cancer. J Clin Oncol 17:2355–2364
39. Dancey JE, Chen HX (2006) Strategies for optimizing combinations of molecularly targeted anticancer agents. Nat Rev Drug Discov 5:649–659
40. Grant S (2008) Co-targeting survival signaling pathways in cancer. J Clin Invest 118:3003–3006
41. Nguyen DM, Chen A, Mixon A, Schrump DS (1999) Sequence-dependent enhancement of paclitaxel toxicity in non-small cell lung cancer by 17-allylamino-17-demethoxygeldanamycin. J Thorac Cardiovasc Surg 118:908–915
42. Blanco E, Bey EA, Khemtong C, Yang S-G, Setti-Guthi J, Chen H, Kessinger CW, Carnvale KA, Bommann WG, Bothman DA, Gao J (2010) β-Lapachone micellar nanotherapeutics for non-small cell lung cancer therapy. Cancer Res 70:3896–3904
43. Kim S-C, Chang E-O, Song I-S, Pai C-M, US Patent 6,322,805, 2001
44. Xiong MP, Yáñez JA, Kwon GS, Davies NM, Forrest ML (2009) A cremophor-free formulation of tanespimycin (17-AAG) using PEO-*b*-PDLLA micelles: Characterization and pharmacokinetics in rats. J Pharm Sci 98:1577–1586
45. Aliabadi HM, Lavasanifar A (2006) Polymeric micelles for drug delivery. Expert Opin Drug Deliv 3:139–162

46. Li P, Zhao L (2002) Cosolubilization of non-polar drugs in polysorbate 80 solutions. Int J Pharm 249:211–217
47. Darwish IA, Florence AT, Saleh AM (1989) Effects of hydrotropic agents on the solubility, precipitation, and protein binding of etoposide. J Pharm Sci 7:577–581
48. Solit DB, Chiosis G (2008) Development and applications of hsp90 inhibitors. Drug Discov Today 13:38–43
49. Ramalingam SS, Egorin MJ, Ramanathan RK, Remick SC, Sikorski RP, Lagattuta TF, Chatta GS, Friedland DM, Stoller RG, Potter DM, Ivy SP, Belani CP (2008) A phase I study of 17-allylamino-17-demethoxygeldanamycin combined with paclitaxel in patients with advanced solid malignancies. Clin Cancer Res 14:3456–3461
50. Solit DB, Basso AD, Olshen AB, Scher HI, Rosen N (2003) Inhibition of heat shock protein 90 function down-regulates akt kinase and sensitizes tumors to taxol. Cancer Res 63:2139–2144
51. Modi S, Stopeck AT, Gordon MS, Mendelson D, Solit DB, Bagatell R, Ma W, Wheler J, Rosen N, Norton L, Cropp GF, Johnson RG, Hannah AL, Hudis CA (2007) Combination of trastuzumab and tanespimycin (17-AAG), KOS-953) is safe and active in trastuzumab-refractory her-2-overexpressing breast cancer: A phase I dose-escalation study. J Clin Oncol 34:5410–5417
52. Nishiyama N, Okazaki S, Cabral H, Miyamoto M, Kato Y, Sugiyama Y, Nishio K, Matsumura Y, Kataoka K (2003) Novel cisplatin-incorporated polymeric micelles can eradicate solid tumors in mice. Cancer Res 63:8977–8983
53. Koizumi F, Kitagawa M, Negishi T, Onda T, Matsumoto S, Hamaguchi T, Matsumura Y (2006) Novel SN-38-incorporating polymeric micelles, NK012, eradicate vascular endothelial growth factor-secreting bulky tumors. Cancer Res 66:10048–10056
54. Hamaguchi T, Matsumura Y, Suzuki M, Shimizu K, Goda R, Nakamura I, Nakatomi I, Yokoyama M, Kataoka K, Kakizoe T (2005) NK105, a paclitaxel-incorporating micellar nanoparticle formulation, can extend in vivo antitumor activity and reduce neurotoxicity of paclitaxel. Br J Cancer 92:1240–1246
55. Matsumura Y, Kataoka K (2009) Preclinical and clinical studies of anticancer agent-incorporating polymer micelles. Cancer Sci 100:572–579
56. Bae Y, Kataoka K (2009) Intelligent polymeric micelles from functional poly(ethylene glycol)-poly(amino acid) block copolymers. Adv Drug Deliv Rev 61:768–784
57. Bae Y, Fukushima S, Harada A, Kataoka K (2003) Design of environment-sensitive supramolecular assemblies for intracellular drug delivery: Polymeric micelles that are responsive to intracellular pH change. Angew Chem Int Ed 42:4640–4643
58. Bae Y, Nishiyama N, Fukushima S, Koyama H, Matsumura Y, Kataoka K (2005) Preparation and biological characterization of polymeric micelle drug carriers with intracellular pH-triggered drug release property: Tumor permeability, controlled subcellular drug distribution, and enhanced in vivo antitumor efficacy. Bioconjugate Chem 16:122–130
59. Bae Y, Jang W-D, Nishiyama N, Fukushima S, Kataoka K (2005) Multifunctional polymeric micelles with folate-mediated cancer cell targeting and pH-triggered drug releasing properties for active intracellular drug delivery. Mol BioSyst 1:242–250
60. Bae Y, Nishiyama N, Kataoka K (2007) In vivo antitumor activity of the folate-conjugated pH-sensitive polymeric micelle selectively releasing adriamycin in the intracellular acidic compartments. Bioconjugate Chem 18:1131–1139
61. Alani AWG, Bae Y, Rao DA, Kwon GS (2010) Polymeric micelles for the pH-dependent controlled, continuous low dose release of paclitaxel. Biomaterials 31:1765–1772
62. Davis ME, Chen Z, Shin DM (2008) Nanoparticle therapeutics: An emerging treatment modality for cancer. Nat Rev Drug Discov 7:771–782

Part III
Targeting and Imaging

Chapter 8
Multifunctional Nanoparticles for Target-Specific Imaging and Therapy

Anita Gianella, Joanna C. Read, David P. Cormode, Zahi A. Fayad, and Willem J.M. Mulder

8.1 Introduction

We have witnessed an explosive growth in the development, preclinical application, and funding of nanoparticle research to improve the management of disease since the early 2000s [1, 2]. This was a result of important advances in nanotechnology and the field of nanoparticle-facilitated drug delivery [3] in general, and the dramatically increased utilization and development of (molecular) imaging technologies that require nanoparticle probes specifically [2]. Molecular imaging [4], which aims to noninvasively visualize processes at the cellular and molecular level, has developed to be an important preclinical diagnostic tool in the twenty-first century. Besides advances in, and the development of, new imaging modalities, the success of molecular imaging highly relies on the development of process-specific probes. Such probes may specifically target a molecular epitope of interest, may be designed or applied to make a cell type of interest "visible," or may attenuate or generate signal as a result of molecular interactions. This latter category is also referred to as activatable probes and is especially, but not exclusively, interesting in combination with optical techniques [5]. For example, activatable quantum dot nanoparticles are applied as Förster resonance energy transfer (FRET)-based biosensors [6]. Target-specific imaging can be realized by functionalization of nanoparticles with ligands that specifically bind an epitope of interest, while cellular imaging can be accomplished by labeling cells with nanoparticle contrast agents ex vivo or in vivo [7].

Simultaneously, the field of nanoparticle-based drug delivery has seen some important advances in the same period [1]. For example, polymeric nanoparticles, including nanoparticles formed from amphiphilic copolymers, were proposed as efficient and biodegradable drug delivery vehicles [3]. These classes of nanoparticles

A. Gianella • J.C. Read • D.P. Cormode • Z.A. Fayad • W.J.M. Mulder (✉)
Translational and Molecular Imaging Institute, Mount Sinai School of Medicine,
One Gustave L. Levy Place, New York, NY 10029, USA
e-mail: willem.mulder@mountsinai.org

S. Svenson and R.K. Prud'homme (eds.), *Multifunctional Nanoparticles for Drug Delivery Applications: Imaging, Targeting, and Delivery*, Nanostructure Science and Technology,
DOI 10.1007/978-1-4614-2305-8_8,

allow for controlled and sustained release of the drug and can be functionalized to increase the specificity of the treatment. Even more sophisticated nanoparticles for drug delivery include nanoparticles whose content is released as function of a physiological or biological trigger, or nanoparticles whose protecting surface can transform once accumulated at the diseased site. One of the latest advances in nanoparticle development for biomedical purposes includes multifunctionality [8]. This may apply to different features, for example, the inclusion of two or more drugs to enhance therapeutic efficacy or specificity or the integration of multiple labels to allow multimodal imaging. In the former case, the nanoparticle design can be such that the release kinetics of the different drugs are independently tuned to the different processes they intervene in [9], while in the latter case, the amount of contrast-generating material may be adjusted to match the sensitivity of the different imaging modalities [10]. Lastly and importantly, nanoparticles can be designed that contain both therapeutic molecules and contrast-generating materials [11]. These so-called theranostic nanoparticle platforms not only allow imaging-guided drug delivery but potentially also allow monitoring of drug release or therapeutic efficacy with noninvasive imaging.

This chapter will discuss some of the latest advances in multifunctional nanoparticle design and application. We will first provide a context by briefly describing the evolution of different nanoparticle categories, ranging from the first manmade class of nanoparticles, i.e., liposomes [12], to very sophisticated hybrid structures that can be employed for multimodal imaging [13], combinatory therapies, or theranostics [2]. Subsequently, we will discuss some recent and important examples of studies where these types of nanoparticle platforms were developed and applied.

8.2 The Evolution of Nanoparticle Development for Biomedical Purposes

8.2.1 Liposomes

The first artificial nanoparticles that were identified are liposomes. They were discovered in the early 1960s by Alec D. Bangham, who found that phospholipids combined with water self-organized into spherical structures because of the amphiphilic character of the lipids [14, 15]. Soon after their discovery, liposomes were suggested for use as a drug carrier vehicle because of their striking biological properties, i.e., they are composed of naturally occurring lipids or derived synthetic lipids, their ability to carry a high payload of water-soluble therapeutic agents, as well as their ability to protect drugs from interactions with plasma proteins, or drug deactivation, while simultaneously enhancing the drugs' circulation half-lives [12]. Liposomes can be defined as spherical, self-closed structures, formed by one or several concentric lipid bilayers with an aqueous phase inside and between the lipid bilayer. Liposomes can vary in size and lamellarity and are therefore subdivided into multilamellar vesicles, large unilamellar vesicles, and small unilamellar vesicles. The latter type, most commonly used for biomedical applications, can be synthesized in a size range of 50–150 nm. Altering their surface properties allows for improved pharmacokinetics and makes specific delivery of liposomes to diseased tissue and into cells possible. These properties also make liposomes excellent candidates to carry or deliver contrast agents for magnetic resonance imaging (MRI), and in the 1980s, the first studies about the use of liposomes as a carrier of MR contrast agents appeared in the literature [16]. Currently, liposomes also serve as contrast agents for optical imaging, nuclear imaging, ultrasound imaging, and computed tomography (CT) [12].

8.2.2 Lipidic Nanoparticles (Micelles and Emulsions)

Besides liposomes, a variety of different other lipid-based nanoparticle platforms have been employed as therapeutic and/or diagnostic agents over the years. Frequently employed platforms include micelles and oil-in-water emulsions. Micelles are composed of amphiphilic molecules that form an aggregate in which the hydrophilic part of these molecules is faced towards the water and the hydrophobic part forms the core. They can be used as carriers for hydrophobic and lipophilic agents. Oil-in-water emulsions, also referred to as microemulsions, are mixtures of water, oil, and an amphiphile (or surfactant), where the amphiphiles serve to stabilize tiny oil droplets in the aqueous environment. The hydrophobic tail groups of the surfactant are embedded in the oil, and the charged headgroups of the surfactant face the water. Microemulsions can be used as carrier vehicles for delivery of hydrophobic drugs in their oil core, whereas lipophilic agents and molecules can be included in the lipid monolayer that can also be used to incorporate target-specific molecules.

8.2.3 Polymeric Nanoparticles

The development of polymeric nanoparticles as long circulating drug carriers has also witnessed a fast expansion after they were first introduced in 1994 [17]. Polymeric nanoparticles exhibit several advantages, including their ability to carry high payloads of water-insoluble drugs and to control release kinetics as well as their biodegradability [1]. Popular synthetic polymers in nanoparticle formulations include poly(amides), poly(amino acids), poly(alkyl-α-cyanoacrylates), poly(esters), poly(orthoesters), poly(urethanes), and poly(acrylamides). Among them, the thermoplastic aliphatic poly(esters) such as poly(lactic acid) (PLA), poly(glycolic acid) (PGA), and, in particular, poly(lactic-*co*-glycolic acid) (PLGA) have been investigated most abundantly due to their excellent biocompatibility and biodegradability.

8.2.4 Inorganic Nanoparticles (Iron Oxide, Quantum Dots, Gold, and Silica)

Another class of important nanoparticles that have seen a fast growth in use in the biomedical field over the past decade comprises inorganic nanocrystals. Some of these have very exciting properties as contrast-generating materials for MRI, CT, and optical imaging. Iron oxide nanoparticles [7] are widely used as contrast in MR imaging but also have other applications in biomedicine, including thermal therapy and drug delivery. They are usually synthesized via a coprecipitation of Fe^{2+} and Fe^{3+} under basic conditions to form superparamagnetic Fe_3O_4 (magnetite). The conditions of synthesis are usually tuned to produce particles that consist of a single domain, are superparamagnetic (i.e., they are magnetized only when an external field is applied), and are smaller than 15 nm. To allow their application to cells and animals, a variety of biocompatible materials, including poly(ethylene glycol) (PEG), lipids, or dextran, have been used as coatings [2]. Iron oxide nanoparticles produce negative contrast in MRI (darkening of the image) due to a reduction in T2* (gradient echo, spin–spin relaxation time), caused by local inhomogeneities in the magnetic field that the particles create. Feridex®, coated with dextran, and Resovist®, coated with carboxydextran, are iron oxide-based agents that have been licensed for clinical use. Both are large aggregates of the coating into which multiple iron cores are embedded. In order to make iron oxides specific, investigators have developed amine-functionalized dextran coatings to allow attachment of targeting ligands and/or other labels, such as fluorophores. Targets such as VCAM-1, E-selectin, and phosphatidylserine-expressing apoptotic cells have been successfully imaged in vivo using this platform [18].

Semiconductor nanocrystals (quantum dots, QDs) have unique optical properties, which make them ideally suited for a number of applications in biomedical imaging, with several important advantages over fluorescent dye molecules [19]. In contrast to fluorescent dyes, the absorption spectrum of a QD is characterized by a very broad band, since any photon with energy equal to or higher than the bandgap is absorbed, while the emission spectrum is rather narrow. The optical properties of QDs can be tuned by a judicious control of composition and size, reaching emission wavelengths spanning from the near-UV to near-infrared (NIR), making QDs particularly suitable for multiplexed imaging. For the visible range CdSe, CdTe, or InP can be used, whereas the NIR can be easily covered by PbSe or PbS nanocrystals. QDs are more stable and brighter than fluorescent dyes. Importantly, the surface of QDs can be easily modified so that new functionalities and properties can be conferred. QDs are potentially cytotoxic when they interact with the cellular environment and are prone to photochemical degradation, albeit to a lesser extent than dyes. These shortcomings can be (partially) overcome by the use of core–shell QDs, suitable coatings, or the use of other materials. High-quality QDs are usually prepared by a chemical synthesis that starts with injecting suitable precursors into a high-boiling point, coordinating solvent. This injection initiates nucleation and subsequent growth of the nanocrystals. Coating these just formed QDs with a shell of a wider bandgap semiconductor yields the so-called core–shell QDs that exhibit a much better stability and lower cytotoxicity.

In recent years, there has been strong interest in the biomedical application of gold nanoparticles [2] due to their unique properties for optical imaging, thermal ablation, and their application as CT contrast agents. Gold nanoparticles may be synthesized with excellent control of core size, and many different biologically compatible coatings such as poly(vinylpyrrolidone) (PVP) or PEG have been applied. The properties of gold nanoparticles used for medical imaging purposes are the strong X-ray attenuation of this element and the above-mentioned array of optical properties, but they can also be used as a scaffold for other contrast-inducing species such as gadolinium. For example, Qian and coworkers were able to detect tumors located in the flank of mice by taking advantage of the surface-enhanced Raman scattering of systemically injected targeted gold nanoparticles [20]. Recently, Cormode and colleagues have shown multicolor CT imaging in mice, using gold nanoparticles and a spectral CT scanner, as we will expound upon later [21].

Silica nanoparticles can be synthesized in a wide range of desired sizes (50–1,000 nm) and have been investigated extensively as model systems to study fundamental colloidal phenomena. Apart from the size and composition, the shape and surface properties of silica can be precisely controlled. In the past decade, silica-based nanoparticles have increasingly been exploited for biomedical applications, including drug and gene delivery, as well as a carrier vehicle for different contrast-generating materials. Interestingly, the ease of incorporation of different chemicals in silica makes it an excellent material for the integration of multiple diagnostically active materials to enable multimodality biomedical imaging [13]. Silica can be synthesized in the presence of cetyltrimethylammonium bromide, which results in mesoporous (honeycombed) particles. These pores can be loaded with therapeutic agents and/or contrast-generating materials. Alternatively, silica can be grown on other nanoparticles, resulting in a silica shell of tunable thickness [22]. This has been accomplished with a variety of nanocrystals, including gold, iron oxide, and quantum dots [23]. The silica shell protects the diagnostically active cores from degradation and allows the size of the particle to be easily increased to the desired size.

8.2.5 *Natural Nanoparticles (Lipoproteins and Viruses)*

In addition to synthetic approaches to create nanoparticles, nature also offers its own nanoparticles. Lipoproteins [24], self-assembled nanoparticles of lipids and apolipoproteins, transport fats throughout the body and are involved in several pathological processes. In cardiovascular disease, for example, high-density lipoprotein (HDL), popularly known as "good cholesterol," is known to have a protective role, while low-density lipoprotein (LDL, "bad cholesterol") is causative of the formation of atherosclerotic plaques. Both HDL and LDL are subject of studies for targeted therapy and/or imaging [24].

Viruses are another example of natural nanoparticles that can be exploited for targeted therapy, i.e., gene transfection, or may be chemically modified to serve as drug or contrast agent vehicles [25, 26]. For diagnostic purposes, the iron-transporting natural nanoparticle ferritin has received considerable interest, especially in relation to MRI [27, 28]. This nanoparticle, composed of protein 24 subunits, contains an approximately 10 nm hydrated iron oxide core that includes thousands of iron atoms and therefore can serve as an MRI reporter gene [28], a natural MRI contrast agent, or can be reconstituted to contain MRI contrast agents such as Gd-DTPA chelates [27].

8.3 Architectural Features of Multifunctional Hybrid Nanoparticles

One of the most interesting features of nanoparticles for their use in biomedicine is the possibility to integrate and combine different materials to build multifunctional hybrid nanostructures [8]. Although this field is young, the number of such hybrid nanoparticles available today is considerable and rapidly growing. A generalized schematic is depicted in Fig. 8.1. It has to be stressed that different variations are possible, and numerous examples that do not resemble this schematic have been developed.

Hybrid nanoparticles may consist of an inorganic core material, for example, silica, iron oxide, or gold, which is subsequently coated with organic materials of synthetic origin, for example, block copolymers or natural materials, for example, lipids [30]. Combinations of hydrophobic oil cores carrying nanocrystalline materials that are stabilized by amphiphilic surfactants are also possible [31]. Combinations of

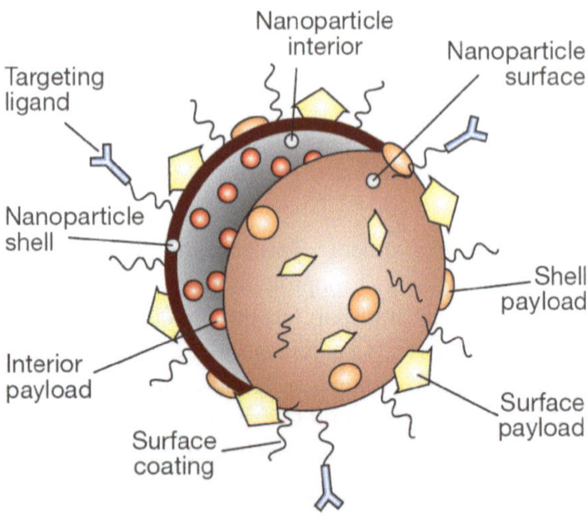

Fig. 8.1 Schematic depicting the typical features and buildup of multimodal nanoparticles. Reproduced with permission from [29]

synthetic and natural molecules and/or nanoparticles are under investigation as well [25], for example, viruses [26] or lipoproteins [32] that are chemically modified for drug delivery or diagnostic purposes. Polymer/lipid hybrid nanoparticles represent a class that combines advantages of lipidic platforms with those of polymeric platforms, including recently introduced PLGA-lecithin-PEG core–shell nanoparticles [33, 34]. Generally, all the abovementioned nanoparticle platforms consist of a core that can contain a payload that is covered by a shell. The latter may have distinctively different properties from the core and therefore may be used to incorporate certain molecules that cannot be integrated in the nanoparticle core. Notably, the surface of most nanoparticles can be designed to be readily modified, which is a valuable feature that can be exploited not only to include diagnostic and therapeutic molecules but also to alter the surface to increase biocompatibility and bioapplicability. Such coatings include hydrophilic polymers like PEG and allow the conjugation of ligands, including antibodies, antibody fragments, sugars, peptides, peptidomimetics, and small molecules [2].

8.4 Examples of Multifunctional Nanoparticle Platforms

8.4.1 Oil-in-Water Emulsions for Combined Diagnosis and Therapy

Nanoparticle platforms for simultaneous drug delivery and diagnostic purposes offer a number of unique advantages that add to our understanding of the targeting, efficacy, mode of action, and specificity of the therapy, especially in a preclinical setting. This field of diagnostic therapy is popularly referred to as “theranostics” and has the ultimate goal to improve and personalize nanotherapeutic approaches. Lanza, Wickline, and colleagues have published a number of theranostic studies in which perfluorocarbon microemulsions were used for simultaneous drug delivery and imaging [35–39]. For example, Winter et al. have shown the application of fumagillin-loaded microemulsions to inhibit atherosclerotic plaque angiogenesis in rabbit models of atherosclerosis and cancer, while using MRI as readout for therapeutic efficacy [36–39]. In a proof-of-concept study, the nanoparticles were functionalized with an angiogenesis-specific targeting ligand, and their neovascularization-reducing efficacy was extensively studied. It was shown that in a rabbit model of atherosclerosis, with abundant angiogenesis, the $\alpha v\beta 3$ integrin-targeted nanoparticles could be applied for noninvasive MR molecular imaging plaque neovascularization. In a subsequent study, a similar approach was applied, albeit in this case the nanoparticles were also loaded with fumagillin. It was shown by nanoparticle-facilitated MRI that rabbits treated with these nanoparticles displayed severe inhibition of neovascularization in their atherosclerotic plaques. In a follow-up study, it was shown that this nanoparticle-induced angiogenesis reduction could be sustained by consecutive statin therapy (Fig. 8.2).

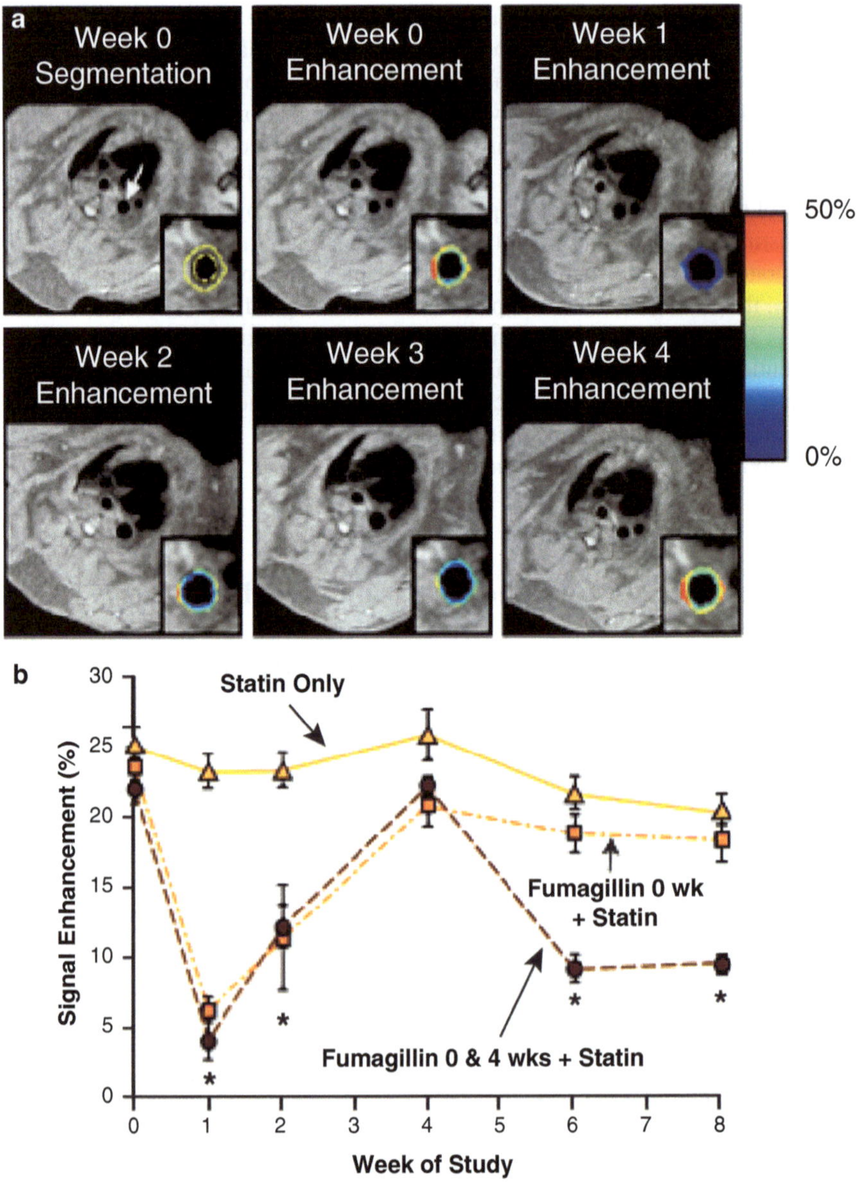

Fig. 8.2 (**a**) MR images of the aortic vessel wall (outlined in *yellow*) before and up to 4 weeks after fumagillin nanoparticle treatment. The color-coded overlay of signal enhancement (%) shows patchy areas of high angiogenesis. At week 1, the signal enhancement has clearly decreased due to the antiangiogenic effect of targeted fumagillin treatment and gradually increased to the pretreatment level. (**b**) Enhancement in rabbits receiving atorvastatin alone (*triangles*) or in conjunction with one (*squares*) or two (*circles*) doses of targeted fumagillin nanoparticles. The combination of two fumagillin doses and statin produced a sustained decrease in angiogenesis. Reproduced with permission from [36]

8.4.2 Hybrid Structures of Inorganic Nanocrystals and Lipoproteins

Recently, Cormode et al. modified HDL particles to create endogenous nanoparticle–inorganic material hybrid composites [32]. In addition to modifying the phospholipid coating to provide contrast for medical imaging, as was shown previously by Frias and colleagues [40], a method was developed to modify the hydrophobic core for the same purpose. The original hydrophobic HDL core of triglycerides and cholesteryl esters was replaced by different nanocrystals, i.e., gold nanoparticles, iron oxide nanoparticles, and quantum dots to produce a broad range of novel contrast agents for multimodality imaging. As appropriate, fluorescent and paramagnetic lipids were included in the phospholipid corona of the particles to render all the nanocrystal HDL formulations at least MRI and fluorescence active. Negative stain TEM, protein analysis, cholesterol efflux, and lipid-exchange studies proved the nanoparticles to be very similar to native HDL. In vivo, it was observed that these nanoparticles trafficked throughout the body as individual entities. In vitro experiments revealed that these particles were abundantly taken up by macrophages, as evidenced by confocal laser scanning microscopy, fluorescence imaging, cell pellet MR, and CT imaging, as well as TEM. In vivo experiments using the imageable HDL nanoparticles were performed with the apolipoprotein E knockout (apoE-KO) mouse model of atherosclerosis. It was observed that the different versions of the nanocrystal core HDL particles trafficked to and accumulated in atherosclerotic plaques. These observations were corroborated by confocal microscopy of aortic sections as well as computed tomography and fluorescence images of intact aorta specimen.

In a recently published study, Cormode and colleagues further explored the potential of this nanoparticle platform [21]. As mentioned above, the gold core version (Au-HDL, Fig. 8.3a) has excellent properties to serve as a contrast agent for CT. In the study, atherosclerotic mice were intravenously injected with Au-HDL to visualize atherosclerotic plaques and with an iodine contrast agent to visualize the vasculature. Subsequently, the animals were imaged by conventional CT (Fig. 8.3b) and spectral CT (Fig. 8.3c), a novel CT approach that allows multicolor imaging of different elements. This unique approach showed, for the first time, that it is possible to apply spectral CT to specifically discern different atherosclerotic plaque components.

In addition to in vivo imaging, nanocrystal core HDL also exhibits unique properties for live cell imaging. Skajaa et al. developed QD-HDL that was additionally labeled with the near-infrared dye Cy5.5, which allowed them to investigate important biological features of lipoproteins, including lipid-exchange dynamics, cellular interactions, and nanoparticle disassembly via Förster resonance energy transfer (FRET) [41].

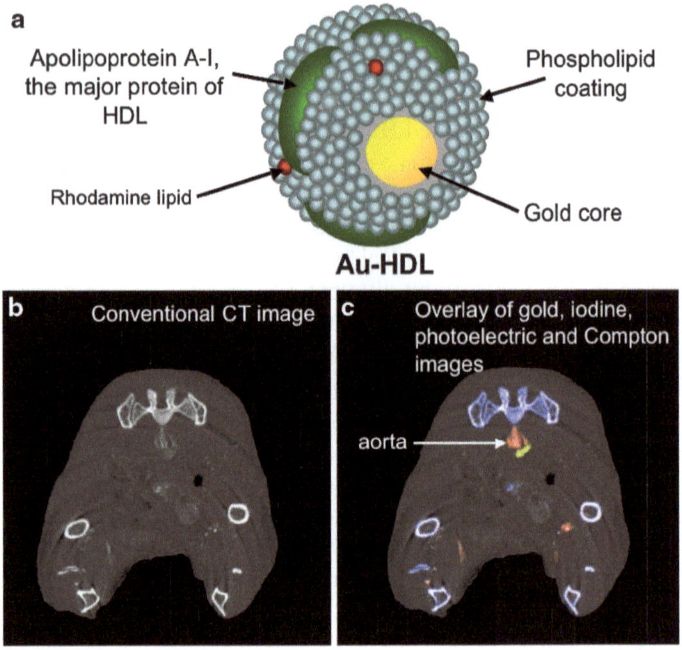

Fig. 8.3 (**a**) Schematic representation of a gold core high-density lipoprotein nanoparticle (Au-HDL). (**b**) A conventional CT image of the abdomen of an atherosclerotic apoE-KO mouse after the administration of an iodine-based vascular agent and the plaque-specific agent, Au-HDL. (**c**) The same image as in (**b**) acquired by spectral CT visualizes the different elements in color codes and allows the differentiation of the bones (*blue*), the aorta (*red*), and the atherosclerotic plaque (*yellow*). Reproduced with permission from [21]

8.4.3 siRNA Delivery and Imaging with Iron Oxide Nanoparticles

A variety of difficulties have limited the application of nucleic acids as a therapeutic strategy. Among those difficulties are degradation of the nucleotides, poor bioavailability, and biodistribution upon intravenous administration, as well as problems that are associated with diminished activity in the cells of interest due to compartmentalization. Nanoparticulate formulations may have significant advantages to deal with these limitations and therefore have been applied extensively in this field of research. Most recently, studies have appeared where the nucleotide delivery vehicle was additionally labeled for visualization with diagnostic imaging techniques.

Medarova et al. reported a study where they developed and employed a dextran-coated iron oxide nanoparticle that had a near-infrared fluorescent dye, a translocation peptide, and a small interfering RNA specific for green fluorescent protein (GFP) mRNA covalently conjugated to the nanoparticle (Fig. 8.4a) [42]. This design allows

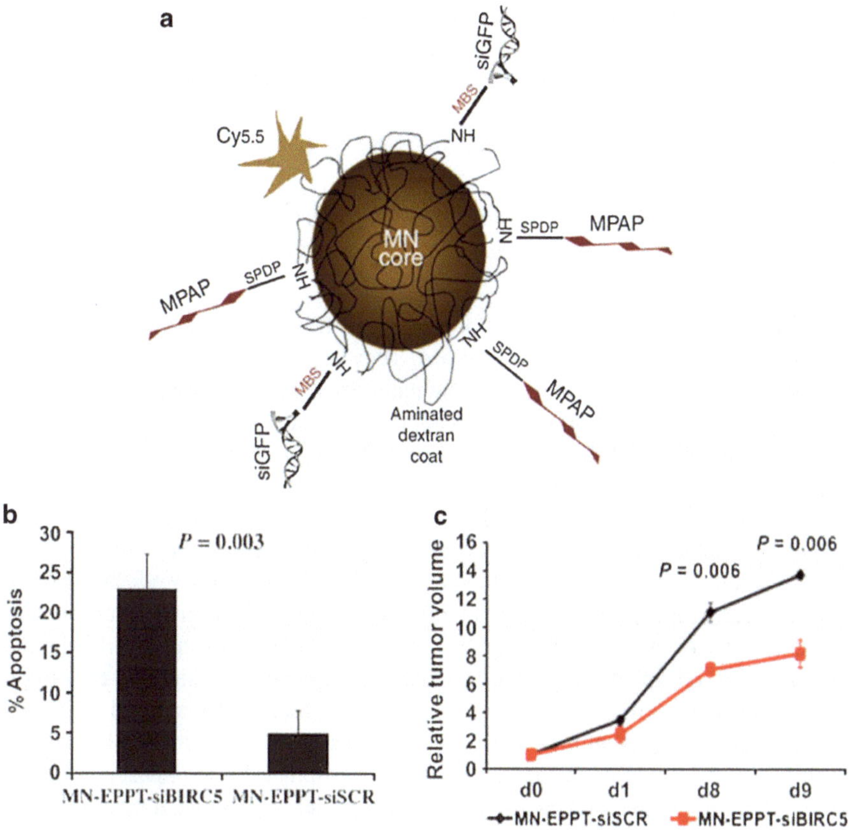

Fig. 8.4 (**a**) Schematic depiction of a multifunctional iron oxide-based MR nanoparticle that is labeled with the near-infrared dye Cy5.5 and functionalized with siRNA specific to GFP as well as with cell penetrating peptides. Reproduced with permission from [42]. (**b**) A nanoparticle similar to the one in (**a**) that was equipped with the antiapoptotic gene BIRC5 induced tumor apoptosis and (**c**) significantly inhibited tumor growth. Reproduced with permission from [43]

the nanoparticle to be visualized with MRI and near-infrared fluorescent imaging to be efficiently taken up by cells and to subsequently silence GFP expression. The authors showed, with MRI and NIRF imaging, that upon intravenous administration, this nanoparticle accumulated in the tumors of mice that had two tumor types inoculated on the flank, of which one tumor expressed red fluorescent protein and one tumor expressed green fluorescent protein. Using in vivo optical imaging, it was shown that GFP expression was selectively silenced 48 h after probe administration, while RFP expression remained unaffected. This study convincingly demonstrated how the combination of nanotechnology, imaging, and genetics generates valuable insights and allows investigators to visualize nanoparticle delivery and subsequent silencing in live animals. In a subsequent study from the same authors, it was demonstrated that upon intravenous injection of a similar nanoparticle, equipped with siRNA that targets the tumor-specific antiapoptotic gene BIRC5, tumor necrosis and apoptosis could be induced (Fig. 8.4b), which resulted in a significant decrease in tumor growth rate (Fig. 8.4c) [43].

8.4.4 *Multimodal Liposomes for Imaging-Guided Treatment of Atherosclerotic Plaque Inflammation*

As stated earlier in this chapter, liposomes can be considered the first artificial nanoparticles for biomedical applications. They are the most widely investigated, developed, and clinically explored nanoparticles [12]. Interestingly, and despite their mature developmental state, novel applications and modifications of liposomes are still being explored. In the early 1990s, PEGylation of liposomal nanoparticles induced a tremendous boost in their applicability, while in the past 10 years the field of (molecular) imaging has caused a renewed interest in the application of liposomes. Metselaar and colleagues developed prednisolone phosphate-loaded and long circulating liposomes for the treatment of inflammatory diseases. Strong therapeutic effects were observed in animal models of rheumatoid arthritis, multiple sclerosis, and cancer [44–46]. The approach of combining an FDA-approved drug, like prednisolone phosphate, and long circulating liposomes to amplify pleiotropic effects and enhance therapeutic effect is simple but effective. Lobatto and colleagues reported another application of the above anti-inflammatory treatment. They applied multifunctional, multimodal, and prednisolone phosphate-loaded liposomes to treat atherosclerotic plaque inflammation in a rabbit model [47]. Since atherosclerosis is a systemic disease of the arteries, the evaluation of valid endpoints is a challenge, especially in a longitudinal study design. A methodology that was comprised of a two-pronged approach was developed (Fig. 8.5). First, long circulating liposomes were applied to deliver glucocorticoids (prednisolone phosphate) to atherosclerotic plaques via the EPR effect, thereby increasing their efficacy and lowering the dose. These liposomes were also labeled with Gd-DTPA and rhodamine-functionalized lipids to enable their visualization with MRI and optical techniques. Second, noninvasive multimodality clinical imaging methods, i.e., MRI, PET, and CT were employed to monitor drug delivery and to evaluate the response after treatment. To corroborate the observations, endpoints determined in vivo were also established histologically *ex vivo*. MRI, fluorescence imaging, and microscopy revealed the accumulation of liposomes in the atherosclerotic lesions, mostly colocalized with macrophages. FDG-PET, applied to evaluate vessel wall inflammation, and DCE-MRI revealed an unprecedented therapeutic effect lasting almost 2 weeks after a single dose of the multifunctional liposomes. In addition to the therapeutic efficacy of prednisolone phosphate liposomes for atherosclerosis, this study also showed the validity of using imaging endpoints.

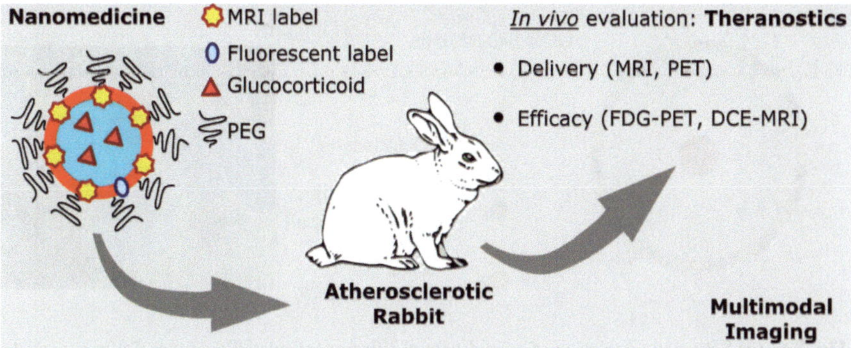

Fig. 8.5 Overview of a nanotheranostics study in atherosclerotic rabbits. Liposomal nanomedicine carrying glucocorticoids in its aqueous lumen and paramagnetic and fluorescent amphiphiles in its lipid bilayer (*left*). The nanotherapy was evaluated in a rabbit model of atherosclerosis (*middle*). Targeting and therapeutic efficacy was evaluated in a longitudinal fashion with multimodal imaging (*right*). Reproduced with permission from [47]

8.4.5 Lipid-Coated Silica Quantum Dot Hybrid Structures for Multimodal Imaging

Despite the promises of silica for biomedical applications, a serious drawback of these inorganic nanoparticles is their inherently low biocompatibility. To address this issue, Koole et al. recently developed a novel method to obtain hydrophobic silica nanoparticles, coated with a physically adsorbed monolayer of PEGylated phospholipids [30]. This highly flexible coating method allows, next to the inclusion of PEGylated lipids, the incorporation of many other lipid species, for example, paramagnetic lipids for MRI and functional lipids to achieve target specificity. The cytotoxicity and pharmacokinetics of these paramagnetic and PEG lipid-coated, quantum dot-containing silica nanoparticles were studied by van Schooneveld et al. [48]. A variety of imaging techniques were employed, and the results were compared with those obtained with bare silica nanoparticles. Compared to bare silica, lipid-coated silica nanoparticles exhibited a prolonged circulation half-life, while analysis of tissue sections of mice sacrificed 24 h after the administration revealed that both particles accumulated in the liver and spleen. Interestingly, the bare silica particles were also found to accumulate in the lungs of the animals, most likely caused by the aggregation of the bare silica particles upon intravenous administration, a phenomenon that was not observed for the lipid-coated nanoparticles. The enhanced biocompatibility of such lipid-coated silica nanoparticles was independently confirmed in a recent study by Hu and Gao [49]. Because of their complex and heterogeneous morphology, these organic–inorganic hybrid nanoparticles are of great interest to be studied with advanced electron microscopy techniques. A detailed study of the different elements in this material was done with electron microscopy combined with electron energy-loss spectroscopy (EELS). For the first time, it was reported that detailed quantitative information about the composition of a complex nanoparticle can be extracted (Fig. 8.6) [50].

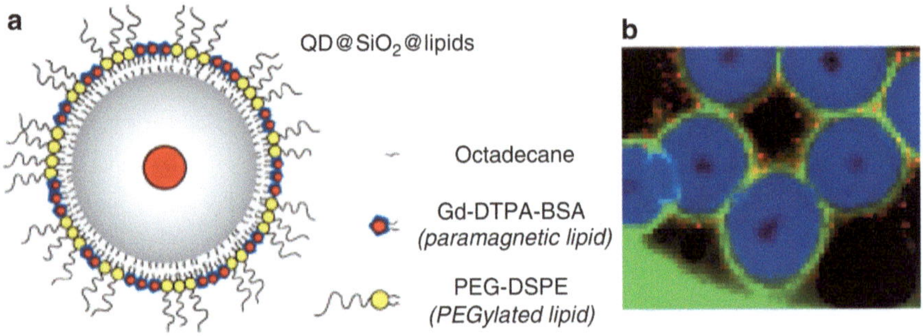

Fig. 8.6 (**a**) Schematic depiction of a lipid-coated silica-encapsulated quantum dot nanoparticle. The lipid layer is comprised of Gd-containing, paramagnetic amphiphiles and PEGylated phospholipids. (**b**) High-angle annular dark-field electron microscopy of the nanoparticle in (**a**). *Red*=Gd; *Blue*=Si; *Green*=C. Reproduced with permission from [50]

8.5 Perspective and Conclusions

In this chapter, we first outlined the nanoparticle developments that led to the current state-of-the-art, multifunctional nanoparticle platforms. Interestingly, the more traditional nanoparticle platforms are presently still under development, while new types of nanoparticles are being developed at an unprecedented pace. So-called multifunctional hybrid nanoparticles currently are a popular topic of investigation, and many variations are possible, for example, nanoparticles that are composed of natural and synthetic components or nanoparticles that have an inorganic or polymeric core that is stabilized and functionalized by phospholipids. We reviewed some interesting nanoparticle platforms, some of them developed in our own laboratory, but it has to be stressed that numerous other examples have appeared in the literature in recent years. We primarily focused our selection on nanoparticle platforms that have also been extensively evaluated in animal models, are utilized in a number of studies, and were independently evaluated by other researchers.

We have shown the unique advantages multifunctional nanoparticles offer for diseases such as atherosclerosis and cancer. Because of their multifunctional character, they may be applied to inhibit or even regress disease as well as to function as diagnostic agents or even as readouts for therapeutic efficacy. In addition, the multifunctionality also allows improved biocompatibility, biodegradability, and, importantly, specificity.

Although important developments have been made in the past decade, the clinical translation of nanoparticles in general and multifunctional nanoparticles specifically is still hampered by regulatory and financial limitations. The complexity of nanoparticle-based agents inhibits the approval for human use, while at the same time the costs associated with such products may not outweigh the benefits. Nevertheless, many aspects remain to be investigated in preclinical studies and will further contribute to the maturation of this exciting and fast growing field.

References

1. Farokhzad OC, Langer R (2009) Impact of nanotechnology on drug delivery. ACS Nano 3(1):16–20
2. Cormode DP, Skajaa T, Fayad ZA, Mulder WJ (2009) Nanotechnology in medical imaging: probe design and applications. Arterioscler Thromb Vasc Biol 29(7):992–1000
3. Shi J, Votruba AR, Farokhzad OC, Langer R (2010) Nanotechnology in drug delivery and tissue engineering: from discovery to applications. Nano Lett 10(9):3223–3230
4. Weissleder R, Mahmood U (2001) Molecular imaging. Radiology 219(2):316–333
5. Mulder WJ, Griffioen AW, Strijkers GJ, Cormode DP, Nicolay K, Fayad ZA (2007) Magnetic and fluorescent nanoparticles for multimodality imaging. Nanomedicine 2(3):307–324
6. Medintz IL, Uyeda HT, Goldman ER, Mattoussi H (2005) Quantum dot bioconjugates for imaging, labelling and sensing. Nat Mater 4(6):435–446
7. Bulte JW, Kraitchman DL (2004) Iron oxide MR contrast agents for molecular and cellular imaging. NMR Biomed 17(7):484–499
8. Jarzyna PA, Gianella A, Skajaa T, Knudsen G, Deddens LH, Cormode DP et al (2010) Multifunctional imaging nanoprobes. Wiley Interdiscip Rev Nanomed Nanobiotechnol 2:138–150
9. Sengupta S, Eavarone D, Capila I, Zhao G, Watson N, Kiziltepe T et al (2005) Temporal targeting of tumour cells and neovasculature with a nanoscale delivery system. Nature 436:568–572
10. Mulder WJ, Koole R, Brandwijk RJ, Storm G, Chin PT, Strijkers GJ et al (2006) Quantum dots with a paramagnetic coating as a bimodal molecular imaging probe. Nano Lett 6(1):1–6
11. Lanza GM, Winter P, Caruthers S, Schmeider A, Crowder K, Morawski A et al (2004) Novel paramagnetic contrast agents for molecular imaging and targeted drug delivery. Curr Pharm Biotechnol 5(6):495–507
12. Torchilin VP (2005) Recent advances with liposomes as pharmaceutical carriers. Nat Rev Drug Discov 4(2):145–160
13. Mulder WJ, Strijkers GJ, van Tilborg GA, Cormode DP, Fayad ZA, Nicolay K (2009) Nanoparticulate assemblies of amphiphiles and diagnostically active materials for multimodality imaging. Acc Chem Res 42(7):904–914
14. Bangham AD, Standish MM, Watkins JC (1965) Diffusion of univalent ions across the lamellae of swollen phospholipids. J Mol Biol 13(1):238–252
15. Bangham AD, Standish MM, Watkins JC, Weissmann G (1967) The diffusion of ions from a phospholipid model membrane system. Protoplasma 63(1):183–187
16. Mulder WJ, Strijkers GJ, van Tilborg GA, Griffioen AW, Nicolay K (2006) Lipid-based nanoparticles for contrast-enhanced MRI and molecular imaging. NMR Biomed 19(1):142–164
17. Gref R, Minamitake Y, Peracchia MT, Trubetskoy V, Torchilin V, Langer R (1994) Biodegradable long-circulating polymeric nanospheres. Science 263:1600–1603
18. McCarthy JR, Kelly KA, Sun EY, Weissleder R (2007) Targeted delivery of multifunctional magnetic nanoparticles. Nanomedicine 2(2):153–167
19. Michalet X, Pinaud FF, Bentolila LA, Tsay JM, Doose S, Li JJ et al (2005) Quantum dots for live cells, in vivo imaging, and diagnostics. Science 307:538–544
20. Qian X, Peng XH, Ansari DO, Yin-Goen Q, Chen GZ, Shin DM et al (2008) In vivo tumor targeting and spectroscopic detection with surface-enhanced Raman nanoparticle tags. Nat Biotechnol 26(1):83–90
21. Cormode DP, Roessl E, Thran A, Skajaa T, Gordon RE, Schlomka JP et al (2010) Atherosclerotic plaque composition: analysis with multicolor CT and targeted gold nanoparticles. Radiology 256(3):774–782
22. van Schooneveld MM, Cormode DP, Koole R, van Wijngaarden JT, Calcagno C, Skajaa T et al (2010) A fluorescent, paramagnetic and PEGylated gold/silica nanoparticle for MRI, CT and fluorescence imaging. Contrast Media Mol Imaging 5(4):231–236
23. Koole R, Mulder WJ, van Schooneveld MM, Strijkers GJ, Meijerink A, Nicolay K (2009) Magnetic quantum dots for multimodal imaging. Wiley Interdiscip Rev Nanomed Nanobiotechnol 1(5):475–491

24. Skajaa T, Cormode DP, Falk E, Mulder WJ, Fisher EA, Fayad ZA (2010) High-density lipoprotein-based contrast agents for multimodal imaging of atherosclerosis. Arterioscler Thromb Vasc Biol 30(2):169–176
25. Cormode DP, Jarzyna PA, Mulder WJ, Fayad ZA (2009) Modified natural nanoparticles as contrast agents for medical imaging. Adv Drug Deliv Rev 62(3):329–338
26. Huang X, Bronstein LM, Retrum J, Dufort C, Tsvetkova I, Aniagyei S et al (2007) Self-assembled virus-like particles with magnetic cores. Nano Lett 7(8):2407–2416
27. Aime S, Frullano L, Geninatti CS (2002) Compartmentalization of a gadolinium complex in the apoferritin cavity: a route to obtain high relaxivity contrast agents for magnetic resonance imaging. Angew Chem Int Ed 41(6):1017–1019
28. Genove G, DeMarco U, Xu H, Goins WF, Ahrens ET (2005) A new transgene reporter for in vivo magnetic resonance imaging. Nat Med 11(4):450–454
29. Mulder WJ, Cormode DP, Hak S, Lobatto ME, Silvera S, Fayad ZA (2008) Multimodality nanotracers for cardiovascular applications. Nat Clin Pract Cardiovasc Med 5(Suppl 2):S103–S111
30. Koole R, van Schooneveld MM, Hilhorst J, Castermans K, Cormode DP, Strijkers GJ et al (2008) Paramagnetic lipid-coated silica nanoparticles with a fluorescent quantum dot core: a new contrast agent platform for multimodality imaging. Bioconjug Chem 19(12):2471–2479
31. Jarzyna PA, Skajaa T, Gianella A, Cormode DP, Samber DD, Dickson SD et al (2009) Iron oxide core oil-in-water emulsions as a multifunctional nanoparticle platform for tumor targeting and imaging. Biomaterials 30(36):6947–6954
32. Cormode DP, Skajaa T, van Schooneveld MM, Koole R, Jarzyna P, Lobatto ME et al (2008) Nanocrystal core high-density lipoproteins: a multimodality contrast agent platform. Nano Lett 8(11):3715–3723
33. Chan JM, Zhang L, Yuet KP, Liao G, Rhee JW, Langer R et al (2009) PLGA-lecithin-PEG core-shell nanoparticles for controlled drug delivery. Biomaterials 30(8):1627–1634
34. Zhang L, Chan JM, Gu FX, Rhee JW, Wang AZ, Radovic-Moreno AF et al (2008) Self-assembled lipid–polymer hybrid nanoparticles: a robust drug delivery platform. ACS Nano 2(8):1696–1702
35. Lanza GM, Winter PM, Caruthers SD, Hughes MS, Hu G, Schmieder AH et al (2010) Theragnostics for tumor and plaque angiogenesis with perfluorocarbon nanoemulsions. Angiogenesis 13(2):189–202
36. Winter PM, Caruthers SD, Zhang H, Williams TA, Wickline SA, Lanza GM (2008) Antiangiogenic synergism of integrin-targeted fumagillin nanoparticles and atorvastatin in atherosclerosis. JACC Cardiovasc Imaging 1(5):624–634
37. Winter PM, Schmieder AH, Caruthers SD, Keene JL, Zhang H, Wickline SA et al (2008) Minute dosages of alpha(nu)beta3-targeted fumagillin nanoparticles impair Vx-2 tumor angiogenesis and development in rabbits. FASEB J 22(8):2758–2767
38. Winter PM, Neubauer AM, Caruthers SD, Harris TD, Robertson JD, Williams TA et al (2006) Endothelial alpha(v)beta3 integrin-targeted fumagillin nanoparticles inhibit angiogenesis in atherosclerosis. Arterioscler Thromb Vasc Biol 26(9):2103–2109
39. Winter PM, Morawski AM, Caruthers SD, Fuhrhop RW, Zhang H, Williams TA et al (2003) Molecular imaging of angiogenesis in early-stage atherosclerosis with alpha(v)beta3-integrin-targeted nanoparticles. Circulation 108(18):2270–2274
40. Frias JC, Williams KJ, Fisher EA, Fayad ZA (2004) Recombinant HDL-like nanoparticles: a specific contrast agent for MRI of atherosclerotic plaques. J Am Chem Soc 126(50):16316–16317
41. Skajaa T, Zhao Y, van den Heuvel DJ, Gerritsen HC, Cormode DP, Koole R et al (2010) Quantum dot and Cy5.5 labeled nanoparticles to investigate lipoprotein biointeractions via Förster resonance energy transfer. Nano Lett 10(12):5131–5138
42. Medarova Z, Pham W, Farrar C, Petkova V, Moore A (2007) In vivo imaging of siRNA delivery and silencing in tumors. Nat Med 13(3):372–377
43. Kumar M, Yigit M, Dai G, Moore A, Medarova Z (2010) Image-guided breast tumor therapy using a small interfering RNA nanodrug. Cancer Res 70(19):7553–7561
44. Banciu M, Schiffelers RM, Fens MH, Metselaar JM, Storm G (2006) Anti-angiogenic effects of liposomal prednisolone phosphate on B16 melanoma in mice. J Control Release 113(1):1–8

45. Metselaar JM, Wauben MH, Wagenaar-Hilbers JP, Boerman OC, Storm G (2003) Complete remission of experimental arthritis by joint targeting of glucocorticoids with long-circulating liposomes. Arthritis Rheum 48(7):2059–2066
46. Schmidt J, Metselaar JM, Wauben MH, Toyka KV, Storm G, Gold R (2003) Drug targeting by long-circulating liposomal glucocorticosteroids increases therapeutic efficacy in a model of multiple sclerosis. Brain 126(Pt 8):1895–1904
47. Lobatto ME, Fayad ZA, Silvera S, Vucic E, Calcagno C, Mani V et al (2010) Multimodal clinical imaging to longitudinally assess a nanomedical anti-inflammatory treatment in experimental atherosclerosis. Mol Pharm 7(6):2020–2029
48. van Schooneveld MM, Vucic E, Koole R, Zhou Y, Stocks J, Cormode DP et al (2008) Improved biocompatibility and pharmacokinetics of silica nanoparticles by means of a lipid coating: a multimodality investigation. Nano Lett 8(8):2517–2525
49. Hu X, Gao X (2010) Silica-polymer dual layer-encapsulated quantum dots with remarkable stability. ACS Nano 4(10):6080–6086
50. van Schooneveld MM, Gloter A, Stephan O, Zagonel LF, Koole R, Meijerink A et al (2010) Imaging and quantifying the morphology of an organic–inorganic nanoparticle at the subnanometre level. Nat Nanotechnol 5(7):538–544

Chapter 9
Polymer-Modified Nanoparticles as Targeted MR Imaging Agents

Stephen G. Boyes, Misty D. Rowe, Chia-Chih Chang, Talia J. Sanchez, Wilasinee Hatakeyama, Natalie J. Serkova, Priya N. Werahera, and Fernando J. Kim

9.1 Introduction

Despite recent advances in the understanding of fundamental cancer biology, many of these advances have not translated into clinical applications via the development of new therapeutics [1]. Arguably, the main reasons for this are the inability to administer therapeutics so that they reach the desired target without causing collateral damage and the lack of tumor-specific, noninvasive imaging modalities to determine the efficacy and selectivity of cancer treatments [2]. As such, a major focus of current cancer research is the development of therapeutic systems that demonstrate the synergistic goals of increasing the efficacy per dose of any therapeutic or imaging agent, increasing the targeted selectivity of the system, and introducing the ability to overcome any biological barriers that may stop the device from reaching its desired target [2–4]. The technology that offers the greatest potential in addressing these needs is nanotechnology, and perhaps, the best example of nanotechnology related to cancer treatment is the use of nanovectors. Nanovectors are solid or hollow multifunctional organic or inorganic nanoparticles, which can be

S.G. Boyes (✉) • M.D. Rowe • C.-C. Chang • T.J. Sanchez • W. Hatakeyama
Department of Chemistry and Geochemistry, Colorado School of Mines, Golden, CO 80401, USA
e-mail: sboyes@mines.edu

N.J. Serkova
Department of Anesthesiology, Biomedical MRI/MRS Cancer Center Core, University of Colorado Denver Anschutz Medical Campus, Aurora, CO 80045, USA

P.N. Werahera
Department of Pathology, University of Colorado Denver Anschutz Medical Campus, Aurora, CO 80045, USA

F.J. Kim
Department of Surgery – Urology, Denver Health Medical Center, Denver, CO 80204, USA

S. Svenson and R.K. Prud'homme (eds.), *Multifunctional Nanoparticles for Drug Delivery Applications: Imaging, Targeting, and Delivery*, Nanostructure Science and Technology, DOI 10.1007/978-1-4614-2305-8_9,

functionalized with therapeutics, targeting ligands, and imaging agents [5–8]. Ideally, nanovectors will be used for the in vivo, noninvasive visualization of early stage molecular markers for cancer, the targeted delivery of therapeutic agents, and the trapping and suppression of lesions before they reach either lethal or malignant phenotype, with little to no simultaneous loss of quality of life [1]. In conjunction to the focus on cancer therapeutics, the development of targeted diagnostic imaging agents has become a major interest for the clinical application of nanotechnology [1, 9, 10]. Because of the versatility and control over the nanostructures that can be formed, nanoparticles have been used as imaging agents for numerous imaging techniques including magnetic resonance imaging (MRI), computed X-ray tomography (CT), and optical imaging [11, 12]. Examples of nanoparticles that have been used as imaging agents include iron oxide nanoparticles [5, 13], silicon and silica nanoparticles [14, 15], gold (Au) nanoparticles and nanoshells [16–20], and quantum dots [21, 22]. Nanoparticle-based diagnostic agents with targeting capabilities show advantages over conventional imaging agents such as increased retention time, specificity, and enhanced imaging capabilities [12, 23–25]. The addition of active targeting ligands such as antibodies has increased the applicability of these novel imaging agents due to their ability to induce receptor-mediated endocytosis by diseased cells, preferential to healthy cells, through biomolecular recognition. While these systems offer much promise, there are a number of challenges that must be overcome before they can be applied in a clinical setting. These challenges include issues such as cytotoxicity, stability, and the inability to incorporate all of the components required of an effective targeted multimodal imaging agent. Arguably, polymer-modified inorganic nanoparticles represent the most promising structures to meet the above challenges. Research into the use of inorganic nanoparticles as nanovectors has highlighted the use of nanoparticles that possess properties useful for molecular imaging. For example, iron oxide nanoparticles behave as negative T_2-contrast agents in MRI, and Au nanostructures can be observed using either CT or confocal microscopy [26]. The use of nanoparticles as the imaging component of multifunctional nanovectors is beneficial due to the fact that a large concentration of imaging agent can be delivered to the desired location per targeting biorecognition event. However, in order to produce a nanoparticle-based imaging agent that can be easily translated to clinical application, it is important that the nanodevice be useful for application with common and widely utilized diagnostic imaging instrumentation, such as MRI, and that the nanodevice can be modified so as to render the nanostructure biocompatible while providing a means of biomolecular targeting and disease therapy.

Superparamagnetic iron oxide nanoparticles (SPIOs) have garnered a majority of the focus on nanoparticle contrast agents for MRI, which has led to current testing of ferucarbotran (Resovist®, approved in Japan, Europe, and Australia) and ferumoxtran in phases II and III clinical trials in the USA, while ferumoxsil (GastroMARK® and Lumirem®) and ferumoxide (Feridex® and Endorem®) are approved for clinical use. Unfortunately, negative contrast agents such as SPIOs suffer from a series of drawbacks, including limited in vivo cell tracking (MRI cannot distinguish the negative contrast agent from other signal voids), and negative contrast agents are limited by partial volume effects [27–29]. Recently, focus on nanoparticle-based MRI contrast agents has shifted to gadolinium (Gd)-based nanoparticles [27, 30–36] due to their ability to act as a positive contrast agent and due to their relationship to currently employed MRI contrast agents, such as gadopentetate dimeglumine (Magnevist®) and gadobenate dimeglumine (MultiHance®), which are based on Gd chelates. This attention on Gd-based nanoparticles has been mostly driven by limitations with the conventional contrast agents based on Gd chelates, such as low concentrations of Gd per molecule, short retention times in vivo due to contrast agent dimensions, limited biostability, and difficulty in functionalization to enable use in more complex diagnostic devices. These Gd-based nanoparticles have been synthesized with both inorganic and organometallic compounds of Gd such as gadolinium oxide, gadolinium phosphate, gadolinium fluoride, gadolinium hexanedione and acetylacetenoate mixed with emulsifying wax, and most recently, nanoscale metal-organic frameworks (MOFs) [27, 31–37]. While Gd-based nanoparticles exhibit relaxivities significantly higher than typical Gd chelates and also provide a contrast agent with higher molecular weight for improved retention times and a high concentration of Gd^{3+} ions per contrast agent particle, their application has been limited due to the difficulty in producing nanoparticles that are biocompatible, stable, and have specific surface functionality [27, 32].

One of the most interesting Gd nanoparticle systems is the nanoscale MOFs, which are constructed from Gd^{3+} ions and organic bridging ligands, such as 1,4-benzenedicarboxylic acid (1,4-BDC), as they have demonstrated exceptional MRI capabilities [31, 38, 39]. However, in order to take advantage of Gd nanoparticles as nanoscale contrast agents for MRI, focus has shifted to overcoming some of their inherent limitations by developing methods to surface-modify the nanoparticles [33, 36]. Current surface modification methods have yielded limited success, as they have resulted in instabilities in the coatings due to their noncovalent nature, poorly defined surfaces, insufficient control over surface functionality, or reduced imaging capabilities due to masking of the underlying Gd nanoparticle. As such, the search for a surface modification technique that provides control over surface functionality and architecture, along with the ability to produce a stable structure without diminishing the inherent imaging properties, represents a significant challenge for researchers. Surface modification of Gd-based MOFs through covalent attachment of well-defined polymers offers a means of modifying and/or tuning the relaxation properties of the nanoparticles, incorporating a higher degree of functionality, and increasing their in vivo stability and biocompatibility.

Herein, we discuss a novel surface modification procedure developed in our research group, allowing the attachment of well-defined polymers synthesized via reversible addition-fragmentation chain transfer (RAFT) polymerization through reduction of the thiocarbonylthio end group under basic conditions to form thiolates, and further attachment through coordination chemistry to the Gd MOF nanoparticles [40, 41]. In vitro MRI was employed to determine the relaxivities of the novel polymer-modified Gd MOF nanoparticles in comparison to the clinically employed contrast agents, Magnevist® and MultiHance®. Furthermore, Gd MOF nanoparticles were surface-modified by the covalent attachment of well-defined

RAFT copolymers containing a targeting ligand and antineoplastic agent to produce a novel theragnostic nanodevice, with bimodal diagnostic imaging capabilities. Finally, tailoring either the size and shape of the nanoparticles or the functionality and thickness of the polymer coating provided a means of tuning the MRI characteristics of these novel polymer-modified positive contrast nanoparticle agents.

9.2 Experimental Section

9.2.1 Materials and Characterization

All chemicals were purchased from Sigma-Aldrich, unless otherwise noted. H-glycine-arginine-glycine-aspartate-serine-NH_2 (GRGDS-NH_2) peptide motif was purchased from AnaSpec. Epidermal growth factor antibody (EGFR) was purchased from Invitrogen. *N*-isopropyacrylamide (NIPAM) was doubly recrystallized in hexanes before use. Azobisisobutyronitrile (AIBN) was doubly recrystallized from methanol prior to use. Triethylamine was distilled under pressure and stored in the freezer prior to use. Styrene (Sty) was filtered over basic alumina oxide and then stored in a freezer prior to use. All other chemicals, unless otherwise discussed, were reagent grade and used as received. S-1-dodecyl S′-(α,α-dimethylacetic acid) trithiocarbonate (DATC) was prepared via Lai et al.'s literature procedure [42]. *N*-(2-hydroxypropyl) methacrylamide (HPMA) was synthesized by a procedure in the literature [43]. A discussion of the characterization of polymers, Gd MOF nanoparticles, and RAFT polymer-modified Gd MOF nanoparticles is described in great detail in the recent literature [40, 41].

9.2.2 Homopolymer Synthesis via RAFT Polymerization

A range of homopolymers including poly(*N*-isopropyacrylamide) (PNIPAM), poly[(2-hydroxypropyl)methacrylamide] (PHPMA), and poly(styrene) (PSty) were synthesized by RAFT polymerization techniques. For example, in order to produce a PNIPAM homopolymer with a $M_{n,\mathrm{theoretical}}$ of 5,500 g/mol, NIPAM (10.07 g, 8.90×10^{-2} mol), *N*,*N*-dimethylformamide (DMF) (50 mL), and DATC (0.488 g, 1.34×10^{-3} mol) were added to a 150-mL Schlenk flask equipped with a stir bar. The flask was sealed, and the solution was gently degassed for 45 min and then left under a high-purity nitrogen atmosphere. The flask was allowed to stir at room temperature until the monomer and DATC were completely dissolved. To a second 150-mL Schlenk flask equipped with a stir bar was added AIBN (0.0204 g, 1.24×10^{-4} mol). This flask was sealed with a rubber septum, subjected to three evacuation-nitrogen purge cycles, and left under a nitrogen atmosphere. The monomer solution was then transferred via cannula to the initiator-containing flask.

The reaction was then heated for 1 h at 60°C, after which the polymer was isolated from the solution by evaporating residual solvent under vacuum at 40°C overnight. Polymer was then purified via precipitation to remove residual monomer. Specific RAFT polymerization conditions for each of the other monomer systems can be found in the literature [40].

9.2.3 Copolymer Synthesis via RAFT Polymerization

A random copolymer of PHPMA-*co*-poly(*N*-methacryloxysuccinimide)-*co*-poly (fluorescein *O*-methacrylate) (PHPMA-*co*-PmNAOS-*co*-PFMA) was synthesized with 15 wt% monomer in a mixture of DMF and tert-butanol at 80°C via RAFT polymerization techniques utilizing a 10:1 molar ratio of DATC, as the RAFT agent, to AIBN, as the initiator. HPMA comprised the majority of the copolymer backbone, with NAOS being incorporated into the backbone at 15 wt% loading to provide a reactive sight for attachment of targeting ligands. FMA, at approximately 0.5 wt% loading, was added allowing for fluorescent tagging of the copolymer for subsequent imaging capabilities in fluorescent microscopy. For example, in order to produce PHPMA-*co*-PmNAOS-*co*-PFMA with a 15 wt% loading of mNAOS, tert-butanol (15 mL), HPMA (3.00 g, 20.8 mmol), FMA (0.0300 g, 0.0749 mmol), DATC (0.100 g, 0.274 mmol), and AIBN (0.00300 g, 0.0183 mmol) were added to a 2-neck 150-mL Schlenk flask equipped with a stir bar and a 100-mL addition funnel. Anhydrous DMF (30 mL) and mNAOS (0.450 g, 2.46 mmol) were added to the addition funnel equipped with a bubbler, which was then sealed. Each solution was gently degassed for 45 min and then left under a high-purity nitrogen atmosphere. The flask was allowed to stir at room temperature until all of the solid reactants were completely dissolved. The reaction was then heated to 80°C, at which time the mNAOS solution was added dropwise via the addition funnel. The polymerization was allowed to proceed for 24 h, after which the polymer was isolated from the solution by double precipitation into a 50:50 vol/vol solution of acetone and ethyl ether.

9.2.4 Attachment of Targeting Ligands to RAFT Copolymer

The PHPMA-*co*-PmNAOS-*co*-PFMA copolymers were subsequently reacted through a condensation reaction of the succinimide groups with primary amine groups of the targeting ligand. Reactions were carried out at room temperature in deuterated *N,N*-dimethylsulfoxide (DMSO) with 0.01 M triethylamine and stirring at room temperature for 24 h. Unreacted targeting ligand was then removed via silica column chromatography and aqueous/solvent extraction. GRGDS-NH_2 and anti-EGFR were incorporated into the copolymer backbone at 9 and 5 wt% loading capacity, respectively. Detailed reaction conditions for the attachment of the GRGDS-NH_2-targeting ligand are discussed in the literature [41].

9.2.5 Synthesis of Gd MOF Nanoparticles

Gd MOF nanoparticles were prepared via a reverse microemulsion synthesis reported by Reiter and coworkers [31]. Briefly, a water-to-surfactant ratio (*w*) of 10 was employed for the reverse microemulsion system containing gadolinium(III) chloride ($GdCl_3$) and the bridging ligand, 1,4-BDC (0.075 M). Cetyltrimethyl ammonium bromide (CTAB) (0.05 M) was employed as the surfactant, with hexanol as a cosurfactant in a heptane oil phase. The resulting solution was allowed to stir for 24 h at room temperature. Unreacted reagents were removed from the Gd MOF nanoparticles through repeated centrifugation and resuspension in ethanol (2×) and water (2×), followed by drying.

9.2.6 Synthesis of Gd MOF Nanoparticles Including Hydrotrope

Gd MOF nanoparticles were prepared exactly as mentioned in the previous section except for the inclusion of varying weight percentages of sodium salicylate (NaSal) to the reverse microemulsion system. Briefly, a water-to-surfactant ratio of 10 was employed for the reverse microemulsion system containing $GdCl_3$ and the bridging ligand, 1,4-BDC (0.075 M). CTAB (0.05 M) was employed as the surfactant, with hexanol as a cosurfactant in a heptane oil phase. The resulting solution was allowed to stir for 5–10 min at room temperature until the solution became clear. The NaSal was weighed according to the desired weight percentage and was subsequently added to the clear solution. The solution was then allowed to stir for 24 h at room temperature. Unreacted reagents were removed from the Gd MOF nanoparticles through repeated centrifugation and resuspension in ethanol (2×) and water (2×), followed by drying.

9.2.7 Surface Modification of Gd MOF Nanoparticles with RAFT Homopolymers and Copolymers

In a typical experiment, RAFT polymer (0.1 g) was added to 25 mL of anhydrous DMF in a 150-mL Schlenk flask equipped with a stir bar and then sealed with a rubber septum. The RAFT polymer solution was purged with high-purity nitrogen and then subsequently left under a nitrogen atmosphere. The RAFT agent-terminated polymer was then converted to a thiolate-terminated polymer through aminolysis, by the addition of hexylamine (0.4 mL, 0.075 M) and stirring for 1 h at room temperature. Gd MOF nanoparticles (0.01 g) were suspended in an additional 25 mL of DMF in a second 150-mL Schlenk flask equipped with a stir bar and then sealed with a rubber septum. The nanoparticle solution was then purged with high-purity nitrogen for 30 min and was subsequently left under a nitrogen atmosphere.

The polymer solution was transferred via cannula to the Gd MOF nanoparticle solution. The resulting solution was allowed to stir for 24 h at room temperature under a nitrogen atmosphere. Untethered polymer was removed from the polymer-modified Gd MOF nanoparticles through repeated centrifugation and resuspension in DMF (2×) and ethanol (2×), followed by drying.

9.3 Results and Discussion

9.3.1 Polymer-Modified Nanoparticles as Targeted Imaging Agents

Despite the tremendous potential that nanoparticles offer in the thriving area of nanomedicines, their modification to provide critical properties, such as multimodal imaging, biocompatibility, and biomolecular targeting, is of utmost importance. Developing a method to provide reproducible, well-defined, and stable surface modification of nanoparticles is critical. RAFT polymerization techniques offer one of the most versatile routes to enhancing the properties of high surface area nanoparticle structures through modification of nanoparticle surfaces with well-defined RAFT polymers. RAFT polymerization is arguably the most versatile living radical polymerization (LRP) technique with respect to polymerization conditions, along with the ability to produce well-defined, low polydispersity index (PDI) polymers with both simple and complex architectures and a high degree of end group control [44, 45]. RAFT polymerization shows great promise in the synthesis of multifunctional polymers due to the versatility of monomer selection and its functional group tolerance. It has been widely employed for the preparation of highly specialized materials for advanced biomedical applications, such as antibody and small interfering ribonucleic acid (siRNA)–polymer conjugates, controlled drug delivery vehicles, and bioconjugation [46–52]. Due to the well-defined nature of the RAFT polymerization technique, another advantage of RAFT polymers is the presence of a thiocarbonylthio group on the end of each polymer chain. Literature has shown that the thiocarbonylthio end groups can be reduced to a thiol in the presence of a nucleophile such as a primary amine or sodium borohydride [53, 54]. Thiols have been shown to react strongly with a variety of metal surfaces, such as gold and silver [54], and the surface of semiconducting nanoparticles, such as CdSe nanoparticles; therefore, RAFT polymerization is uniquely placed as one of the premier polymerization techniques to prepare polymers for surface functionalization of a wide range of both planar and nanoparticle substrates [53–56]. Specifically, gold nanorods and Gd nanoparticles have gained interest due to their potential application as biomedical imaging agents in dark field microscopy and MRI, respectively [26, 27, 30–33, 35–37, 39, 57–59].

To date, there are few reports of the utilization of the RAFT polymerization technique to produce well-defined polymers that allow for the surface modification of Gd

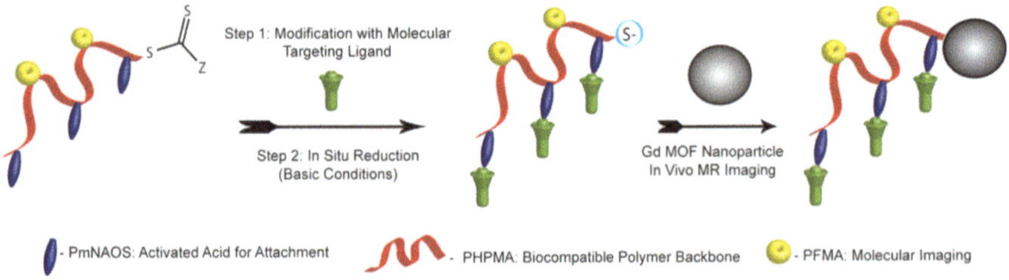

Scheme 9.1 General route to modification of Gd MOF nanoparticles with biomolecular targeted RAFT copolymers

MOF nanoparticles to produce polymer-modified positive contrast nanoparticle agent platforms for MRI [40, 41]. Herein, we discuss a novel surface modification procedure developed in our research, allowing the attachment of well-defined polymers synthesized *via* RAFT polymerization through reduction of the thiocarbonylthio end group under basic conditions to form thiolates, and further attachment through coordination chemistry to the Gd MOF nanoparticles (Scheme 9.1). In vitro MRI was employed to determine the relaxivities of the novel polymer-modified Gd MOF nanoparticles in comparison to the clinically employed contrast agents, Magnevist® and MultiHance®. Furthermore, tailoring either the size and shape of the nanoparticles or the functionality and thickness of the polymer coating provided a means of tuning the MRI characteristics of these novel polymer-modified positive contrast nanoparticle agents. Finally, Gd MOF nanoparticles were surface-modified by the covalent attachment of well-defined RAFT copolymers containing a targeting ligand to produce a novel targeted diagnostic nanodevice, with bimodal imaging capabilities.

9.3.2 Gd MOF Nanoparticles

The use of Gd MOF nanoparticles as positive contrast agents for MRI should provide several advantages over the clinically employed Gd chelates, such as enhanced imaging through magnetic resonance, increased biostability, and longer in vivo retention [27, 30, 31, 33–37]. Gd-based contrast agents largely reduce the longitudinal relaxation time (T_1) and longitudinal relaxivity (r_1) and are called positive contrast agents, where the relaxivity value, r, is defined as the inverse of the relaxation time with respect to the contrast agent concentration [27, 60]. The ratio of the transverse relaxivity (r_2) to r_1 is used to provide information about the contrast agent, where r_2/r_1 values below 2 show brightening in T_1-weighted images, yielding a positive contrast agent [27]. As there is a preference for the use of positive contrast agents at the clinical level due to their wider dynamic range, contrast agents based on Gd^{3+} chelates are the most widely used.

In this research, Gd MOF nanoparticles were synthesized as described in the literature [31, 32, 61]. The Gd MOF nanoparticles were characterized thoroughly by employing transmission electron microscopy (TEM), attenuated total reflectance

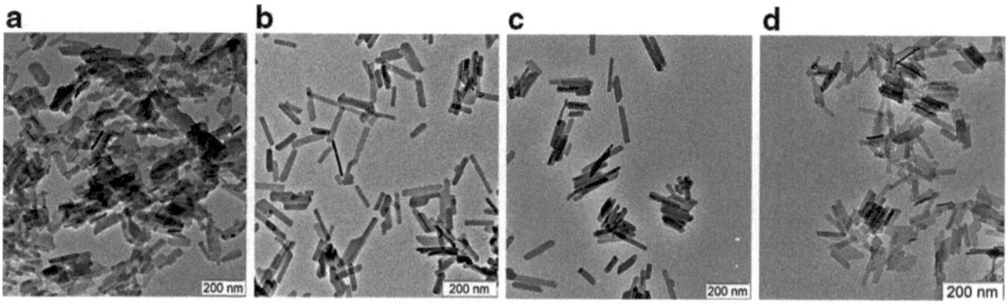

Fig. 9.1 Transmission electron microscopy of Gd MOFs synthesized via the traditional method, utilizing (**a**) no hydrotrope at a water-to-surfactant ratio (w)=10, (**b**) 135 μM NaSal at w=10, (**c**) 540 μM NaSal at w=10, and (**d**) 135 μM NaSal at w=7

Table 9.1 Dimensions of Gd MOFs synthesized with the addition of NaSal as a hydrotrope at various water-to-surfactant ratios (w)

	NaSal		Length (nm)			Width (nm)			Aspect ratio
W	μM	wt%[a]	Average[b]	Std Dev[b]	%RSD[b]	Average[b]	Std Dev[b]	%RSD[b]	Average[b]
10	–	–	120	60	50%	50	20	50%	3
10	67.5	0.05	120	40	33%	23	5	22%	5
7	135	0.1	80	20	25%	17	4	24%	5
10	135	0.1	110	40	33%	25	6	22%	4
10	270	0.2	160	50	31%	28	7	23%	6
10	405	0.3	150	50	33%	24	5	23%	6
10	540	0.4	160	40	25%	26	5	23%	6

[a]Wt% is calculated with respect to CTAB

[b]A minimum of 500 particles were measured and averaged from at least five separate batches of Gd MOF nanoparticles

Fourier transform infrared spectroscopy (ATR-FTIR), and thermogravimetric analysis (TGA). Figure 9.1a and Table 9.1 show the average dimensions of the synthesized Gd MOF nanoparticles to be 120 nm in length and 50 nm in width. The ATR-FTIR spectrum showed a characteristic out-of-plane = C–H aromatic stretch at 725 cm^{-1}, symmetric carboxylate stretch at 1,400 cm^{-1}, and an asymmetric carboxylate stretch at 1,540 cm^{-1}, along with peaks at 2,855, 2,925, and 3,065 cm^{-1}, which are attributed to the –C–H stretching vibrations of the 1,4-BDC bridging ligand, and 3,460 cm^{-1} which was attributed to the –OH stretch of the water ligand. Additionally, TGA was employed and confirmed the empirical formula of the Gd MOF nanoparticles to be $Gd(1,4\text{-}BDC)_{1.5}(H_2O)_{2,}$ as discussed in the literature [31].

9.3.3 *Effect of NaSal on Gd MOF Nanoparticle Synthesis*

The use of NaSal in normal micelles containing the surfactant CTAB has shown a transition from spherical/ellipsoidal micelles to wormlike micelles at varying concentration, indicating that there is an influence of NaSal on the micelle structure as a function of hydrotrope concentration [62, 63]. The same reported transition from ellipsoidal to wormlike micelles is expected for reverse microemulsion systems and potentially provides a mechanism to control particle size and reduce particle size distribution. Once again, Gd MOF nanoparticles were synthesized as described in the literature; however, NaSal was incorporated into the final microemulsion during combining of the two miniemulsions of $GdCl_3$ and 1,4-BDC. The effect of varying the weight percentage of NaSal on the Gd MOF nanoparticle size and size distribution can be seen in Fig. 9.1 and Table 9.1. Synthesis of Gd MOF nanoparticles with 135 μM of NaSal at a water-to-surfactant ratio of $w = 10$ compared to Gd MOF nanoparticles synthesized without a hydrotrope at the same *w*-value demonstrated comparable average particle lengths of 110 nm and 120 nm, respectively (Fig. 9.1a, b). However, the standard deviation of particle length decreased from 60 nm (without NaSal) to 40 nm (with NaSal), which is a percentage deviation drop from 50% to 33% (Table 9.1). The diameter of the nanoparticles also shows a decrease in percentage standard deviation when synthesized with NaSal, from 50% to 22%. With an increase in the concentration of NaSal added to the Gd MOF nanoparticle synthesis, at a constant *w*-value, the length appears to increase to a maximum value of approximately 160 nm (Fig. 9.1c and Table 9.1), while the diameter of the nanoparticles decreases to a value of approximately 25 nm. These results demonstrate that there is an increase in the aspect ratio of the nanoparticles from 3 to 6 with an increasing NaSal concentration. Based on these observations, it appears as though the addition of NaSal to the reverse microemulsion results in the formation of wormlike micelle shapes, most likely due to the NaSal simultaneously screening the surfactant head groups and increasing the end cap energy for surfactant and hydrotrope to align linearly along the axial interface [62, 63]. In addition to studying the effect of NaSal concentration on nanoparticle size and shape, the effect of changing the *w*-value at a fixed NaSal concentration of 135 μM was also examined. A decrease in the *w*-value from ten to seven at a fixed concentration of NaSal resulted in a decrease in both the length and width of the nanoparticles to 80 nm and 17 nm, respectively (Fig. 9.1d and Table 9.1). There was also a significant improvement in the percentage standard deviation of the nanoparticle length from 50% to 25%, indicating more uniform nanoparticle structures (Table 9.1).

9.3.4 Surface Modification of Gd MOF Nanoparticles with RAFT Homopolymers

As DATC is employed as the RAFT agent in the formation of RAFT polymers in this study, each of the polymers produced should have trithiocarbonate-terminated chains. Surface modification of Gd MOF nanoparticles was achieved by the "grafting to" technique, which involved an initial aminolysis, using hexylamine, of the trithiocarbonate end group of the RAFT polymers to a thiolate functionality under inert and basic conditions (Scheme 9.1). Subsequently, it is hypothesized that the thiolate-terminated homopolymer was covalently attached to the nanoparticle surface through a coordination reaction between the thiolate end group moiety and vacant orbitals on the Gd^{3+} ions at the surface of the Gd MOF nanoparticles. Using this procedure, Gd MOF nanoparticles were modified with various RAFT homopolymers, including PNIPAM, PHPMA, and PSty. It should be noted that after polymer deposition and prior to characterization or use, the nanoparticles were washed several times with a good solvent for the polymer, via repeated washing and centrifugation steps, to remove any untethered polymer from the system. Prior to attachment of the RAFT polymers to the Gd MOF nanoparticles, proton nuclear magnetic resonance spectroscopy (^{1}H NMR), ultraviolet–visible (UV–vis) spectroscopy, and ATR-FTIR were used to verify the conversion of the trithiocarbonate end group to a thiolate upon addition of the hexylamine. In each case, the aminolysis reaction resulted in approximately quantitative conversion of the RAFT polymer end groups to thiolates, as had been observed in the literature for similar reactions [53], with minimal dimerization confirmed by ^{1}H NMR and gel permeation chromatography (GPC). Surface modification of the Gd MOF nanoparticles with RAFT homopolymers was characterized through both TEM (Table 9.2) and ATR-FTIR spectroscopy. The Gd MOF nanoparticles were first modified with a series of different molecular weight PNIPAM homopolymers synthesized via RAFT polymerization. ATR-FTIR was employed to confirm the addition of the PNIPAM homopolymer onto the surface of the Gd MOF nanoparticles: several of the characteristic stretches of the free PNIPAM homopolymer, including a broad N–H stretch above 3,300 cm^{-1} and a small N–H bend at 1,640 cm^{-1}, indicating the presence of the acrylamide functionality; an increase in intensity of the $-CH_2$ stretching and C–H stretching vibrations between 2,800 and 3,000 cm^{-1} due to backbone methylenes; a peak at 1,720 cm^{-1} assigned to the carbonyl stretch of the amide; and a stretch at 1,380 cm^{-1} attributed to the addition of $-CH_3$ and isopropyl groups display good transference to the polymer-modified Gd MOF nanoparticles when compared to the unmodified Gd MOF nanoparticle. TEM images show a relatively uniform coating of the PNIPAM homopolymer around the Gd MOF nanoparticles after deposition with an average coating thickness of about 11 nm.

Other RAFT homopolymers such as PHPMA and PSty synthesized via RAFT polymerization techniques were also employed in the successful modification of Gd MOF nanoparticles. In each case, TEM showed relatively uniform coatings on the surface of the nanoparticles after deposition. Polymer coating thicknesses for these samples are shown in Table 9.2. Additionally, upon modification of the Gd MOF nanoparticles, several of the respective characteristic stretches of each of the free homopolymers displayed good transference to the polymer-modified Gd MOF nanoparticles when compared to the unmodified Gd MOF nanoparticles [40].

Table 9.2 Molecular weight and coating properties of RAFT-synthesized homopolymers

Polymer coating	M_n (g/mol)		PDI [d,e]	Coating thickness (nm) [f]	Grafting density (chain/nm^2)	
	Theory [a]	Experimental [b,c]			Theory [g]	Experimental [h,i]
PNIPAM	5,500	5,700	1.23	4.2±0.3	0.1849	0.1395
	8,700	8,600	1.12	7.0±0.4	0.1591	0.1037
	17,100	17,800	1.12	10.9±1.0	0.1070	0.0873
PHPMA	5,100	5,300	1.44	2.4±1.1	0.1847	0.1379
	9,000	10,200	1.30	3.5±0.7	0.0957	0.0993
	19,700	19,400	1.24	7.3±0.6	0.0729	0.0638
PSTY	4,900	4,800	1.15	1.6±0.3	0.1660	0.1839
	9,400	9,000	1.15	5.7±0.9	0.1341	0.0960
	18,900	15,300	1.13	8.7±0.9	0.1033	0.0815

[a] $M_{n,\text{theoretical}}$ is defined as the theoretical number average molecular weight, and each RAFT polymerization was calculated using the equation M_n = (molecular weight of RAFT agent) + (molecular weight of monomer) × ([monomer]$_0$/[RAFT agent]$_0$) × (monomer conversion)
[b] Values determined by gel permeation chromatography
[c] Values calculated by end group analysis using proton nuclear magnetic resonance spectroscopy (^{1}H NMR)
[d] Values determined by matrix-assisted laser desorption ionization time-of-flight mass spectrometry (MALDI-TOFMS)
[e] PDI was calculated using the equation PDI = (M_w/M_n)
[f] Polymer coating thickness was determined by TEM and is an average of ten measurements
[g] Theoretical grafting densities were calculated from equations discussed in the literature, using an average Gd MOF nanoparticle length of 122 nm and width of 53 nm, along with the experimental molecular weight and bulk density of each polymer
[h] Experimental grafting densities were calculated using an average Gd MOF nanoparticle lengtht of 122 nm and width of 53 nm, the experimental molecular weight of each polymer, a bulk density of 2.529 g/cm^3 for the Gd MOF nanoparticles, along with the mass of polymer per mass of Gd MOF nanoparticle determined by TGA
[i] Percent relative standard deviations were determined to be <7% for the experimental grafting density calculations

9.3.5 *Effect of Polymer Molecular Weight on Coating Thickness and Grafting Density*

The different PNIPAM, PHPMA, and PSty samples were also used to examine the effect of molecular weight on grafting density, coating thickness, and relaxivity. As can be seen in Table 9.2, there is a definite trend of increasing polymer coating thickness with molecular weight of each of the three polymers. For example, the average thickness of the PNIPAM homopolymer with an $M_{n,\text{experimental}}$ of 5,700 g/mol is approximately 4.2 ± 0.3 nm. The PNIPAM homopolymer with an $M_{n,\text{exp}}$ equal to 8,600 g/mol increased the polymer coating thickness to 7.0 ± 0.4 nm, while an $M_{n,\text{exp}}$ of 17,800 g/mol further raised the thickness to 10.9 ± 1.0 nm. Similar results were seen for the PHPMA samples. For example, the average thickness of the PHPMA homopolymer with an $M_{n,\text{exp}}$ of 5,327 g/mol is approximately 2.4 ± 1.1 nm. The PHPMA homopolymer with an $M_{n,\text{exp}}$ equal to 10,281 g/mol increased the polymer coating thickness to 3.5 ± 0.7 nm, while an $M_{n,\text{exp}}$ of 5,327 g/mol further raised the thickness to 7.3 ± 0.6 nm. Finally, the average thickness of the PSty homopolymer with an $M_{n,\exp}$ of 4,802 g/mol is approximately 1.6 ± 0.3 nm. The PSty homopolymer with an $M_{n,\exp}$ equal to 8,972 g/mol increased the polymer coating thickness to 6.7 ± 0.9 nm, while an $M_{n,\exp}$ of 15,245 g/mol further raised the thickness to 8.7 ± 0.9 nm (Table 9.2). This trend suggests the ability to tailor the polymer coating thickness by simply changing the molecular weight characteristics of the RAFT polymer used for modification.

Table 9.2 shows both the theoretical and experimental grafting density values for each of the polymer-modified Gd MOF nanoparticle samples, calculated from equations discussed in the literature [40]. As can be seen in Table 9.2, the theoretical and experimental grafting densities for each of the polymers used for surface modification of the Gd MOF nanoparticles correlated quite well, with nearly all of the experimental calculations being within 25% of their theoretical grafting densities. For example, the PHPMA homopolymer with a molecular weight of 5,327 g/mol employed to modify Gd MOF nanoparticles provided comparable theoretical and experimental polymer grafting densities of 0.1847 chain/nm^2 and 0.1379 chain/nm^2, respectively. In each case, the grafting densities are relatively high for use of a "grafting to" technique, as most of the samples have values around 0.1 chains/nm^2, indicating modification of the Gd MOF nanoparticles with each polymer in the "brush" regime [64–66]. However, although the values are quite high, similar experimental grafting densities have been documented in the literature for the "grafting to" technique [67–70]. Despite this high grafting density, a definite trend was seen, which showed a decrease in the grafting density with increased molecular weight of the grafted polymer (Table 9.2). For example, as the molecular weight of the PHPMA increased from 5,327 to 10,281 g/mol, the experimental grafting density decreased from 0.1379 to 0.0993 chains/nm^2. The experimental grafting density further decreased to 0.0638 chains/nm^2 with modification of the Gd MOF nanoparticles using the PHPMA homopolymer with an experimental molecular weight of 19,370 g/mol. This trend of decreasing grafting density with increasing polymer molecular weight has been discussed extensively in the literature with the "grafting to" technique and is a result of limited diffusion of polymer chains to reactive sites on the nanoparticle surface due to increased steric hindrance of polymer chains that are already attached to the surface [64–66]. As such, as the molecular weight of the chains increases, it becomes more difficult for chains to diffuse to the surface, providing lower polymer grafting densities.

9.3.6 *In Vitro Imaging Properties of RAFT Homopolymer-Modified Gd MOF Nanoparticles*

In order to provide information about the clinical imaging viability of the polymer-modified Gd MOF nanoparticles as positive contrast nanoparticle agents, in vitro MRI was employed to determine relaxation properties of the unmodified and polymer-modified Gd MOF nanoparticles. Table 9.3 compares the relaxivity values

Table 9.3 Relaxivity properties of Gd MOF nanoparticles modified with RAFT-synthesized homopolymers

Contrast agent	r_1 (s^{-1} mM^{-1})[a]	r_2 (s^{-1} mM^{-1})[a]	r_2/r_1
Magnevist®	13.44	21.40	1.59
MultiHance®	19.45	30.44	1.57
Gd MOF nanoparticles	9.86	17.94	1.82
PNIPAM (5,700 g/mol)-modified Gd MOF nanoparticles	20.27	29.73	1.47
PNIPAM (8,600 g/mol)-modified Gd MOF nanoparticles	46.99	64.10	1.36
PNIPAM (17,800 g/mol)-modified Gd MOF nanoparticles	62.51	79.90	1.28
PHPMA (5,300 g/mol)-modified Gd MOF nanoparticles	17.81	25.77	1.45
PHPMA (10,200 g/mol)-modified Gd MOF nanoparticles	32.94	44.85	1.36
PHPMA (19,400 g/mol)-modified Gd MOF nanoparticles	105.36	129.63	1.23
PSty (4,800 g/mol)-modified Gd MOF nanoparticles	1.17	14.16	12.10
PSty (9,000 g/mol)-modified Gd MOF nanoparticles	1.20	25.75	21.46
PSty (15,300 g/mol)-modified Gd MOF nanoparticles	3.91	123.40	31.56

[a] Longitudinal relaxivity (r_1) and transverse relaxivity (r_2) values, calculated as the reciprocal values of the longitudinal relaxation time (T_1) and transverse relaxation time (T_2), respectively, of each of the contrast agents were determined with a 1.5-T scanner with samples diluted in deionized ultrafiltered water by acquiring signal intensity (I) measurements via region-of-interest analysis of the samples for all pulse sequences with T_1 and T_2 values being calculated using: $I_i = I_{0,I}(1 - \exp^{(-t/T_i)})$

collected on a 1.5 T scanner for unmodified Gd MOF nanoparticles, each of the polymer-modified Gd MOF nanoparticles, and the clinically employed contrasts agents, Magnevist® and MultiHance®. The r_1 values demonstrate that both the unmodified and polymer-modified Gd MOF nanoparticles, with the exception of the PSty-modified Gd MOF nanoparticles, result in a large shortening of the T_1 relaxation time and, thus, behave as positive contrast agents. Of particular note is the fact that the polymer modification of the Gd MOF nanoparticles with the PHPMA and PNIPAM RAFT homopolymers demonstrated significantly higher relaxivity values in comparison to both the unmodified Gd MOF nanoparticles and the clinically employed small molecule contrast agents. This phenomenon is attributed to increased water retention by the hydrophilic RAFT homopolymer matrices attached to the surface of the Gd MOF nanoparticles. The increased water retention allows for more favorable interactions between the water protons and the free orbitals of the Gd^{3+}-containing MOF nanoparticle, thus enhancing T_1 relaxation shortening effects. For example, when the highly hydrophilic PHPMA homopolymer, with a molecular weight of 19,370 g/mol, was employed for the modification of the Gd MOF nanoparticles, r_1 and r_2 values of 105.36 s^{-1} mM^{-1} and 129.63 s^{-1} mM^{-1}, respectively, were determined. These values are over ten times higher than the observed relaxivities of the unmodified Gd MOF nanoparticles and six times higher than the values for the clinically employed Magnevist® and MultiHance®. Furthermore, in each case, the polymer modification of the Gd MOF nanoparticles provided a much lower r_2/r_1 value in comparison to both the unmodified Gd MOF nanoparticles and the clinically employed contrast agents, which is advantageous for their use as clinical positive contrast agents. The PSty-modified Gd MOF nanoparticles showed very low r_1 values and very large r_2 values, which provided relaxivity ratio values more than an order of magnitude higher than the other polymer-modified samples. For instance, in comparison to the PHPMA homopolymer (19,370 g/mol)-modified Gd MOF nanoparticles, which showed an r_2/r_1 value of 1.23, the PSty samples with a comparable molecular weight of 15,245 g/mol yielded an r_2/r_1 value of 31.56 (Table 9.3). This difference was attributed to decreased water retention due to the hydrophobic nature of the PSty surface-modified Gd MOF nanoparticles, which minimizes interactions between the Gd^{3+} and water molecules and thus lengthens the T_1 relaxation times in comparison to the other systems.

In order to determine the effect of polymer molecular weight on the relaxation properties of the polymer-modified Gd MOF nanoparticle, three different molecular weights of PHPMA, PNIPAM, and PSty were used to surface-modify the nanoparticle constructs. As can be seen in Table 9.3, both PNIPAM and PHPMA-modified Gd MOF nanoparticles showed a trend of enhanced r_1 values and r_2/r_1 values with a respective increase in number average molecular weight of the polymer. For instance, as the PNIPAM molecular weight was increased from 5,700 to 8,600 g/mol, the longitudinal relaxivity increased from 20.27 to 46.99 s^{-1} mM^{-1}, with a decreased r_2/r_1 value of 1.36 from 1.47. Furthermore, as the molecular weight of the PNIPAM was increased to 17,800 g/mol, the longitudinal relaxivity value increased to 62.51 s^{-1} mM^{-1}, while the r_2/r_1 value decreased to 1.28, which is substantially improved in comparison to the clinically employed contrast agents and unmodified Gd MOF nanoparticles. Increases in the molecular weight of the PHPMA used for modification of the Gd MOF nanoparticles showed similar enhanced r_1 values. In contrast, an increase in the molecular weight of the PSty from 4,800 to 15,300 g/mol resulted in a limited change in the r_1 values, but a large increase in the r_2 values from 14.16 to 123.40 s^{-1} mM^{-1}, which is nearly ten times higher than that of the unmodified Gd MOF nanoparticles, and Magnevist® and MultiHance® (Table 9.3). This apparent decrease in the r_1 values was attributed to the substantial hydrophobic coating of PSty, which prevents interactions of water with vacant orbitals on the Gd^{3+}.

9.3.7 Surface Modification of Gd MOF Nanoparticles with Multifunctional RAFT Polymers

Next, a highly functional random copolymer of PHPMA-*co*-PmNAOS-*co*-PFMA was synthesized *via* RAFT polymerization employing the RAFT agent, DATC.

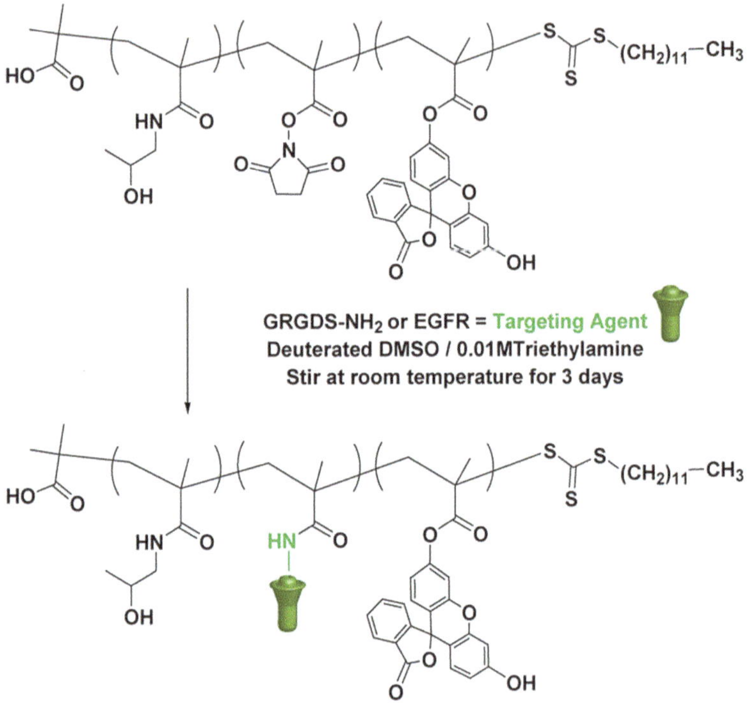

Scheme 9.2 General modification of RAFT copolymer with biomolecular targeting agents

The HPMA monomer was chosen because of its extensive use in polymers for bio-based applications and the overall biocompatibility of PHPMA [71], which comprises the major component of the prepared random copolymers in most cases. The mNAOS monomer was incorporated into the copolymer as a site for the attachment of different targeting ligands, as shown in Scheme 9.2. It has been widely reported that primary amines and alcohols will react readily with the succinimide functionality present on the mNAOS monomer [46, 72]. As each of the targeting ligands chosen to be used in this work contains an available primary amine, the targeting and/or chemotherapeutic loading can be easily modified by introducing the mNAOS monomer at different weight percentages into the copolymer. However, for the purpose of this study, the content of the mNAOS monomer was fixed at 15 wt%. The addition of the FMA monomer into the copolymer at 0.5 wt% allowed for the introduction of a fluorescence moiety, providing in vitro cellular level imaging and making the final polymer-modified Gd MOF nanoparticle bimodal with respect to imaging. RAFT polymerization is arguably the best living radical polymerization technique for the polymerization of functional monomers, such as HPMA, mNAOS, and FMA [44, 73, 74]. The experimental M_n of PHPMA-*co*-PmNAOS-*co*-PFMA (at 15 wt% mNAOS) was 10,100 g/mol, which is within 10% of the theoretical M_n of 10,900 g/mol. Furthermore, ^{1}H NMR confirmed the experimental weight percentage of PmNAOS to be within 2 wt% of the corresponding theoretical value.

Finally, due to the controlled nature of RAFT polymerizations, the PDI for the copolymer synthesized was 1.12, indicating a very narrow molecular weight distribution, and the GPC curve indicated a monomodal molecular weight distribution. Additionally, since the copolymerization of HPMA and mNAOS monomers was allowed to proceed to moderate conversions before the addition of the FMA monomer, ^{1}H NMR confirmed the copolymer backbone had approximately a random structure with blocky PFMA characteristic near the chain end.

Once the random copolymer was synthesized, the ability to incorporate a targeting ligand onto the polymer backbone via reaction with the mNAOS segments was investigated. The peptide GRGDS-NH_2 and antibody EGFR were chosen as the targeting ligands. The GRGDS-NH_2 ligand was chosen due to its ability to target the $\alpha_V\beta_3$ integrin, which is expressed in angiogenic vasculature in a range of cancerous tumors [75], while the anti-EGFR was chosen due to its ability to selectively target the epidermal growth factor receptor on a variety of cancer cells [76]. Attachment of the GRGDS-NH_2 or anti-EGFR peptide motifs to the copolymer was achieved via a condensation reaction between the succinimide group on the mNAOS monomer and the primary amine present on each motif in the presence of 0.01 M triethylamine (Scheme 9.2). Successful attachment of the targeting ligand was qualitatively confirmed by ^{1}H NMR spectroscopy. The ^{1}H NMR studies also provided information regarding critical characteristics of the multifunctional copolymers, including the number of targeting ligands per polymer chain. After surface modification of the Gd MOF nanoparticles with each of the targeted RAFT copolymers, the average grafting density was calculated to be about 22,500 chains/nm^2. Using the grafting density, the number of targeting agents per particle was calculated to be about 30×10^4 molecules GRGDS-NH_2 and 20×10^4 molecules of anti-EGFR, respectively.

The surface modification of nanoparticles to incorporate advanced functionality and biocompatibility is a critical step in the development of the next generation of nanoscale-targeted imaging agents. In a similar procedure as before, modification of the Gd MOF nanoparticles was achieved via initial aminolysis, using hexylamine, of the trithiocarbonate end group of the RAFT copolymers to the thiolate functionality under inert and basic conditions. Both ATR-FTIR and ^{1}H NMR spectroscopy confirmed near quantitative reduction of the trithiocarbonate end groups of the RAFT copolymer to the thiolate with the addition of hexylamine, followed by subsequent attachment of the thiolate end group to the Gd MOF nanoparticle surface. As discussed earlier, after polymer deposition and prior to characterization, the nanoparticles were washed several times with a good solvent for the polymer to remove untethered polymer from the system. Successful modification of the Gd MOF nanoparticles with the RAFT copolymer was characterized through both TEM and ATR-FTIR spectroscopy. The TEM image indicates a relatively uniform coating of polymer around the Gd MOF nanoparticles after deposition, with an average thickness of approximately 8 nm. ATR-FTIR was utilized to confirm the addition of the polymer onto the nanoparticles without the loss of the copolymer functionality. Several characteristic stretches of the free copolymer, including the carbonyl stretch at 1,735 cm^{-1}, succinimide stretch at 1,650 cm^{-1}, and methylene stretches from 2,810 to 3,000 cm^{-1}, confirm good transference to the polymer-modified Gd MOF nanoparticles when compared to the unmodified Gd MOF nanoparticles.

9.3.8 *In Vitro Imaging Capabilities of Multifunctional RAFT Copolymer-Modified Gd MOF Nanoparticles*

One of the primary requirements of any nanoscale-targeted imaging agent is the ability to image the construct using standard clinical techniques. The PHPMA-*co*-PmNAOS-*co*-PFMA copolymer-modified Gd MOF nanoparticles prepared in this work provide the specific advantage of multimodal imaging capability. The incorporation of the FMA monomer into the RAFT copolymer allows for cellular level imaging via fluorescence microscopy, while the Gd MOF nanoparticle acts as a positive contrast agent for MRI, providing diagnostic imaging at the clinical level. The presence of fluorescence and MRI in one nanoparticle is of high interest because it combines the sensitivity of the fluorescence component with the high degree of spatial resolution of MRI [34, 39, 58]. The clinical imaging viability of the copolymer-modified Gd MOF nanoparticles as a positive contrast agent was determined through in vitro MR imaging. The relaxivity values demonstrate that both the unmodified and the PHPMA-*co*-PmNAOS-*co*-PFMA copolymer-modified Gd MOF nanoparticles result in a large decrease in the r_1 values and, thus, behave as positive contrast agents. Of particular note is the fact that the copolymer-modified Gd MOF nanoparticles increased the longitudinal relaxivity ($r_1 = 40.3$ s^{-1} mM^{-1}) fourfold compared to the values determined for the unmodified Gd MOF nanoparticles ($r_1 = 9.86$ s^{-1} mM^{-1}), and two- and threefold compared to the clinical contrast agents, MultiHance® ($r_1 = 19.45$ s^{-1} mM^{-1}) and Magnevist® ($r_1 = 13.44$ s^{-1} mM^{-1}), respectively. As discussed before, this phenomenon is attributed to increased water retention by the hydrophilic RAFT copolymer matrix attached to the surface of the Gd MOF nanoparticles, thus enhancing T_1 relaxation shortening effects. This data demonstrates the feasibility of achieving clinically useful T_1 shortening effects with these novel multifunctional, polymer-modified Gd MOF nanoparticles.

9.3.9 *Molecular Targeting of Multifunctional Copolymer-Modified Gd MOF Nanoparticles*

In addition to an imaging component, one of the other essential requirements of a successful targeted imaging nanodevice is the presence of a molecular targeting component or ligand to increase the target selectivity of the system and, in the case of nanoparticles, to take advantage of the large amount of imaging agent that can be delivered to the desired location per targeting biorecognition event, thus reducing the amount of Gd^{3+} required for effective MR imaging [2]. The synthesized multifunctional, copolymer-modified Gd MOF nanoparticles have been designed to incorporate dual targeting components. The first targeting component is incorporated by tailoring the PHPMA-*co*-PmNAOS-*co*-PFMA copolymer with an active targeting ligand, either GRGDS-NH_2 or anti-EGFR. The second potential route of targeting is passive targeting through the enhanced permeability and retention (EPR) effect. This tumor-targeting mechanism results from the leaky vasculature present in most tumors, resulting in preferential extravasation and protracted lodging of particles of a particular size [77]. The EPR effect can be taken advantage of due to the controllable size of the Gd MOF nanoparticles.

To investigate the ability of the polymer-modified Gd MOF nanoparticles to be utilized as theragnostic devices for cancer, using either the active GRGDS-NH_2 ligand or anti-EGFR, cell flow cytometry and in vitro targeted MRI experiments were completed. Two head and neck squamous cell carcinomas were chosen for these experiments, UMSCC10 and UMSCC2. UMSCC10 was utilized as the negative control because there should be no active targeting mechanisms available in this cell line. On the other hand, UMSCC2 cells were used as the positive control because they overexpress $\alpha_v\beta_3$ integrins, which are targeted with GRGDS-NH_2, and they show an increased number of EGFR at their vasculature, which can be targeted utilizing the anti-EGFR moiety. Cell flow cytometry experiments proved to be quite successful for both targeted Gd MOF nanoparticle samples. As was noted above, FMA was incorporated into the RAFT copolymer backbone to allow for optical imaging, in this case, allowing for a quantitative measurement of targeting by the Gd MOF nanoparticles. For example, after the untargeted Gd MOF nanoparticles, or Gd MOF nanoparticles modified with GRGDS-NH_2 (Fig. 9.2a) or anti-EGFR were incubated for 1 h with the negative control cells, UMSCC10, a nominal amount of less than 30% of fluorescence shift was visualized, showing minimal nonspecific uptake of any of these nanoparticles by the cells. It is quite possible that this small shift in fluorescence is due to passive targeting strictly based on the Gd MOF nanoparticle size. However, when the GRGDS-NH_2-modified Gd MOF nanoparticles were incubated with UMSCC2 cells, an increase of almost 100% of fluorescence shift was seen in comparison to the untreated cells (Fig. 9.2b). This data not only confirms a cellular level-specific interaction but also shows that nearly all of the Gd MOF nanoparticles are labeled with the GRGDS-NH_2 targeting ligand. This shift in fluorescence is expected to be caused by the increased number of $\alpha_v\beta_3$ integrins on or within the UMSCC2 cells interacting with the GRGDS-NH_2 peptide on the Gd MOF nanoparticles. In addition, Gd MOF nanoparticles modified with the anti-EGFR ligand showed similar results. Again, a gated increase of nearly 100% confirmed both specificity of the anti-EGFR-modified Gd MOFs towards the UMSCC2 cell line and uniformly labeled nanoparticle constructs.

Finally, in vitro MRI experiments allowed determination of enhanced imaging capabilities with incorporation of a molecular targeting moiety. T_1-weighted MRI was performed using a 4.7-T Bruker PharmaScan instrument. For each experiment, Gd MOF nanoparticle samples were incubated with either UMSCC10 or UMSCC2 cell lines for 1 h in phosphate-buffered saline (PBS). Incubation dishes were washed several times with PBS to remove unattached Gd MOF nanoparticles before the cells were pelletized in microcentrifuge tubes. In vitro MRI was then utilized to measure T_1 relaxation times of each sample to determine if there was enhanced imaging as a result of incorporation of active targeting Gd MOF nanoparticles. A baseline sample of UMSCC10 and UMSCC2 cell pellets was prepared as above

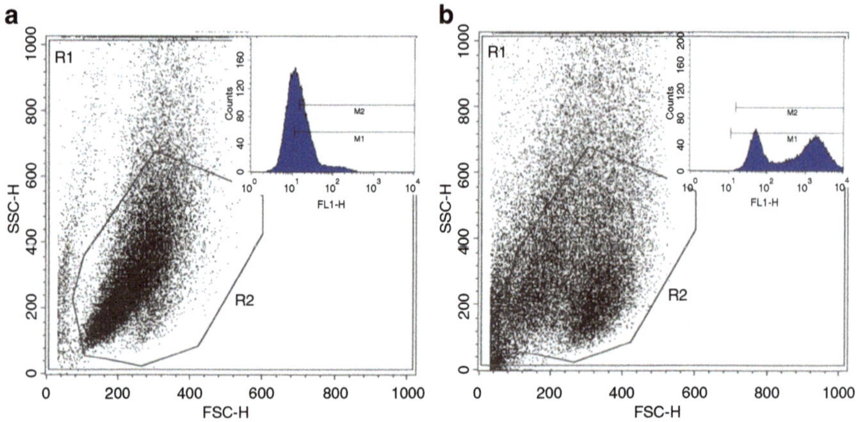

Fig. 9.2 In vitro cell flow cytometry for GRGDS-NH_2-modified Gd MOF nanoparticles, incubated with (**a**) UMSCC10 cell line and (**b**) UMSCC2 cell line. The peak shift observed in the *inset* in (**b**) demonstrates attachment of the nanoparticles to the targeted cells (main plots: *y*-axis—SSC-H = side light scatter; *x*-axis—FSC-H = forward light scatter)

Table 9.4 In vitro MRI relaxation experimental data

	T1 relaxation (ms)	
Contrast agent	UMSCC10(−)	UMSCC2 (+)
Control (cells only)	971	1,046
Untargeted Gd MOF nanoparticles	984	978
GRGDS-NH2-modified Gd MOF nanoparticles	940	912
Anti-EGFR-modified Gd MOF nanoparticles	979	822

without the addition of Gd MOF nanoparticles. As with the cell flow cytometry studies, when the untargeted Gd MOF nanoparticles, or Gd MOF nanoparticles modified with GRGDS-NH_2 or anti-EGFR were incubated with the negative control UMSCC10 cell line, there were only nominal changes in the T_1 relaxation times in comparison to the baseline (Table 9.4). As suggested above, these small changes could be due to either a small amount of nonspecific uptake or, more likely, due to a passive targeting mechanism. However, when the GRGDS-NH_2-targeted Gd MOF nanoparticles were incubated with the positive UMSCC2 cell line, a T_1 relaxation rate of 912 ms was seen, which is a significant enhancement from that of the baseline. Furthermore, the anti-EGFR-modified nanoparticles showed an even larger change of T_1 relaxation time from 822 to 1,046 ms in comparison to the baseline measurements (Table 9.4). These results not only demonstrate the capability of targeted Gd MOF nanoparticles to be employed as positive nanoparticle-based contrast agents in MRI but also show the flexibility of our constructs to be utilized in targeted imaging and treatment of a range of different advanced diseases.

9.4 Future Perspective

It is becoming increasingly more evident from the number of publications, conference symposia, and clinical trials that nanotechnology will play a pivotal role in the development of the next generation of diagnostics and therapeutics. The development of nanoparticles for use as imaging agents has grown dramatically in the past decade. This development has been driven by tremendous advances in chemistry, physics, nanotechnology, molecular biology, and radiology and has resulted in a wide range of different formulations to diagnose a multitude of diseases. Of particular interest recently has been the development of Gd-based or Gd-labeled nanoparticles for use as MRI contrast agents. The use of Gd in the nanoparticle construct is particularly interesting as it provides a positive MRI contrast agent. While SPIO nanoparticles have previously received significant attention as a MRI contrast agent, they are a negative contrast agent and have several drawbacks, including limited in vivo cell tracking (MRI cannot distinguish the negative contrast agent from other signal voids), and negative contrast agents are limited by partial volume effects. Positive contrast nanoparticles based upon Gd demonstrate tremendous potential as their use enables the loading of the nanoparticle with large quantities of Gd, dramatically improving the relaxivity when compared to small molecule Gd chelates. In addition, the ability to control the nanoparticle size allows for passive targeting and a decrease of the tumbling rate of the contrast agent which, again, improves the relaxivity. Finally, new surface modification techniques have allowed for the incorporation of specific molecular targeting properties and improved biocompatibility, both of which will significantly improve the specificity of the contrast agent, leading to earlier diagnosis. Even though remarkable steps have been taken in the development of nanoparticle-based diagnostic devices, the advances on the horizon are even more exciting. Researchers are already developing multimodal imaging agents to allow for tracking of tumor cells, provide guidance to surgeons, and enable quick and accurate analysis to ensure complete tumor removal. The use of stimuli-responsive polymers on the surface of nanoparticle imaging agents will allow for the development of "smart" imaging agents that are capable of responding directly to their surrounding environment. For example, pH-responsive polymers may allow for metabolic imaging and also for improved specificity in cancer diagnosis. While the strides made in developing nanoparticles for use as a positive MRI contrast agents have been extraordinary, there is still a long way to go before these nanodevices reach the clinic. The vast majority of current reports on Gd-based nanoparticles as MRI contrast agents have focused on in vitro studies. The translation of the nanodevices from the lab to the clinic will require comprehensive in vivo studies, investigating the pharmacological and toxicological properties of potential nanodevices. This will require multidisciplinary research teams consisting of chemists, engineers, molecular biologists, radiologists, and clinicians to optimize the size, shape, surface properties, and functionality of the nanodevices in order to optimize in vivo efficacy and safety. Although the required work to develop nanoparticles for clinical use as positive MRI contrast agent may seem vast, the benefits of these devices are immense and potentially offer unparalleled effectiveness in the diagnosis and treatment of disease and improve the quality of life for patients.

9.5 Conclusions

The development of nanoscale imaging agents for the diagnosis of cancer represents one of the primary targets of the general field of nanomedicine. Despite the fact that the incredible potential of these devices is widely recognized, their clinical application has yet to be realized due to poor design and manufacturing techniques. The research presented here has developed a method for the surface modification of Gd MOF nanoparticles with well-defined, highly functional RAFT polymers and demonstrated their applicability towards use as positive contrast nanoparticle agents in MRI. Specifically, a range of RAFT homopolymers, PNIPAM, PHPMA, and PSty, and novel multifunctional copolymers of PHPMA-*co*-PmNAOS-*co*-PFMA were synthesized by employing the RAFT agent, DATC. PHPMA-*co*-mPNAOS-*co*-PFMA copolymers were successfully combined with a targeting ligand, GRGDS-NH_2 or anti-EGFR, to form a multifunctional, polymeric nanoparticle. Following preparation of the RAFT polymers, quantitative aminolysis of the trithiocarbonate end groups in basic conditions provided thiolate end groups which allowed for the direct attachment of the polymers to the surface of the Gd nanoparticle MOFs. Successful modification was attributed to thiolate attachment through vacant orbitals on the Gd^{3+} ions at the surface of the Gd MOF nanoparticles. To evaluate the potential of the RAFT polymer-modified nanoparticles as a targeted diagnostic nanodevice, MRI, fluorescence imaging, and cell flow cytometry studies were performed. This research demonstrated that the molecular weight and chemical properties of polymers used to modify Gd MOF nanoparticles are intimately connected with their T_1 relaxation rates and that variation in these properties provides a means to tune the relaxivity of the nanoparticles. In the MRI studies, the unmodified Gd MOF nanoparticles showed an r_1 value comparable to the clinically employed contrast agents, Magnevist® and MultiHance®, while the majority of the RAFT polymer-modified Gd MOF nanoparticles displayed significantly enhanced r_1 values in comparison to those clinical contrast agents. Furthermore, tailoring the chemical and physical properties of the RAFT polymers used for the surface modification of Gd MOF nanoparticles has shown the ability to tailor and tune the r_1 values, thus providing greatly enhanced T_1 relaxation values in comparison to the unmodified structure and clinically used small molecule contrast agents. The incorporation of FMA into the multifunctional copolymer constructs provided polymer-modified Gd MOF nanoparticles that were successfully used as a bimodal diagnostic imaging device for both MR and fluorescence imaging. Additionally, the use of fluorescence imaging in conjunction with cell flow cytometry demonstrated that the polymer-modified targeted nanoparticles have a cellular level-specific interaction, and also that nearly all of the Gd MOF nanoparticles are labeled with the GRGDS-NH_2 targeting ligand.

Finally, in vitro MRI experiments demonstrated the capability of targeted Gd MOF nanoparticles to be employed as positive, nanoparticle-based contrast agents in MRI and in addition demonstrated the flexibility of our constructs to be utilized in targeted imaging and treatment of a range of different advanced diseases. By taking advantage of advancements in nanotechnology and polymer science, this research has effectively prepared new nanoscale multifunctional devices with tumor targeting and diagnostic imaging capability.

References

1. Ferrari M (2005) Cancer nanotechnology: opportunities and challenges. Nat Rev Cancer 5:161–171
2. Allen TM (2002) Ligand-targeted therapeutics in anticancer therapy. Nat Rev Drug Discov 2:750–763
3. Jain RK (1998) The next frontier of molecular medicine: delivery of therapeutics. Nat Med 4:655–657
4. Yezhelyev MV, Gao X, Xing Y, Al-Hajj A, Nie S, O'Regan RM (2006) Emerging use of nanoparticles in diagnosis and treatment of breast cancer. Lancet Oncol 7:657–667
5. Nasongkla N, Bey E, Ren J, Al H, Khemtong C, Guthi JS, Chin S-F, Sherry AD, Boothman DA, Gao J (2006) Multifunctional polymeric micelles as cancer-targeted, MRI-ultrasensitive drug delivery systems. Nano Lett 6:2427–2430
6. Panchapakesan B (2005) Nanotechnology: part 2 – tiny technology – tremendous therapeutic potential. Oncol Issues 20:20–23, November/December
7. Park K (2007) Nanotechnology: what it can do for drug delivery. J Control Release 120:1–3
8. Wickline SA, Lanza GM (2003) Nanotechnology for molecular imaging and targeted therapy. Circulation 107:1092–1095
9. Wang X, Yang L, Chen Z, Shin DM (2008) Application of nanotechnology in cancer therapy and imaging. CA Cancer J Clin 58:97–110
10. Salamanca-Buentello F, Persad DL, Court EB, Martin DK, Daar AS, Singer PA (2005) Nanotechnology and the developing world. PLoS Med 2:383–386
11. Cheon J, Lee J-H (2008) Synergistically integrated nanoparticles as multimodal probes for nanobiotechnology. Acc Chem Res 41:1630–1640
12. Sharma P, Brown S, Walter G, Santra S, Moudgil B (2006) Nanoparticles for bioimaging. Adv Colloid Interface Sci 123–126:471–485
13. Zhang Y, Shang M (2004) Self-assembled coatings on individual monodisperse magnetite nanoparticles for efficient cellular uptake. Microdevices 6:33–40
14. Cohen MH, Melnik K, Boiasrki A, Ferrari M, Martin FJ (2003) Microfabrication of silicon-based nanoporous particulates for medical applications. Biomed Microdevices 5:253–259
15. He XX, Li J (2003) Bioconjugated nanoparticles for DNA protection from cleavage. J Am Chem Soc 125:7168–7169
16. Chithrani BD, Chan WCW (2007) Elucidating the mechanism of cellular uptake and removal of protein-coated gold nanoparticles of different sizes and shapes. Nano Lett 7:1542–1550
17. Daniel M-C, Astruc D (2004) Gold nanoparticles: assembly, supramolecular chemistry, quantum-size-related properties, and applications toward biology, catalysis, and nanotechnology. Chem Rev 104:293–346
18. Everts M, Saini V, Leddon JL, Kok RJ, Stoff-Khalili M, Preuss MA, Millican CL, Perkins G, Brown JM, Bagaria H, Nikles DE, Johnson DT, Zharov VP, Curiel DT (2006) Covalently linked Au nanoparticles to a viral vector: potential for combined photothermal and gene cancer therapy. Nano Lett 6:587–591

19. Kim D, Park S, Lee JH, Jeong YY, Jon S (2007) Antibiofouling polymer-coated gold nanoparticles as a contrast agent for in vivo X-ray computed tomography imaging. J Am Chem Soc 129:7661–7665
20. Su C-H, Sheu H-S, Lin C-Y, Huang C-C, Lo Y-W, Pu Y-C, Weng J-C, Shieh D-B, Chen J-H, Yeh C-S (2007) Nanoshell magnetic resonance imaging contrast agents. J Am Chem Soc 129:2139–2146
21. Bharali DJ, Lucey DW, Jayakumar H, Pudavar HE, Prasad PN (2005) Folate-receptor-mediated delivery of InP quantum dots for bioimaging using confocal and two-photon microscopy. J Am Chem Soc 127:11364–11371
22. Loo C, Lowery A, Halas N, West J, Drezek R (2005) Immunotargeted nanoshells for integrated cancer imaging and therapy. Nano Lett 5:709–711
23. McCarthy JR, Weissleder R (2008) Multifunctional magnetic nanoparticles for targeted imaging and therapy. Adv Drug Deliv Rev 60:1241–1251
24. Sanvicens N, Marco MP (2008) Multifunctional nanoparticles – properties and prospects for their use in human medicine. Trends Biotechnol 26:425–433
25. Smith AM, Duan H, Mohs AM, Nie S (2008) Bioconjugated quantum dots for in vivo molecular and cellular imaging. Adv Drug Deliv Rev 60:1226–1240
26. Huang X, El-Sayed IH, Qian W, El-Sayed MA (2006) Cancer cell imaging and photothermal therapy in the near-infrared region by using gold nanorods. J Am Chem Soc 128:2115–2120
27. Hifumi H, Yamaoka S, Tanimoto A, Citterio D, Suzuki K (2006) Gadolinium-based hybrid nanoparticles as a positive MR contrast agent. J Am Chem Soc 128:15090–15091
28. Artemov D, Bhujwalla ZM, Bulte JWM (2004) Magnetic resonance imaging of cell surface receptors using targeted contrast agents. Curr Pharm Biotechnol 5:485–494
29. Rinck PA, Bjørnerud A (2001) Magnetic resonance in medicine. Wiley-Blackwell, New York
30. Reynolds CH, Annan N, Beshah K, Huber JH, Shaber SH, Lenkinske RE, Wortman JA (2000) Gadolinium-loaded nanoparticles: new contrast agents for magnetic resonance imaging. J Am Chem Soc 122:8940–8945
31. Rieter WJ, Taylor KML, An H, Lin W, Lin W (2006) Nanoscale metal-organic frameworks as potential multimodal contrast enhancing agents. J Am Chem Soc 128:9024–9025
32. Rieter WJ, Taylor KML, Lin W (2007) Surface modification and functionalization of nanoscale metal-organic frameworks for controlled release and luminescence sensing. J Am Chem Soc 129:9852–9853
33. Oyewumi MO, Mumper RJ (2002) Engineering tumor-targeted gadolinium hexanedione nanoparticles for potential applications in neutron capture therapy. Bioconjug Chem 13:1328–1335
34. Bridot J-L, Faure A-C, Laurent S, Riviere C, Billotey C, Hiba B, Janier M, Josserand V, Coll J-L, Vander Elst L, Muller R, Roux S, Perriat P, Tillement O (2007) Hybrid gadolinium oxide nanoparticles: multimodal contrast agents for in vivo imaging. J Am Chem Soc 129:5076–5084
35. Oyewumi MO, Liu S, Moscow JA, Mumper RJ (2003) Specific association of thiamine-coated gadolinium nanoparticles with human breast cancer cells expressing thiamine transporters. Bioconjug Chem 14:404–411
36. Oyewumi MO, Yokel RA, Jay M, Coakley T, Mumper RJ (2004) Comparison of cell uptake, biodistribution and tumor retention of folate-coated and PEG-coated gadolinium nanoparticles in tumor-bearing mice. J Control Release 95:613–626
37. Evanics F, Diamente PR, van Veggel FCJM, Stanisz GJ, Prosser RS (2006) Water-soluble GdF_3 and GdF_3/LaF_3 nanoparticles – physical characterization and NMR relaxation properties. Chem Mater 18:2499–2505
38. Gobin AM, Lee MH, Halas NJ, James WD, Drezek RA, West JL (2007) Near-infrared resonant nanoshells for combined optical imaging and photothermal cancer therapy. Nano Lett 7:1929–1934
39. Taylor KML, Jin A, Lin W (2008) Surfactant-assisted synthesis of nanoscale gadolinium metal-organic frameworks for potential multimodal imaging. Angew Chem Int Ed Engl 47:7722–7725

40. Rowe MD, Chang C-C, Thamm DH, Kraft SL, Harmon JF Jr, Vogt AP, Sumerlin BS, Boyes SG (2009) Tuning the magnetic resonance imaging properties of positive contrast agent nanoparticles by surface modification with RAFT polymers. Langmuir 25:9487–9499
41. Rowe MD, Thamm DH, Kraft SL, Boyes SG (2009) Polymer-modified gadolinium metal-organic framework nanoparticles used as multifunctional nanomedicines for the targeted imaging and treatment of cancer. Biomacromolecules 10:983–993
42. Lai JT, Filla D, Shea R (2002) Functional polymers from novel carboxyl-terminated trithiocarbonates as highly efficient RAFT agents. Macromolecules 35:6754–6756
43. Kopecek J, Bazilova H (1973) Poly[N-(2-hydroxypropyl)methacrylamide]. 1. Radical polymerization and copolymerization. Eur Polym J 9:7–14
44. Perrier S, Takolpuckdee P, Westwood J, Lewis DM (2004) Versatile chain transfer agents for reversible addition fragmentation chain transfer (RAFT) polymerization to synthesize functional polymer architectures. Macromolecules 37:2709–2717
45. Le TP, Moad G, Rizzardo E, Thang SH (1998) PCT Int Appl WO 98 01478 A1 980115
46. Yanjarappa MJ, Gujraty KV, Joshi A, Saraph A, Kane RS (2006) Synthesis of copolymers containing an active ester of methacrylic acid by RAFT: controlled molecular weight scaffolds for biofunctionalization. Biomacromolecules 7:1665–1670
47. Hong C-Y, Pan C-Y (2006) Direct synthesis of biotinylated stimuli-responsive polymer and diblock copolymer by RAFT polymerization using biotinylated trithiocarbonate as RAFT agent. Macromolecules 39:3517–3524
48. Nguyen TL, Tey SY, Pourgholami MH, Morris DL, Davis TP, Barner-Kowollik C, Stenzel MH (2007) Synthesis of semi-biodegradable crosslinked microspheres for the delivery of 1,25 dihydroxyvitamin D3 for the treatment of hepatocellular carcinoma. Eur Polym J 43:1754–1767
49. Scales CW, Huang F, Li N, Vasilieva YA, Ray J, Convertine AJ, McCormick CL (2006) Corona-stabilized interpolyelectrolyte complexes of SiRNA with nonimmunogenic, hydrophilic/cationic block copolymers prepared by aqueous RAFT polymerization. Macromolecules 39:6871–6881
50. Zelikin AN, Such GK, Postma A, Caruso F (2007) Poly(vinylpyrrolidone) for bioconjugation and surface ligand immobilization. Biomacromolecules 8:2950–2953
51. Li M, De P, Gondi SR, Sumerlin BS (2008) Responsive polymer-protein bioconjugates prepared by RAFT polymerization and copper-catalyzed azide-alkyne click chemistry. Macromol Rapid Commun 29:1172–1176
52. De P, Li M, Gondi SR, Sumerlin BS (2008) Temperature-regulated activity of responsive polymer–protein conjugates prepared by grafting-from via RAFT polymerization. J Am Chem Soc 130:11288–11289
53. Sumerlin BS, Lowe AB, Stroud PA, Zhang P, Urban MW, McCormick CL (2003) Modification of gold surfaces with water-soluble (co)polymers prepared via aqueous reversible addition-fragmentation chain transfer polymerization. Langmuir 19:5559–5562
54. Hotchkiss JW, Lowe AB, Boyes SG (2007) Surface modification of gold nanorods with RAFT synthesized polymers. Chem Mater 19:6–13
55. Zhang Q, Gupta S, Emrick T, Russell TP (2006) Surface-functionalized CdSe nanorods for assembly in diblock copolymer templates. J Am Chem Soc 128:3898–3899
56. Lowe AB, Sumerlin BS, Donovan MS, McCormick CL (2002) Facile preparation of transition metal nanoparticles stabilized by well-defined (co)polymers synthesized via aqueous reversible addition-fragmentation chain transfer polymerization. J Am Chem Soc 124:11562–11563
57. Hartman KB, Laus S, Bolskar RD, Muthupillai R, Helm L, Toth E, Merbach AE, Wilson LJ (2008) Gadonanotubes as ultrasensitive pH-smart probes for magnetic resonance imaging. Nano Lett 8:415–419
58. Kim JS, Rieter WJ, Taylor KML, An H, Lin W, Lin W (2007) Self-assembled hybrid nanoparticles for cancer-specific multimodal imaging. J Am Chem Soc 129:8962–8963
59. Allen M, Bulte JWM, Liepold L, Basu G, Zywicke HA, Frank JA, Young M, Douglas T (2005) Paramagnetic viral nanoparticles as potential high-relaxivity magnetic resonance contrast agents. Magn Reson Med 54:807–812

60. Caravan P, Ellison JJ, McMurry TJ, Lauffer RB (1999) Gadolinium(III) chelates as MRI contrast agents: structure, dynamics, and application. Chem Rev 99:2293–2352
61. Lin W, Rieter WJ, Taylor KML (2009) Modular synthesis of functional nanoscale coordination polymers. Angew Chem Int Ed 48:650–658
62. Rodrigues RK, da Silva MA, Sabadini E (2008) Worm-like micelles of CTAB and sodium salicylate under turbulent flow. Langmuir 24:13875–13879
63. Shukla A, Rehage H (2008) Zeta potentials and Debye screening lengths of aqueous viscoelastic surfactant solutions (cetyltrimethylammonium bromide/sodium salicylate system). Langmuir 24:8507–8513
64. Boyes SG, Granville AM, Baum M, Akgun B, Mirous BK, Brittain WJ (2004) Recent Advances in the synthesis and rearrangement of block copolymer brushes. In: Advincula RC, Brittain WJ, Caster KC, Ruhe J (eds) Polymer brushes. Weinheim, Wiley-VCH Verlag GmbH & Co, pp 151–165
65. Boyes SG, Granville AM, Baum M, Akgun B, Mirous BK, Brittain WJ (2004) Polymer brushes-surface immobilized polymers. Surf Sci 570:1–12
66. Jordan R (ed) (2006) Surface initiated polymerization I and II: advances in polymer science. Springer, Berlin, p 214
67. Liu Y, Klep V, Zdyrko B, Luzinov I (2004) Polymer grafting via ATRP initiated from macroinitiator synthesized on surface. Langmuir 20:6710–6718
68. Motornov M, Sheparovych R, Katz E, Minko S (2008) Chemical gating with nanostructured responsive polymer brushes: mixed brush versus homopolymer brush. ACS Nano 2:41–52
69. Muthukrishnan S, Erhard DP, Mori H, Muller AHE (2006) Synthesis and characterization of surface-grafted hyperbranched glycomethacrylates. Macromolecules 39:2743–2750
70. Gao H, Matyjaszewski K (2007) Synthesis of molecular brushes by "grafting onto" method: combination of ATRP and click reactions. J Am Chem Soc 129:6633–6639
71. Vihola H, Laukkanen A, Valtola L, Tenhu H, Hirvonen J (2005) Cytotoxicity of thermosensitive polymers poly(*N*-isopropylacrylamide), poly(*N*-vinylcaprolactam) and amphiphilically modified poly(*N*-vinylcaprolactam). Biomaterials 26:3055–3064
72. Favier A, D'Angosto F, Charreyre M-T, Pichot C (2004) Synthesis of N-acryloxysuccinimide copolymers by RAFT polymerization, as reactive building blocks with full control of composition and molecular weight. Polymer 45:7821–7830
73. Rizzardo E, Chiefari J, Mayadunne RTA, Moad G, Thang SH (2000) Synthesis of defined polymer by reversible addition-fragmentation chain transfer (the RAFT process). ACS Symp Ser 768:278
74. Lowe AB, McCormick CL (2007) Reversible addition–fragmentation chain transfer (RAFT) radical polymerization and the synthesis of water-soluble (co)polymers under homogeneous conditions in organic and aqueous media. Prog Polym Sci 32:283–351
75. Ye Y, Bloch S, Xu B, Achilefu S (2006) Design, synthesis, and evaluation of near infrared fluorescent multimeric RGD peptides for targeting tumors. J Med Chem 49:2268–2275
76. Ciardiello F, Tortora G (2003) Epidermal growth factor receptor (EGFR) as a target in cancer therapy: understanding the role of receptor expression and other molecular determinants that could influence the response to anti-EGFR drugs. Eur J Cancer 39:1348–1354
77. Brigger I, Dubernet C, Couvreur P (2002) Nanoparticles in cancer therapy and diagnosis. Adv Drug Deliv Rev 54:631–651

Part IV
Diagnostic and Therapy ("Theranostics")

Chapter 10
Injectable Multistage Nanovectors for Enhancing Imaging Contrast and Directed Therapy

Biana Godin, Rita E. Serda, Xuewu Liu, and Mauro Ferrari

10.1 Introduction: Injectable Nanovector Taxonomy

The concept of a "magic bullet," envisioned by Paul Ehrlich in the beginning of the twentieth century [1], was initially considered futuristic and unfeasible. However, the advent of molecular targets has altered this view. Currently, remarkable advances in understanding the pathological processes and identifying molecular signatures of various diseases have been made, enabling us to design specific and efficient targeted therapeutics. As an example, in just one decade, the FDA has approved over 30 molecularly targeted anticancer drugs [2]. However, albeit an indisputable therapeutic potential at the molecular level, achieving successful clinical translation of these agents is frequently challenging. Physicochemical properties of some of these agents significantly impact their effective administration. As an example, the polycyclic nature of certain drugs, such as paclitaxel, makes them practically insoluble in aqueous environments [3]. On the other hand, recently discovered, highly potent bioactive substances, such as small-interfering RNA (siRNA), rapidly degrade in blood. Both characteristics make these drugs unacceptable for intravenous administration. A variety of obstacles further lie in the presence of multiple biological barriers, preventing the administered drug or imaging agent from reaching the target tissue. As a result, the distribution of these agents is highly unspecific, with only 1 in 10,000–100,000 molecules reaching their target tissue. Overcoming these problems is the fundamental driving force behind the concept of nanotherapeutic drug delivery. In other words, nanovectors are being developed and investigated as carriers for personalized therapeutic and imaging contrast agents based on the simultaneous, anticipated advantages of homing to the diseased site (e.g., cancer lesions, atherosclerotic plaque). Being at the interface of the dimensions of molecules, the

B. Godin (✉) • R.E. Serda • X. Liu • M. Ferrari (✉)
The Methodist Hospital Research Institute, Department of Nanomedicine,
6670 Bertner, St., R7-122, Houston, TX 77030, USA
e-mail: bgodin@tmhs.org; mferrari@tmhs.org

S. Svenson and R.K. Prud'homme (eds.), *Multifunctional Nanoparticles for Drug Delivery Applications: Imaging, Targeting, and Delivery*, Nanostructure Science and Technology, DOI 10.1007/978-1-4614-2305-8_10, © Springer Science+Business Media, LLC 2012

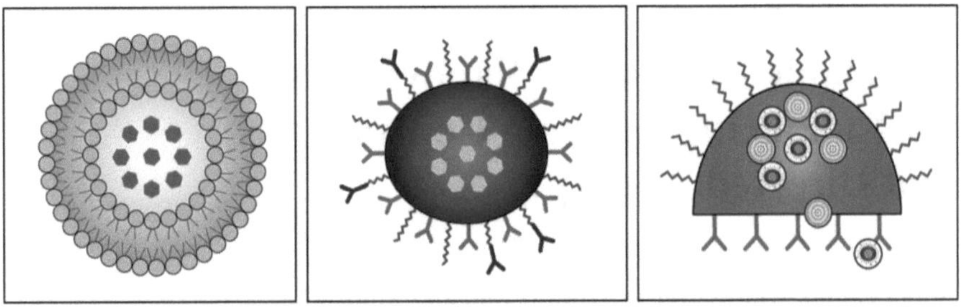

Fig. 10.1 Schematic presentation of three generations of nanovectors: (*Left*) First-generation nanovectors, as the currently clinical liposomes, comprise a container (double phospholipid bilayer membrane) and an active principle (*dots*). They localize in the tumor by enhanced permeation and retention (EPR); (*middle*) Second-generation nanovectors further possess the ability for the targeting of their therapeutic action via antibodies and other biomolecules, remote activation, or responsiveness to environment; (*right*) Third-generation nanovectors such as multistage agents are capable of more complex functions, such as time-controlled deployment of multiple waves of active nanoparticles, deployed across different biological barriers and with different subcellular targets. Reproduced with permission from [10]

biological building blocks, and cells, the biologically relevant functional structures, nanocarriers can be designed for overcoming the variety of obstacles, so-called bio-barriers, located between the administration site and the target organ. Historically, oncology represents the field of medicine to which nanotechnology made the most prominent contributions. Beginning with the FDA approval of liposomal doxorubicin in the mid-1990s for use against Kaposi's sarcoma [4–6], a variety of nanocarrier-based drug delivery systems have been conceived and are in different stages of development, including particles with various compositions, physicochemical characteristics, geometry, and surface functionalizations [7, 8]. The library of particles, generated by all the possible combinations, is gigantic, and clear considerations should be taken when developing carriers for specific drugs or conditions.

A general taxonomy can be applied to nanovectors, dividing them into three main generations. These three generations are schematically presented in Fig. 10.1 [4, 8–10]. According to this classification, the first generation of nanovectors describes carriers that localize to the lesion through passive mechanisms. Among these carriers, the liposomes still remain the main representatives [11], utilizing the enhanced permeability of the disease-associated vasculature and localizing into the desired site through the so-called enhanced permeability and retention (EPR) mechanism proposed by Maeda [12, 13]. Other nanocarriers in this generation include polymer–drug conjugates [14], polymeric micelles [15], and dendrimers [16]. Some surface modifications are frequently seen in nanovectors of this subclass, e.g., attachment of a poly(ethylene glycol) (PEG). This modification produces "stealth" nanovectors with substantially prolonged circulation time and, as a result, more likely passive tumor homing [11, 17–19]. The second generation of nanovectors describes systems with additional advanced functionalities, such as (1) attachment of recognition moieties on the nanovector surface specific to the disease "zip codes" or (2) a possibility for active/triggered release of the payload at the diseased location through the use of physical remote energy [20–24]. This category generally presents a progressive evolution of the first generation of nanovectors with emerging degrees of sophistication. Liposomes and other antibody-targeted nanoparticles have been the most investigated example of the second generation [4, 6, 8, 11, 20, 22, 25–27]. A variety of other targeting moieties besides antibodies are under extensive investigation worldwide, including ligands, aptamers, small peptides, and phage-display peptides binding to specific target cell surface markers or surface markers expressed in the disease microenvironment [28]. Examples of other nanocarriers in the first and second generations include metal nanoparticles for use in diagnostics and therapy [29, 30], albumin-bound paclitaxel nanoparticles approved for use in metastatic breast cancer [31], drug–polymer construct dendrimers, and polymeric micelles [25, 32–36].

The third generation of nanovectors encompasses therapeutic and diagnostic multicomponent and multifunctional constructs with logic-embedded functions. These systems aim at overcoming a variety of obstacles in order for the therapeutic and diagnostic agents to efficiently reach their target, representing a paradigm shift when compared to the previous two generations of nanovectors [27]. The third-generation multicomponent carriers are comprised of a number of nanoengineered components with the advantage of decoupling functions such as biorecognition, cytotoxicity, and biobarrier avoidance to separate nanocomponents acting in a synergistic pre-programmable and sequential manner, encoded in the properties of the material.

Multistage nanovectors (MSV), an emblematic system for the third-generation nanocarriers, are comprised of nanoporous silicon particles that utilize their unique particle geometry and other physical characteristics in concert with active biological targeting moieties to efficiently solve sequential mission-critical challenges and to deliver therapeutic payloads of nanoparticles, loaded into the porous silicon structure, to the disease loci [6, 27, 37]. In MSV carriers, each component (or stage) performs part of the journey from the site of administration toward the target lesion, negotiating one or more biological barriers and adding a degree of targeting selectivity in the process. The geometry of the "mothership" nanoporous silicon particles, or first-stage particles, is mathematically optimized to exhibit superior margination, firm cellular adhesion [38, 39], and internalization [40, 41] properties. Using photolithographic techniques and bioconjugation methods, we have fabricated a large array of first-stage particles with specific surface characteristics to meet the criteria chosen by predictive design maps. These biodegradable and biocompatible first-stage vectors are loaded with the second-stage nanovectors, which can essentially be any of the above-mentioned first- or second-generation vectors [6, 37]. The release profiles of the second-stage vector from the first-stage particle can be finely tuned to take place at different times and through different paths. For example, particles can be intracellularly internalized [42, 43] and tailor-made to deliver their payloads to different subcellular structures. In this chapter, we will describe studies related to the MSV, focusing on fabrication, intracellular multisite trafficking, advanced therapeutic systems, and contrast agents.

10.2 Fabrication of the Multistage Nanovectors

As mentioned above, the concept of a multistage delivery system requires fabrication of proper injectable porous particles with designed geometry and pore morphology as the first-stage carriers to load, transport, and on-site release the second-stage nanoparticles [37]. Such particles should be able to provide extensive surface area to hold second-stage nanometer-sized delivery carriers such as liposomes, carbon nanotubes, and other organic or inorganic nanoparticles. Porous silicon, a biodegradable nanomaterial, has been explored for numerous biomedical applications, based on its readily modified physicochemical and biophysical properties [43–47]. Owing to its advantageous, fully biodegradability nature, in addition to the full set of sophisticated tools from the silicon micromachining industry, porous silicon particles are emerging as promising injectable drug delivery agents, thus making porous silicon an ideal candidate for multistage delivery vectors.

Until recently, porous silicon was used in the format of powdered materials obtained by ultrasonic fracture [48, 49] or ball milling [47] of electrochemically etched porous silicon films. The resulting particles were characterized by their irregular shape and polydispersed size even though different subsequent sorting strategies were applied. In the multistage system, precisely controlled size and shape of the first-stage particles would be desired for highly efficient directed delivery. Size and shape of nanovectors have been shown to fundamentally determine several properties of the particle that are relevant for drug delivery, such as flow dynamics, margination, degradation rate, and cell uptake [40, 50]. Based on mathematical design, nonspherical vectors are favorable compared to their spherical counterparts. Thus, integration of top-down fabrication approaches of porous silicon production with photolithography patterning, originated in the silicon semiconductor industry, has been adapted to produce particles with defined geometry for biomedical applications. Silicon technology has been well established in terms of production, characterization, and translation into biomedical nanotechnologies. Scalability, precision, and reproducibility are characteristics of the silicon micromachining processes that will be extremely valuable when translated into clinical applications.

Our group has introduced a series of combinatorial surface and bulk silicon micromachining methods within the framework of industry-standard, lithography-based microfabrication. We developed a series of protocols to fabricate porous silicon particles by combination of photolithography and electrochemical porosification of silicon. These protocols allow us to fabricate particulates of essentially any desired nonspherical shape, size (dimensions from 50 nm to hundreds of microns), cylindrical nanopore diameter (range between 5 and 150 nm), and porosity. The fabrication protocols consist of two steps: (1) formation of nanoporous silicon film and (2) engineering the particles from porous films. The geometry of porous silicon particles is precisely defined by changing the photolithographic mask, while the porosity and pore size are controlled by parameters including silicon doping, electric current, and concentration of hydrogen fluoride (HF) etching solutions.

A porous silicon film is produced through top-down porosification strategies as opposed to the synthesized "bottom-up" silica nanomaterials, the last process referring

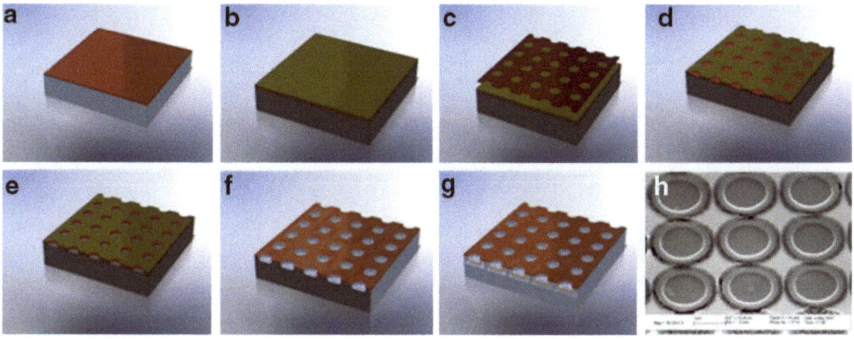

Fig. 10.2 (**a**) Si_3N_4 film on Si wafer; (**b**) photoresist coating; (**c**) photo process; (**d**) pattern development; (**e**) RIE dry etch; (**f**) Si trench etching; (**g**) 2-step electrochemical etch to make porous silicon particles; and (**h**) SEM image of etched porous silicon particles

to the self-assembly of silica sol–gel with polymeric templates as structure-directing agents. The commonly used top-down porosification method consists of electrochemical etching of single-crystalline silicon wafers in an aqueous solution of HF [51]. A simplest setup of this approach consists of a cathode immersed in the solution, the silicon wafer as the anode, and a power supply providing a constant current, with the porosification process sustained as long as an electrical current is applied. The main characteristics of the porous silicon such as porosity, pore size, and morphology are determined by the current density, the type and concentration of dopant in silicon, and the concentration of the etchant solution [52]. The most common solution used for the production of porous silicon is a mixture of HF, H_2O, and ethanol. Ethanol does not take an active part in the porosification but is employed merely as a surface-active cosolvent [52]. A top-down lithography process following the top-down porosification process can be applied to produce discoidal particles. If the porosification process is performed after lithographic patterning, hemispherical particles are produced. Thus, the integration of the top-down lithography process and the top-down porosification can provide the flexibility necessary to fabricate monodisperse, porous silicon particles with tailored physical attributes as well as the uniformity and reproducibility, which are critical for both the proper performance of the system and the safety profile required for regulatory approval and clinical effectiveness.

A variety of porous silicon particles have been fabricated, and their biomedical applications have been investigated [53], including hemispherical, discoidal, and cylindrical porous silicon particles. With only one type of silicon wafer, such as P-type and 0.005 ohm-cm, silicon particles with pore size ranging from 5 to 150 nm, and porosity from 40% to 90% can be obtained by controlling the current and etchant [53]. Figure 10.2 shows a typical fabrication process of our featured hemispherical porous silicon particles [37] by patterning an array of circles on a dielectric layer (silicon nitride, Si_3N_4) on the surface of a silicon wafer, followed by a two-step electrochemical etch in a HF solution. Briefly, heavily doped p++ type (100) wafers with resistivity of 0.005 ohm-cm are used as the substrate. A 100-nm layer of low stress silicon nitride is deposited by a low-pressure chemical vapor deposition system (Fig. 10.2a). Then standard photolithography is used to pattern an

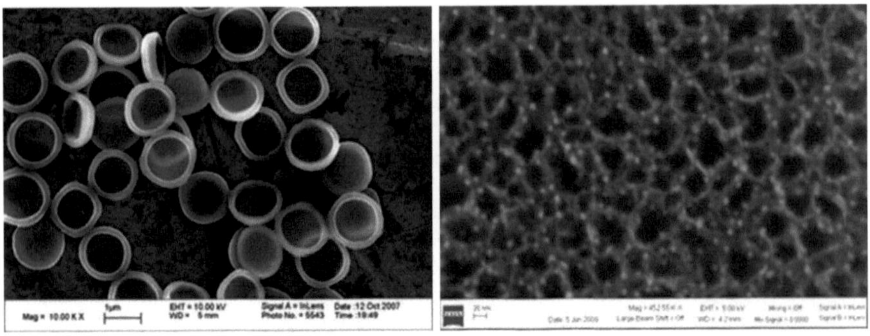

Fig. 10.3 (*Left*) SEM image of microfabricated hemispherical, porous silicon particles; (*right*) pores of mesoporous silicon particles loaded with gold nanoparticles

array of circles, using an EVG 620 aligner (Fig. 10.2b–d). The silicon nitride is selectively removed by reactive-ion etching (RIE) (Fig. 10.2e), and the trenches of desired depth are etched into silicon on the exposed area (Fig. 10.2f). The silicon nitride on the back side of the wafer is then removed by RIE. After the photoresist is removed, the wafer is assembled in a homemade Teflon cell for electrochemical etching. The nanopores are formed in the mixture of HF and ethanol (1:3 v/v) with an applied current density of 16 mA/cm^2 for 20 s (for 1.6 μm particles), controlled by a programmable power working at the constant current mode supply. A high-porosity layer is then formed by applying the current density of 400 mA/cm^2 for 6 s (Fig. 10.2g). Figure 10.2h shows an example SEM image of etched porous silicon particles on the silicon wafer. Because of the nonuniform current density distribution, the resulting hemispherical-shaped particles have aligned cylindrical nanopores at the center and radial distributed pores around the rings [37]. After removing the silicon nitride layer by HF, a clean porous silicon particle array is kept on the substrate, allowing thorough washing and various surface modifications. These particles are released in isopropyl alcohol (IPA) by ultrasound for 1 min.

Figure 10.3 shows an example SEM image of fabricated hemispherical MSV particles. In this MSV system, 5-nm gold nanoparticles are covalently bound to the pore walls of porous silicon particles, as shown in the right panel. Applying porous silicon particles as the first-stage carrier in the MSV systems, a wide range of biological studies are carried out in our laboratory to evaluate biodistribution, therapeutic, and diagnostic characteristics as will be further discussed in this chapter.

10.3 Intracellular Delivery and Multisite Trafficking of the MSV

Larger particles (>500 nm in diameter) are internalized by cells via actin-mediated processes known as macropinocytosis and phagocytosis [54]. It is believed that cationic properties of particles enhance interactions with cell surfaces containing negative glycocalyx moieties [55]. The extracellular matrix is reported to form a gel on the cell membrane, consisting of sulfated glycosaminoglycans and polysaccharide acids, both creating negative surface properties. Additionally, integral membrane proteins, including lipids and glycoproteins, the latter bearing sialic acid residues, both contribute to a negative surface potential. J774 macrophages have been shown to rapidly internalize cationic, multilamellar colloids (200–300 nm) with negligible binding to neutral colloids [55]. We have shown that J774 macrophages associate with both positive and negative first-stage silicon particles [41]. Interestingly, serum opsonization of cationic MSV particles alters their surface potential, leading to a net negative charge on the particle surface [56]. Therefore, physiologically relevant adhesion of particles to cell surfaces appears to be mediated largely by interactions between particle-bound serum components with their respective cell surface receptors rather than purely by surface charge. Each unique particle formulation binds a distinct repertoire of proteins, leading to different affinities for various cell populations and a means for achieving targeted delivery.

With respect to porous silicon microparticles, following cellular adhesion, first-stage particles are internalized either by macropinocytosis or phagocytosis [41]. Both processes depend on actin polymerization, with the resulting vacuole termed the phagosome. During the initial phase of cellular internalization, pseudopods extend from the cell surface, creating a phagocytic cup beneath the porous silicon particle. The pseudopods extend further, wrapping around the particle and pulling it into the cell. Phagocytosis is reported to differ qualitatively, depending on the type of receptors mediating cell surface adhesion [57]. For example, mitogens and cytokines activate complement receptors and mediate cell adhesion of complement-bound particulates with the attached particulate "sinking into the cell," resulting in the formation of more loosely associated vesicles [58]. Fc-gamma receptor-mediated adhesion leads to internalization mediated by "zipper-like" growth of the cell membrane around the particulate and leads to the formation of a vesicle whose membrane is tightly associated with the particulate. Under serum-free conditions, low-porosity first-stage silicon particles (1.6 and 3.2 μm in diameter), anionic and cationic, were shown to be internalized into cellular vacuoles of endothelial cells with membranes bound tightly around the microparticle [59].

In association with F-actin are molecular motor proteins, with myosin II accumulating in the actin cup [60]. Other motor proteins, including myosin I, V, and IX, colocalize with evolving phagosomes and play a role in trafficking of phagosomes to the perinuclear region of the cell. Cellular migration occurs along microtubules and is mediated by members of the kinesin and dynein motor families [61]. Transport along microtubules is disrupted by the presence of microtubule-dissociating agents, such as nocodazole. Endothelial cells (human microvascular endothelial cells; HMVEC) incubated with first-stage silicon particles in the absence (a) or presence (b) of 150 nM nocodazole are shown in Fig. 10.4 [62]. The cells were stained using DRAQ5 and FITC-conjugated antibody, specific for nuclear staining and α-tubulin, respectively. Accumulation of microparticles in the perinuclear region of the cell 4 h after introduction was decreased by 80% in the presence of nocodazole, supporting involvement of microtubules in the transport of microparticle-loaded endosomes.

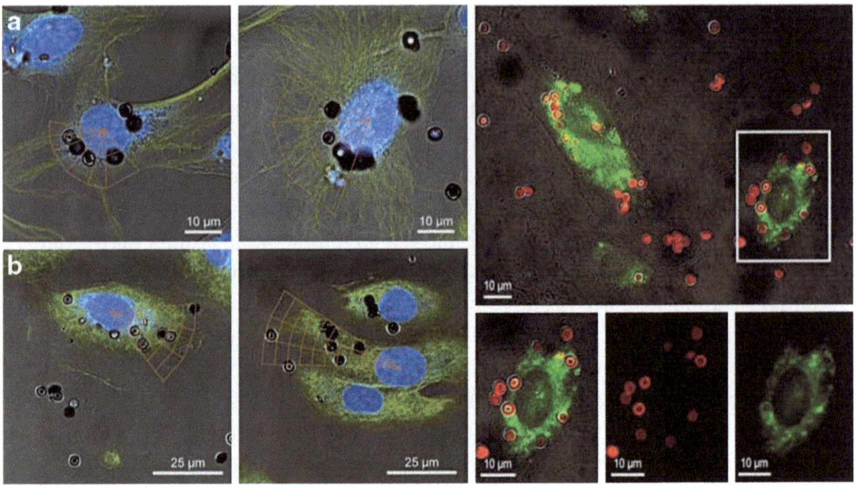

Fig. 10.4 (*Left*) Microtubule-mediated trafficking of porous silicon first-stage particles. Confocal micrographs of endothelial cells with internalized first-stage particles (*black objects*) in the absence (**a**) and presence (**b**) of nocodazole. Cells are labeled with FITC-conjugated antitubulin antibody and DRAQ5 nuclear dye; (*right*) endolysosomal trafficking of first-stage nanocarriers. Endothelial cells expressing NPC1-GFP were incubated with Dylight 594-labeled 3.2-μm silicon microparticles for 4 h at 37°C. Colocalization of the fluorescent signals indicated lysosomal localization of the internalized microparticles. Reproduced with permission from [62]

Proteins that associate transiently with the maturing endosome/phagosome include early endosomal autoantigen 1 (EEA1) and Rab 5 [63]. These proteins associate with the newly formed compartments (early endosome/phagosome), while proteins such as Lamp1 and NPC1 (Niemann–Pick C1) associate with more mature endosomes and lysosomes [64]. HMVECs, transfected with green fluorescent protein (GFP) fused to NPC1, were incubated with Dylight 594-labeled 3.2-μm silicon microparticles for 4 h (Fig. 10.4c). In confocal images, rings of GFP–NPC1 surround the internalized microparticles, with microparticles predominately located in the perinuclear region of the cell. Thus, endosomes containing porous silicon microparticles mature along the endolysosomal pathway. Unlabeled microparticles were similarly located in lysosomes at the indicated time.

We recently reported on the assembly of MSV consisting of first-stage discoidal porous silicon particles (50 nm pores; 3.2 μm in diameter), loaded with second-stage iron oxide nanoparticles [65]. Nanoparticles were loaded into the porous matrix by capillary action (i.e., dry microparticles were incubated with a concentrated solution of nanoparticles). Retention of cationic nanoparticles in the anionic oxidized silicon matrix was based on electrostatic interactions. Similar to the first-stage unloaded porous silicon particles, the MSV were rapidly internalized by both endothelial cells and macrophages under serum-free conditions.

Through a process known as "maturation," which may involve the fusion of multiple phagosomes and endosomes, the early vacuole develops into a "sorting endosome" [66]. This early endosome is the organelle in which sorting of cargo and membrane receptors occurs. Acquisition of a new array of proteins enables the division of endosomal components into different regions of the endosome. This sorting process is accompanied by the formation of tubular regions and regions rich in multivesicular bodies (MVBs) [66]. MVBs that fuse with the cell membrane release the smaller intraluminal vesicles, known as exosomes, into the extracellular environment. Exosome membranes contain proteins and lipids, and the lumen contains proteins, RNA, and other constituents selected for secretion from the cell. The exosomes are 50–90 nm in diameter [67] and may represent a means for intercellular communication between cells, independent of cell-to-cell contact. This process may be a mechanism for rapid propagation of information.

The size and surface chemistry of particles are known to modulate cellular uptake, intracellular trafficking, and cytotoxicity of particles. The impact of varying the surface chemistry of second-stage iron oxide nanoparticles on their intracellular release from first-stage silicon particles and on intracellular trafficking of the nanoparticles was recently studied [68]. Nanoparticles were either coated with PEGylated amine groups or chitosan. Chitosan, a biodegradable polysaccharide composed of random β-(1–4)-linked D-glucosamine and *N*-acetyl-D-glucosamine units, was chosen as a coating for the iron oxide nanoparticles because of its biocompatibility, high charge density, and reports of its potential use for intracytoplasmic delivery of drugs and nanoparticles. Iron oxide nanoparticles coated with chitosan concentrations ranging from 0.01 to 1.0 mg/ml showed a steady increase in bound chitosan, reaching a plateau at higher concentrations. At 1.0 mg/ml of chitosan and higher, 10 μg of chitosan was bound to 5 μg of nanoparticles, yielding a mass ratio of 1:2 nanoparticles to chitosan. Zeta potential (surface charge) and dynamic light scattering (size) measurements were consistent with the finding that the amount of bound chitosan on iron oxide nanoparticles remained unchanged at concentrations of chitosan of 1 mg/ml and higher. Additionally, the change in iron oxide nanoparticle zeta potential, from −27.4 to 26.1 mV that accompanied chitosan coating of carboxylated iron oxide nanoparticles, supported surface coating with positively charged chitosan.

Two hours after the introduction of the MSV, internalized MSV were present in large vacuoles. When the first-stage silicon nanocarriers were heavily loaded with nanoparticles, several nanoparticles were seen in the cytoplasm at 24 h; however, the majority of chitosan-coated nanoparticles remained associated with the first-stage particles. At 48 h, the iron oxide nanoparticles were spatially removed from the silicon particle, and no membrane was detected around the released nanoparticles. It is thought that release of chitosan-coated nanoparticles may result from protonation of the primary amines of the glucosamine residues as the endosome matures and becomes more acidic, leading to a high charge density, membrane destabilization, and release of the nanoparticles.

In contrast to chitosan-coated iron oxide nanoparticles, PEGylated amine nanoparticles were not released into the cytoplasm [65, 68]. As the cationic nanoparticles were released from the silicon first-stage nanocarriers (24 h), they aggregated into regions rich in multivesicular bodies (MVBs; Fig. 10.5a). Six days after the introduction of the MSV to macrophages, ultrastructural examination revealed both

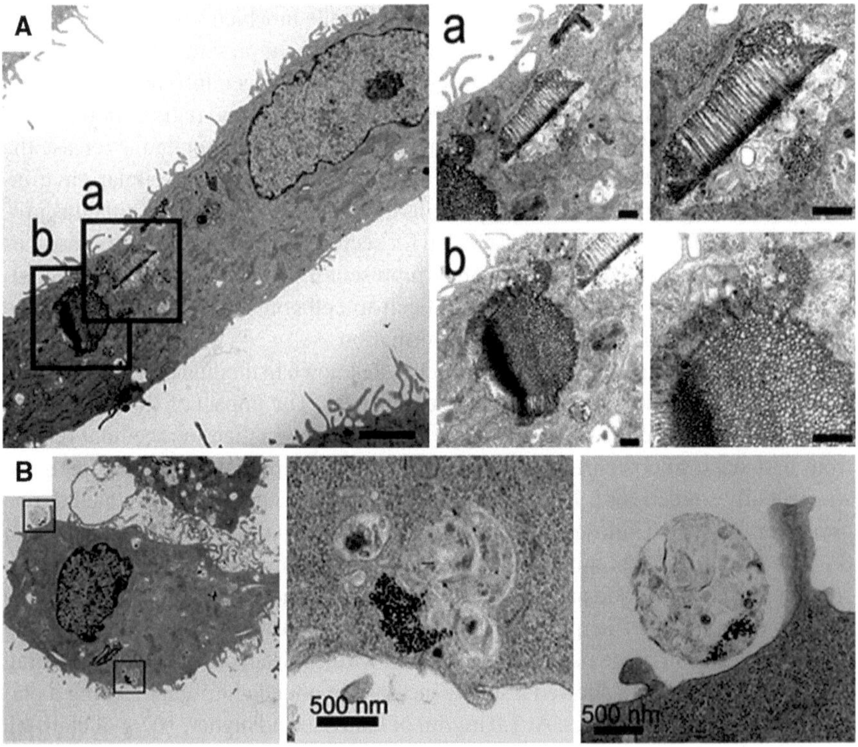

Fig. 10.5 Intracellular partitioning and cellular secretion of vesicles containing iron oxide nanoparticles. (**a**) TEM micrographs of macrophage-internalized MSVs 24 h after introduction to cells. Two internalized MSVs are shown in the image to the left in *boxed regions* and amplified in the images to the right. Clusters of second-stage iron oxide nanoparticles, released from the silicon first-stage particles, are visible in the amplified micrographs. (**b**) TEM micrographs of a macrophage 6 days after cellular uptake of MSVs (*left*, 6 k; *middle*, 25 k; and *right*, 50 k magnification). The cell to the left is further amplified in images to the right to show both an internalized (*middle*) and a secreted (*right*) vesicle containing second-stage iron oxide nanoparticles. Reproduced with permission from [68]

intracellular and extracellular vesicles, containing MVBs and nanoparticles (Fig. 10.5b, bottom row). The porous silicon carrier particle remained in distinct membrane-bound compartment. The membrane-bound compartments containing second-stage nanoparticles were approximately 1.5 μm in diameter. In the secreted vesicle shown in Fig. 10.5b, 10 nm iron oxide nanoparticles are visible in the region of the vesicle adjacent to the plasma membrane.

In support of cellular secretion of vesicles containing iron oxide nanoparticles and reports of eukaryotic cell-derived vectors containing iron oxide nanoparticles [69], as well as studies demonstrating exocytosis of nanoparticles in animal cells [70–72], the iron content from the supernatant of cells treated with the MDS was determined. Quantization of iron in the cell culture media, based on a Prussian blue assay, was consistent with release of both 15 and 30 nm iron oxide nanoparticles from macrophages on all days tested. The amount of iron oxide nanoparticles released from cells treated with 15 nm particles was greater than that of cells treated with 30 nm particles, indicating greater release of the smaller iron oxide nanoparticles, which is consistent with findings reported by other investigators [70].

One of the advantages of multistage, multiparticle delivery systems is the ability of a single delivery vehicle to be "intracellularly partitioned" into discrete regions of the endosome and trafficked to unique locations. By controlling the surface of the vectors, particles can simultaneously or sequentially reach diverse intracellular locations for multiple independent or synergistic effects. Alternatively, the delivery cargo can be a combination of imaging and therapeutic agents, creating theranostics for real-time monitoring of drug delivery or therapeutic efficacy. The ability of cells to communicate with other cells in a contact-independent manner through secretion and uptake of vesicles containing protein, RNA, lipid, and nanoparticle-delivered signals allows for rapid signal propagation, expanding the impact of the therapeutic agent to the entire lesion microenvironment.

10.4 MSV for Imaging and Diagnostics: Contrast Agents with Enhanced Efficiency

Various strategies were employed to design MSV with diagnostic and imaging capabilities. These include loading with gadolinium carbon nanotubes and iron oxide nanoparticles for magnetic resonance imaging (MRI) [65, 73], gold nanoparticles and nanoshuttles for computed tomography (CT) and RAMAN spectroscopy [10], and attaching fluorescent probes to particle surfaces for near-infrared (NIR) imaging [74]. Below, we will describe two studies for enhancing MRI contrast efficiency.

10.4.1 MSV Loaded with Gadolinium Carbon Nanotubes

It is commonly agreed that MRI currently presents one of the most powerful, noninvasive diagnostic imaging techniques used in clinics and in research, with close to 30 million MRI procedures performed in the USA annually. Early diagnosis, treatment, and prognosis of many diseases currently rely on the superior MRI resolution. Elements that contain an odd number of protons and neutrons have a property called "spin" [75]. Spin causes a magnetic moment in the direction of the spin axes. MRI images are obtained as a result of the magnetization of nuclear spins of water protons in tissues and organs, producing longitudinal and transverse relaxation times. Since hydrogen nuclei have strong magnetic moments and exist in abundance in the body, they make good targets for imaging. The magnetic moments align in the presence of a magnetic field. Introduction of radiofrequency energy alters the orientation of the spin. The return of the spin to the equilibrium state is known as relaxation, and the presence of contrast agents alters the relaxation rate. Paramagnetic metal ions or chemical contrast agents (CAs) have been widely used for improving the MRI sensitivity based on their ability to decrease the relaxation time of water protons. Gadolinium-based contrast agents are the most widely used in clinics; however, Gd^{3+} ions in aqueous solution are highly toxic and have to be chelated. While minimizing their toxicity, chelation significantly decreases the number of coordination sites available for water proton exchange (8–9 sites for free Gd^{3+} compared to 1–2 sites for Gd^{3+}-chelate compounds), resulting in reduced contrast enhancement (relaxivity). In addition, almost all of the clinically used CAs are extracellular fluid (ECF) space agents with low blood circulation times (few minutes) and without tissue selectivity and cellular uptake, which limit their contrast enhancement even more. Generally, clinically used CAs have relaxivities smaller than 4 mM^{-1} s^{-1} at a magnetic field strength of 1.41 T. We proposed a new category of nanoconstructs for MRI contrast enhancement by loading Gd-based carbon nanoparticles (carbon nanotubes and fullerenes) into the nanoporous structure of discoidal or hemispherical MSV (Fig. 10.6) [73].

The ability of a paramagnetic material to act as a MRI contrast agent is expressed in terms of its relaxivity (ri). This can be described as the change in the relaxation rate (1/Ti s^{-1}) of water protons per mM concentration of the CAs and can be calculated using the expression ri = (1/Ti−1/Tid)/[CA], where Ti is the relaxation time in the presence of the CAs, Tid is the relaxation time in the absence of CA, and [CA] is the concentration of theGd^{3+} ions present in solution (mM). The loaded MSV were examined for their longitudinal relaxation properties using a relaxometer at 1.41 T and 37°C. All prepared nanoconstructs showed a high increase in longitudinal proton relaxivity, with values of up to 40 times higher than those of currently used contrast agents. It was suggested that the enhancement in MRI performance was attributed to the geometrical confinement of Gd-CAs into nanopores of MSV, which would influence the paramagnetic behavior of the Gd^{3+} ions by altering both the inner and outer sphere contributions to the longitudinal relaxivity. In summary, geometrical confinement of CAs encapsulated in the pores of MSV was found to (1) reduce the ability of CAs to tumble, (2) decrease the mobility of the water molecules, and (3) favor clustering and mutual interactions among the loaded CAs, thus enhancing the imaging characteristics of the resulting MSV [73].

10.4.2 MSV Loaded with Iron Oxide Nanoparticles as Negative Contrast Agents for MR Imaging

Iron oxide-based contrast agents have a greater impact on transverse, spin–spin relaxation (T2, decrease *x–y* component), compared to longitudinal, spin–lattice relaxation (T1, realignment with external magnetic field), leading to areas of negative contrast. The impact of size is important when considering the biodistribution of iron oxide nanoparticles. Particles less than 20 nm are reported to have longer circulation times due to reduced uptake by the reticuloendothelial system (RES),

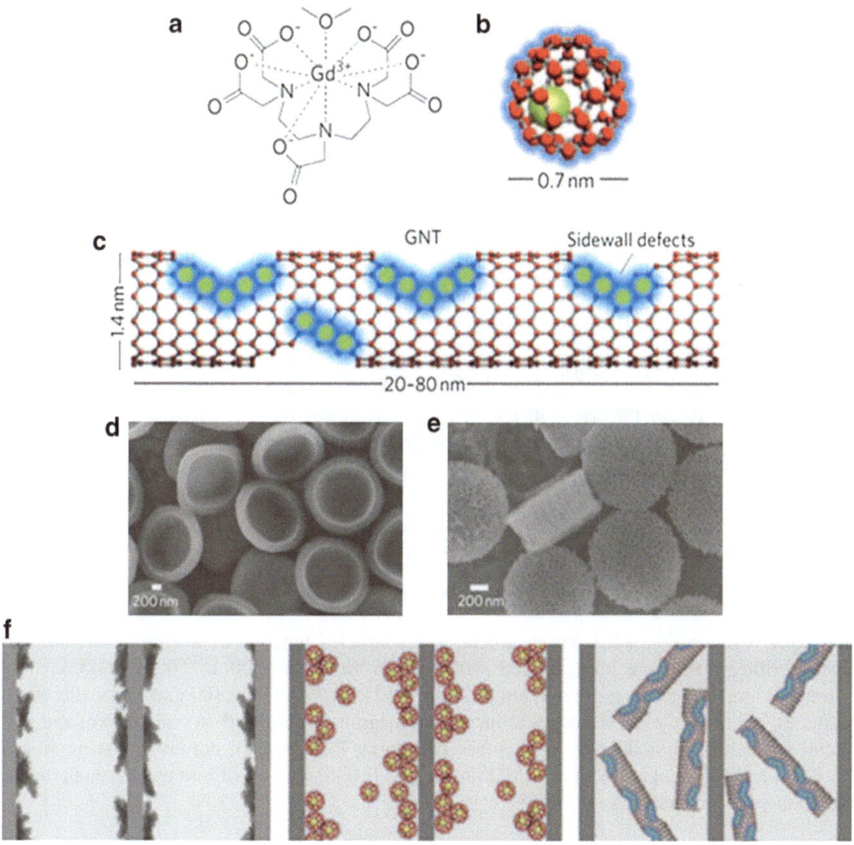

Fig. 10.6 Schematic representations of the new MRI nanoconstructs. (**a**) Commercial contrast agent Magnevist® (MAG); (**b**) Gadofullerenes (GF); and (**c**) debundled Gd–carbon nanotubes (GNT) that can be loaded into mesoporous silicon MSV of different size and shape. SEM micrographs of the (**d**) quasi-hemispherical (1.6 μm in diameter; 1.0 μm in thickness) and (**e**) discoidal (1.0 μm in diameter; 0.4 μm in thickness) MSV particles. (**f**) Cartoons showing MAG, GF, and GNT entrapped within the porous structure of the MSV. The geometrical confinement of the Gd-based CA within the nanopores of the MSV enhances the *T* contrast. Reproduced with permission from [73]

allowing for more uptake by the lymphatic system and by bone marrow [76]. However, particles smaller than 8 nm undergo greater renal elimination, narrowing the window of opportunity. Encapsulation of iron oxide nanoparticles into a porous silicon matrix is one means of transporting large numbers of nanoparticles to common targeted locations [68]. High-resolution scanning electron microscope (SEM) images in Fig. 10.7a, b show large numbers of iron oxide nanoparticles (10 and 30 nm), loaded throughout the porous silicon matrix.

Phantoms were prepared for magnetic resonance imaging by suspending porous silicon microparticles, either free or loaded with variable levels of iron oxide nanoparticles, into a bed of agarose, housed in a NMR tube. Magnetic resonance images

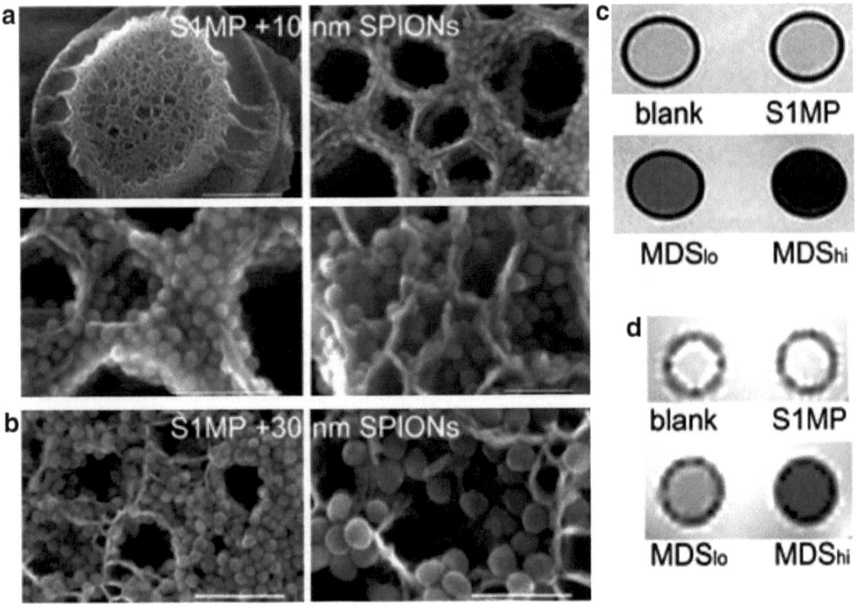

Fig. 10.7 Iron oxide-loaded porous silicon microparticles function as negative MRI contrast agents. Silicon microparticles, loaded with 10 nm (**a** 25 k, 300 k, 500 k, 600 k; *bars* 1 μm,100 nm,100 nm, 50 nm) or 30 nm (**b** 200 k, 450 k; *bars* 200 nm, 100 nm) iron oxide nanoparticles, are shown in SEM images at increasing magnifications. (**c, d**) Axial spin- (**c**) and gradient- (**d**) echo MR images of NMR tubes containing PBS (blank), porous first-stage silicon particles, and MSV loaded with low (MDSlo) or high (MDShi) levels of iron oxide nanoparticles. Reproduced with permission from [68]

of the phantoms in Fig. 10.7c represent samples containing PBS (phosphate buffered saline) alone, unloaded porous silicon microparticles in PBS (S1MP), and silicon microparticles loaded with varying levels of iron oxide nanoparticles (MDShi, MDSlo) in PBS. Significant negative contrast is seen in the porous silicon containing iron oxide nanoparticles compared to unloaded silicon microparticles. Phantom models containing the MSV shortened T2 and T2* relaxation times in a manner dependent on iron oxide concentration, with higher levels of iron oxide nanoparticles (MDShi) yielding greater contrast than lesser loads (MDSlo).

10.5 From Rational Design to In Vivo MSV Distribution and Therapeutic Efficacy

The ultimate goal of rationally designing nanovectors is to optimize accumulation in the biological target through modification of governing engineered parameters, which affect the series of events encountered by the particle on its way from the site of administration to the intended site of action. In this regard, there are multiple factors which can affect the transport of MSV carriers, including margination in the bloodstream, interaction with the vascular walls through specific ligand–receptor or nonspecific (e.g., electrostatic) interactions, and cell internalization. Margination, the term used in physiology to describe the lateral drift of leukocytes and platelets from the center of the blood vessels toward the endothelial walls, is one of the fundamental events in the intravascular "journey." Rational design of this property allows a close contact between the circulating particles and the vessel walls, which is required for targeting vascular endothelium and for extravasation based on the EPR concept [38, 39]. Unlike spherical particles, nonspherical particles exhibit more complex motions with tumbling and rolling, which can be exploited to control their margination dynamics without any need for lateral external forces. The longitudinal (drag) and lateral (lift) forces, as well as the torque exerted by the flowing blood, depend on the particle geometry and orientation. In general, elongated or disc-shaped particles were shown to marginate and interact better with the cell surface [38], as demonstrated by flow chamber in vitro experiments [77, 78]. There are a number of different mechanisms governing the behavior of particles in vivo, as shown in Fig. 10.8. Our recent in vivo studies have demonstrated that particles of different shape biodistribute differently in vivo [79]. As an example, both discoidal and hemispherical particles tended to accumulate at the tumor site in higher concentration than spherical ones. It was also shown that size affects the biodistribution of particles, with smaller particles accumulating at the tumor site better than larger ones. Discoidal particles with diameter of 600 nm were found to highly accumulate in the tumor (up to 10% of injected dose/g tumor) (unpublished data).

Several therapeutic agents were encapsulated in liposomes and loaded into MSV carriers. One drug that was successfully encapsulated in liposomes was annamycin, a non-cross-resistant anthracycline [80]. The annamycin pre-liposome lyophilized powder contains phospholipids (dimyristoylphosphatidylcholine and dimyristoylphosphatidylglycerol at a 7:3 molar ratio), annamycin (lipid–drug at a ratio 50:1 w/w), and Tween 20. The surfactant in the formulation allows for better solubilization of the drug, shortening the reconstitution step, as well as serves as a means to form nanosize carriers without destroying the liposomal structure [81]. Since the drug possesses native fluorescence in the red region similar to doxorubicin, we have shown loading of annamycin liposomes into the multistage particles using flow cytometry, which resulted in the shift in the mean fluorescent intensity from 3 to 1,285 AU.

Another agent successfully tested with the multistage delivery system belongs to the class of siRNA therapeutics. The discovery of RNAi a decade ago by Fire et al. [82] opened up an exciting field of cancer therapeutics with vast clinical application, theoretically being able to silence any cancer-related gene pathway. However, safe and effective delivery of siRNA therapeutics has been the major bottleneck in the translation of this technology to clinical applications. "Naked" siRNAs have an extremely fast degradation profile and are subjected to immune system activation and recognition. Therefore, liposomes were intensively investigated as an attractive delivery system for siRNA therapeutics, offering protection from extensive degradation and inhibition of harmful nonspecific binding to normal tissues. In vitro and

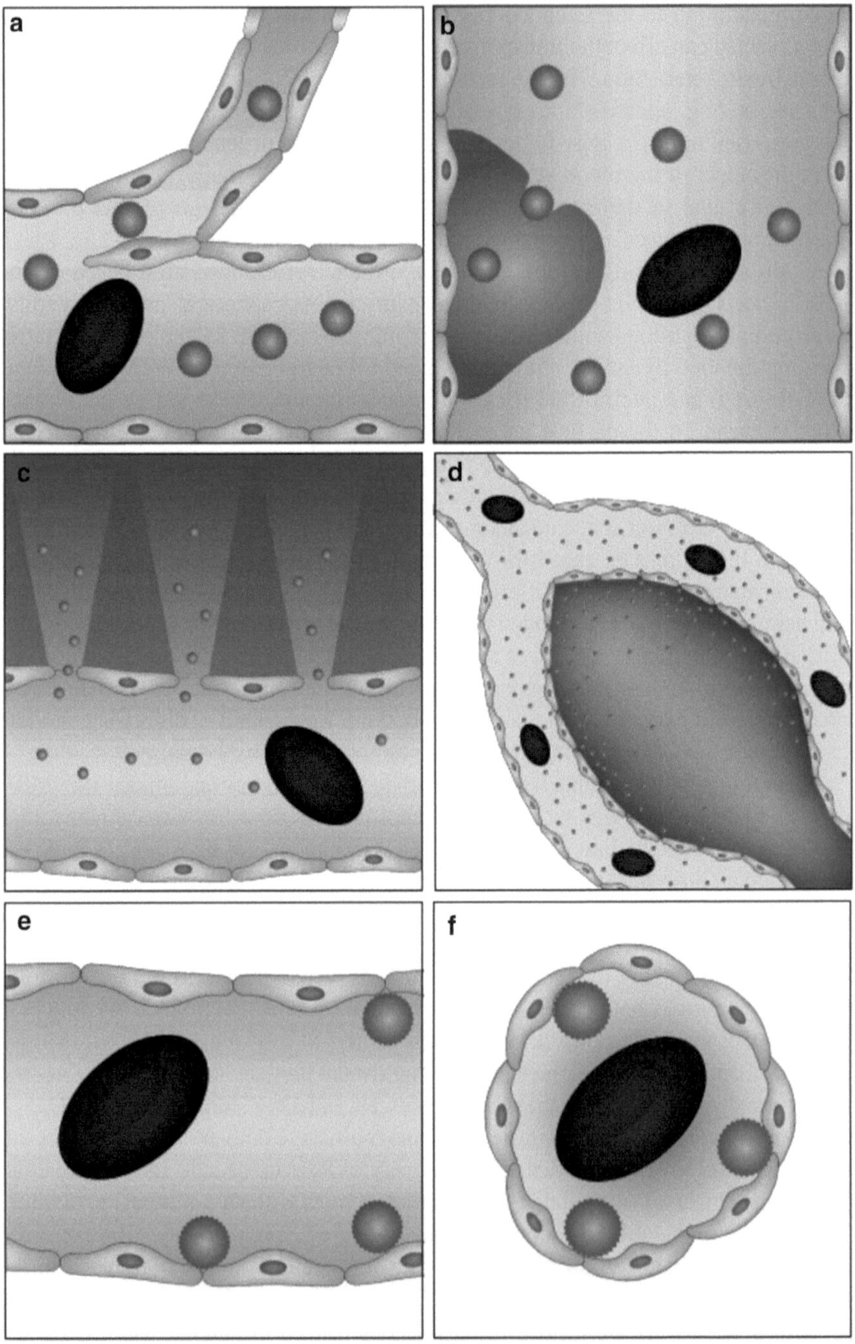

Fig. 10.8 Mechanisms of particle sequestration from the circulation after intravenous injection (**a**) entrapment in small capillaries; (**b**) engulfment by phagocytic cells; (**c**) extravasation through fenestrated endothelium; (**d**) excretion through the kidneys glomeruli; (**e**) and (**f**) adhesion to the blood vessel walls. Reproduced with permission from [79]

in vivo studies have demonstrated improved siRNA delivery to melanoma, lung cancer, breast cancer, and ovarian cancer when liposomal carriers were used [83, 84]. Recently, we examined MSV loaded with neutral dioleoylphosphatidylcholine (DOPC) nanoliposomes, containing siRNA targeted against the EphA2 oncoprotein, which is overexpressed in most malignancies, including ovarian cancer. The study was conducted in two independent orthotopic mouse models of ovarian cancer. Mice bearing SKOV3ip1 ovarian tumors were given an i.v. injection of S1MP, loaded with either single-dose EphA2-siRNA-DOPC (5 μg EphA2-siRNA) or triple-dose EphA2-siRNA-DOPC (15 μg EphA2-siRNA). A single administration of the high dose of S1MP-EphA2-siRNA-DOPC resulted in >80% reduction of EphA2 expression for at least 21 days, while the control single injection of EphA2-siRNA liposomes alone downregulated EphA2 expression to a much lesser extent and for only 5–6 days. Sustained gene silencing in the tumor was also supported by immunohistochemical analysis, showing EphA2 downregulation for 28 days (Fig. 10.9). The therapeutic potential of a single administration with S1MP-EphA2-siRNA-DOPC (15 μg EphA2-siRNA) was compared to six siRNA-DOPC doses given twice a week for 3 weeks (5 μg EphA2-siRNA per injection, for a total administered dose of 30 μg EphA2-siRNA over 3 weeks). A significant reduction of tumor burden was achieved with a single injection of S1MP-EphA2-siRNA-DOPC. This effect was comparable to the one obtained to six repeated i.v. injections of liposomal EphA2-siRNA for a total of twice the siRNA dosage. This magnificent finding was followed by significantly reduced angiogenesis and cell proliferation when compared to free liposomes (Fig. 10.10) [85]. Other therapeutic agents that were successfully loaded into the multistage drug delivery system include liposomal and micellar paclitaxel and doxorubicin. These systems are currently under investigation in in vitro and in vivo efficacy studies in various cancer models.

It is noteworthy that the biocompatibility of MSV was tested and confirmed in a number of in vitro and in vivo studies. The first-stage carriers were shown to degrade in physiological environments in vitro and in vivo [85, 86]. In vitro studies in endothelial cells and macrophages have shown that, following the interaction with MSV, the cells do not produce inflammatory cytokines (33 cytokines were tested) and exhibit normal cell cycle [59, 74, 87]. No release of biochemical markers and plasma cytokines was observed in in vivo studies examining the effect of MSV acute and subchronic administration in healthy mice [88].

10.6 Conclusions and Future Perspectives

To summarize, the last few decades have witnessed the discovery of a great arsenal of highly specific pharmacological agents. However, there are a large number of unresolved issues, which, in general, live in symbiosis with new and intriguing opportunities. The major challenges to the systemically administrated therapeutics are all related to overcoming the multiplicity of biological and biophysical barriers. The third-generation nanosystems, and in particular MSV described above, are only

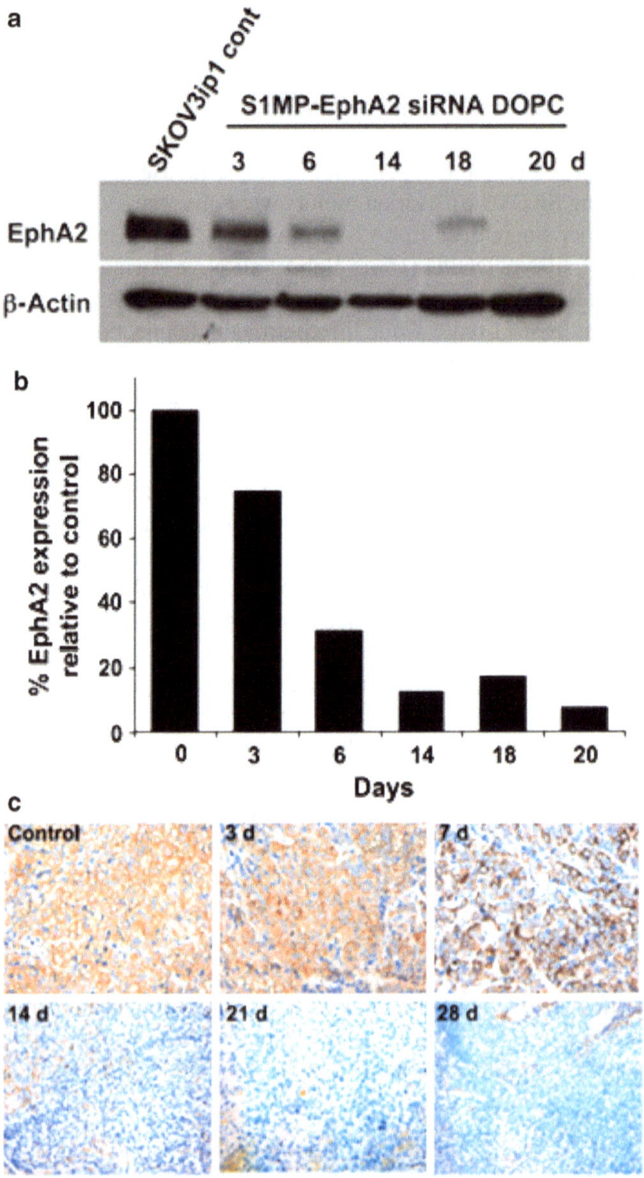

Fig. 10.9 Systemic delivery of siRNA-DOPC using S1MP results in long-lasting in vivo gene silencing. The mice (three mice per time point), bearing SKOV3ip1 orthotopic ovarian tumors, were injected with S1MP-EphA2-siRNA-DOPC or left nontreated. (**a**) The tumors were harvested at the indicated time points for Western blot to measure EphA2 expression levels; (**b**) densitometric analysis was performed to normalize EphA2 expression by β-actin; and (**c**) immunohistochemical analysis of EphA2 expression in the SKOV1ip3 tumor. Reproduced with permission from [85]

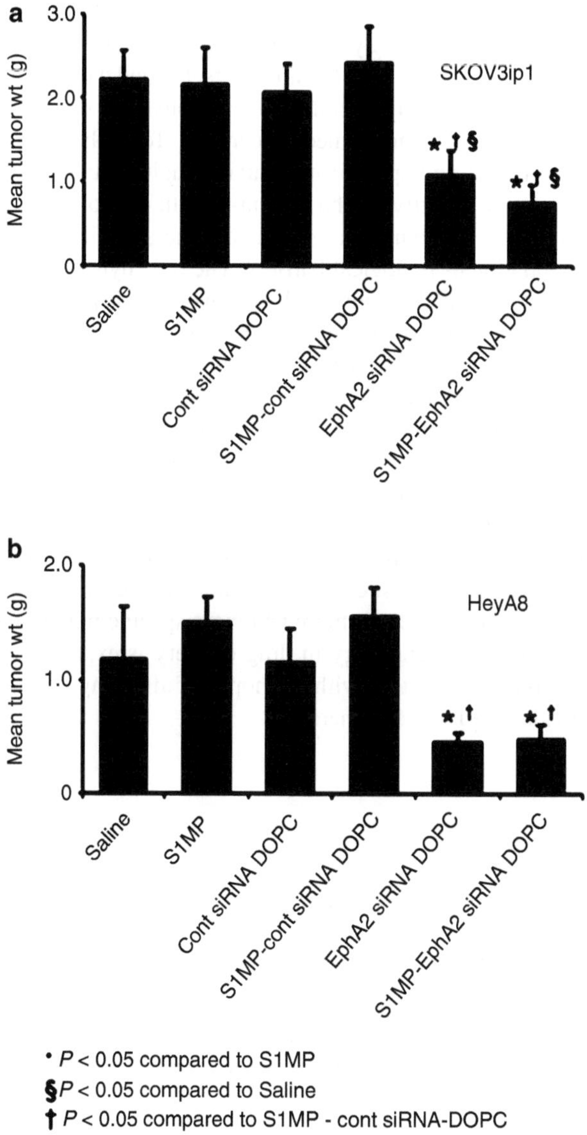

Fig. 10.10 Therapeutic efficacy of sustained EphA2-siRNA-DOPC delivery by MSV (S1MP). Nude mice were injected i.p. with SKOV3ip1 or HeyA8 cells and randomly allocated to one of six treatment groups ($n = 10$) (a) saline, (b) S1MP, (c) nonsilencing control siRNA-liposomes (DOPC), (d) S1MP-nonsilencing control siRNA-DOPC, (e) EphA2-siRNA-DOPC, and (f) S1MP-EphA2-siRNA-DOPC. SiRNA-DOPC was i.v. injected biweekly at a dose of 5 μg siRNA. S1MP-EphA2-siRNA-DOPC was injected in a single administration in 3 weeks at a dose of 15 μg siRNA. The mice were injected with saline biweekly in the rest of treatment period. When control animals (saline- and nonsilencing siRNA-treated mice) began to appear moribund (4–5 weeks after cell injection), all animals in an experiment were sacrificed and mouse weight, tumor weight (wt), tumor number, ascites volume, and tumor location were recorded. Columns, mean tumor weight from SKOV3ip1 (**A**) or HeyA8 cells (**B**); *bars*: SD. Reproduced with permission from [85]

a first step toward the development of a general, modular, synergistic system that can systematically and sequentially address the biological barriers. We foresee the development of further generations of nanovectors, beyond the current three subcategories that will enable highly specific and personalized therapy through "by-design" biodistribution of nanoengineered carriers. To enable personalization of therapy, highly efficient and specific molecular imaging techniques are required. These will guide us in identifying the patient-specific pathologies, microenvironmental abnormalities, molecular targets, and time-specific disease alterations required for negotiation of biological barriers. The time dynamics of the evolution of the lesion do not necessarily require a change in cytotoxic payload—the response to the evolution of the lesion and its microenvironment may be built in the individualization of carrier. With their exquisite control of geometry, surface chemistry, and overall design parameters, nanovectors are great candidates for time-sensitive, pre-programmable, and logic-embedded release of therapeutic and imaging agents. Additional challenges and opportunities are targeting multiple cellular organelles by various therapeutic agents and manipulation of the cellular microenvironment.

Presently, the majority of nanoscale drug delivery vehicles continue to enable the use of conventional medications by providing longer circulation times, greater tolerability, and improved disease site delivery, factors that result in better patient outcomes. While future drugs become more and more specific in their mechanisms of action, the future of nanotechnology in drug delivery is expected to shift toward logic-embedded functionalization with the hopes of affording the patient the most promising individualized mode of therapy.

References

1. Winau F, Westphal O, Winau R (2004) Paul Ehrlich – in search of the magic bullet. Microbes Infect 6(8):786–789
2. Blagosklonny MV (2004) Analysis of FDA approved anticancer drugs reveals the future of cancer therapy. Cell Cycle 3(8):1035–1042
3. Hatefi A, Amsden B (2002) Camptothecin delivery methods. Pharm Res 19(10):1389–1399
4. Heath JR, Davis ME (2008) Nanotechnology and cancer. Annu Rev Med 59:251–265
5. Wang MD et al (2007) Nanotechnology for targeted cancer therapy. Expert Rev Anticancer Ther 7(6):833–837
6. Riehemann K et al (2009) Nanomedicine – challenge and perspectives. Angew Chem Int Ed 48(5):872–897
7. Wagner V et al (2006) The emerging nanomedicine landscape. Nat Biotechnol 24(10):1211–1217
8. Ferrari M (2005) Cancer nanotechnology: opportunities and challenges. Nat Rev Cancer 5(3):161–171
9. Sanhai WR et al (2008) Seven challenges for nanomedicine. Nat Nanotechnol 3(5):242–244
10. Godin B et al (2010) An integrated approach for the rational design of nanovectors for biomedical imaging and therapy. Adv Genet 69:31–64
11. Torchilin VP (2005) Recent advances with liposomes as pharmaceutical carriers. Nat Rev Drug Discov 4(2):145–160
12. Hashizume H et al (2000) Openings between defective endothelial cells explain tumor vessel leakiness. Am J Pathol 156(4):1363–1380

13. Maeda H (2001) The enhanced permeability and retention (EPR) effect in tumor vasculature: the key role of tumor-selective macromolecular drug targeting. Adv Enzyme Regul 41:189–207
14. Duncan R (2003) The dawning era of polymer therapeutics. Nat Rev Drug Discov 2(5):347–360
15. Sutton D et al (2007) Functionalized micellar systems for cancer targeted drug delivery. Pharm Res 24(6):1029–1046
16. Lee CC et al (2005) Designing dendrimers for biological applications. Nat Biotechnol 23(12):1517–1526
17. Harris JM, Chess RB (2003) Effect of pegylation on pharmaceuticals. Nat Rev Drug Discov 2(3):214–221
18. Maeda H, Bharate GY, Daruwalla J (2009) Polymeric drugs for efficient tumor-targeted drug delivery based on EPR-effect. Eur J Pharm Biopharm 71(3):409–419
19. Duncan R (2006) Polymer conjugates as anticancer nanomedicines. Nat Rev Cancer 6(9):688–701
20. Brannon-Peppas L, Blanchette JO (2004) Nanoparticle and targeted systems for cancer therapy. Adv Drug Deliv Rev 56(11):1649–1659
21. Kale AA, Torchilin VP (2007) "Smart" drug carriers: PEGylated TATp-modified pH-sensitive liposomes. J Liposome Res 17(3–4):197–203
22. Farokhzad OC, Langer R (2009) Impact of nanotechnology on drug delivery. ACS Nano 3(1):16–20
23. Souza GR, Staquicini FI, Christianson DR, Ozawa MG, Miller JH, Pasqualini R, Arap W. Combinatorial targeting and nanotechnology applications. Biomed Microdevices. 2010; 12(4):597–606
24. Juweid M et al (1992) Micropharmacology of monoclonal antibodies in solid tumors: direct experimental evidence for a binding site barrier. Cancer Res 52(19):5144–5153
25. Nakanishi T et al (2001) Development of the polymer micelle carrier system for doxorubicin. J Control Release 74(1–3):295–302
26. Kukowska-Latallo JF et al (2005) Nanoparticle targeting of anticancer drug improves therapeutic response in animal model of human epithelial cancer. Cancer Res 65(12):5317–5324
27. Ferrari M (2010) Frontiers in cancer nanomedicine: directing mass transport through biological barriers. Trends Biotechnol 28(4):181–188
28. Hajitou A, Pasqualini R, Arap W (2006) Vascular targeting: recent advances and therapeutic perspectives. Trends Cardiovasc Med 16(3):80–88
29. Gindy ME, Prud'homme RK (2009) Multifunctional nanoparticles for imaging, delivery and targeting in cancer therapy. Expert Opin Drug Deliv 6(8):865–878
30. Latorre M, Rinaldi C (2009) Applications of magnetic nanoparticles in medicine: magnetic fluid hyperthermia. P R Health Sci J 28(3):227–238
31. Chuang VT, Kragh-Hansen U, Otagiri M (2002) Pharmaceutical strategies utilizing recombinant human serum albumin. Pharm Res 19(5):569–577
32. Ho DH et al (1986) Clinical pharmacology of polyethylene glycol-L-asparaginase. Drug Metab Dispos 14(3):349–352
33. Vasey PA et al (1999) Phase I clinical and pharmacokinetic study of PK1 [N-(2-hydroxypropyl) methacrylamide copolymer doxorubicin]: first member of a new class of chemotherapeutic agents-drug-polymer conjugates. Cancer Research Campaign Phase I/II Committee. Clin Cancer Res 5(1):83–94
34. Meerum Terwogt JM et al (2001) Phase I clinical and pharmacokinetic study of PNU166945, a novel water-soluble polymer-conjugated prodrug of paclitaxel. Anticancer Drugs 12(4):315–323
35. Matsumura Y et al (2004) Phase I clinical trial and pharmacokinetic evaluation of NK911, a micelle-encapsulated doxorubicin. Br J Cancer 91(10):1775–1781
36. Kim TY et al (2004) Phase I and pharmacokinetic study of Genexol-PM, a cremophor-free, polymeric micelle-formulated paclitaxel, in patients with advanced malignancies. Clin Cancer Res 10(11):3708–3716
37. Tasciotti E et al (2008) Mesoporous silicon particles as a multistage delivery system for imaging and therapeutic applications. Nat Nanotechnol 3(3):151–157

38. Decuzzi P, Ferrari M (2008) Design maps for nanoparticles targeting the diseased microvasculature. Biomaterials 29(3):377–384
39. Decuzzi P, Ferrari M (2010) Modulating cellular adhesion through nanotopography. Biomaterials 31(1):173–179
40. Decuzzi P, Ferrari M (2008) The receptor-mediated endocytosis of nonspherical particles. Biophys J 94(10):3790–3797
41. Serda RE et al (2009) The association of silicon microparticles with endothelial cells in drug delivery to the vasculature. Biomaterials 30(13):2440–2448
42. Serda RE et al (2009) Quantitative mechanics of endothelial phagocytosis of silicon microparticles. Cytometry A 75(9):752–760
43. Anglin EJ et al (2008) Porous silicon in drug delivery devices and materials. Adv Drug Deliv Rev 60(11):1266–1277
44. Cohen MH et al (2003) Microfabrication of silicon-based nanoporous particulates for medical applications. Biomed Microdevices 5(3):253–259
45. Nashat AH, Moronne M, Ferrari M (1998) Detection of functional groups and antibodies on microfabricated surfaces by confocal microscopy. Biotechnol Bioeng 60(2):137–146
46. Salonen J et al (2008) Mesoporous silicon in drug delivery applications. J Pharm Sci 97(2):632–653
47. Salonen J et al (2005) Mesoporous silicon microparticles for oral drug delivery: loading and release of five model drugs. J Control Release 108(2–3):362–374
48. Meade SO et al (2004) Porous silicon photonic crystals as encoded microcarriers. Adv Mater 16(20):1811–1814
49. Park JH et al (2009) Biodegradable luminescent porous silicon nanoparticles for in vivo applications. Nat Mater 8(4):331–336
50. Decuzzi P et al (2005) A theoretical model for the margination of particles within blood vessels. Ann Biomed Eng 33(2):179–190
51. Canham LT, INSPEC (Information service) (1987) Properties of porous silicon, xviii. INSPEC, London, 405 p
52. Zhang XG (2004) Morphology and formation mechanisms of porous silicon. J Electrochem Soc 151(1):C69–C80
53. Chiappini C et al (2010) Tailored porous silicon microparticles: fabrication and properties. Chemphyschem 11(5):1029–1035
54. Rupper A, Cardelli J (2001) Regulation of phagocytosis and endo-phagosomal trafficking pathways in *Dictyostelium discoideum*. Biochim Biophys Acta 1525(3):205–216
55. Chenevier P et al (2000) Interaction of cationic colloids at the surface of J774 cells: a kinetic analysis. Biophys J 79(3):1298–1309
56. Serda RE et al (2011) Proteomic analysis of serum opsonins impacting biodistribution and cellular association of porous silicon microparticles. Mol Imaging 10(1):43–55
57. Aderem A, Underhill DM (1999) Mechanisms of phagocytosis in macrophages. Annu Rev Immunol 17:593–623
58. Kaplan G (1977) Differences in the mode of phagocytosis with Fc and C3 receptors in macrophages. Scand J Immunol 6(8):797–807
59. Serda RE et al (2009) Mitotic trafficking of silicon microparticles. Nanoscale 1(2):250–259
60. Olazabal IM et al (2002) Rho-kinase and myosin-II control phagocytic cup formation during CR, but not FcgammaR, phagocytosis. Curr Biol 12(16):1413–1418
61. Ma S, Fey P, Chisholm RL (2001) Molecular motors and membrane traffic in *Dictyostelium*. Biochim Biophys Acta 1525(3):234–244
62. Ferrati S et al (2010) Intracellular trafficking of silicon particles and logic-embedded vectors. Nanoscale 2:1512–1520
63. Barysch SV et al (2009) Sorting in early endosomes reveals connections to docking- and fusion-associated factors. Proc Natl Acad Sci USA 106(24):9697–9702
64. Clague MJ (1998) Molecular aspects of the endocytic pathway. Biochem J 336:271–282
65. Serda RE et al (2010) Cellular association and assembly of a multistage delivery system. Small 6(12):1329–1340

66. Scott CC, Botelho RJ, Grinstein S (2003) Phagosome maturation: a few bugs in the system. J Membr Biol 193(3):137–152
67. Fevrier B et al (2005) Exosomes: a bubble ride for prions? Traffic 6(1):10–17
68. Serda RE et al (2010) Logic-embedded vectors for intracellular partitioning, endosomal escape, and exocytosis of nanoparticles. Small 6(23):2691–2700
69. Wilhelm C et al (2008) Intracellular trafficking of magnetic nanoparticles to design multifunctional biovesicles. Small 4(5):577–582
70. Chithrani BD, Chan WC (2007) Elucidating the mechanism of cellular uptake and removal of protein-coated gold nanoparticles of different sizes and shapes. Nano Lett 7(6):1542–1550
71. Walczak P et al (2007) Applicability and limitations of MR tracking of neural stem cells with asymmetric cell division and rapid turnover: the case of the shiverer dysmyelinated mouse brain. Magn Reson Med 58(2):261–269
72. Panyam J, Labhasetwar V (2003) Dynamics of endocytosis and exocytosis of poly(D, L-lactide-co-glycolide) nanoparticles in vascular smooth muscle cells. Pharm Res 20(2):212–220
73. Ananta JS et al (2010) Geometrical confinement of gadolinium-based contrast agents in nanoporous particles enhances T1 contrast. Nat Nanotechnol 5(11):815–821
74. Tasciotti E et al (2011) Near-infrared imaging method for the in vivo assessment of the biodistribution of nanoporous silicon particles. Mol Imaging 10(1):56–68
75. Puddephat M (2011) Principles of magnetic resonance imaging. http://www.mikepuddephat.com/Page/1603/Principles-of-magnetic-resonance-imaging. Accessed 9 May 2011
76. Fukuda Y et al (2006) Superparamagnetic iron oxide (SPIO) MRI contrast agent for bone marrow imaging: differentiating bone metastasis and osteomyelitis. Magn Reson Med Sci 5(4):191–196
77. Gentile F et al (2008) The effect of shape on the margination dynamics of non-neutrally buoyant particles in two-dimensional shear flows. J Biomech 41(10):2312–2318
78. Gentile F et al (2008) The margination propensity of spherical particles for vascular targeting in the microcirculation. J Nanobiotechnol 6:9
79. Decuzzi P et al (2010) Size and shape effects in the biodistribution of intravascularly injected particles. J Control Release 141(3):320–327
80. Booser DJ et al (2002) Phase II study of liposomal annamycin in the treatment of doxorubicin-resistant breast cancer. Cancer Chemother Pharmacol 50(1):6–8
81. Zou Y, Priebe W, Perez-Soler R (1996) Lyophilized preliposomal formulation of the non-cross-resistant anthracycline annamycin: effect of surfactant on liposome formation, stability and size. Cancer Chemother Pharmacol 39(1–2):103–108
82. Fire A et al (1998) Potent and specific genetic interference by double-stranded RNA in *Caenorhabditis elegans*. Nature 391:806–811
83. Mangala LS et al (2009) Liposomal siRNA for ovarian cancer. Methods Mol Biol 555:29–42
84. Whitehead KA, Langer R, Anderson DG (2009) Knocking down barriers: advances in siRNA delivery. Nat Rev Drug Discov 8(2):129–138
85. Tanaka T et al (2010) Sustained small interfering RNA delivery by mesoporous silicon particles. Cancer Res 70(9):3687–3696
86. Godin B et al (2010) Tailoring the degradation kinetics of mesoporous silicon structures through PEGylation. J Biomed Mater Res A 94(4):1236–1243
87. Serda RE et al (2011) Multi-stage delivery nano-particle systems for therapeutic applications. Biochim Biophys Acta 1810(3):317–329
88. Tanaka T et al (2010) In vivo evaluation of safety of nanoporous silicon carriers following single and multiple dose intravenous administrations in mice. Int J Pharm 402(1–2):190–197

Chapter 11
Targeted, Multifunctional Hydrogel Nanoparticles for Imaging and Treatment of Cancer

Yong-Eun Koo Lee and Raoul Kopelman

11.1 Introduction

Cancer is the second most common cause of death in the United States, accounting for nearly one of every four deaths [1]. Yet, diagnosis and treatment of cancer remains a challenge due to limitations of the existing medical modalities. The current standard practice of cancer diagnosis consists of invasive tissue biopsies that can provide information about its histological type, classification, grade, and potential aggressiveness. It is useful to identify a fully developed cancer but not always helpful to detect the premalignant or early lesions. Early stage cancer is treatable, in general, and therefore its detection can improve the cure rate. To date, modern imaging techniques, such as computed tomography (CT), positron emission tomography (PET), ultrasound, and magnetic resonance imaging (MRI), are the emerging standards for the detection of cancers. However, these imaging scans, even with current image contrast agents, are neither tumor sensitive nor selective enough, so as to replace biopsy or to detect invisible cancer cells at early stages of the disease. The typical role of the imaging modality is to locate and stage the neoplasm and visualize the tumor before biopsy or at the time of surgery [2].

Current cancer treatment modalities include surgery and/or nonsurgical therapy such as chemotherapy, photodynamic therapy (PDT), and radiation therapy. Surgery is the primary cancer treatment but is invasive and, often not efficient due to incomplete resection. Note that often, neoplastic tissue is virtually indistinguishable from normal tissue, and sometimes unresectable due to the location of the tumor. Nonsurgical therapy usually leads to various side effects because of its nonselectivity. Chemotherapy is the primary nonsurgical therapy, but the present status of chemotherapy

Y.-E. Koo Lee • R. Kopelman (✉)
Department of Chemistry, The University of Michigan, 930 N. University Avenue, Ann Arbor, MI 48109, USA
e-mail: kopelman@umich.edu

S. Svenson and R.K. Prud'homme (eds.), *Multifunctional Nanoparticles for Drug Delivery Applications: Imaging, Targeting, and Delivery*, Nanostructure Science and Technology, DOI 10.1007/978-1-4614-2305-8_11,

is far from satisfactory. Most of the chemotherapeutic drugs are highly toxic and nonspecific, affecting not only cancer cells but also healthy cells. A majority of these drugs are low-molecular-weight compounds that diffuse rapidly and are distributed evenly within the body. The drugs are often cleared too quickly from the bloodstream, by the kidneys, or through immune recognition, demanding a higher dose treatment. As a result, chemotherapy often causes significant collateral organ damage and, consequently, severe systemic side effects. Moreover, a majority of the drugs for cancer treatment are hydrophobic with very poor solubility in aqueous solution, requiring solubilizers that can also contribute to the toxicity of the administered drug. There is another problem faced by cancer chemotherapy due to multidrug resistance (MDR) of the cancer cells. MDR prevents a wide variety of chemotherapeutic drugs from being delivered into cancer cells due to the very effective drug efflux system, P-glycoprotein, or MDR-associated protein [3, 4]. PDT is a relatively new therapeutic method that allows localized treatment by light-activating the PDT drugs (photosensitizers) that are delivered to a treatment spot, either locally or systemically, to produce singlet oxygen and other reactive oxygen species (ROS). PDT is more selective and less toxic than other chemotherapeutic drugs but limited by the tissue penetration depth of the excitation light used and the level of oxygen. Moreover, PDT efficacy is also limited by inefficient delivery of the drugs as the PDT drugs are mostly hydrophobic small molecules. Radiation therapy (or radiotherapy) uses ionizing radiation to treat malignant cancer cells. The principal initial reaction is the radiolysis of the abundant water molecules present in the tumor tissues, creating the OH radical which subsequently reacts with oxygen or other organic molecules to create other ROS and short-lived organic radicals. As the oxygen plays a key role in propagating the initial damage, radiation therapy is usually inefficient in hypoxic condition. Radiation therapy may pose the risk of secondary malignancy in the irradiated area as well as other side effects by damaging normal, healthy cells near the cancer. To circumvent such problems, radiosensitizers, drugs that make tumor cells more sensitive to radiation therapy, are sometimes used. Also, radionuclides (or radioisotopes) are used as an ionizing radiation source, instead of an external beam source. However, the therapeutic efficacy of both radiosensitizers and radionuclides is limited by their delivery efficiency.

Because of the above-mentioned existing problems regarding cancer imaging and treatment, it was hoped that a new breakthrough in technology would result in more efficient diagnosis and treatment of cancer. The molecularly targeted approach, with advances in finding cancer-specific molecular signatures or markers, has emerged as a new paradigm in cancer imaging and therapy. Here, the medical intervention is "targeted" to molecular signatures, such as certain proteins, that are specifically expressed on the surface of the malignant cancer cell or neovasculature; thus, in principle, it affects only diseased tissues and keeps normal tissues undisturbed [5]. Targeted therapy can drastically increase the therapeutic index, with smaller therapeutic doses and with reduced side effects stemming from the toxicity of drugs. Targeted molecular imaging can achieve a dramatic enhancement of image contrast, enabling detection of cancer at the earliest possible time, and may even perform pathological characterization as biopsy does. It turned out that such targeting could be amplified by the use of nanoparticle drugs [6–7]. Another emerging new paradigm ("multifunctionality") is to combine or closely integrate the imaging and therapy so as to achieve an image-guided therapy, i.e., a single agent detects the diseased tissue, treats the tissue on-site, and monitors prognosis. Such an multifunctional approach is indeed expected to revolutionize cancer treatment, allowing individual-based diagnosis, therapy, and monitoring, i.e., personalized medicine.

Nanotechnology has been a powerful tool for validating and enabling the above two emerging new paradigms of cancer intervention because of the size, versatile engineerability, and nontoxic nature of nanoparticles [8–12].

First, the size of nanoparticles may enable passive targeting of nanoparticles into a tumor, due to the "enhanced permeability and retention" (EPR) effect [6], in which leaky neovasculatures and a lack of lymphatic drainage of the tumor lead to macromolecular accumulations. *Second*, the surfaces of nanoparticles can be engineered with moieties that selectively bind to molecular signatures of cancer, enabling molecular targeting. Often, the nanoparticles are also surface-engineered with poly(ethylene glycol) (PEG) for longer plasma circulation time, thus further enhancing tumor accumulation of the nanoparticles [8–12]. *Third*, the nanoparticles can be engineered for controlled loading and releasing of single or multiple drugs and/or of image contrast agents, thus enabling synergistic multimodal therapy/imaging as well as image-guided therapy. *Fourth*, most of the nanoparticle matrixes are biocompatible and nontoxic, thus reducing immunogenicity and side effects of the drug as well as increasing the maximum tolerated dose of the drug, by masking the drug molecule (i.e., entrapping it within the nanoparticles). *Fifth*, the nanoparticles can significantly enhance the delivery of drugs that are normally excluded from tumors by the MDR of cancer cells. Thus, the advent of nanoparticle drugs enabled the design of multifunctional, targeted, biocompatible, long circulating nanoplatforms that combine imaging and therapy [8–12]. The schematics of a targeted, multifunctional nanoparticle are summarized in Fig. 11.1. As imaging is often associated with diagnostics, these are sometimes termed "theranostic," even though an important characteristic of the dual or multifunctional modality is the combination of therapy and of the monitoring and guidance, by imaging, of that therapy. The first suggested multifunctional nanoplatforms were based on a hydrogel matrix, which easily enables the multifunctionality due to its chemical flexibility [8–12].

Among the diverse classes of nanomaterials used so far in cancer therapy and imaging, hydrogel nanoparticles have drawn considerable interest because of additional advantages stemming from unique characteristics of the hydrogel, such as high aqueous suspendability and swelling in water, that can be modulated by environmental changes. This chapter aims to cover the use of hydrogel nanoparticles in cancer therapy and imaging. It includes an overview on hydrogel nanoparticles (hydrogel types and preparation methods), their intrinsic and engineered properties for cancer therapy and imaging, and examples of their applications. It should be noted that hydrogel shell nanoparticles with metal or with a metallic oxide core are not covered here.

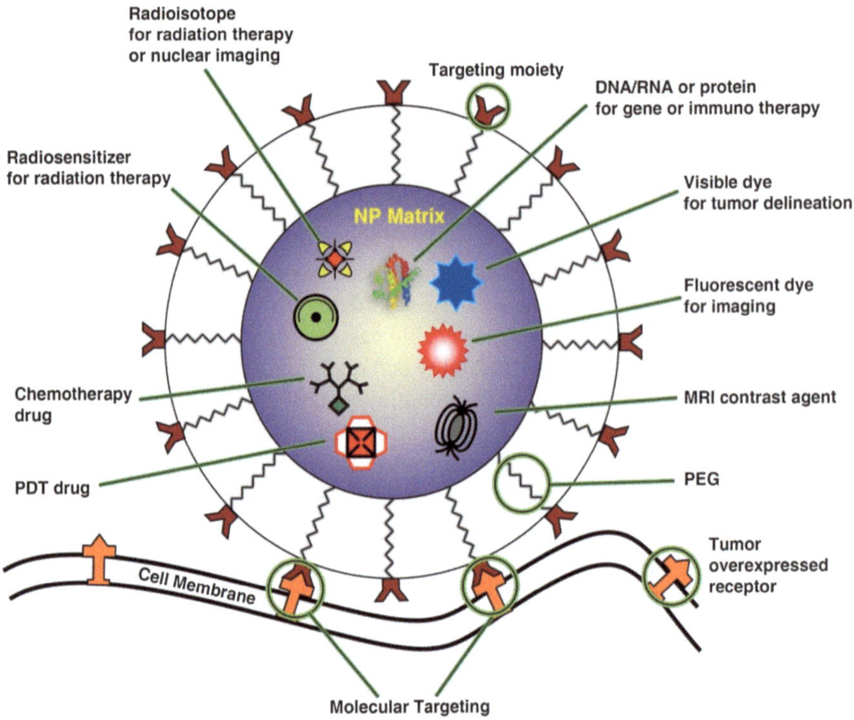

Fig. 11.1 Schematic diagram of a targeted multifunctional nanoparticle with therapeutic and imaging options that have been used for cancer detection and therapy. In reality, no more than 2–3 modalities per nanoplatform would be used, in addition to the targeting and "cloaking" with PEG molecules

11.2 Hydrogel Nanoparticles: A Brief Overview

Hydrogels are three-dimensional polymeric networks, physically or covalently cross-linked, that are water-insoluble but can absorb a large amount of water or biological fluids. They can swell/deswell while keeping their shapes due to the presence of the physical or covalent cross-links in the hydrogel network. Hydrogels are typically soft and elastic, resulting in tissue-like mechanical properties. Hydrogels are also biocompatible and injectable. Their water-absorbing ability is attributed to the presence of hydrophilic groups in the polymers that form hydrogel structures [13]. Hydrogels often contain functional groups such as amine, carboxyl, hydroxyl, and aldehyde, which can be utilized for covalent linkage to other molecules/segments, for example, hydrophobic groups, drugs, contrast agents, and targeting moieties. The swelling of a hydrogel can be explained in terms of the concept of ion osmotic swelling pressure and ion exchange kinetics [14]. The key factors in controlling the swelling properties are identified as polymer morphology (structure, porosity) and polymer composition (lengths, ionic character and hydrophilicity of

Table 11.1 Examples of polymer types used for hydrogel nanoparticles in cancer applications

Natural polymers	Polysaccharides • *Anionic*: Alginate, hyaluronan (also called hyaluronic acid or hyaluronate), xanthene gum • *Cationic*: Chitosan • *Neutral*[a]: Pullulan, agarose, dextran, cellulose Proteins/polypeptides • Collagen, gelatin, albumin, polylysine, elastin-like polypeptide (ELP)[b]
Synthetic polymers	• Poly(acrylic acid) • Poly(acrylamide), poly(*N*-isopropylacrylamide), poly(2-hydroxyethyl methacrylate), poly[2-(*N*,*N*-diethylammo)ethyl methacrylate] • Poly(ethylene oxide) (PEO)[c], poly(propylene oxide) (PPO), PEO-PPO-PEO block copolymers (or Pluronics®), poly(oligo(ethylene oxide) monomethyl ether methacrylate) • Poly(vinyl alcohol) • Poly(vinyl pyrrolidone) • Poly(*N*-vinylformamide) • Poly(ethyleneimine) • Poly(*N*-vinyl caprolactam), block copolymer of methoxy poly(ethylene glycol)–poly(caprolactone)

[a] Neutral polysaccharides are sometimes modified with acidic/basic moieties to change into anionic/cationic forms

[b] ELP is made of Val-Pro-Gly-Xaa-Gly pentapeptide repeat (where the "guest residue" Xaa is any amino acid except Pro), derived from a structural motif found in mammalian elastin [19]

[c] PEO refers to high-molecular-weight poly(ethylene glycol), typically >20,000 g/mol

linear chains, and cross-linkers) as well as factors affecting ionization equilibrium, such as pH, ionic strength, nature of counterions, and buffer composition. Swelling/deswelling of hydrogels can be engineered so as to respond to various physical, chemical or biochemical stimuli. Because of these properties, hydrogels are especially suitable for in vivo biomedical applications. The examples of biomedical applications include drug/image contrast agent delivery systems [13, 15, 16], biosensors [17] as well as tissue engineering [18].

11.2.1 Types of Hydrogel

Hydrogels are composed of hydrophilic polymers of natural or synthetic origins, or a combination of both, with or without additional cross-linkers. Table 11.1 shows examples of natural and synthetic polymers that have been used to prepare nanoparticles for cancer applications. Natural polymers are intrinsically biodegradable, either via enzymes or hydrolysis, and possess a high content of functional groups, including hydroxyl-, amino-, and carboxyl groups. These functional groups are utilized for cross-linking the polymers into hydrogels as well as for additional conjugation with a variety of groups, such as tumor-targeting moieties. Synthetic polymers are usually nonbiodegradable. However, degradable synthetic hydrogels can be made by incorporating cross-linkers or groups that are degradable by hydrolysis, by enzymatic digestions, or by environmental parameters such as pH.

11.2.2 Hydrogel Nanoparticle Preparation Methods

Hydrogels are prepared either by physically/chemically cross-linking of polymers, or by polymerization of monomers. The physical cross-linkage is based on noncovalent, attractive forces between the polymer chains, which include charge–charge (ionic) interactions, hydrophobic interactions, or hydrogen bonding. The chemical cross-linkage is based on covalent bonds via a cross-linking agent that is typically a molecule with at least two coupling or polymerizable functional moieties. Hydrogel nanoparticles, i.e., hydrogels in size from 10 to 1,000 nm, have been made by a variety of methods as listed below.

11.2.2.1 Reverse Microemulsion Polymerization/Gelation

Reverse micelles consist of water-in-oil droplets separated by surfactant-rich films. The reverse micellar droplets are highly monodisperse, and the droplet sizes are usually well controlled and less than 20 nm in diameter [20]. Therefore, this method can produce ultrafine nanoparticles with narrow size distribution, in a larger amount as compared to the other methods described below that are based on surfactant-free synthesis. However, use of organic solvent and surfactants is disadvantageous. The size of the nanoparticles is affected by the factors responsible for interdroplet interaction such as interfacial rigidity of the droplets, size of the droplets, temperature, and concentration of the hydrophilic materials in the aqueous core of the droplets [21]. The nanoparticles can be produced either from polymerization of monomer mixtures—including both monomers for linear polymers and cross-linking monomers—or from gelation of a mixture of linear polymers using either gel-inducing counterions (for instance, Ca^{2+} for alginate) or coupling molecules such as glutaraldehyde and carbodiimide. Nanoparticles made of poly(acrylamide) [15], poly(vinyl pyrrolidone) [22], and their derivatives have been prepared from monomer polymerization, and their size is typically less than 100 nm. The cross-linker monomers are typically molecules with a double bond on each side, which include nondegradable ones such as *N*,*N*′-methylene-*bis*-acrylamide as well as biodegradable ones such as glycerol dimethacrylate and poly(ethylene glycol) dimethacrylate. The nanoparticles made from polymers through chemical cross-linking or gelation are larger than those made from monomer polymerization, but their sizes are typically still less than 300 nm. The examples include chitosan nanoparticles [23], nanoparticles based on hyaluronic acid chemically cross-linked with α,β-poly(aspartylhydrazide) [24], disulfide-linked hyaluronan nanoparticles [25], and alginate nanoparticles [26].

11.2.2.2 Homogeneous Gelation in Water

In this approach, the gelation is generally conducted in dilute solution in order to prevent macroscopic gelation that leads to bulk hydrogels. The size distribution of the nanoparticles is usually broader than that of nanoparticles prepared from reverse micelle gelation, and often, aggregates are formed.

Ionic Gelation

In this method, the nanoparticles are formed in water—usually without surfactants—from physical cross-linking of the polymers based on electrostatic interaction, for instance, between two oppositely charged polymers or between charged polymers and counterions. For the nanoparticles prepared from two polymers, cationic chitosan is mixed with anions such as sodium tripolyphosphate (TPP) [27, 28] or polyanionic polymers that include polyelectrolytes such as poly(ethylene imine) (PEI) [29] and anionic polysaccharides such as hyaluronan (a weak polyanion) or heparin (a strong polyanion) [30, 31]. For the nanoparticles prepared from polymers and counterions, alginate nanoparticles are a representative example [32]. Alginate–chitosan composite nanoparticles were also prepared from formation of dilute alginate gel with Ca^{2+} ions, followed by ionic complexation with chitosan [33, 34]. The size of the nanoparticles is typically larger (300–600 nm) than those made by reverse microemulsion method. The size is affected by the molecular weight of the linear polymers that are used.

Physical Cross-Linking by Hydrophobic Association (Polymeric Micelles)

Amphiphilic polymers bearing hydrophobic substituents can self-assemble in water, by hydrophobic interactions, into nanoparticles with hydrophobic core and hydrophilic shell. The examples of amphiphilic polymers include hydroxyethylcellulose modified with various fatty acids [35] and cellulose acetate derivatives [36]; pullulan modified with cholesterol [37, 38]; hyaluronan grafted with various hydrophobic groups such as tetradecylamine [39], poly(lactic-*co*-glycolic acid) (PLGA) [40] or PEG-PLGA [41]; PEG-poly(caprolactone) [42]; poly(vinyl alcohol) (PVA) grafted with PLGA [43]; PEO cross-linked with PEI [44]; and dextran grafted with polylactide [45]. The nanoparticles are prepared either by simple sonication of the polymers in water [37, 38] or, more often, by solvent displacement (or desolvation) method. In solvent displacement method, the polymers are dissolved in an organic solvent, and the polymer solution is subsequently emulsified into an aqueous solution, often using emulsifying agents like gelatin, PVA, polysorbate 80, and Poloxamer® 188. The solvent is then displaced/removed either by evaporation [35, 41–44] or by dialysis [36, 39, 40, 45] so as to produce a nanoparticle suspension. The size of the self-assembled nanoparticles varies with the degree of hydrophobic groups on the hydrophilic polysaccharide and the molecular weight of the polysaccharide [46]. The average size ranges from 20 to 550 nm.

Monomer Polymerization

In this method, monomers forming linear chains and cross-linking monomers are copolymerized in water containing surfactant but without water-immiscible organic solvent, unlike the reverse micelle polymerization. Nanoparticles of poly(*N*-isopropylacrylamide) or its copolymer with poly(allylamine) or poly(acrylic acid) were prepared [47]. The average size of nanoparticles decreased with increasing surfactant concentration (350–120 nm at room temperature) [47]. Nanoparticles of poly(*N*-vinyl caprolactam) were prepared this way as well. Their size was 200 nm at 20°C without drugs but 360 nm when encapsulated with drugs [48].

Covalent Cross-Linking of Polymers

In this method, polymers are chemically bound using coupling agents. As an example, covalently linked chitosan nanoparticles of less than 100 nm in diameter (as measured by DLS) were prepared in surfactant-free water by covalently linking the amino groups of linear chitosan chains using cross-linker molecules such as PEG dicarboxylic acid [49] and ethylenediaminetetraacetic dianhydride [50]. The nanoparticles' size was affected by the molecular weight of chitosan and the ratio of cross-linking [49]. In addition, covalently cross-linked gelatin nanoparticles were prepared using glutaraldehyde, for which a water–acetone mixture was used in a two-step solvent displacement procedure [51].

11.2.2.3 Spray Drying

In this method, solutions and suspensions of drugs and polymers are prepared and then atomized to fine droplets using small-diameter nozzles. The droplets from a nozzle were accelerated across a voltage gradient or by the presence of a hot drying gas flow and dried through evaporation during the transit in the gas flow. This method is a rapid and high throughput production of particles, which are typically in the micron size range. However, recently, elastin-like polypeptide nanoparticles of 300–400 nm [19] and albumin nanoparticles of 130 nm [52] were prepared this way.

11.3 Applications of Hydrogel Nanoparticles in Cancer Therapy and Imaging

Hydrogel nanoparticles have excellent properties as a delivery vehicle for cancer drugs or image contrast agents due to synergistic advantages from combination of hydrogel properties and nanoscale size, as summarized in Table 11.2. The optimal size for cancer imaging and treatment is believed to be 10–100 nm, although the

Table 11.2 Advantages of hydrogel nanoparticles as a delivery vehicle for cancer drugs/image contrast agents

Properties	Origin of the properties	Resultant advantages
Biocompatibility and low toxicity	Properties of hydrogel matrix materials High water-absorbing properties of hydrogel	Reduced side effects from drugs/contrast agents Reduced unnecessary premature degradation of the drugs/contrast agents
High payload of drugs or image contrast agents per nanoparticle	Interaction between drugs/contrast agents and hydrogel matrix Size of the nanoparticles compared to that of drugs/contrast agents	High image contrast enhancement or high therapeutic efficiency
Loading of multiple drugs and/or image contrast agents	Interaction between drugs/contrast agents and hydrogel matrix Size of the nanoparticles compared to that of drugs/contrast agents	Image-guided therapy or personalized medicine
Controlled release of drugs	Interaction between drugs/contrast agents and hydrogel matrix Swelling/deswelling or structure rupturing of hydrogel in response to various environmental changes or external stimuli	High therapeutic efficacy Reduce systemic toxicity (or side effects) of drugs
Hydrophilicity	High water-absorbing properties of hydrogel	Injectable formulation
Low nonspecific protein binding	Properties of hydrogel matrix materials Surface modification of nanoparticles with more hydrophilic segments or PEG	Long plasma circulation that can help increase high tumor accumulation of the drugs or contrast agents
Efficient targeting to tumor	Size of the nanoparticles enabling EPR effect Nanoparticles' surface modification with tumor-specific moieties using functional groups of hydrogel	High image contrast enhancement or high therapeutic efficiency Reduced side effects Reduced dose of drugs/contrast agents
Biodegradation with safe biodegradable products	Intrinsic properties of polymers that constitute hydrogel or cross-linkers of the hydrogel	Bioelimination within a reasonable time period

upper limit is still controversial. Nanoparticles with a size greater than 10 nm can avoid kidney clearance, resulting in prolonged and elevated levels in the blood stream, and those with a size smaller than 100 nm can penetrate deep into tissues without being trapped by the phagocytes [53].

Hydrogel nanoparticles have been loaded with drugs and/or image contrast agents, and often with active targeting moieties for cancer therapy and imaging, using the methods described below. Loading of drugs/contrast agents is carried out during the synthesis of hydrogel nanoparticles or after the nanoparticles are formed. The loading mechanisms include entrapment within the matrix by steric constriction or binding with the matrix either chemically through covalent bonds or physically through electrostatic interactions, hydrogen bond formation, and/or hydrophobic interactions. Charge or hydrophobicity of the hydrogel matrix can be engineered by grafting (conjugating) charged or hydrophobic groups to the hydrogel. Loading during the synthesis may have higher loading efficiency because both entrapment by steric restriction and physical/chemical interaction mechanisms are applied. However, harsh synthetic conditions such as use of surfactant, organic solvents, or radical polymerization, sometimes compromise the integrity and, consequently, efficiency of the drugs/contrast agents. This compromise occurs mostly when the agents are made from delicate biological molecules like proteins and nucleic acids. Harsh conditions can sometimes also reduce the efficiency of low-molecular-mass agents. When the loading is done only through physical entrapment and physical interactions, the loaded drugs/contrast agents may be released from the nanoparticles by simple diffusion or by external stimuli if the hydrogel matrix is properly designed. Often, the loaded agents are released before the nanoparticles reach their target area in vivo, thus reducing the efficiency of treatment and possibly resulting in side effects, although these side effects are greatly reduced compared to naked drugs/contrast agents. Postsynthesis loading by physical interactions does not cause chemical damage to the drugs/contrast agents, but the nanoparticles prepared by this method are more prone to prematurely releasing the loaded agents during in vivo applications. When the loading mechanism involves chemical bonding between the drugs/contrast agents and the nanoparticle matrix, there should be no leaching-related problems during systemic in vivo circulation for both cases, the loading during the synthesis case or postloading case. However, covalently linked drugs/contrast agents may reduce therapeutic/image contrast efficiency of the original, nonlinked agents.

Surface modification of hydrogel nanoparticles with active targeting moieties is typically done by conjugation of amine or carboxyl groups available on the nanoparticles' surface with similar functional groups available on targeting moieties using bifunctional cross-linking agents [12, 54]. Actively targeted nanoparticles were reported to have enhanced binding affinity and specificity over targeted molecular drug and contrast agents due to the multiple numbers of targeting ligands packed on their surface, a "multivalency effect" [7, 55].

11.3.1 Hydrogel Nanoparticles for Therapy

Hydrogel nanoparticles have been applied for chemotherapy, gene therapy, immunotherapy, photodynamic therapy, and radiotherapy. The active agents in these therapies, chemical drugs, immunomodulating proteins such as cytokines, DNAs, antisense oligonucleotides, and siRNA, should directly interact with cancer cells and, consequently, should be released from the nanoparticles at the target site. However, release is not necessary for PDT and radiotherapy as the cytotoxic agents for these therapies are ROS, not the loaded drugs/agents themselves.

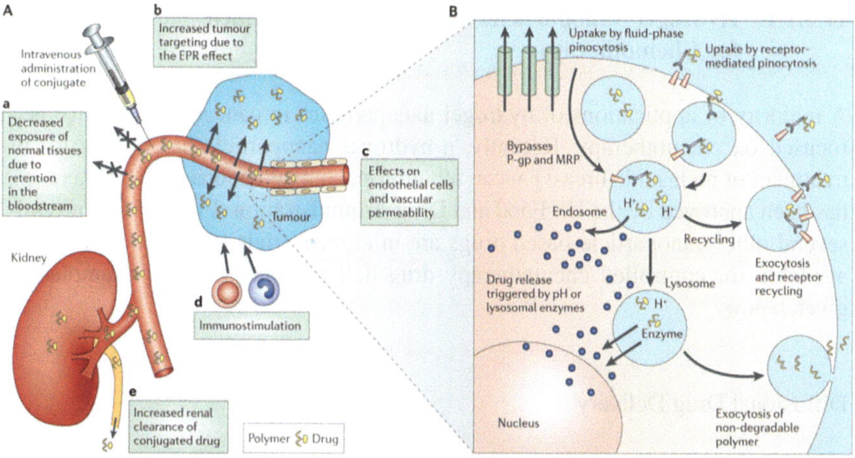

Fig. 11.2 (**A**) Drug delivery pathway of intravenously injected hydrogel nanoparticles or hydrophilic polymer-drug conjugates. (**B**) Mechanisms of cellular uptake of nanoparticles and subsequent intracellular drug release. Reprinted with permission from [62]

Ideal delivery vehicles should (1) retain high dose of drugs or cytotoxic agents, (2) reach only tumor cells and tumor metastases and not healthy tissue, (3) release the drugs or cytotoxic agents only at the tumor site and not during circulation, and (4) be cleared from the body in a reasonable time period. Hydrogel nanoparticles have properties that could meet these requirements (see Table 11.2). Furthermore, some hydrogels—so-called environment-sensitive hydrogels, stimuli-sensitive gels, responsive gels, or smart gels—show volumetric transitions in response to external stimuli. These hydrogels undergo physicochemical changes upon environmental changes that affect the movement of water and ions into and from the hydrogel matrix, resulting in volumetric changes (swelling or shrinking). Because of these special properties, hydrogel nanoparticles enable not only time-controlled release of drugs by diffusion but also local burst of drugs at tumor site in response to external stimuli or environmental changes. The available stimuli include chemical or biochemical stimuli such as pH, glucose [56], antigen [57], and redox agents such as glutathione, as well as physical stimuli including temperature [48, 58], electric fields [59], and light [60, 61]. Nanoparticles have been administered into the body by intravenous, oral, intratumoral, and intraperitoneal delivery. Figure 11.2 illustrates detailed mechanisms of hydrogel nanoparticle-based drug delivery after intravenous administration. Hydrogel nanoparticles preferentially accumulate in tumor cells due to their long blood circulation time, EPR effect, and surface-conjugated, tumor-specific targeting moieties. At the tumor site, the nanoparticles enter the cells, and the drugs are released either by diffusion or by environmental changes within cells. The rate of the stimuli-induced, volumetric phase transition for hydrogel nanoparticles is relatively fast. For instance, deswelling of thermoresponsive poly(*N*-isopropylacrylamide) (pNIPAm) hydrogel nanoparticles (100–400 nm in diameter) occurs on the microsecond timescale [63].

11.3.1.1 Hydrogel Nanoparticles as Drug Delivery Vehicle for Chemotherapy

A majority of applications of hydrogel nanoparticles in cancer therapy have been focused on chemotherapy. Recently, a hydrogel nanoparticle-based drug for the treatment of metastatic breast cancer, albumin-paclitaxel nanoparticle (Abraxane®), has been approved by the US Food and Drug Administration (FDA) [64]. Currently, several other nanoparticle-based drugs are in clinical trials [65]. The mechanisms available for controlled chemotherapy drug delivery and selective examples are given below.

Diffusional Drug Delivery

Diffusion plays the predominant role in drug release for hydrogel nanoparticles made of neutral polymers and cross-linkers that are either nondegradable or slowly degradable. Factors affecting diffusion-controlled drug release include concentration of the loaded components, osmotic effect, polymer characteristics (e.g., composition, flexibility, and elasticity), and degree of cross-linking [14]. We note that, to a certain degree, the drug release from these nanoparticles can also be affected by environmental changes. For example, neutral poly(acrylamide) hydrogels cross-linked with a nondegradable cross-linker, when partially hydrolyzed in a solvent such as an acetone–water mixture, undergo discrete and reversible volume transitions upon small changes in temperature, solvent composition, pH, or concentration of an added salt [59]. We also note that hydrogels made of neutral polymers can be converted into environment-sensitive gels by using environment-sensitive, degradable cross-linkers or by modifying the polymer chains with weakly ionizable groups. Examples for this class of hydrogel nanoparticles include the nanoparticles made of poly(acrylamide) [66], poly(vinyl pyrrolidone) [67], Pluronic® F127 (PEO-PPO-PEO triblock copolymer) [68], block copolymer of methoxy poly(ethylene glycol)–poly(caprolactone) (mPEG–PCL) [69], pullulan modified with cholesterol [70], and dextran [71]. These hydrogel nanoparticles have been used to deliver chemotherapy drugs such as cisplatin [66, 69], bleomycin [67], and doxorubicin [68, 70, 71]. Some of these nanoparticles were conjugated with active targeting moieties such as F3 peptide [66] or folic acid [68, 70] so as to enhance cellular binding and uptake by receptor-mediated endocytosis. These nanoparticles showed superior antitumor efficiency with reduced side effects when compared to the corresponding free drugs.

Some of these nanoparticles were found to have high therapeutic efficiency even in multidrug resistant (MDR) cells. Doxorubicin-loaded, self-assembled dextran nanoparticles (~110 nm in hydrodynamic diameter as measured by DLS) showed notable antiproliferative effects against human osteosarcoma cell lines, both in doxorubicin-sensitive (KHOS, U-2OS) and doxorubicin-resistant ($KHOS_{R2}$ and $U\text{-}2OS_{R2}$) cells [71]. Another example consists of F3-peptide-conjugated cisplatin-loaded poly(acrylamide) (PAA) nanoparticles (F3-Cis-NPs) (20–30 nm in diameter in dried state) that were used for vascular-targeted chemotherapy in both murine and human ovarian cancer models. The F3 peptide is known to bind to the angiogenic vasculature within solid tumors as well as to some tumor cells and become internalized, both in cell culture systems and in vivo, by interacting with nucleolin, a cell surface receptor [72, 73]. It was demonstrated, using fluorescently labeled nanoparticles (Fl-NPs), that surface-conjugated F3 peptides enabled the nanoparticles to bind with high specificity to both human ovarian tumor cells and tumor endothelial cells in vitro. The F3-Cis-NPs demonstrated cytotoxic activity against the tumor endothelial cells. In vivo studies with Fl-NPs showed that the F3-targeted nanoparticles bound primarily to tumor endothelial cells. At 100 mg/kg dose, administered to mice bearing highly vascular ID8-VEGF ovarian tumors models, nontargeted Fl-NPs showed little tumor-specific uptake but were found at significant levels in the liver and kidney. In contrast, at the same concentration, F3-Fl-NP showed significant uptake in tumor vessels and some uptake within tumor parenchyma. Therapeutic studies were performed in both flank and orthotopic intraperitoneal (i.p.) murine ovarian tumor models, as well as in human tumor xenograft models. Here, F3-Cis-NPs or cisplatin together with F3-blank nanoparticles (F3-NPs) were administered into tumor-bearing mice by i.v. or i.p. injection, or by a combination of both. The results suggested that vascular exposure is the primary means of therapy, as the addition of i.p. therapy, which could directly target tumor cells, had no added benefit. The F3-Cis-NP treatment at a cisplatin dose of 75 μg/kg demonstrated rapid tumor regression, while the control treatment using cisplatin together with F3-NPs at the dose of 150 μg/kg did not (Fig. 11.3). This treatment was effective in both cisplatin-sensitive and cisplatin-resistant cell lines. The therapy targeting a human vasculature, with F3-Cis-NP, led to near complete loss of all human tumor vessels in a murine model of human tumor vasculature. Moreover, minimal toxicity was noted with F3-Cis-NP treatment as evidenced by stable levels of blood enzymes and complete blood counts in the collected serum from animals 24 h after the last i.v. treatment.

Chemotherapy by Stimuli-Induced Drug Delivery

pH-Responsive Drug Release

Physiological pH varies among different organs/tissues in the body (e.g., pH 2 for stomach, pH 6.2–7.5 for the small intestine, pH 6.9–7.1 for brain, pH 7.2–7.6 for skin) as well as among cellular compartments (pH 5.0–6.8 for endosomes, 4.5–5.5 for lysosomes, 7.2–7.4 for cytosol) [74, 75]. Moreover, most human tumors have been reported to be slightly more acidic than normal tissue with pH values between 6.15 and 7.40 (the latter value representing the mean pH of arterial blood), while normal tissues usually have pH values between 7.0 and 7.4 [75]. Therefore, pH-responsive hydrogel nanoparticles have drawn considerable interest as controllable delivery systems [76].

Hydrogel nanoparticles showing pH-responsive volume changes have been prepared in two ways: (1) use of polymers containing weakly ionizable moieties, such

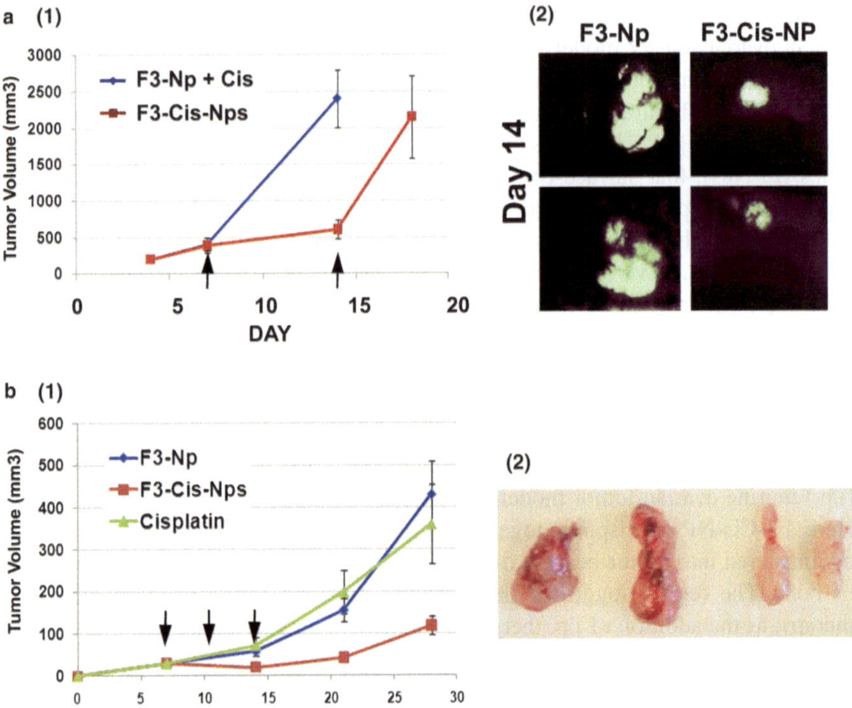

Fig. 11.3 Therapeutic efficacy of F3-Cis-NP against human tumor xenografts. (**a**) A2780 cisplatin-sensitive tumor xenografts treated with F3-NP+cisplatin or F3-Cis-NP (1) tumor growth curves, (2) in vivo fluorescent imaging. (**b**) Cisplatin-resistant SKOV3 tumor xenografts treated with F3-NP+cisplatin or F3-Cis-NP (1) tumor growth curves, (2) gross tumor pathology. F3-Cis-NP-treated tumors were significantly smaller and pale/avascular ($n=5$/group in independent experiments). Reproduced with permission from [66]

as amine or carboxyl groups, for example, chitosan [27, 28, 77–79], poly(acrylic acid-*co*-acrylamide) [80], elastin-like polypeptide (ELP) [19], pullulan modified with acidic/basic moieties [81–83], poly[2-(*N*,*N*-diethylamino)ethyl methacrylate] [84], or (2) use of pH-labile cross-linkers [85, 86]. Polymers containing basic groups (or cationic polymers) ionize at low pH, while polymers containing acidic groups (or anionic polymers) ionize at high pH, and therefore, cationic hydrogels swell at low pH and anionic hydrogels swell at high pH. The degree of swelling of the hydrogels containing these ionizable polymers depends on the degree of ionization and the amount of cross-linkers. For instance, poly(acrylic acid-*co*-acrylamide) hydrogel nanoparticles of 20–80 nm in diameter were prepared from radical polymerization of monomer mixtures of acrylamide and acrylic acid monomer and the nondegradable cross-linker, *N*,*N*-methylene-bis-acrylamide (MBA) [80]. The ratio of acrylamide-to-acrylic acid changed both drug loading and pH-dependent drug release from the nanoparticles. In another study, nanoparticles consisting of a cross-linked core of poly[2-(diethylamino)ethyl methacrylate] surface grafted with PEG

(50–150 nm in average diameter) were prepared [87]. Here, control over mesh size, surface charge, encapsulation efficiency, and in vitro biocompatibility was obtained by varying the cross-linking density. Hydrogel nanoparticles made of pH-labile cross-linkers also showed that the pH-dependent release rate of proteins varies with the amount of the cross-linkers [85, 87]. It should be noted that these pH-responsive hydrogel nanoparticles are usually highly charged. The charged nanoparticles have a much higher opsonization rate than neutral particles [88], resulting in rapid clearance by the reticuloendothelial system (RES). Additional nonionic hydrophilic coatings, such as PEG, may be necessary for these charged nanoparticles to become efficient in in vivo applications.

The pH-responsive delivery has been applied for chemotherapy drugs such as 5-fluorouracil (5-FU) [27, 80], doxorubicin [19, 77, 81, 82, 89], paclitaxel [79], methotrexate disodium [28], epirubicin [83], and cisplatin [78, 84]. For more efficient, tumor-specific delivery, the hydrogel nanoparticles have been decorated with targeting moieties such as folic acid [83, 89], hyaluronic acid [27], and apo-transferrin [28], resulting in significantly higher cellular uptake by the target tumor cells in vitro.

Redox-Responsive Drug Release

Disulfide is a biodegradable group, which can be cleaved to the corresponding thiols by reducing agents. Hydrogel nanoparticles containing disulfide bonds have been developed for preferential release of the encapsulated drugs inside the tumor cells by glutathione (GSH). GSH is a major intracellular reducing agent whose intracellular concentration (1–10 mM) is significantly higher than its extracellular concentrations (up to ~1,000 times) [90]. For example, hydrogel nanoparticles (~225 nm in diameter) made of poly[(oligo(ethylene oxide) monomethyl ether methacrylate] with encapsulated doxorubicin or fluorescent dye rhodamine 6G were prepared [91]. Results obtained from optical fluorescence microscope images and live/dead cytotoxicity assays of HeLa cancer cells suggested that the released doxorubicin molecules penetrated cell membranes, and therefore, could suppress the growth of cancer cells.

Temperature-Responsive Drug Release

Some hydrophilic polymers containing hydrophobic groups have temperature-dependent solubility in aqueous solutions, exhibiting a cloud point (CP) or lower critical solution temperature (LCST). Hydrogels containing these polymers undergo an abrupt shrinking above the LCST, which stems from the hydrophilic–lipophilic balance (HLB) of polymer chains [74]. Temperature-responsive hydrogel have been prepared from poly(*N*-isopropylacrylamide) (PNIPAAm) [16], Pluronics® [58], and more recently from poly(*N*-vinyl caprolactam) (PVCL) [48, 92]. The degree of swelling/squeezing of these hydrogels was reported to depend on the cross-linking density and the number of ionic groups along the backbone [93]. The LCST can also be controlled by the composition of these hydrogels, for instance, the ratio of temperature-responsive polymers and nonresponsive polymers, as shown for copolymeric nanoparticles of poly(methyl methacrylate-*co*-*N*-vinylcaprolactam) [92]. The applications of these hydrogels, however, have been mostly focused on in situ implants rather than on an individual nanoparticle-type delivery system, although a couple of such nanoparticles were designed [92, 93]. Unlike bulk hydrogel, the temperature of the nanoparticles will be quickly equilibrated with the body temperature, which is higher than the LCST of the common temperature-sensitive polymers (e.g., 32°C for PNIPAAm and PVCL), as soon as they are introduced. Therefore, nanoparticles operated on body temperature-based drug release may not be an efficient systemic drug delivery system. For nanoparticles made of hydrogel with LCST higher than body temperature, the drug release may be initiated at the tumor site by external heat. However, inducing a hyperthermic condition in tumors may also be difficult unless either the tumors are located near the surface of the body or the nanoparticles contain additional components that could be remotely heated, for example, by a magnetic field. Recently, hydrogel nanoparticles that can respond not only to temperature but also to pH or GSH have been designed for more efficient drug delivery. For instance, self-assembled nanoparticles (size 40–110 nm) of PNIPAAm-*g*-pullulan with thiol end groups (sulfanylthiocarbonylsulfanyl) were prepared for temperature- and/or GSH-responsive drug release [94], while poly(*N*,*N*-diethylacrylamide-*co*-methacrylic acid) nanoparticles (size 130–250 nm) [95] and PNIPAAm/chitosan nanoparticles (size 50–150 nm) [96] were prepared for temperature- and/or pH-responsive drug release.

11.3.1.2 Hydrogel Nanoparticles as Drug Delivery Vehicle for Gene-/Immunotherapy

The drugs for gene therapy or immunotherapy are biological molecules that are hydrophilic, unlike chemotherapy drugs. These biological molecules are highly charged and subject to fast degradation by serum enzymes, resulting in very low therapeutic efficacy and potential side effects when administered in their "naked" forms. Therefore, the use of efficient and safe delivery vehicles for these biological drugs is essential.

In case of gene therapy, the polyanionic nature of nucleic acids prevents efficient transport across the anionic cell membrane. Cationic polymer-based gene delivery has been used due to the following two reasons [97]: (1) the electrostatic attraction between the cationic polymer and the oppositely charged nucleic acid contributes to condensing DNA into small particles, called polyplexes; and (2) cellular uptake ensues as a result of charge–charge interactions between the cationic polyplex and the anionic surface of the cells, which consequently will enhance the transfection efficiency of the polyplex. Common cationic polymers include natural proteins (gelatin, collagen, albumin, histone, protamine), synthetic polyamino acids (polylysine, polyarginine), polysaccharides (cyclodextrin, chitosan), and other synthetic polymers such as polyesters, polyurethanes, tetraminofullerene, polybrene, PEI, poly(α-(4-aminobutyl)-L-glycolic acid), and cationic dendrimers [97].

However, gene delivery based on these polymers showed low transfection efficiency and also toxicity. For instance, PEI, the most efficient gene transfer polymeric vector, suffers from toxic effects such as apoptosis induction [98].

Hydrogel nanoparticles have been applied to enhance transfection efficiency by engineering PEG, targeting moieties, and redox-responsive cross-linkers in the nanoparticle constructs. Hydrogel nanoparticles made of gelatin [99, 100] were used to deliver plasmid DNAs, and the nanoparticles made of dextran [101, 102], hyaluronan [25], and poly(*N*-isopropylmethacrylamide) (PNIPMA) [103] were used for siRNA delivery. PEGylated gelatin nanoparticles with encapsulated reporter plasmid DNAs, encoding for β-galactosidase (pCMV-β), were prepared to have a size of ~200 nm and a slightly negative surface charge [100]. These nanoparticles efficiently transfected Lewis lung carcinoma (LLC) cells in vitro, as well as in LLC-bearing mice. The same gelatin nanoparticles were further engineered to contain disulfide bonds so as to achieve preferential intracellular release of the encapsulated plasmid DNAs encoding for the soluble form of the extracellular domain of the vascular endothelial growth factor receptor-1 (VEGF-R1 or sFlt-1) [99]. The PEGylated nanoparticles (~300 nm in diameter) induced a higher level of sFlt-1 expression, compared to the unmodified nanoparticles, in the MDA-MB-435 human breast adenocarcinoma cell line in vitro, as well as in female Nu/Nu mice bearing orthotopic MDA-MB-435 breast adenocarcinoma xenografts. PEGylated dextran nanoparticles (100–180 nm in diameter) were used to deliver siRNA to silence either multidrug resistance 1 (MDR-1) [101] or epidermal growth factor receptor (EGFR) genes [102], so as to enhance the chemotherapy drugs' efficiency inside multidrug resistant cancer cells. The nanoparticles caused a significant knockdown of these genes in osteosarcoma cell lines ($KHOS_{R2}$ and $U\text{-}2OS_{R2}$) [101] and in the human hepatoma cell line (HuH-7 EGFP) [102]. Targeted intracellular delivery of siRNA, to silence EGFR genes, was also tried using nanoparticles made of PNIPMA and surface-conjugated YSA peptides (~100 nm in diameter) [103] and nanoparticles made of hyaluronan and disulfide linkages (200–500 nm in diameter) [25]. The YSA peptide (YSAYPDSVPMMS), a 12-amino acid peptide, mimics the ligand ephrin-A1, which binds to the erythropoietin-producing hepatocellular (Eph) A2 receptor that is overexpressed in the neovasculature and in a number of tumor cells, including Hey cells [103]. Hyaluronan is one of the main components of the extracellular matrix, and the hyaluronan receptor CD44 is overexpressed in many types of tumor cells, including colorectal carcinoma (HCT-116) [104, 105]. Therefore, it can serve not only as a hydrogel matrix but also as an active targeting moiety. These actively targeted nanoparticles were readily taken up by the specific tumor cells that overexpress these receptors. For instance, two ovarian cancer cells, Hey cells (EphA2 positive) and SK-OV-3 cells (EphA2 negative), were treated with siRNA-loaded, YSA-targeted PNIPMA nanoparticles [103]. The Hey cells showed decreased EGFR expression levels and significantly increased the sensitivity to docetaxel, while SK-OV-3 cells did not.

Hydrogel nanoparticles have also been used for delivering proteins for tumor vaccine/immunotherapy. The nanoparticles were prepared from poly(acrylamide) (35 nm and 3.5 μm) [106], poly(2-(*N*,*N*-diethylammo)ethyl methacrylate) derivatives (50–200 nm) [107, 108] and cholesterol-bearing hydrophobicized pullulan (CHP) (30–40 nm) [109–111]. The types of delivered protein include protein antigens such as NY-ESO-1 protein [109, 110], model protein antigens such as ovalbumin [85, 106] and albumin [107], lysozyme [108], and cytokines such as IL-12. Currently, one of these hydrogel nanoparticles, CHP nanoparticles containing NY-ESO-1, is in clinical trials, showing that vaccination with the nanoparticles had potent activity for inducing tumoral immune responses against the NY-ESO-1 antigen in cancer patients [110]. The release mechanisms of these proteins were either diffusional [109–111] or pH-sensitive [85, 106–108]. One controversial finding was reported in T-cell activation studies with antigen-loaded poly(acrylamide) (PAA) nanoparticles of two different sizes, 35 nm and 3.5 μm [106]. The in vivo tests of T-cell activation demonstrated that for the pH-labile PAA hydrogel carrier system particle size does not appear to play a significant role in the generation of cytotoxic T-lymphocyte (CTL) immune responses, which is contrary to previous reports in the literature.

11.3.1.3 Hydrogel Nanoparticles for Photodynamic Therapy (PDT)

PDT is a light-stimulated, localized therapy where light activates PDT drugs to react with the surrounding oxygen, producing ROS, mainly singlet oxygen. Note that the release of PDT drugs from the nanoparticle matrix is not necessary for PDT, as the major role of the PDT drugs is not to directly kill cancer cells but to produce ROS. Consequently, PDT drugs have been loaded into nanoparticles not only by physical entrapment, as for chemotherapy drugs, but also by nondegradable covalent linkage. Both loading methods have advantages and disadvantages for PDT. The covalent linkage has advantages in better controlling the drug loading, its distribution and retention, resulting in higher singlet oxygen production (i.e., PDT efficiency) per nanoparticle [54]. Most photosensitizers easily form dimers or aggregates, and usually, these structures have very low or no PDT efficiency [112, 113]. The covalent linkage can prevent dimer formation as individual photosensitizer moieties are separately conjugated to the nanoparticle matrix, and therefore, can be well separated, thus maximizing the singlet oxygen production from each nanoparticle [54]. On the other hand, the covalently linked photosensitizers may have reduced PDT efficiency compared to free photosensitizer. Moreover, the nanoparticles do not generally enter specific subcellular organelles, such as the mitochondria [114], while molecular PDT drugs do. These two factors may lead to reduced PDT efficiency. The physical encapsulation method has advantages and disadvantages in a way opposite to the covalent linkage method. The singlet oxygen production per nanoparticle may be low due to dimer formation or due to drug leaching out of the nanoparticle matrix before the nanoparticles reach the tumors. However, the PDT drug can be released from the nanoparticles inside cells and enter specific subcellular organelles. Hydrogel nanoparticles used for PDT have been made of poly(acrylamide) [10, 11, 54, 115–118], cross-linked albumin [119, 120], marine atelocollagen/xanthene gum [121], and alginate [26, 122].

So far, PAA nanoparticles are the most intensely studied hydrogel nanoparticles for PDT. The size of the PAA nanoparticles is typically in the range of 30–70 nm in diameter [8–12, 54, 115–117, 124], but ultrafine PAA nanoparticles (2–3 nm in diameter) have also been used [118]. PAA nanoparticles have been loaded with methylene blue [54, 115, 116], *meta*-tetra(hydroxyphenyl)chlorin (mTHPC) [118], Photofrin [10–12], and the two-photon absorption photosensitizer, 5,10, 15,20-tetrakis(1-methyl 4-pyridinio) porphyrin tetra(p-toluenesulfonate) (TMPyP) [117]. It was demonstrated that the photodynamic activity of the loaded PDT drugs is well preserved within the inert PAA matrix, even for photosensitizers like methylene blue that can be degraded by plasma enzymes [123] and significantly lose its therapeutic efficacy following systemic administration [115]. These PAA nanoparticles have shown an effective photodynamic activity in cancer cells such as 9L glioma, C6 glioma, F98, and MBA-MD-435 cells. Moreover, F3 peptides have been conjugated to the surfaces of PAA nanoparticles with encapsulated Photofrin [12] and those with covalently linked methylene blue [54], to selectively kill the target tumor cells in vitro [12, 54] as well as in vivo [12].

Other noteworthy hydrogel PDT nanoparticles are alginate nanoparticles that were used for a combination therapy of PDT and chemotherapy. Alginate nanoparticles (62 nm in diameter by DLS), containing both methylene blue and doxorubicin, as well as other control alginate nanoparticles, containing either only methylene blue or only doxorubicin or nothing, were prepared and used to treat one of the multidrug resistant tumor cells, NCI/ADR-RES cells [122]. All the control nanoparticles, except the nanoparticles containing only methylene blue, did not induce cytotoxicity. However, the nanoparticles containing both drugs induced enhanced nuclear accumulation of both drugs, resulting in significantly elevated ROS production upon light exposure compared to the nanoparticles containing only methylene blue. Cytotoxicity was enhanced in the presence of methylene blue, even without light exposure. This observation suggests that nanoparticle-mediated combination of PDT and chemotherapy can lead to improved cytotoxicity in drug-resistant tumor cells.

11.3.1.4 Hydrogel Nanoparticles for Radiation Therapy

Hydrogel nanoparticles have been utilized as a carrier of radiosensitizers or radionuclides. PAA nanoparticles with covalently linked mitomycin C (MMC), a radiosensitizer, were prepared to serve as a nanoparticle-based radiosensitizer [124]. The in vitro tests—48 h incubation of free MMC or MMC nanoparticles with 9L gliosarcoma cells, followed by 10 Gy X-ray irradiation—showed that MMC nanoparticles retained the radiosensitizing activity of MMC but significantly reduced the toxicity of MMC. For radionuclide-containing hydrogel nanoparticles, rhenium-188-labeled pullulan acetate nanoparticles (60–230 nm in diameter) [125] and iodine-131-labeled albumin nanoparticles [126] have been reported. The pullulan acetate nanoparticles had a tendency to aggregate in biological solutions of high ionic strength, and therefore, were able to accumulate in the tumor efficiently after intratumoral injection into mice bearing CT-26 colon cancer. The nanoparticle treatment inhibited tumor growth significantly, compared to the free radioisotope treatment. The albumin nanoparticles were conjugated with fibrinogen, a native ligand containing two RGD peptides, in order to target the $\alpha_{2b}\beta_3$ integrin, a receptor within the tumor vasculature. Interestingly, radiation was used to induce the $\alpha_{2b}\beta_3$ integrin expression in the tumor vasculature, and therefore, to guide the fibrinogen-conjugated nanoparticles to the receptor. Administration of the targeted nanoparticles in combination with radiation resulted in significantly increased tumor regression and growth delay in mice bearing murine B16F0 melanoma, as compared to tumors treated with other controls, i.e., nontargeted nanoparticles with radiation, radiation alone, and targeted or nontargeted nanoparticles without radiation.

11.3.2 Hydrogel Nanoparticles for Imaging: Diagnosis and Tumor Delineation for Surgery

Current modalities of cancer imaging rely on contrast agents mostly made of small molecules that are not specific to the tumor and have short plasma circulation times. Because of these shortcomings, imaging techniques have played only a supplementary role to invasive biopsies in cancer diagnosis. Hydrogel nanoparticles have a high potential to improve the tumor specificity and sensitivities of these imaging modalities so that they can serve as noninvasive diagnostic tools due to their intrinsic and engineerable excellent properties (see Table 11.2). The ultimate goal in this enterprise would be to obviate the need for biopsies.

Hydrogel nanoparticles have been prepared for several cancer imaging modalities. For instance, hydrogel nanoparticles were loaded with iron oxide [10–12, 127] or gadolinium chelate [9, 128] for MRI, technetium-99m (^{99m}Tc) for SPECT or planar scintigraphy [129], and ^{64}Cu for PET [130, 131]. The hydrogel nanoparticle-based contrast agents achieved increased blood circulation time, selective tumor targeting, and a significantly enhanced signal, demonstrating a high promise for sensitive, tumor-specific imaging. For instance, superparamagnetic iron oxide (SPIO) nanocrystals (~10 nm in diameter) were encapsulated in PAA hydrogel nanoparticles (~30–70 nm in diameter). These SPIO–PAA nanoparticles showed very high R2 and R2* relaxivities, approximately fivefold greater than other SPIOs due to their high payload of SPIO [127]. Moreover, control of plasma half-life with surface-conjugated PEG of different lengths [127], and selective tumor targeting with surface-conjugated F3 peptides [12] was observed in in vivo MRI studies of brain tumors in a rat 9L gliosarcoma model. A similarly enhanced outcome for PET was observed with ^{64}Cu-radiolabeled shell-cross-linked hydrogel nanoparticles made of polystyrene (PS) and poly(acrylic acid) (PAAc) block copolymers, grafted with a copper-chelating agent, a lysine derivative of 1,4,7,10-tetraazocyclododecane-*N*,*N*′,*N*″,*N*‴-tetraacetic acid (DOTA) (DOTAlysine) [130, 131]. These nanoparticles showed impressive specific activities (ca. 400 μCi/μg). Furthermore, biodistribution experiments and PET images demonstrated that the blood retention of PEGylated nanoparticles could be tuned, depending on the mPEG grafting density and the nanoparticle surface properties [131].

Hydrogel nanoparticles were also utilized to prepare a color contrast agent for tumor delineation so as to perform visible, color-guided surgery under normal operating room lighting condition [132, 133]. The PAA nanoparticles were loaded with coomassie blue dye and conjugated with F3 peptides to solve the limitations of "naked" dye delineation, i.e., the dyes' lack of specificity and the inability to concentrate an adequate quantity of dye within tumor cells to achieve visual satisfactory contrast. In vitro tests showed that F3-targeted nanoparticles induced ~5 times faster and higher color contrast in 9L gliosarcoma cells (a high nucleolin expressing cell) than nontargeted nanoparticles [132]. In vivo tests in a rat cranial window model with implanted 9L gliosarcoma showed that the tumor delineation by the F3-targeted nanoparticles was quite persistent and increased even at 6 h after tail vein administration, while the tumor delineation by free coomassie blue dye started to diminish after 40 min and almost disappeared at 6-h postadministration [133].

11.3.3 Multifunctional Hydrogel Nanoparticles for Image-Guided Therapy

When imaging contrast agents are combined with drugs, the distribution of drugs and drug dose delivered to diseased tissues can be quantified, and therefore, therapeutic response can be predicted. Such combination can be efficiently realized with nanoparticles. The nanoparticles can be loaded with high payloads of the two agents and delivered selectively to the tumor by passive or active targeting mechanisms, thus enabling image-guided therapy. Due to hydrogel characteristics, hydrogel nanoparticles have a high potential for this integrated imaging and therapy. Hydrogel nanoparticles have been prepared to accommodate the following combinations: chemotherapy with fluorescent imaging [134, 135], PDT with MRI [9–12], and PDT with nuclear imaging [136].

The combination of chemotherapy and fluorescence imaging was realized in two different polysaccharide nanoparticle platforms: hydroxypropylcellulose-poly(acrylic acid) (HPC-PAAc) [134] and glycol-chitosan nanoparticles [135]. The HPC-PAAc nanoparticles (100–150 nm in diameter) are responsive to two environmental stimuli, pH through PAAc and temperature through HPC (LCST 41°C). CdSe quantum dots (QD), which have two emission peaks, at 592 and 741 nm, can be utilized as a fluorescent imaging contrast agent. These QD were in situ immobilized within the HPC-PAAc nanoparticles, based on ionic interaction between the precursor Cd^{2+} ions and the OH groups of the HPC chains. Temozolomide, an anticancer drug, was postloaded into HPC-PAAc nanoparticles for chemotherapy. The developed nanoparticles showed a pH-dependent drug release rate and pH-sensitive photoluminescence (741 nm), demonstrating pH-triggered sustained drug release and their potential use as a pH sensor. The blank nanoparticles were nontoxic, but the drug-loaded nanoparticles showed cytotoxicity in mouse melanoma B16F10 cells, which increased with drug loading. Photoluminescence at 592 nm, an emission relatively insensitive to pH, was used for imaging of B16F10 cells, demonstrating the nanoparticles' imaging ability. Another example is self-assembled nanoparticles made of glycol chitosan bearing hydrophobic cholanic acid moieties [135]. Cisplatin was loaded as a chemotherapy drug, and a near-infrared fluorescence dye, Cy5.5, was loaded as a fluorescent imaging component. These nanoparticles were 300–500 nm in diameter and released the drugs in a sustained manner. In vivo fluorescence imaging in mice bearing squamous cell carcinoma (SCC7) showed that the nanoparticles successfully accumulated in tumor tissues, probably by the EPR effect. The nanoparticles showed higher antitumor efficacy and lower toxicity than free cisplatin, as shown by changes in tumor volume, body weight, and survival rate, as well as by immunohistological TUNEL assay data.

The first suggested examples of a multifunctional hydrogel nanoparticles for combination of imaging and therapy, i.e., the combinations of PDT and MRI or PDT and fluorescence, have been based on PAA nanoparticles as the nanoplatform [8–12]. These nanoparticles combined (1) MRI contrast enhancement by an encapsulated contrast agent (iron oxide or gadolinium chelate), (2) PDT by an encapsulated PDT dye (Photofrin®), (3) plasma residence time control via surface-attached PEG, and (4) specific targeting to tumor sites by targeting units (RGD or F3 peptide). In vivo MRI studies in a rat 9L gliosarcoma model showed significant MRI contrast enhancement with F3-targeted nanoparticles (F3-NPs), compared to controls (nontargeted nanoparticles and blank nanoparticles) [12]. In vivo PDT studies demonstrated that the presence of surface-conjugated F3 peptides significantly enhanced the PDT efficiency, according to therapeutic activity evaluations by Kaplan–Meier survival statistics and diffusion MRI [12] (see Fig. 11.4). Note that in diffusion MRI, the apparent diffusion coefficients (ADC) of water within tissues are evaluated, and the increase in ADC within the tumor tissue corresponds to a loss of tumor cellularity. A single tail-vein injection of tumor-bearing rats with F3-NPs, followed by 7.5 min of red light irradiation through an inserted optical fiber, showed significant improvement in survival rate and in the tumor ADC compared to four different sets of control rats: (1) those that received the same light treatment but with nontargeted nanoparticles or (2) with Photofrin, (3) those that only received the light treatment, and (4) those that received no treatment.

For PDT and nuclear imaging, albumin nanoparticles (100–200 nm in diameter) were linked with a PDT drug, hematoporphyrin (HP), which was labeled with gamma-emitting nuclides (^{99m}Tc) for scintigraphic imaging [136]. It was found that the PDT efficiency of the nanoparticles was dependent on the HP content per nanoparticles when tested in A549 cells. In vitro cellular uptake studies using fluorescently labeled albumin nanoparticles showed enhanced accumulation of the HP-albumin nanoparticles in A549 and CT-26 cancer cell lines as compared to albumin nanoparticles, indicating HP may serve as an active targeting moiety. The nanoparticles also showed increased in vivo tumor accumulation in a CT-26 murine metastatic lung cancer model compared to the free HP drugs. The in vivo scintigraphic imaging of rabbits, administered with the nanoparticles, showed ~15-fold extended biological half-life for ^{99m}Tc-HP-albumin nanoparticles compared to free ^{99m}Tc-HP.

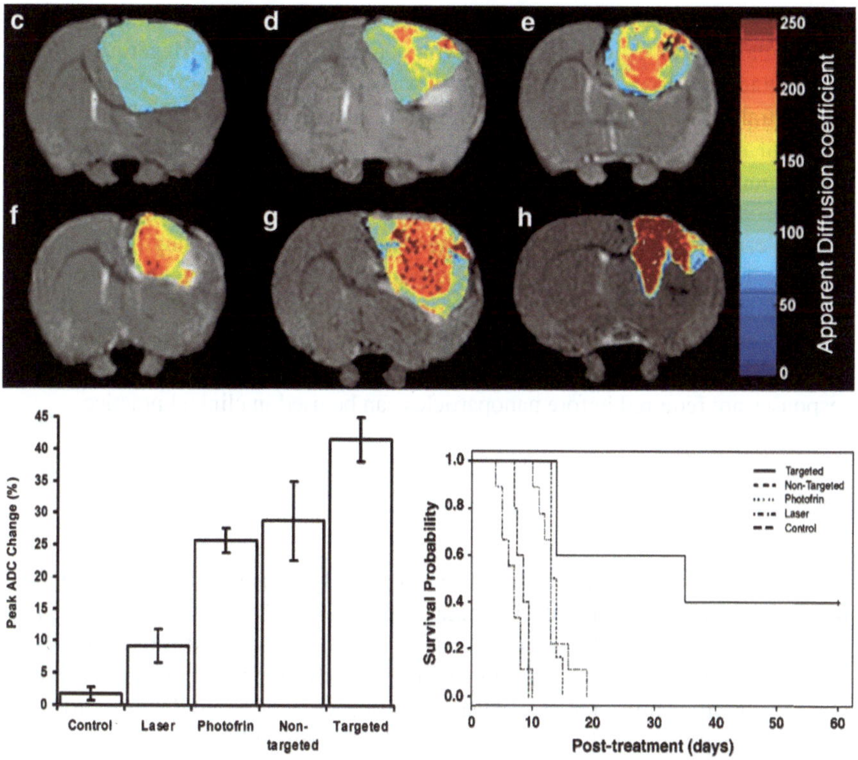

Fig. 11.4 *Top*: ADC color diffusion maps overlaid on top of T2-weighted MR images. (**c**) Nontreated control, (**d**) laser only treated, (**e**) Photofrin and laser treated, (**f**) nontargeted Photofrin nanoparticles and laser treated, and (**g**) targeted Photofrin/F3-NPs laser treated 9L tumors, all at 8-days postlaser treatment (7.5 min *red light*). (**h**) Targeted Photofrin/F3-NPs laser treated tumor at 40-day posttreatment. *Bottom left*: Mean peak percentage change in tumor ADC at day 8 after PDT. *Bottom right*: Kaplan–Meier survival plot. Note: None of the control rats survived to be tested after 20 days. Reproduced with permission from [12]

11.4 Future Perspective

Applications of hydrogel nanoparticles in cancer therapy and imaging have been successful in enhancing therapeutic efficiency and image contrast. Moreover, multifunctional hydrogel nanoparticles, combining therapeutic agents, imaging contrast agents, and specific targeting, all in a single nanoplatform, have already demonstrated the high promise of a new paradigm for cancer treatment, i.e., image-guided therapy that enables individualized (personal) detection, treatment, and monitoring of therapy. Looking ahead, continued research efforts should progress toward further improvement in existing issues such as (1) tumor targeting, (2) retention time of drugs/contrast agents in tumor tissue, (3) tumor-specific drug release, and (4) image-guided therapy. Challenges on these issues include finding cancer biomarkers with higher specificity, improving the efficiency of delivery to target tumor cells or subcellular organelles in combination with optimal drug release-triggering methods, and finding ways to cross the blood–brain barrier (especially for tumors related with the central nervous system). Research efforts should also involve the challenging issues of nanoparticle clearance and toxicity in order to progress toward clinical applications. For instance, the addition of PEG to the nanoparticle surface can prevent opsonization and increase plasma circulation time, but in turn, it delays systemic clearance of the nanoparticles, thereby potentially increasing the risks from toxicity. There has been conflicting evidence and a lack of details regarding the working mechanism of PEG, based on results from various animal models [88]. Further research efforts to understand the controlling mechanisms stemming from particle characteristics (size, geometry, coating), blood contact, and bioelimination responses are required before nanoparticles can be used in clinical practice.

11.5 Conclusions

Hydrogel nanoparticles, made of diverse natural and synthetic hydrophilic polymers, have all the right properties for becoming tumor-specific carriers of drugs and imaging contrast agents. Hydrogel nanoparticles have been prepared for various therapeutic modalities (chemo-, gene, immnuno-, photodynamic, and radiation therapy), and imaging modalities (MRI, nuclear, fluorescence, and visual delineation). They have high payloads of drug/contrast agents per nanoparticle, controllable release characteristics, and the ability of active/passive targeting. Consequently, hydrogel nanoparticles show superior antitumor therapeutic efficiency and reduced side effects when compared to the free drugs. They induce high therapeutic efficacy even in multidrug resistant cells, while generally causing reduced systemic toxicity. Hydrogel nanoparticles also enable significantly enhanced tumor imaging contrast. Importantly, hydrogel nanoparticles have demonstrated their potential for a relatively straightforward construction of nanodevices for image-guided therapy; the latter may revolutionize medicine. Because of these highly promising research outcomes and future perspectives, hydrogel-based nanoparticles are expected to play an important role in cancer diagnosis and therapy, already in the near future.

Acknowledgments This work was supported by NIH grants 1R01EB007977 and R21/R33CA125297 (RK).

References

1. American Cancer Society (2010) Cancer facts & figures 2010. American Cancer Society, Atlanta
2. Nimsky C, Ganslandt O, Kober H, Buchfelder M, Fahlbusch R (2001) Intraoperative magnetic resonance imaging combined with neuronavigation: a new concept. Neurosurgery 48:1082–1089

3. Izquierdo MA, Scheffer GL, Flens MJ, Shoemaker RH, Rome LH, Scheper RJ (1996) Relationship of LRP-human major vault protein to in vitro and clinical resistance to anticancer drugs. Cytotechnology 19:191–197
4. Gatmaitan ZC, Arias IM (1993) Structure and function of P-glycoprotein in normal liver and small intestine. Adv Pharmacol 24:77–97
5. Allen TM (2002) Ligand-targeted therapeutics in anticancer therapy. Nat Rev Cancer 2:750–763
6. Maeda H (2001) The enhanced permeability and retention (EPR) effect in tumor vasculature: the key role of tumor-selective macromolecular drug targeting. Adv Enzyme Regul 41:189–207
7. Hong S, Leroueil PR, Majoros IJ, Orr BG, Baker JR, Holl MMB (2007) The binding avidity of a nanoparticle-based multivalent targeted drug delivery platform. Chem Biol 14:107–115
8. Harrell JA, Kopelman R (2000) Biocompatible probes measure intracellular activity. Biophotonics Int 7:22–24
9. Xu H, Buck SM, Kopelman R, Philbert MA, Brasuel M, Ross B, Rehemtulla A, Festschrift J (2004) Photo-excitation based nano-explorers: chemical analysis inside live cells and photodynamic therapy. Isr J Chem 44:317–337
10. Ross B, Rehemtulla A, Koo Y-EL, Reddy R, Kim G, Behrend C, Buck S, Schneider RJ, Philbert MA, Weissleder R, Kopelman R (2004) Photonic and magnetic nanoexplorers for biomedical use: from subcellular imaging to cancer diagnostics and therapy. Proc SPIE 5331:76–83
11. Kopelman R, Koo YL, Philbert M, Moffat BA, Reddy GR, McConville P, Hall DE, Chenevert TL, Bhojani MS, Buck SM, Rehemtulla A, Ross BD (2005) Multifunctional nanoparticle platforms for in vivo MRI enhancement and photodynamic therapy of a rat brain cancer. J Magn Magn Mater 293:404–410
12. Reddy GR, Bhojani MS, McConville P, Moody J, Moffat BA, Hall DE, Kim G, Koo Y-E, Woolliscroft MJ, Sugai JV, Johnson TD, Philbert MA, Kopelman R, Rehemtulla A, Ross BD (2006) Vascular targeted nanoparticles for imaging and treatment of brain tumors. Clin Cancer Res 12:6677–6686
13. Hamidi M, Azadi A, Rafiei P (2008) Hydrogel nanoparticles in drug delivery. Adv Drug Deliv Rev 60:1638–1649
14. Kashyap N, Kumar N, Kumar MNVR (2005) Hydrogels for pharmaceutical and biomedical applications. Crit Rev Ther Drug Carrier Syst 22:107–149
15. Koo Y-EL, Reddy GR, Bhojani M, Schneider R, Philbert MA, Rehemtulla A, Ross BD, Kopelman R (2006) Brain cancer diagnosis and therapy with nano-platforms. Adv Drug Deliv Rev 58:1556–1577
16. Qiu Y, Park K (2001) Environment-sensitive hydrogels for drug delivery. Adv Drug Deliv Rev 53:321–339
17. Koo Lee Y-E, Smith R, Kopelman R (2009) Nanoparticle PEBBLE sensors in live cells and in vivo. Annu Rev Anal Chem 2:57–76
18. Lee KY, Mooney DJ (2001) Hydrogels for tissue engineering. Chem Rev 101:1869–1879
19. Wu Y, MacKay JA, McDaniel JR, Chilkoti A, Clark RL (2009) Fabrication of elastin-like polypeptide nanoparticles for drug delivery by electrospraying. Biomacromolecules 10:19–24
20. Maitra A (1984) Determination of size parameters of water-aerosol OT-oil reverse micelles from their nuclear magnetic resonance data. J Phys Chem 88:5122–5125
21. Munshi N, De TK, Maitra A (1997) Size modulation of polymeric nanoparticles under controlled dynamics of microemulsion droplets. J Colloid Interface Sci 190:387–391
22. Bharali DJ, Sahoo SK, Mozumdar S, Maitra A (2003) Cross-linked polyvinylpyrrolidone nanoparticles: a potential carrier for hydrophilic drugs. J Colloid Interface Sci 258:415–423
23. Ohya Y, Shiratani M, Kobayashi H, Ouchi T (1994) Release behavior of 5-fluorouracil from chitosan-gel nanospheres immobilizing 5-fluorouracil coated with polysaccharides and their cell specific cytotoxicity. Pure Appl Chem A31:629–642
24. Pitarresi G, Craparo EF, Palumbo FS, Carlisi B, Giammona G (2007) Composite nanoparticles based on hyaluronic acid chemically cross-linked with α, β-polyaspartylhydrazide. Biomacromolecules 8:1890–1898

25. Lee H, Mok H, Lee S, Oh Y-K, Park TG (2007) Target-specific intracellular delivery of siRNA using degradable hyaluronic acid nanogels. J Control Release 119:245–252
26. Khdair A, Gerard B, Handa H, Mao G, Shekhar MPV, Panyam J (2008) Surfactant-polymer nanoparticles enhance the effectiveness of anticancer photodynamic therapy. Mol Pharm 5:795–807
27. Jain A, Jain SK (2008) In vitro and cell uptake studies for targeting of ligand anchored nanoparticles for colon tumors. Eur J Pharm Sci 35:404–416
28. Zhang H, Mardyani S, Chan WCW, Kumacheva E (2006) Design of biocompatible chitosan microgels for targeted pH-mediated intracellular release of cancer therapeutics. Biomacromolecules 7:1568–1572
29. Zhou X, Liu B, Yu X, Zha X, Zhang X, Chen Y, Wang X, Jin Y, Wu Y, Chen Y, Shan Y, Chen Y, Liu J, Kong W, Shen J (2007) Controlled release of PEI/DNA complexes from mannose-bearing chitosan microspheres as a potent delivery system to enhance immune response to HBV DNA vaccine. J Control Release 121:200–207
30. Boddohi S, Moore N, Johnson PA, Kipper MJ (2009) Polysaccharide-based polyelectrolyte complex nanoparticles from chitosan, heparin, and hyaluronan. Biomacromolecules 10:1402–1409
31. Duceppe N, Tabrizian M (2009) Factors influencing the transfection efficiency of ultra low molecular weight chitosan/hyaluronic acid nanoparticles. Biomaterials 30:2625–2631
32. Rajaonarivony M, Vauthier C, Couarraze G, Puisieux F, Couvreur P (1993) Development of a new drug carrier made from alginate. J Pharm Sci 82:912–917
33. Sarmento B, Ribeiro AJ, Veiga F, Ferreira DC, Neufeld RJ (2007) Insulin-loaded nanoparticles are prepared by alginate ionotropic pregelation followed by chitosan polyelectrolyte complexation. J Nanosci Nanotechnol 7:2833–2841
34. Ahmad Z, Sharma S, Khuller GK (2007) Chemotherapeutic evaluation of alginate nanoparticle-encapsulated azole antifungal and antitubercular drugs against murine tuberculosis. Nanomedicine 3:239–243
35. Besheer A, Hause G, Kressler J, Maeder K (2007) Hydrophobically modified hydroxyethyl starch: synthesis, characterization, and aqueous self-assembly into nano-sized polymeric micelles and vesicles. Biomacromolecules 8:359–367
36. Hornig S, Heinze T (2008) Efficient approach to design stable water-dispersible nanoparticles of hydrophobic cellulose esters. Biomacromolecules 9:1487–1492
37. Akiyoshi K, Deguchi S, Moriguchi N, Yamaguchi S, Sunamoto J (1993) Self-aggregates of hydrophobized polysaccharides in water. Formation and characteristics of nanoparticles. Macromolecules 26:3062–3068
38. Kuroda K, Fujimoto K, Sunamoto J, Akiyoshi K (2002) Hierarchical self-assembly of hydrophobically modified pullulan in water: gelation by networks of nanoparticles. Langmuir 18:3780–3786
39. Choi KY, Lee S, Park K, Kim K, Park JH, Kwon IC, Jeong SY (2008) Preparation and characterization of hyaluronic acid-based hydrogel nanoparticles. J Phys Chem Solids 69:1591–1595
40. Lee H, Ahn C-H, Park TG (2009) Poly[lactic-co-(glycolic acid)]-grafted hyaluronic acid copolymer micelle nanoparticles for target-specific delivery of doxorubicin. Macromol Biosci 9:336–342
41. Yadav AK, Mishra P, Mishra AK, Mishra P, Jain S, Agrawal GP (2007) Development and characterization of hyaluronic acid-anchored PLGA nanoparticulate carriers of doxorubicin. Nanomedicine 3:246–257
42. Yadav AK, Mishra P, Jain S, Mishra P, Mishra AK, Agrawal GP (2008) Preparation and characterization of HA–PEG–PCL intelligent core–corona nanoparticles for delivery of doxorubicin. J Drug Target 16:464–478
43. Westedt U, Kalinowski M, Wittmar M, Merdan T, Unger F, Fuchs J, Schäller S, Bakowsky U, Kissel T (2007) Poly(vinyl alcohol)-graft-poly(lactide-co-glycolide) nanoparticles for local delivery of paclitaxel for restenosis treatment. J Control Release 119:41–51
44. Vinogradov SV, Tatiana KTK, Kabanov AV (2002) Nanosized cationic hydrogels for drug delivery: preparation, properties and interactions with cells. Adv Drug Deliv Rev 54:135–147

45. Nagahama K, Mori Y, Ohya Y, Ouchi T (2007) Biodegradable nanogel formation of polylactide-grafted dextran copolymer in dilute aqueous solution and enhancement of its stability by stereocomplexation. Biomacromolecules 8:2135–2141
46. Oh JK, Lee DI, Park JM (2009) Biopolymer-based microgels/nanogels for drug delivery applications. Prog Polym Sci 34:1261–1282
47. Huang G, Gao J, Hu ZB, John JVS, Ponder BC, Moro D (2004) Controlled drug release from hydrogel nanoparticle networks. J Control Release 94:303–311
48. Vihola H, Laukkanen A, Hirvonen J, Tenhu H (2002) Binding and release of drugs into and from thermosensitive poly(N-vinyl caprolactam) nanoparticles. Eur J Pharm Sci 16:69–74
49. Bodnar M, Hartmann JF, Borbely J (2006) Synthesis and study of cross-linked chitosan-N-poly(ethylene glycol) nanoparticles. Biomacromolecules 11:3030–3036
50. Shen X, Zhang L, Jiang X, Hu Y, Guo J (2007) Reversible surface switching of nanogel triggered by external stimuli. Angew Chem Int Ed 46:7104–7107
51. Coester CJ, Langer K, Von Briesen H, Kreuter J (2000) Gelatin nanoparticles by two step desolvation a new preparation method, surface modifications and cell uptake. J Microencapsul 17:187–193
52. Maham A, Tang Z, Wu H, Wang J, Lin Y (2009) Protein-based nanomedicine platforms for drug delivery. Small 5:1706–1721
53. Davis ME, Chen Z, Shin DM (2008) Nanoparticle therapeutics: an emerging treatment modality for cancer. Nat Rev Drug Discov 7:771–782
54. Hah HJ, Kim G, Koo Lee Y-E, Orringer DA, Sagher O, Philbert MA, Kopelman R (2011) Methylene blue-conjugated hydrogel nanoparticles and tumor-cell targeted photodynamic therapy. Macromol Biosci 11:90–99
55. Montet X, Funovics M, Montet-Abou K, Weissleder R, Josephson L (2006) Multivalent effects of RGD peptides obtained by nanoparticle display. J Med Chem 49:6087–6093
56. Kataoka K, Miyazaki H, Bunya M, Okano T, Sakurai Y (1998) Totally synthetic polymer gels responding to external glucose concentration: their preparation and application to on-off regulation of insulin release. J Am Chem Soc 120:12694–12695
57. Miyata T, Asami N, Uragami T (1999) A reversibly antigen-responsive hydrogel. Nature 399:766–769
58. Bromberg LE, Ron ES (1998) Temperature-responsive gels and thermogelling polymer matrices for protein and peptide delivery. Adv Drug Deliv Rev 31:197–221
59. Tanaka T, Nishio I, Sun S-T, Ueno-Nishio S (1982) Collapse of gels in an electric field. Science 218:467–469
60. Suzuki A, Tanaka T (1990) Phase transition in polymer gels induced by visible light. Nature 346:345–347
61. Mamada A, Tanaka T, Kungwatchakun D, Irie M (1990) Photoinduced phase-transition of gel. Macromolecules 23:1517–1519
62. Duncan R (2006) Polymer conjugates as anticancer nanomedicines. Nat Rev Cancer 6:688–701
63. Wang J, Gan D, El-Sayed MA (2001) Temperature-jump investigations of the kinetics of hydrogel nanoparticle volume phase transitions. J Am Chem Soc 123:11284–11289
64. http://www.abraxane.com/
65. Service RF (2010) Nanotechnology: nanoparticle Trojan horses gallop from the lab into the clinic. Science 330:314–315
66. Winer I, Wang S, Koo Lee Y-E, Fan W, Gong Y, Burgos-Ojeda D, Spahlinger G, Kopelman R, Buckanovich RJ (2010) F3-targeted cisplatin-hydrogel nanoparticles as an effective therapeutic that targets both murine and human ovarian tumor endothelial cells in vivo. Cancer Res 70:8674–8683
67. Guowei D, Adriane K, Chen X, Jie C, Yinfeng L (2007) PVP magnetic nanospheres: biocompatibility, in vitro and in vivo bleomycin release. Int J Pharm 328:78–85
68. Huang S-J, Sun S-L, Feng T-H, Sung K-H, Lui W-L, Wang L-F (2009) Folate-mediated chondroitin sulfate-Pluronic® 127 nanogels as a drug carrier. Eur J Pharm Sci 38:64–73

69. Li X, Li R, Qian X, Ding Y, Tu Y, Guo R, Hu Y, Jiang X, Guo W, Liu B (2008) Superior antitumor efficiency of cisplatin-loaded nanoparticles by intratumoral delivery with decreased tumor metabolism rate. Eur J Pharm Biopharm 70:726–734
70. Hidaka M, Kanematsu T, Ushio K, Sunamoto J (2006) Selective and effective cytotoxicity of folic acid-conjugated cholesteryl pullulan hydrogel nanoparticles complexed with doxorubicin in in vitro and in vivo studies. J Bioact Compat Polym 21:591–602
71. Susa M, Iyer AK, Ryu K, Hornicek FJ, Mankin H, Amiji MM, Duan ZF (2009) Doxorubicin loaded polymeric nanoparticulate delivery system to overcome drug resistance in osteosarcoma. BMC Cancer 9:399
72. Christian S, Pilch J, Akerman ME, Porkka K, Laakkonen P, Ruoslahti E (2003) Nucleolin expressed at the cell surface is a marker of endothelial cells in angiogenic blood vessels. J Cell Biol 163:871–878
73. Huang Y, Shi H, Zhou H, Song X, Yuan S, Luo Y (2006) The angiogenic function of nucleolin is mediated by vascular endothelial growth factor and nonmuscle myosin. Blood 107:3564–3571
74. Liechty WB, Kryscio DR, Slaughter BV, Peppas NA (2010) Polymers for drug delivery systems. Annu Rev Chem Biomol Eng 1:149–173
75. Vaupel P, Kallinowski F, Okunieff P (1989) Blood flow, oxygen and nutrient supply, and metabolic microenvironment of human tumors: a review. Cancer Res 49:6449–6465
76. Lee ES, Gao Z, Bae YH (2008) Recent progress in tumor pH targeting nanotechnology. J Control Release 132:164–170
77. Park JH, Kwon S, Lee M, Chung H, Kim JH, Kim YS, Park RW, Kim IS, Seo SB, Kwon IC, Jeong SY (2006) Self-assembled nanoparticles based on glycol chitosan bearing hydrophobic moieties as carriers for doxorubicin: in vivo biodistribution and anti-tumor activity. Biomaterials 27:119–126
78. Kim J-H, Kim Y-S, Park K, Lee S, Nam HY, Min KH, Jo HG, Park JH, Choi K, Jeong SY, Park R-W, Kim I-S, Kim K, Kwon IC (2008) Antitumor efficacy of cisplatin-loaded glycol chitosan nanoparticles in tumor-bearing mice. J Control Release 127:41–49
79. Zhao ZM, He M, Yin LC, Bao J, Shi L, Wang B, Tang C, Yin C (2009) Biodegradable nanoparticles based on linoleic acid and poly(beta-malic acid) double grafted chitosan derivatives as carriers of anticancer drugs. Biomacromolecules 10:565–572
80. Ray D, Mohapatra DK, Mohapatra RK, Mohanta GP, Sahoo PK (2008) Synthesis and colon-specific drug delivery of a poly(acrylic acid-co-acrylamide)/MBA nanosized hydrogel. J Biomater Sci Polym Ed 19:1487–1502
81. Na K, Lee KH, Bae YH (2004) pH-sensitivity and pH-dependent interior structural change of self-assembled hydrogel nanoparticles of pullulan acetate/oligo-sulfonamide conjugate. J Control Release 97:513–525
82. Na K, Lee ES, Bae YH (2007) Self-organized nanogels responding to tumor extracellular pH: pH-dependent drug release and in vitro cytotoxicity against MCF-7 cells. Bioconjug Chem 18:1568–1574
83. Zhang H-Z, Li X-M, Gao F-P, Liu L-R, Zhou Z-M, Zhang Q-Q (2010) Preparation of folate-modified pullulan acetate nanoparticles for tumor-targeted drug delivery. Drug Deliv 17:48–57
84. Xu PS, Van Kirk EA, Murdoch WJ, Zhan YH, Isaak DD, Radosz M, Shen YQ (2006) Anticancer efficacies of cisplatin-releasing pH-responsive nanoparticles. Biomacromolecules 7:829–835
85. Murthy N, Xu M, Schuck S, Kunisawa J, Shastri N, Frechet JM (2003) A macromolecular delivery vehicle for protein-based vaccines: acid-degradable protein-loaded microgels. Proc Natl Acad Sci USA 100:4995–5000
86. Shi L, Khondee S, Linz TH, Berkland C (2008) Poly(N-vinylformamide) nanogels capable of pH-sensitive protein release. Macromolecules 41:6546–6554
87. Fisher OZ, Peppas NA (2009) Polybasic nanomatrices prepared by UV-initiated photopolymerization. Macromolecules 42:3391–3398

88. Owens DE, Peppas NA (2006) Opsonization, biodistribution, and pharmacokinetics of polymeric nanoparticles. Int J Pharm 307:93–102
89. Fan L, Li F, Zhang H, Yukun Wang Y, Cheng C, Li X, Gu C-H, Qian Yang Q, Wu H, Zhang S (2010) Co-delivery of PDTC and doxorubicin by multifunctional micellar nanoparticles to achieve active targeted drug delivery and overcome multidrug resistance. Biomaterials 31:5634–5642
90. Schafer FQ, Buettner GR (2001) Redox environment of the cell as viewed through the redox state of the glutathione disulfide/glutathione couple. Free Radic Biol Med 30:1191–1212
91. Oh JK, Siegwart DJ, Lee H, Sherwood G, Peteanu L, Hollinger JO, Kataoka K, Matyjaszewski K (2007) Biodegradable nanogels prepared by atom transfer radical polymerization as potential drug delivery carriers: synthesis, biodegradation, in vitro release, and bioconjugation. J Am Chem Soc 129:5939–5945
92. Shah S, Pal A, Rajiv Gude R, Devi S (2010) Synthesis and characterization of thermo-responsive copolymeric nanoparticles of poly(methyl methacrylate-co-N-vinylcaprolactam). Eur Polym J 46:958–967
93. van den Brom CR, Anac I, Roskamp RF, Retsch M, Jonas U, Menges B, Preece JA (2010) The swelling behaviour of thermoresponsive hydrogel/silica nanoparticle composites. J Mater Chem 20:4827–4839
94. Morimoto N, Qiu X-P, Winnik FM, Akiyoshi K (2008) Dual stimuli-responsive nanogels by self-assembly of polysaccharides lightly grafted with thiol-terminated poly(N-isopropylacrylamide) chains. Macromolecules 41:5985–5987
95. Ma LW, Liu MZ, Liu HL, Chen J, Cui D (2010) In vitro cytotoxicity and drug release properties of pH- and temperature-sensitive core-shell hydrogel microspheres. Int J Pharm 385:86–91
96. Fan L, Wu H, Zhang H, Li F, Yang T-H, Gu C-H, Yang Q (2008) Novel super pH-sensitive nanoparticles responsive to tumor extracellular pH. Carbohydr Polym 73:390–400
97. Nezhadi SH, Choong PFM, Lotfipour F, Dass CR (2009) Gelatin-based delivery systems for cancer gene therapy. J Drug Target 17:731–738
98. Patil SD, Rhodes DG, Burgess DJ (2005) DNA-based therapeutics and DNA delivery systems: a comprehensive review. AAPS J 7:E61–E77
99. Kommareddy S, Amiji M (2007) Antiangiogenic gene therapy with systemically administered sFlt-1 plasmid DNA in engineered gelatin-based nanovectors. Cancer Gene Ther 14:488–498
100. Kaul G, Amiji M (2005) Tumor-targeted gene delivery using poly(ethylene glycol)-modified gelatin nanoparticles: in vitro and in vivo studies. Pharm Res 22:951–961
101. Susa M, Iyer AK, Ryu K, Choy E, Hornicek FJ, Mankin H, Milane L, Amiji MM, Duan Z (2010) Inhibition of ABCB1 (MDR1) expression by an siRNA nanoparticulate delivery system to overcome drug resistance in osteosarcoma. PLoS One 5:e10764
102. Naeye B, Raemdonck K, Remaut K, Sproat B, Demeester J, De Smedt SC (2010) PEGylation of biodegradable dextran nanogels for siRNA delivery. Eur J Pharm Sci 40:342–351
103. Dickerson EB, Blackburn WH, Smith MH, Kapa LB, Lyon LA, McDonald JF (2010) Chemosensitization of cancer cells by siRNA using targeted nanogel delivery. BMC Cancer 10:10
104. Day AJ, Prestwich GD (2002) Hyaluronan-binding proteins: tying up the giant. J Biol Chem 277:4585–4588
105. Ossipov DA (2010) Nanostructured hyaluronic acid-based materials for active delivery to cancer. Expert Opin Drug Deliv 7:681–703
106. Cohen JA, Beaudette TT, Tseng WW, Bachelder EM, Mende I, Engleman EG, Frechet JMJ (2009) T-cell activation by antigen-loaded pH-sensitive hydrogel particles in vivo: the effect of particle size. Bioconjug Chem 20:111–119
107. Fisher O, Kim T, Dietz S, Peppas NA (2009) Enhanced core hydrophobicity, functionalization and cell penetration of polybasic nanomatrices. Pharm Res 26:51–60
108. Hu Y, Litwin T, Nagaraja AR, Kwong B, Katz J, Watson N, Irvine DJ (2007) Cytosolic delivery of membrane-impermeable molecules in dendritic cells using pH-responsive core-shell nanoparticles. Nano Lett 7:3056–3064

109. Hasegawa K, Noguchi Y, Koizumi F, Uenaka A, Tanaka M, Shimono M, Nakamura H, Shiku H, Gnjatic S, Murphy R, Hiramatsu Y, Old LJ, Nakayama E (2006) In vitro stimulation of CD8 and CD4 T cells by dendritic cells loaded with a complex of cholesterol-bearing hydrophobized pullulan and NYESO-1 protein: identification of a new HLA-DR15-binding CD4 T-cell epitope. Clin Cancer Res 12:1921–1927
110. Kawabata R, Wada H, Isobe M, Saika T, Sato S, Uenaka A, Miyata H, Yasuda T, Doki Y, Noguchi Y, Kumon H, Tsuji K, Iwatsuki K, Shiku H, Ritter G, Murphy R, Hoffman E, Old LJ, Monden M, Nakayama E (2007) Antibody response against NY-ESO-1 in CHPNY-ESO-1 vaccinated patients. Int J Cancer 120:2178–2184
111. Shimizu T, Kishida T, Hasegawa U, Ueda Y, Imanishi J, Yamagishi H, Akiyoshi K, Otsuji E, Mazda O (2008) Nanogel DDS enables sustained release of IL-12 for tumor immunotherapy. Biochem Biophys Res Commun 367:330–335
112. Severino D, Junqueira HC, Gugliotti M, Gabrielli DS, Baptista MS (2003) Influence of negatively charged interfaces on the ground and excited state properties of methylene blue. Photochem Photobiol 77:459–468
113. Tanielian C, Heinrich G (1995) Effect of aggregation on the hematoporphyrin-sensitized production of singlet molecular oxygen. Photochem Photobiol 61:131–133
114. Buytaert E, Dewaele M, Agostinis P (2007) Molecular effectors of multiple cell death pathways initiated by photodynamic therapy. Biochim Biophys Acta 1776:86–107
115. Tang W, Xu H, Park EJ, Philbert MA, Kopelman R (2008) Encapsulation of methylene blue in polyacrylamide nanoparticle platforms protects its photodynamic effectiveness. Biochem Biophys Res Commun 369:579–583
116. Tang W, Xu H, Kopelman R, Philbert MA (2005) Photodynamic characterization and in vitro application of methylene blue-containing nanoparticle platforms. Photochem Photobiol 81:242–249
117. Gao D, Agayan RR, Xu H, Philbert MA, Kopelman R (2006) Nanoparticles for two-photon photodynamic therapy in living cells. Nano Lett 6:2383–2386
118. Gao D, Xu H, Philbert MA, Kopelman R (2007) Ultrafine hydrogel particles: synthetic approach and therapeutic application in living cells. Angew Chem 46:2224–2227
119. Chen K, Preuß A, Hackbarth S, Wacker M, Langer K, Röder B (2009) Novel photosensitizer-protein nanoparticles for photodynamic therapy: photophysical characterization and in vitro investigations. J Photochem Photobiol B 96:66–74
120. Rodrigues MMA, Simioni AR, Primo FL, Siqueira-Moura MP, Morais PC, Tedesco AC (2009) Preparation, characterization and in vitro cytotoxicity of BSA-based nanospheres containing nanosized magnetic particles and/or photosensitizer. J Magn Magn Mater 321:1600–1603
121. Deda DK, Uchoa AF, Carita E, Baptista MS, Toma HE, Araki K (2009) A new micro/nano-encapsulated porphyrin formulation for PDT treatment. Int J Pharm 376:76–83
122. Khdair A, Handa H, Mao G, Panyam J (2009) Nanoparticle-mediated combination chemotherapy and photodynamic therapy overcomes tumor drug resistance in vitro. Eur J Pharm Biopharm 71:214–222
123. Gabrielli D, Belisle E, Severino D, Kowaltowski AJ, Baptista MS (2004) Binding, aggregation and photochemical properties of methylene blue in mitochondrial suspensions. Photochem Photobiol 79:227–232
124. Koo Lee Y-E, Kopelman R (2010) Multifunctional nanoparticles for targeted imaging and therapy of cancer. In: Bao Y, Dattelbaum AM, Tracy JB, Yin Y (eds) Multifunctional nanoparticle systems—coupled behavior and applications. Mat Res Soc Sym Proc 1257, 1257-O07-02
125. Song H-C, Na K, Park KH, Shin C-H, Bom H-S, Kang D, Kim S, Lee ES, Lee DH (2006) Intratumoral administration of rhenium-188-labeled pullulan acetate nanoparticles (PAN) in mice bearing CT-26 cancer cells for suppression of tumor growth. J Microbiol Biotechnol 16:1491–1498
126. Hallahan D, Geng L, Qu S, Scarfone C, Giorgio T, Donnelly E, Gao X, Clanton J (2003) Integrin-mediated targeting of drug delivery to irradiated tumor blood vessels. Cancer Cell 3:63–74

127. Moffat BA, Reddy GR, McConville P, Hall DE, Chenevert TL, Kopelman RR, Philbert M, Weissleder R, Rehemtulla A, Ross BD (2003) A novel polyacrylamide magnetic nanoparticle contrast agent for molecular imaging using MRI. Mol Imaging 2:324–332
128. Ma H, Shiraishi K, Minowa T, Kawano K, Yokoyama M, Hattori Y, Maitani Y (2010) Accelerated blood clearance was not induced for a gadolinium-containing PEG-poly(L-lysine)-based polymeric micelle in mice. Pharm Res 27:296–302
129. Banerjee T, Singh AK, Sharma RK, Maitra AN (2005) Labeling efficiency and biodistribution of Technetium-99m labeled nanoparticles: interference by colloidal tin oxide particles. Int J Pharm 289:189–195
130. Sun G, Hagooly A, Xu J, Nystrom AM, Li ZC, Rossin R, Moore DA, Wooley KL, Welch MJ (2008) Facile, efficient approach to accomplish tunable chemistries and variable biodistributions for shell cross-linked nanoparticles. Biomacromolecules 9:1997–2006
131. Sun G, Xu J, Hagooly A, Rossin R, Li Z, Moore DA, Hawker CJ, Welch MJ, Wooley KL (2007) Strategies for optimized radiolabeling of nanoparticles for in vivo PET Imaging. Adv Mater 19:3157–3162
132. Orringer DA, Koo Y-EL, Chen T, Kim G, Hah H, Xu H, Wang S, Keep R, Philbert MA, Sagher O, Kopelman R (2009) In vitro characterization of a targeted, dye-loaded nanodevice for intraoperative tumor delineation. Neurosurgery 64:965–972
133. Orringer DA, Sagher O, Kopelman R, Koo YE (2010) Dye-loaded nanoparticles. US patent US2010/0098637
134. Wu W, Aiello M, Zhou T, Berliner A, Banerjee P, Zhou S (2010) In-situ immobilization of quantum dots in polysaccharide-based nanogels for integration of optical pH-sensing, tumor cell imaging, and drug delivery. Biomaterials 31:3023–3031
135. Kim JH, Kim YS, Kim S, Park JH, Kim K, Choi K, Chung H, Jeong SY, Park RW, Kim IS, Kwon IC (2006) Hydrophobically modified glycol chitosan nanoparticles as carriers for paclitaxel. J Control Release 111:228–234
136. Yang SG, Chang JE, Shin B, Park S, Na K, Shim CK (2010) ^{99m}Tc-hematoporphyrin linked albumin nanoparticles for lung cancer targeted photodynamic therapy and imaging. J Mater Chem 20:9042–9046

Chapter 12
Theranostic Polymeric Micelles for Cancer Imaging and Therapy

Gang Huang, Chalermchai Khemtong, Erik A. Bey, David A. Boothman, Baran D. Sumer, and Jinming Gao

12.1 Introduction

Cancer has surpassed cardiovascular diseases as the leading cause of deaths among men and women under 85 years of age. In the United States alone, a total of 1.5 million new cancer cases and over half a million deaths are expected in 2010 [1]. Current cancer treatments include surgical resection, radiation therapy, and chemotherapy. Recent progress in cancer diagnosis and treatments has resulted in decreased mortality rates in the past few years. However, despite many exciting advances, most current therapeutic modalities still lack specificity and are further limited by undesirable toxic side effects as well as high rates of tumor recurrence.

Nanotechnology has the potential to offer paradigm-shifting solutions to improve the outcome of cancer diagnosis and therapy [2, 3]. Nanoscale therapeutic and diagnostic agents (10–100 nm in diameter) provide unique pharmacological properties such as prolonged blood circulation times and reduced systemic toxicity compared to traditional formulations [4]. They offer high payloads of anticancer drugs and diagnostic agents, efficient cell uptake, and passive/active targeting at tumor sites.

G. Huang • C. Khemtong • J. Gao (✉)
Departments of Pharmacology,
Harold C. Simmons Comprehensive Cancer Center, University of Texas
Southwestern Medical Center at Dallas, Dallas, TX 75390, USA
e-mail: jinming.gao@utsouthwestern.edu

D.A. Boothman • E.A. Bey
Department of Pharmocology and Radiation Oncology,
Harold C. Simmons Comprehensive Cancer Center, University of Texas
Southwestern Medical Center at Dallas, Dallas, TX 75390, USA

B.D. Sumer
Department of Pharmocology and Radiation Oncology,
Harold C. Simmons Comprehensive Cancer Center, University of Texas
Southwestern Medical Center at Dallas, Dallas, TX 75390, USA

S. Svenson and R.K. Prud'homme (eds.), *Multifunctional Nanoparticles for Drug Delivery Applications: Imaging, Targeting, and Delivery*, Nanostructure Science and Technology,
DOI 10.1007/978-1-4614-2305-8_12,

Moreover, recent advances in molecular imaging have made possible a number of imaging tools to study and characterize biological events in cancer cells and tumors at molecular levels. In addition to their therapeutic applications, nanoscale formulations have also been heavily explored as potential molecular imaging agents. Combination of both therapeutic and diagnostic capabilities into one nanoscale particulate has inspired a novel concept of "theranostics," in which the research is focused on development of new multifunctional nanoparticles for diagnosis and treatment of diseases [5–17]. Compared to traditional formulations of drugs and imaging agents, theranostic nanomedicines combine diagnostic and therapeutic moieties into one entity. The unique physical and chemical properties of the nanoscale formulations also allow for an attachment of an antifouling layer and disease-specific ligands for homing purposes. Different chemistries can be carried out to achieve additional functionalities, such as responsiveness to heat and acidic conditions. More in-depth reviews on these discoveries have been discussed elsewhere [18–22]. To date, a number of different nanoscale systems have been proposed and investigated as theranostic agents for the detection and treatment of cancer, atherosclerosis, and many other diseases [23–35]. Commonly studied nanosystems include, but are not limited to, polymeric micelles, liposomes, dendrimers, inorganic particles, and polymer–drug conjugates. Among them, polymeric micelles have been extensively investigated as a complementary and unique nanotherapeutic system for cancer diagnosis and treatment due to their excellent biocompatibility, prolonged circulation times, and facile chemical functionalization. In this chapter, we review recent advances in the development of theranostic polymeric micelles for cancer imaging and therapeutic applications. First, an overview of theranostic nanomedicine for cancer diagnosis and treatment will be given, followed by a discussion of polymeric micellar formulations and their properties. Finally, several representative examples of reported theranostic micelles for cancer applications will be presented.

12.2 Theranostic Nanomedicine

12.2.1 What Is Theranostic Nanomedicine?

Cancer is a highly heterogeneous disease, including multiple cell phenotypes in a single tumor. This heterogeneity is further complicated by the complex tumor microenvironment consisting of tumor microvasculature, extracellular matrix, and infiltrating inflammatory cells whose compositions change over time and space. This biological complexity leads to ineffective treatments as cancer cells can evade monotherapies, develop adaptive resistance in response to treatments, and grow and reconstitute more tumor tissues in the body. The traditional "one-size-fits-all" anticancer drugs often fail to generate highly effective treatment outcomes, eventually leading to resistance. A shift in treatment regimens is usually required to increase therapeutic efficacy and circumvent such "induced" resistance. On the other hand, severe side effects resulting from the toxicity of anticancer drugs are also a major drawback of traditional therapeutic formulations.

"Theranostics" was originally used as a term to describe the process of diagnostic therapy for individual patients, testing them for possible reactions to new medication and tailoring the treatment in response to the test results [36]. Rapid progress in nanotechnology has allowed for the development of highly integrated systems with incorporated imaging and therapeutic functionalities. The integrated diagnostic and imaging functions can potentially allow for image-guided therapy of tumors, where the diagnostic function can theoretically allow for non-invasive imaging of drug targeting, controlled release of anticancer drugs in tumor tissues, and posttherapy assessment of drug efficacy. This functional integration can potentially address the limitations of tumor heterogeneity and adaptive resistance by providing well-informed design of therapeutic regimen to achieve "personalized medicine" of cancer.

12.2.2 Cancer Targets

It is widely accepted that nanomedicine can take advantage of the leaky tumor vessels described as the enhanced permeability and retention (EPR) effect to extravasate into tumor tissues [37]. The accumulated nanoparticles can release their drug payload within the vicinity of tumor cells, which, in turn, reduces toxicity to healthy tissues. However, such passive targeting strategies often lack specificity, and their efficiencies are highly dependent on the tumor leakiness, which can be highly variable among tumors or even within the same tumor type [38]. Moreover, cancer cells can still develop resistance through systems like multiple-drug resistance (MDR) [39]. In an attempt to overcome these limitations, a series of novel cancer-specific targets have been investigated using a number of ligands reported to bind to tumor-specific targets with high specificity and affinity [40–43]. The availability of targeting ligands prompted an intense research effort to develop nanoparticles conjugated with the ligands for tumor-targeted delivery of nanomedicines [44]. The incorporation of targeting ligands allows for specific binding of nanoparticles to cancer cells or tumor endothelia within tumor tissues, which can be internalized into cells before the delivered drug payload is released.

Targeting ligands with high affinity and specificity are generally identified through mass screening of compounds or phage libraries. The ligands are commonly antibodies, small organic molecules, peptides, or aptamers. For targeted nanomedicine, known targeting ligands with well-characterized binding specificities and affinities are often chosen for developing cancer-targeted theranostic nanomedicines. For example, Torchilin et al. conjugated the monoclonal antibody (mAb) 2C5 to the corona of *p*-nitrophenylcarbonyl poly(ethylene oxide)-phosphatidylethanolamine (*p*NP-PEO-PE) micelles [45]. Compared to micelles without 2C5 or free drugs, the targeted, paclitaxel (PTX)-loaded micelles exhibited a superior binding ability to human breast cancer MCF-7 cells in vitro [46]. ^{111}In-labeled 2C5-encoded micelles also showed significantly higher accumulation in Lewis lung carcinoma (LLC) tumors compared to unlabelled micelles in vivo. Park and coworkers functionalized doxorubicin (DOX)-containing poly(ethylene glycol)-poly(lactic-*co*-glycolic acid) (PEG-PLGA) micelles with folic acid and were able to show significantly increased uptake and cytotoxicity in KB cancer cells that express folate receptors. DOX was found to be more efficacious against KB cells in both in vitro and in vivo studies when delivered by folate-conjugated DOX micelles compared to unmodified micelles [43]. Our research group conjugated cyclic (Arg-Gly-Asp-D-Phe-Lys) (cRGDfK) peptide to the surface of PEG-poly(L-lactic acid) (PEG-PLA) micelles by thiol–maleimide linkage and investigated the targeting efficiency of angiogenic tumors overexpressing $\alpha_v\beta_3$ integrins. Confocal laser scanning microscopy showed 30-fold higher accumulation of cRGDfK-encoded micelles compared to unmodified micelles [47].

Results from several research groups discussed above suggested that cancer-specific nanomedicines help improve the efficiency of delivering anticancer drugs, with increased compound accumulation at tumor sites compared to control nanoparticles lacking targeting ligands. This observation confirms the tremendous potential of cancer-targeted nanomedicines for cancer treatments in the future.

12.2.3 Cancer Molecular Imaging

Molecular and cellular imaging has become an important tool in cancer drug discovery and development [48–50]. With a number of available molecular imaging modalities, biological events in cancer cells can be visualized and characterized [48, 51–53]. It is highly desirable that such imaging functionality be incorporated in nanoparticles containing a highly potent anticancer drug. The imaging functionalities of nanoparticles allow for the detection of tumor tissue, as well as monitoring the treatment efficiency of the nanoparticles. A variety of non-invasive imaging modalities, including magnetic resonance imaging (MRI), positron emission tomography (PET), computed tomography (CT), ultrasound, and optical imaging, have been widely utilized in the development of theranostic nanomedicine. These molecular imaging systems offer the opportunity to visualize and locate tumor tissue in the body, while simultaneously characterizing the molecular information in the tumor such as expression and activity of particular molecules, cells that influence tumors, and responses to therapeutic drugs. Chen and coworkers reported peptide-labeled quantum dots (QD) for in vivo targeting and imaging of tumor vasculature [54]. A near-infrared (NIR) QD705 was conjugated with c(RGDyK) that can target $\alpha_v\beta_3$ integrin expression in the tumor microvasculature. U87MG glioblastoma tumors in mice were detected by fluorescence imaging, with maximum intensity achieved 6 h after an i.v. administration of QD705-RGD into the mice. The fluorescence intensity of tumors treated with targeted quantum dots was excellent compared to tumors treated with quantum dots lacking targeting ligands. This approach provided a powerful NIR fluorescence (NIRF) probe for non-invasive detection of tumor vasculature in vivo by optical imaging. In another study, Li et al. reported a chelator-free [^{64}Cu]-CuS nanoparticle as an efficient radiotracer for both PET imaging and photothermal ablation of tumors [55]. PEG- or citrate-coated [^{64}Cu]-CuS nanoparticles (~11 nm in diameter) showed excellent tumor accumulation through the EPR effect as confirmed by PET imaging. The destruction of tumors was achieved by NIR laser irradiation ($\lambda = 808$ nm) of U87 tumor tissue in mice, injected intratumorally or intravenously with CuS nanoparticles. These theranostic nanoprobes provide potential applications for both PET imaging and photothermal therapy.

12.2.4 Current Nanomedicine Platforms

There are several different types of nanoparticles being studied as cancer theranostic nanomedicines. Polymer conjugates and dendrimers require functionalizable chemical groups on the drug or imaging agents so that drugs or imaging agents can be conjugated to the carriers [56]. Functional groups that can undergo facile chemical transformations, such as acid-labile or enzyme-degradable groups, are generally desirable for the conjugation of anticancer drugs because these groups can preferentially be cleaved at tumor sites for drug release [57]. The relatively small size of polymer conjugates and dendrimers, however, leads to short circulation times compared to other types of nanoparticles. With average sizes of <10 nm, polymer conjugates and dendrimers can cross the glomerular basement membrane and are cleared through the kidneys [39]. Liposomes, on the other hand, have considerably prolonged blood circulation times, and therefore, effective passive targeting to tumors through the EPR effect. Liposomes are usually used as carriers for hydrophilic therapeutics such as DNA, protein, and water-soluble drugs, as the core of liposomes is hydrophilic in nature. Although hydrophobic agents can be encapsulated within the lipid bilayer membranes of liposomes, the loading capacity is very limited, and such encapsulation leads to poor stability of liposomes [58]. Inorganic nanoparticles have also emerged as novel theranostic nanoplatforms [49]. This class of nanoparticles has not, however, been approved to date for clinical use. The innate toxicity, non-degradability, and low drug-loading capacity of inorganic nanoparticles have also been discussed as major concerns for future clinical translations of these nanoparticles, which somewhat limits the development of inorganic nanoparticles as theranostic nanomedicine [9].

Compared to other nanomedicine platforms, polymeric micelles have several advantages in cancer diagnostic and therapeutic applications (Fig. 12.1). Their small size (10–100 nm) can be easily controlled by varying the lengths of hydrophobic and hydrophilic blocks of copolymers. In contrast, this size range is difficult to achieve for liposomes due to intrinsic limitations on the curvature of the lipid bilayer [59]. Polymeric micelles have significantly prolonged blood half-lives ($t_{1/2}$) [60, 61] because their hydrophilic PEG shells can effectively prevent protein binding and uptake by the reticuloendothelial system (RES). Micelles (typically MW = 10^3–10^6 kDa) are above the threshold limit (~40 kD) [62] of glomerular filtration in the kidneys, preventing quick renal clearance, which is a major problem for small molecular drugs. The prolonged blood circulation of PEG-stabilized micelles permits passive targeting of micelles to solid tumors via the EPR effect. Both imaging agents and anticancer drugs can be easily encapsulated physically or chemically in

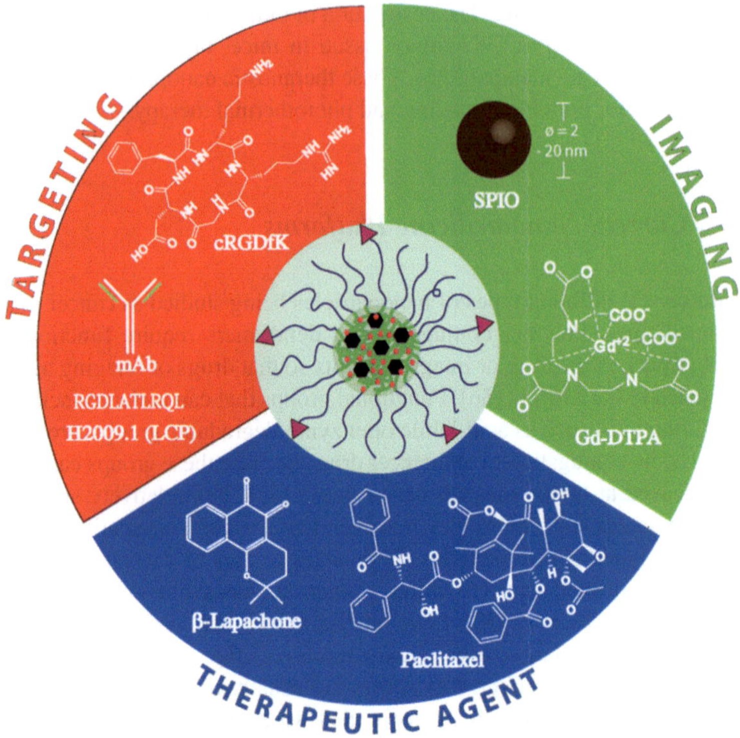

Fig. 12.1 Schematic illustration of theranostic micelle nanomedicine that incorporates imaging, targeting, and therapeutic functions in one system

micelles with high loading capacities. Furthermore, active targeting of micelles to specific tissues can be achieved by functionalization of the hydrophilic PEG shell with cancer-specific ligands.

12.3 Physical Properties and Polymeric Micelle Formulations

12.3.1 Micelle Formulations

Polymeric micelles were first introduced as drug delivery vehicles in the early 1980s by Helmut Ringsdorf [63, 64]. They are composed of amphiphilic block copolymers containing distinguished hydrophobic and hydrophilic segments. The distinct chemical nature of the two blocks results in thermodynamic phase separation in aqueous solution, forming nanoscopic supramolecular core-shell structures. During the micellization process in an aqueous medium, hydrophobic blocks associate to form a micelle core, whereas the hydrophilic blocks form the shell that stabilizes the water-insoluble core in aqueous media. This unique architecture enables the micelle core to serve as a nanoscopic depot for therapeutic or imaging agents and the shell as biospecific surfaces for targeting applications.

A variety of copolymers have been used for micelle formulations. For clinical applications, biocompatibility and biodegradability of the polymers used to form micelles are two critical factors for determining the chemical composition of copolymers. For hydrophilic segments, the most commonly used polymer is poly(ethylene glycol) (PEG) with a molecular weight of 2–15 kD [65]. PEG is water soluble, nontoxic, and uncharged. PEG can reduce the interfacial energy in an aqueous environment, prevent undesirable aggregation, and minimize clearance by protein adsorption in the body. Therefore, PEG-containing shells allow prolonged circulation times of the micelles and increase the opportunity for micelles to reach targeted tumor sites. Other hydrophilic polymers such as poly(*N*-vinyl pyrrolidone) (PVP) [66] or poly(*N*-isopropylacrylamide) (PNIPA) [67–69] have also been used as corona for the micelles. The choice of hydrophobic block depends on the thermodynamic and kinetic stability of the micelles. The thermodynamic stability is controlled by molecular interactions in a micellar system, while the kinetic stability depends on the rate of disassembly of hydrophobic blocks in micelle cores that, in turn, depends on block length and physical state of a core. Polyesters and polyamides are the most commonly used materials. These hydrophobic polymers can undergo hydrolytic and enzyme-catalyzed degradations. Other types of polymers, such as polyethers, have also been used as core materials for micelles.

In recent years, pH-responsive micelles have attracted a lot of interest due to the potential of pH-triggered drug release at tumor sites. It has been widely reported that tumors are acidic with lower pH values (as low as 5.7) than normal tissues (pH 7.4) [70]. In addition, when these micelles are internalized by cancer cells through endocytosis, they can be sequestered into an early endosome (pH ~6.0) and later into acidic lysosomes (pH 5.0–5.5). The acid nature within the tumor tissue and during micelle endocytosis provides a great inspiration for the development of pH-responsive micelles that can release their encapsulated contents upon exposure into acidic environments. A few examples of polymers that are sensitive to acidic environments include, but are not limited to, poly(L-histidine) [22, 71], poly(β-amino ester) [72, 73], and poly(2-(diisopropylamino) ethyl methacrylate) (PDPA) [74–76]. These polymers are hydrophobic at neutral pH and become highly protonated at acidic pH, which leads to disassembly of the micelles.

12.3.2 Critical Micelle Concentration and Micelle Stability

The concentration at which amphiphilic copolymer chains self-assemble in solution to form micelles is known as the critical micelle concentration (CMC). It is a critical parameter that determines whether the polymer solutions compose of unimers or polymeric micelles. Above CMC, the polymer exists as micelles in equilibrium with a small fraction of single chains, while below CMC, the copolymers exist in solution as unimers. Self-assembly of amphiphilic block copolymers is a thermodynamically driven and reversible process. At room temperature, the typical range of CMC for PEG-polyester micelles is around 10^{-7}–10^{-6} M. It is a significant challenge to achieve adequate thermodynamic stability of micelles when micelles are administered intravenously in vivo due to the extreme dilution by circulating blood. If the polymer concentration in circulation is too low, the micelles can prematurely dissociate, resulting in early release of encapsulated drugs before the micelles reach the tumor target. On the other hand, if the copolymer concentration is above a critical threshold, onset of micelle aggregation and precipitation can occur inside blood vessels. The CMC of an amphiphilic block copolymer is primarily controlled by the length of its hydrophobic block. Incorporation of hydrophobic drugs or solutes in micelle cores can further decrease the CMC. The kinetic stability of polymer micelles also depends on the size of hydrophobic block, as well as the physical state of micelle core, such as the glass transition temperature (T_g). Micellar cores with T_g above physiological temperature may survive with minimal aggregation for many hours and even days upon dilution below CMC [77, 78]. Other micelles with "soft" cores (e.g., Pluronics®, poly(ethylene oxide)-*co*-poly(propylene oxide)-*co*-poly(ethylene oxide) triblock copolymers) dissociate into unimers within minutes under the same conditions.

12.3.3 Polymeric Micelles in Clinical Trials

Several polymeric micelle formulations have been evaluated in preclinical and clinical trials for solubilization and controlled delivery of anticancer drugs. NK911 are DOX-encapsulated micelles from a PEG-*block*-poly(L-aspartate) copolymer that showed favorable pharmacokinetic pattern for passive drug targeting. NK911 nearly tripled the half-life of free DOX from 48 min to 2.3 h and improved the clearance kinetics of the drug [79]. The Pluronic micelle formulation of DOX (SP1049C) is in Phase II clinical trial. The formulation has good encapsulation efficiency and shows partial response in some patients, but the pharmacokinetic parameters are similar to those of free DOX in human [80]. Delivery of paclitaxel (PTX) encapsulated in polymeric micelles has also been successfully developed. Genexol-PM, the PEO-*b*-poly(D,L-lactide)-based micelle formulation of PTX, significantly increased the drug's solubility [81]. NK105 is another example of PTX-encapsulating polymeric micelles that have been tested for clinical efficacy. The formulation was developed using PEG-*b*-poly(4-phenyl-1-butanoate)-L-aspartamide as copolymer and has shown 86-fold increase in the AUC of PTX in plasma and strong antitumor activity in C-26 tumor-bearing mice [82]. The most recent PEG-*b*-poly(glutamic acid) micelle formulation with cisplatin (NC-6004) reduced nephrotoxicity and neurotoxicity in rats and demonstrated higher antitumor efficacy compared to free cisplatin. This formulation is currently in Phase I clinical trials [83].

All micellar formulations discussed above take advantage of the leaky microvasculature in tumors and deliver encapsulated drug by passive targeting mechanisms. It is suggested that such drug delivery regimen has low drug accumulation with inept specificity to the disease. Moreover, they also lack any means to visualize tumor accumulation events of the micelles and to monitor the progress of treatments. In light of these limitations, attempts to incorporate imaging capabilities into therapeutic micelles have been reported. The ideal theranostic polymeric micelles would allow for the highly specific delivery of potent anticancer drugs to the tumor sites while providing visualization tools to monitor the accumulation of the drug and the efficiency of the treatment regimen.

12.4 Case Examples of Theranostic Micelles Against Cancer

A number of anticancer agents have been incorporated in theranostic micelles to achieve the desired therapeutic efficacy, including small molecular anticancer drugs such as DOX and PTX, small interfering RNA (siRNA), and some protein drugs. Novel anticancer drugs, such as β-lapachone, have also been encapsulated inside the core of micelles and shown outstanding antitumor efficacy in vivo [84, 85]. Several excellent articles have reviewed the development of small molecular drug-based nanomedicine [16, 86–88]. In this section, we will primarily focus on theranostic polymeric micelles with different imaging modalities and provide a few highlighted examples of multifunctional micellar systems in cancer diagnosis and therapy.

12.4.1 Micelles with Magnetic Resonance Imaging Functions

MRI has been a vital tool for cancer diagnosis because of its excellent spatial and temporal resolutions. However, MRI's poor sensitivity underlies the necessity of contrast agents for cellular and molecular imaging. MRI contrast agents produce image contrast by affecting relaxation properties of water protons, producing images with distinct brightness or darkness depending upon image-weighting parameters [89]. The two most common MRI contrast agents are T_1 and T_2 agents that produce image contrast by shortening T_1 and T_2 times of water protons, respectively. T_1 agents often generate positive image contrast by increasing longitudinal relaxation rates of surrounding water protons, while T_2 agents often generate negative contrast by increasing transverse relaxation rates of water.

T_1 agents are commonly gadolinium(III)-based chelating complexes. Gd(III) ions have seven unpaired electrons, making them strongly paramagnetic. Gd(III) is able to coordinate to diethylenetriamine backbones modified with carboxylic acids. Recently, Kataoka and coworkers reported a multifunctional polymeric micelle system with both MR imaging and cancer therapeutic functionalities [27]. In this study, PEG-*b*-poly(L-glutamic acid) (PEG-P(Glu)) micelles were used to encapsulate gadolinium-diethylenetriamine pentaacetic acid (Gd-DTPA) complex and a parent compound of the anticancer drug oxaliplatin, DACHPt (R,R-*trans*-1,2-diaminocyclohexane) platinum(II), in the micellar core by reversible complexation. The results showed

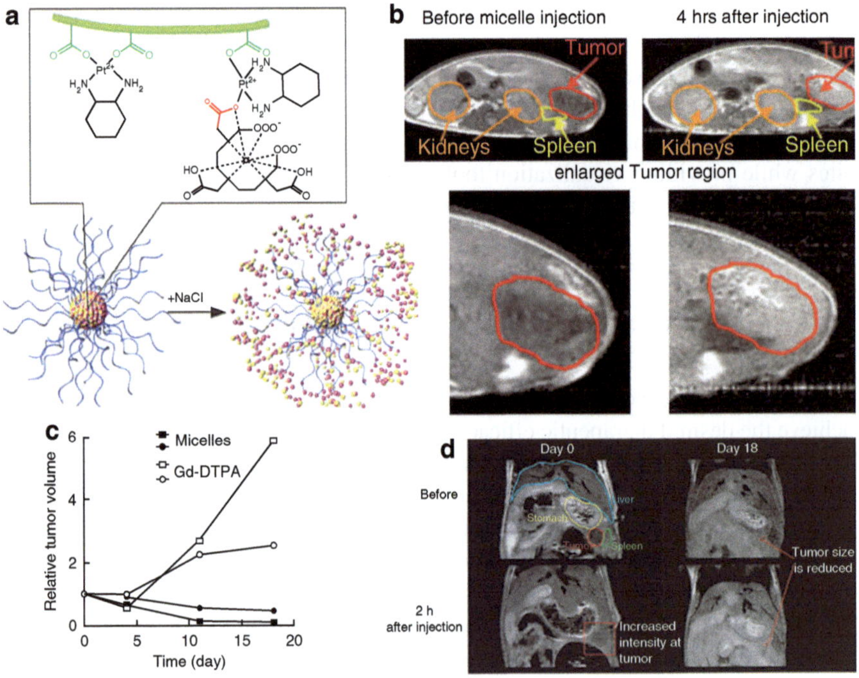

Fig. 12.2 (**a**) Schematic of Gd-DTPA-/DACHPt-loaded micelles and release of Pt and Gd complexes from micelles in physiological environment. (**b**) In vivo T1-weighted MRI images of orthotopic pancreatic cancer (BxPC3) before and after i.v. injection of Gd-DTPA-/DACHPt-loaded micelles at 5 μmol/kg Gd-DTPA. (**c**) In vivo antitumor efficacy of Gd-DTPA-/DACHPt-loaded micelles on orthotopic pancreatic tumor xenografts assessed by volumetric MRI. (**d**) T1-weighted MR images of mice bearing pancreatic tumor at day 0 and 18 after treated with Gd-DTPA-/DACHPt-loaded micelles. Reproduced with permission from [27]

no release of the drug and imaging agent in distilled water, while a sustained release of both agents was observed under physiological conditions (Fig. 12.2a). Compared to free Gd-DTPA and DACHPt, micelles containing these agents significantly increased blood circulation times. In addition, the micelles also enhanced tumor accumulation by 27.7 times for DACHPt at 24 h and over 100 times for Gd-DTPA at 4 h. After an intravenous administration of Gd-DTPA-/DACHPt-loaded micelles to mice bearing orthotopic pancreatic tumors, T_1-weighted MR images showed specific contrast enhancement at the tumor area for over 4 h. The amount of Gd-DTPA at the tumor site delivered by the micelles was seven times higher than the accumulation obtained from free Gd-DTPA administration (Fig. 12.2b). The antitumor activity of Gd-DTPA-/DACHPt-loaded micelles was also evaluated by MRI. The mice treated with micelles at 8 mg/kg Pt equivalent had a significant tumor growth inhibition 18 days after the micelle injection (Fig. 12.2c). Moreover, increased MR signal intensity was also observed in the tumor treated with these micelles (Fig. 12.2d). The colocalization of Gd-DTPA and DACHPt is valuable as MRI can be used to

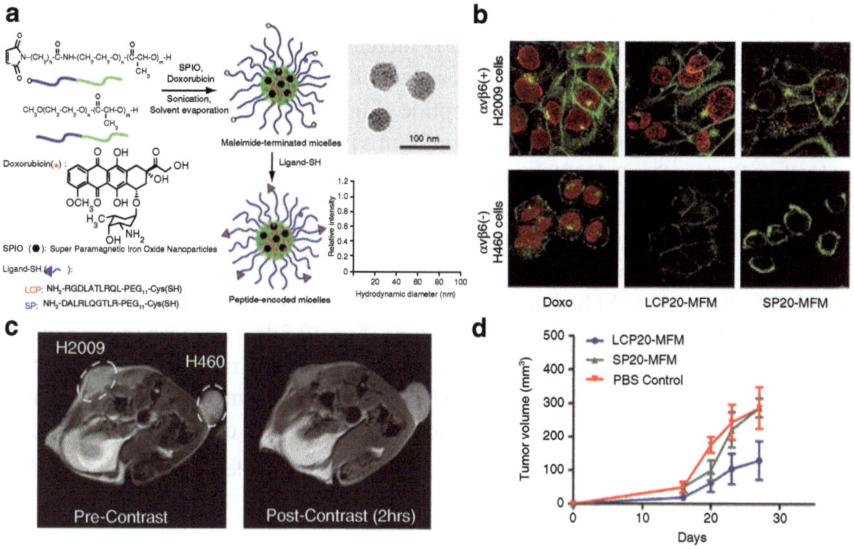

Fig. 12.3 (**a**) Schematic formation of multifunctional micelles with DOX and SPIO loading in the micellar core and peptide conjugation at the micelle surface. Transmission electron microscopy (TEM) and dynamic light scattering (DLS) illustrated the clustering of SPIO nanoparticles (6 nm in diameter) in each micellar particle with a diameter of ~50 nm. (**b**) Confocal laser scanning microscopy of DOX fluorescence in $\alpha_v\beta_6$-expressing H2009 and $\alpha_v\beta_6$-negative H460 cells treated with free DOX, LCP-encoded micelles, and SP-encoded micelles. (**c**) T_2-weighted images of female SCID mouse bearing both H2009 and H460 tumor xenografts before and 2 h after i.v. injection of LCP-encoded micelles on a 4.7 T MRI scanner. (**d**) Evaluation of antitumor efficacy of LCP-encoded MFM in female SCID mice bearing H2009 subcutaneous tumor xenografts at different time points when treated with LCP- and SP-encoded MFM at 4 mg DOX/kg and PBS control. (**a**) and (**b**) reproduced with permission from [26]

visualize the drug distribution within the tumors, allowing for real-time monitoring of the treatment efficacy.

T_2 contrast agents mainly consist of superparamagnetic iron oxide (SPIO) nanoparticles that produce much stronger magnetic susceptibility, affecting a larger number of water molecules to yield higher sensitivity. Our group recently reported the development of multifunctional micelles (MFM) based on PEG-PLA copolymers with integrated MR imaging and therapeutic delivery capabilities [26]. Anticancer drug DOX and SPIO nanoparticles were loaded inside the micellar core to render therapeutic and MR imaging functions, respectively. The surface of the micelles was encoded with the lung cancer-targeting peptide H2009.1 (LCP, with the sequence of RGDLATLRQL), which specifically binds to highly expressed $\alpha_v\beta_6$ integrins in many human non-small cell lung carcinomas (NSCLC) (Fig. 12.3a). For comparison, scrambled peptide (SP) was also conjugated to the surface of micelles. The resulting micelles showed an average diameter of 60–70 nm with narrow size distribution. This SPIO-clustered polymeric micelle design showed much higher T_2 relaxivity than the commercial Federix I.V. sample [90] and decreased the

MR detection limit to subnanomolar concentrations. Demonstrated by confocal laser scanning microscopy, a significantly increased amount of LCP-encoded MFM was observed in $\alpha_v\beta_6$(+) H2009 lung cancer cells over $\alpha_v\beta_6$(−) H460 cells, as well as control micelles with scrambled peptide (Fig. 12.3b). We further evaluated the imaging efficacy of lung tumor xenografts in vivo with injection of LCP-encoded micelles. Both $\alpha_v\beta_6$ (+) H2009 and $\alpha_v\beta_6$ (−) H460 lung tumors were grown on each flank of a female SCID mouse. T_2-weighted images of mice showed that signal intensity of H2009 tumors decreased 2 h after micelle injection, while H460 tumor still kept the same signal intensity (Fig. 12.3c). Preliminary data in vivo also showed that LCP-encoded MFM had much better tumor growth inhibition over SP-encoded micelles and PBS control after four injections (Fig. 12.3d). The mean tumor volume of mice treated with LCP-encoded MFM was only 130 ± 57 mm^3 at day 27, which is two times less than SP-encoded MFM (287 ± 28 mm^3) and PBS control (290 ± 62 mm^3). These results demonstrate the feasibility of combining a multifunctional design of micelle nanomedicine for integrated diagnosis and therapeutic treatment of lung cancer.

12.4.2 *Micelles with Optical Imaging Modality*

Optical imaging is one of the most common modalities used in cancer research. Optical imaging utilizes photons emitted from fluorescent or bioluminescent probes, which is a relatively inexpensive and quick analytical tool. However, this modality has very poor tissue penetration and high background noise due to the tissue autofluorescence and light-absorbing components (hemoglobin, deoxyhemoglobin, and water). Compared to visible light, near-infrared (NIR) light with wavelengths ranging from 700 to 900 nm has advantages of reduced autofluorescence, low tissue scattering, and better tissue penetration, which is highly desirable for imaging applications in vivo [91–93].

Hsiue and coworkers recently developed multifunctional micelles for cancer chemotherapy and imaging by mixing graft copolymer poly(2-hydroxylethyl methacrylate-histidine)-*g*-poly-(D,L-lactide) (PHEMA-*g*-PLA) with diblock copolymer PEG-PLA [30]. Diagnosis and targeting moieties were modified by linking the NIR dye Cy5.5 and folic acid to the end of PEG-PLA. Anticancer drug DOX was encapsulated inside the core of the micelles (Fig. 12.4a). Micelle accumulation in a HeLa tumor was evaluated in vivo with real-time NIR imaging. An analysis comparing the fluorescent intensity in the tumor with the intensity of the whole body of the mice injected with folate micelles revealed twofold higher fluorescence intensity at the tumor site compared to the intensity of the whole body (Fig. 12.4b). Antitumor efficacy in vivo also demonstrated that folate-encoded micelles inhibited HeLa tumor volume by up to 80% at day 25, much more effective than found for mice treated with free DOX or passive targeting micelles (Fig. 12.4c). With the help of NIR imaging, these multifunctional micelles can be utilized to develop combined therapy and diagnosis for oncology.

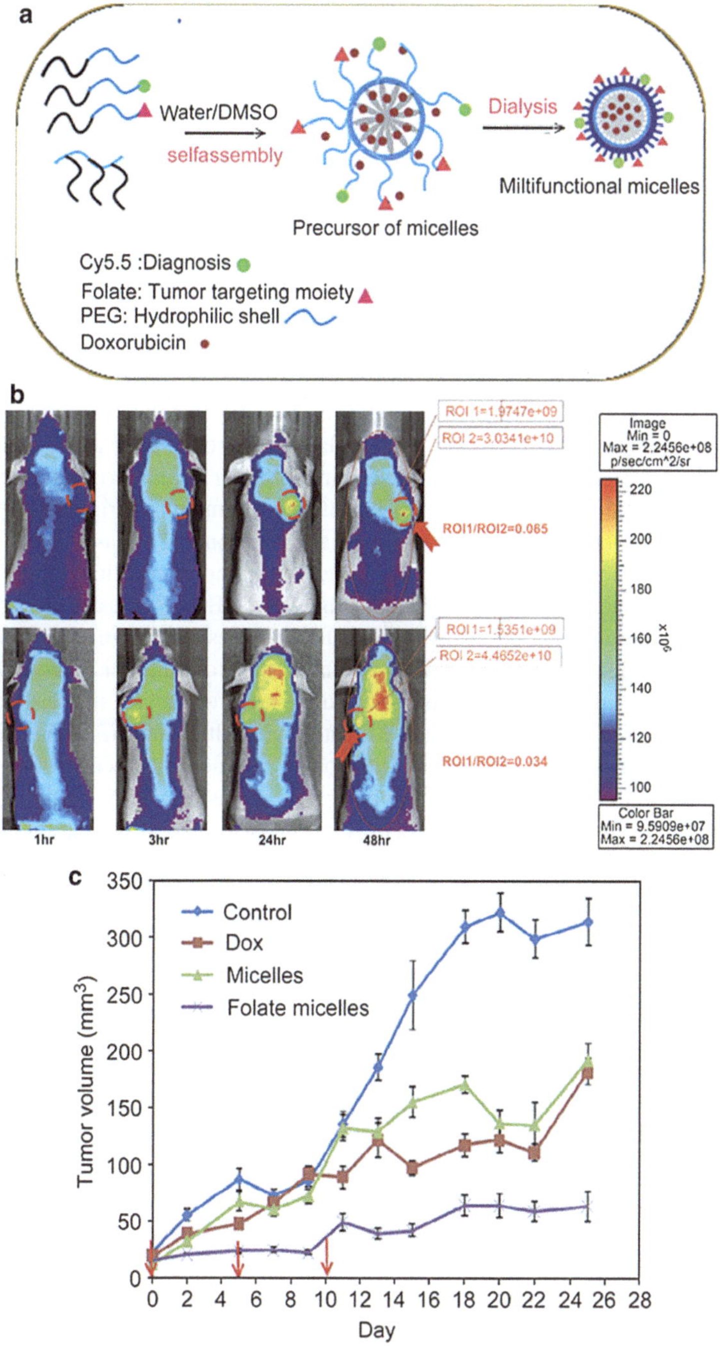

Fig. 12.4 (**a**) Schematic illustration of preparation of folate-encoded, DOX-loaded multifunctional micelles. (**b**) In vivo optical fluorescence imaging of HeLa tumor xenografts administered with folate-encoded micelles. (**c**) Tumor growth inhibition curve of mice bearing subcutaneous HeLa tumors treated with different micelles at 5 mg/kg DOX dosage. Reproduced with permission from [30]

12.4.3 Micelles with Ultrasound Modality

Ultrasound is one of the most commonly used clinical imaging modalities due to its safety, low cost, and simplicity. When a transducer with high-frequency sound waves is applied against skin, the images can be obtained from the sound waves reflected from internal organs [94]. Nanoparticle-based contrast agents can improve imaging by introducing gas with different acoustic properties from that of tissues [95].

Rapoport et al. described novel nanoparticles composed of polymeric micelles and echogenic microbubbles that utilized ultrasound for targeted drug delivery and tumor imaging [29]. DOX-containing PEG-PLLA micelles and nanodroplets of perfluoropentane (PFP), stabilized by an outer layer of block copolymer, were able to permeate through the defective tumor vasculature and accumulate at the tumor site. Those extravasated nanobubbles coalesced in the tumor volume to form microbubbles that produce strong echo signals by ultrasound imaging. Triggered by ultrasound, the microbubbles went through inertial cavitation and released encapsulated drug from polymer micelles (Fig. 12.5a). After intratumoral injection of 0.5% PEG-PLLA/2% perfluoropentane microbubbles in MDA-MB-231 breast cancer-bearing mice, strong ultrasound contrast was generated and preserved in tumors for several days (Fig. 12.5b). Upon i.v. administration of this formulation, MDA-MB-231 tumors without ultrasound treatment showed a pattern of growth similar to control tumors, while the tumors in the group treated with ultrasound (3 MHz for 30 s) demonstrated significant regression. Ultrasound treatment of microbubbles, accumulated in tumor tissue, triggered the controlled release of DOX, resulting in the observed tumor regression. This system is a prime example of utilization of theranostic nanomedicine for cancer therapy and treatment monitoring using ultrasound.

12.5 Summary and Future Perspectives

The field of cancer nanotechnology has rapidly expanded in recent years and continues to progress. Theranostic polymeric micelles have the potential to incorporate multiple functionalities, including cell targeting, ultrasensitive imaging, and therapeutic treatment within one nanoparticle platform to overcome biological complexity and various therapeutic challenges during cancer chemotherapy. However, a number of questions and concerns still need to be addressed before the field can pass beyond its infancy. First, a more detailed understanding of the pharmacokinetic and toxicological properties of these nanoplatforms is needed. There is a general lack of data regarding structure–property relationships of nanoparticles with respect to their size and surface properties and ultimate interactions with biological systems in vivo. A general lack of vigorous safety studies and difficulty associated with scale-up manufacturing are also significant barriers hampering the rapid translation of nanotherapeutics into the clinic. In addition, the dilemma of mismatch between doses required for imaging and therapy has to be carefully considered. Finally, successful development of theranostic nanomedicine builds on the premise and necessity

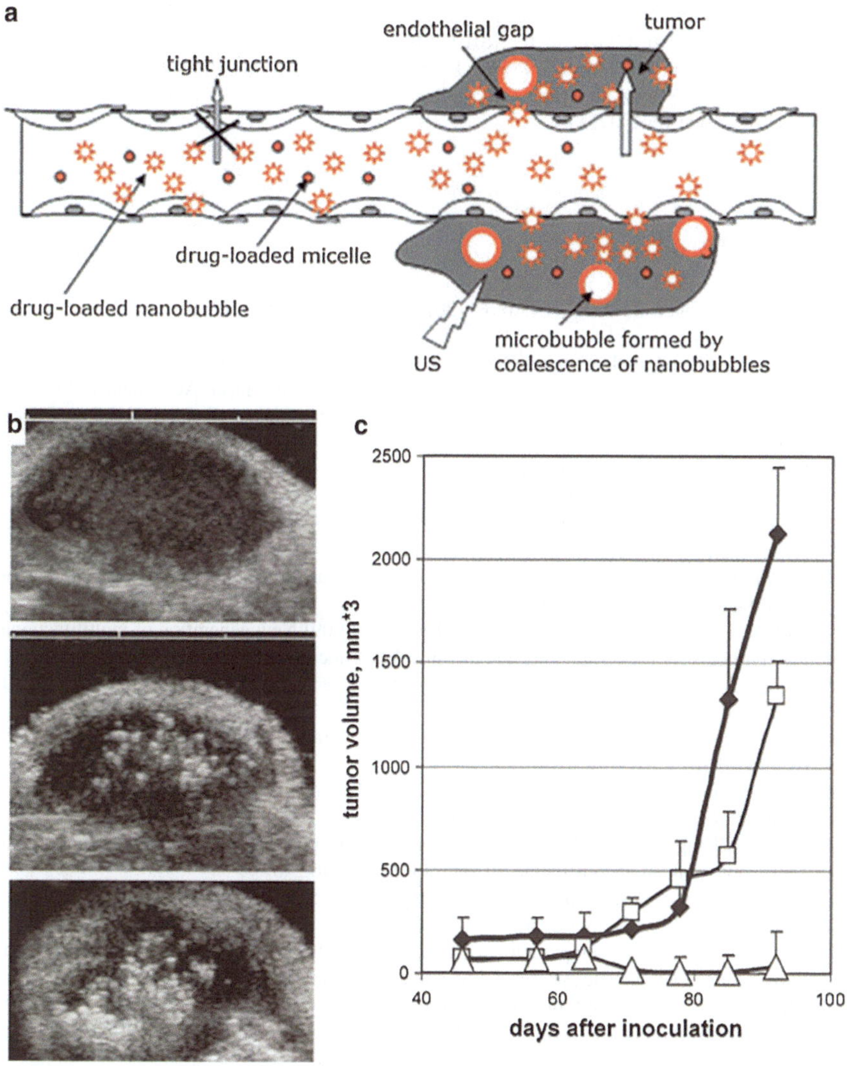

Fig. 12.5 (**a**) Schematic representation of drug targeting through the defective tumor microvasculature using echogenic polymeric micelle system. (**b**) Ultrasound images of MDA-MB231 breast tumor xenografts in mice before and 4 h after injection of 100 ml 0.5% PEG-PLLA/2% perfluoropentane microbubble formulation. (**c**) Antitumor efficacy of tumor-bearing mice treated with microbubble-encapsulated DOX under ultrasound. Reproduced with permission from [29]

of synergistic, interdisciplinary collaborations among chemists, bioengineers, imaging physicists, biologists, and physicians. Although we are still far from Paul Ehrlich's prediction of a "magic bullet," with advances in cancer biology and explosive developments in materials science and imaging technology, it is feasible that we can break through the current threshold and enter a new era of personalized nanomedicine for combined cancer diagnosis and therapy.

Acknowledgment This work is supported by the National Cancer Institute R21-005394, R01-122994, and R01-129011 to JG, R01-CA102792 to DAB, and UL1 RR024982 award to BDS. GH is supported by Susan G. Komen Foundation Postdoctoral Fellowship (PDF0707216). CK is supported by a DOD Breast Cancer Research Program Multidisciplinary Postdoctoral Award (W81XWH-06-1-0751).

References

1. Jemal A, Siegel R, Xu J, Ward E (2010) Cancer statistics 2010. CA Cancer J Clin 60:277–300
2. Farrell D, Alper J, Ptak K, Panaro NJ, Grodzinski P, Barker AD (2010) Recent advances from the National Cancer Institute Alliance for Nanotechnology in Cancer. ACS Nano 4:589–594
3. Ferrari M (2005) Cancer nanotechnology: opportunities and challenges. Nat Rev Cancer 5:161–171
4. Moghimi SM, Hunter AC, Murray JC (2001) Long-circulating and target-specific nanoparticles: theory to practice. Pharmacol Rev 53:283–318
5. Blanco E, Kessinger CW, Sumer BD, Gao J (2009) Multifunctional micellar nanomedicine for cancer therapy. Exp Biol Med (Maywood) 234:123–131
6. Khemtong C, Kessinger CW, Gao J (2009) Polymeric nanomedicine for cancer MR imaging and drug delivery. Chem Commun 3497–3510
7. Park K, Lee S, Kang E, Kim K, Choi K, Kwon C (2009) New generation of multifunctional nanoparticles for cancer imaging and therapy. Adv Funct Mater 19:1–14
8. Sumer B, Gao J (2008) Theranostic nanomedicine for cancer. Nanomedicine (Lond) 3:137–140
9. Xie J, Lee S, Chen X (2010) Nanoparticle-based theranostic agents. Adv Drug Deliv Rev 62:1064–1079
10. Cho K, Wang X, Nie S, Chen ZG, Shin DM (2008) Therapeutic nanoparticles for drug delivery in cancer. Clin Cancer Res 14:1310–1316
11. Fang C, Zhang M (2010) Nanoparticle-based theragnostics: integrating diagnostic and therapeutic potentials in nanomedicine. J Control Release 146:2–5
12. Janib SM, Moses AS, Mackay JA (2010) Imaging and drug delivery using theranostic nanoparticles. Adv Drug Deliv Rev 62:1052–1063
13. Lucignani G (2009) Nanoparticles for concurrent multimodality imaging and therapy: the dawn of new theragnostic synergies. Eur J Nucl Med Mol Imaging 36:869–874
14. McCarthy JR (2009) The future of theranostic nanoagents. Nanomedicine (Lond) 4:693–695
15. McCarthy JR, Weissleder R (2008) Multifunctional magnetic nanoparticles for targeted imaging and therapy. Adv Drug Deliv Rev 60:1241–1251
16. Sutton D, Nasongkla N, Blanco E, Gao J (2007) Functionalized micellar systems for cancer targeted drug delivery. Pharm Res 24:1029–1046
17. Torchilin VP (2007) Targeted pharmaceutical nanocarriers for cancer therapy and imaging. AAPS J 9:E128–E147
18. Rapoport N (2007) Physical stimuli-responsive polymeric micelles for anti-cancer drug delivery. Prog Polym Sci 32:962–990
19. Ulbrich K, Subr V (2004) Polymeric anticancer drugs with pH-controlled activation. Adv Drug Deliv Rev 56:1023–1050
20. Motornov M, Roiter Y, Tokarev I, Minko S (2010) Stimuli-responsive nanoparticles, nanogels and capsules for integrated multifunctional intelligent systems. Prog Polym Sci 35:174–211
21. Smith AE, Xu XW, Mccormick CL (2010) Stimuli-responsive amphiphilic (co)polymers via RAFT polymerization. Prog Polym Sci 35:45–93
22. Lee ES, Gao Z, Bae YH (2008) Recent progress in tumor pH targeting nanotechnology. J Control Release 132:164–170

23. Bagalkot V, Zhang L, Levy-Nissenbaum E et al (2007) Quantum dot-aptamer conjugates for synchronous cancer imaging, therapy, and sensing of drug delivery based on bi-fluorescence resonance energy transfer. Nano Lett 7:3065–3070
24. Choi Y, Weissleder R, Tung CH (2006) Selective antitumor effect of novel protease-mediated photodynamic agent. Cancer Res 66:7225–7229
25. Diagaradjane P, Shetty A, Wang JC et al (2008) Modulation of in vivo tumor radiation response via gold nanoshell-mediated vascular-focused hyperthermia: characterizing an integrated anti-hypoxic and localized vascular disrupting targeting strategy. Nano Lett 8:1492–1500
26. Guthi JS, Yang SG, Huang G et al (2010) MRI-visible micellar nanomedicine for targeted drug delivery to lung cancer cells. Mol Pharm 7:32–40
27. Kaida S, Cabral H, Kumagai M et al (2010) Visible drug delivery by supramolecular nanocarriers directing to single-platformed diagnosis and therapy of pancreatic tumor model. Cancer Res 70:7031–7041
28. Medarova Z, Pham W, Farrar C, Petkova V, Moore A (2007) In vivo imaging of siRNA delivery and silencing in tumors. Nat Med 13:372–377
29. Rapoport N, Gao Z, Kennedy A (2007) Multifunctional nanoparticles for combining ultrasonic tumor imaging and targeted chemotherapy. J Natl Cancer Inst 99:1095–1106
30. Tsai HC, Chang WH, Lo CL et al (2010) Graft and diblock copolymer multifunctional micelles for cancer chemotherapy and imaging. Biomaterials 31:2293–2301
31. Yang J, Lee CH, Ko HJ et al (2007) Multifunctional magneto-polymeric nanohybrids for targeted detection and synergistic therapeutic effects on breast cancer. Angew Chem Int Ed 46:8836–8839
32. Saravanakumar G, Kim K, Park JH, Rhee K, Kwon IC (2009) Current status of nanoparticle-based imaging agents for early diagnosis of cancer and atherosclerosis. J Biomed Nanotechnol 5:20–35
33. Wickline SA, Neubauer AM, Winter PM, Caruthers SD, Lanza GM (2007) Molecular imaging and therapy of atherosclerosis with targeted nanoparticles. J Magn Reson Imaging 25:667–680
34. Reddy GR, Bhojani MS, McConville P et al (2006) Vascular targeted nanoparticles for imaging and treatment of brain tumors. Clin Cancer Res 12:6677–6686
35. Kim K, Kim JH, Park H et al (2010) Tumor-homing multifunctional nanoparticles for cancer theragnosis: simultaneous diagnosis, drug delivery, and therapeutic monitoring. J Control Release 146:219–227
36. Warner S (2004) Diagnostics plus therapy = theranostics. Scientist 18:38–39
37. Matsumura Y, Maeda H (1986) A new concept for macromolecular therapeutics in cancer chemotherapy: mechanism of tumoritropic accumulation of proteins and the antitumor agent smancs. Cancer Res 46:6387–6392
38. Jain RK (1994) Barriers to drug delivery in solid tumors. Sci Am 271:58–65
39. Peer D, Karp JM, Hong S, Farokhzad OC, Margalit R, Langer R (2007) Nanocarriers as an emerging platform for cancer therapy. Nat Nanotechnol 2:751–760
40. Colombo M, Corsi F, Foschi D et al (2010) HER2 targeting as a two-sided strategy for breast cancer diagnosis and treatment: outlook and recent implications in nanomedical approaches. Pharm Res 62:150–165
41. Farokhzad OC, Cheng J, Teply BA et al (2006) Targeted nanoparticle-aptamer bioconjugates for cancer chemotherapy in vivo. Proc Natl Acad Sci USA 103:6315–6320
42. Jeong YI, Seo SJ, Park IK et al (2005) Cellular recognition of paclitaxel-loaded polymeric nanoparticles composed of poly(gamma-benzyl L-glutamate) and poly(ethylene glycol) diblock copolymer endcapped with galactose moiety. Int J Pharm 296:151–161
43. Yoo HS, Park TG (2004) Folate receptor targeted biodegradable polymeric doxorubicin micelles. J Control Release 96:273–283
44. Mahmud A, Xiong XB, Aliabadi HM, Lavasanifar A (2007) Polymeric micelles for drug targeting. J Drug Target 15:553–584
45. Torchilin VP, Lukyanov AN, Gao Z, Papahadjopoulos-Sternberg B (2003) Immunomicelles: targeted pharmaceutical carriers for poorly soluble drugs. Proc Natl Acad Sci USA 100:6039–6044

46. Gao Z, Lukyanov AN, Chakilam AR, Torchilin VP (2003) PEG-PE/phosphatidylcholine mixed immunomicelles specifically deliver encapsulated taxol to tumor cells of different origin and promote their efficient killing. J Drug Target 11:87–92
47. Nasongkla N, Bey E, Ren J et al (2006) Multifunctional polymeric micelles as cancer-targeted, MRI-ultrasensitive drug delivery systems. Nano Lett 6:2427–2430
48. Weissleder R, Pittet MJ (2008) Imaging in the era of molecular oncology. Nature 452:580–589
49. Cheon J, Lee JH (2008) Synergistically integrated nanoparticles as multimodal probes for nanobiotechnology. Acc Chem Res 41:1630–1640
50. Pan D, Lanza GM, Wickline SA, Caruthers SD (2009) Nanomedicine: perspective and promises with ligand-directed molecular imaging. Eur J Radiol 70:274–285
51. Ayyagari AL, Zhang X, Ghaghada KB, Annapragada A, Hu X, Bellamkonda RV (2006) Long-circulating liposomal contrast agents for magnetic resonance imaging. Magn Reson Med 55:1023–1029
52. Wang Y, Ye F, Jeong EK, Sun Y, Parker DL, Lu ZR (2007) Noninvasive visualization of pharmacokinetics, biodistribution and tumor targeting of poly[*N*-(2-hydroxypropyl)methacrylamide] in mice using contrast enhanced MRI. Pharm Res 24:1208–1216
53. Rosenblum LT, Kosaka N, Mitsunaga M, Choyke PL, Kobayashi H (2010) In vivo molecular imaging using nanomaterials: general in vivo characteristics of nano-sized reagents and applications for cancer diagnosis. Mol Membr Biol 27:274–285
54. Cai W, Shin DW, Chen K et al (2006) Peptide-labeled near-infrared quantum dots for imaging tumor vasculature in living subjects. Nano Lett 6:669–676
55. Zhou M, Zhang R, Huang M et al (2010) A chelator-free multifunctional [(64)Cu]CuS nanoparticle platform for simultaneous micro-PET/CT imaging and photothermal ablation therapy. J Am Chem Soc 132:15351–15358
56. Gillies ER, Frechet JM (2005) Dendrimers and dendritic polymers in drug delivery. Drug Discov Today 10:35–43
57. Gopin A, Ebner S, Attali B, Shabat D (2006) Enzymatic activation of second-generation dendritic prodrugs: conjugation of self-immolative dendrimers with poly(ethylene glycol) via click chemistry. Bioconjug Chem 17:1432–1440
58. Liu J, Lee H, Huesca M, Young A, Allen C (2006) Liposome formulation of a novel hydrophobic aryl-imidazole compound for anti-cancer therapy. Cancer Chemother Pharmacol 58:306–318
59. Enoch HG, Strittmatter P (1979) Formation and properties of 1000-A-diameter, single-bilayer phospholipid vesicles. Proc Natl Acad Sci USA 76:145–149
60. Kataoka K, Kwon GS, Yokoyama M, Okano T, Sakurai Y (1993) Block copolymer micelles as vehicles for drug delivery. J Control Release 24:119–132
61. Kwon GS, Yokoyama M, Okano T, Sakurai Y, Kataoka K (1994) Enhanced tumor accumulation and prolonged circulation times of micelle-forming poly(ethylene oxide-aspartate) block copolymer-adriamycin conjugates. J Control Release 28:334–335
62. Delgado C, Francis GE, Fisher D (1992) The uses and properties of PEG-linked proteins. Crit Rev Ther Drug Carrier Syst 9:249–304
63. Pratten MK, Lloyd JB, Horpel G, Ringsdorf H (1985) Micelle-forming block copolymers-pincytosis by macrophages and interaction with model membranes. Makromol Chem Macromol Chem Phys 186:725–733
64. Gros L, Ringsdorf H, Schupp H (1981) Polymer anti-tumor agents on a molecular and on a cellular level. Angew Chem Int Ed 20:305–325
65. Torchilin VP (2001) Structure and design of polymeric surfactant-based drug delivery systems. J Control Release 73:137–172
66. Benahmed A, Ranger M, Leroux JC (2001) Novel polymeric micelles based on the amphiphilic diblock copolymer poly(*N*-vinyl-2-pyrrolidone)-block-poly(D,L-lactide). Pharm Res 18:323–328
67. Chung JE, Yokoyama M, Aoyagi T, Sakurai Y, Okano T (1998) Effect of molecular architecture of hydrophobically modified poly(N-isopropylacrylamide) on the formation of thermoresponsive core-shell micellar drug carriers. J Control Release 53:119–130

68. Chung JE, Yokoyama M, Okano T (2000) Inner core segment design for drug delivery control of thermo-responsive polymeric micelles. J Control Release 65:93–103
69. Chung JE, Yokoyama M, Yamato M, Aoyagi T, Sakurai Y, Okano T (1999) Thermo-responsive drug delivery from polymeric micelles constructed using block copolymers of poly(*N*-isopropylacrylamide) and poly(butylmethacrylate). J Control Release 62:115–127
70. Martin GR, Jain RK (1994) Noninvasive measurement of interstitial pH profiles in normal and neoplastic tissue using fluorescence ratio imaging microscopy. Cancer Res 54:5670–5674
71. Lee ES, Na K, Bae YH (2005) Super pH-sensitive multifunctional polymeric micelle. Nano Lett 5:325–329
72. Lynn DM, Amiji MM, Langer R (2001) pH-responsive polymer microspheres: rapid release of encapsulated material within the range of intracellular pH. Angew Chem Int Ed 40:1707–1710
73. Shenoy D, Little S, Langer R, Amiji M (2005) Poly(ethylene oxide)-modified poly(beta-amino ester) nanoparticles as a pH-sensitive system for tumor-targeted delivery of hydrophobic drugs. 1. In vitro evaluations. Mol Pharm 2:357–366
74. Giacomelli C, Le Men L, Borsali R et al (2006) Phosphorylcholine-based pH-responsive diblock copolymer micelles as drug delivery vehicles: light scattering, electron microscopy, and fluorescence experiments. Biomacromolecules 7:817–828
75. Licciardi M, Craparo EF, Giammona G, Armes SP, Tang Y, Lewis AL (2008) In vitro biological evaluation of folate-functionalized block copolymer micelles for selective anti-cancer drug delivery. Macromol Biosci 8:615–626
76. Du J, Tang Y, Lewis AL, Armes SP (2005) pH-sensitive vesicles based on a biocompatible zwitterionic diblock copolymer. J Am Chem Soc 127:17982–17983
77. Kwon GS, Okano T (1999) Soluble self-assembled block copolymers for drug delivery. Pharm Res 16:597–600
78. Kang N, Perron ME, Pru'dhomme RE, Zhang Y, Gaucher G, Leroux JC (2005) Stereocomplex block copolymer micelles: core-shell nanostructures with enhanced stability. Nano Lett 5:315–319
79. Matsumura Y, Hamaguchi T, Ura T et al (2004) Phase I clinical trial and pharmacokinetic evaluation of NK911, a micelle-encapsulated doxorubicin. Br J Cancer 91:1775–1781
80. Danson S, Ferry D, Alakhov V et al (2004) Phase I dose escalation and pharmacokinetic study of pluronic polymer-bound doxorubicin (SP1049C) in patients with advanced cancer. Br J Cancer 90:2085–2091
81. Kim TY, Kim DW, Chung JY et al (2004) Phase I and pharmacokinetic study of Genexol-PM, a cremophor-free, polymeric micelle-formulated paclitaxel, in patients with advanced malignancies. Clin Cancer Res 10:3708–3716
82. Hamaguchi T, Matsumura Y, Suzuki M et al (2005) NK105, a paclitaxel-incorporating micellar nanoparticle formulation, can extend in vivo antitumour activity and reduce the neurotoxicity of paclitaxel. Br J Cancer 92:1240–1246
83. Uchino H, Matsumura Y, Negishi T et al (2005) Cisplatin-incorporating polymeric micelles (NC-6004) can reduce nephrotoxicity and neurotoxicity of cisplatin in rats. Br J Cancer 93:678–687
84. Blanco E, Bey EA, Dong Y et al (2007) Beta-lapachone-containing PEG-PLA polymer micelles as novel nanotherapeutics against NQO1-overexpressing tumor cells. J Control Release 122:365–374
85. Blanco E, Bey EA, Khemtong C et al (2010) Beta-lapachone micellar nanotherapeutics for non-small cell lung cancer therapy. Cancer Res 70:3896–3904
86. Duncan R (2003) The dawning era of polymer therapeutics. Nat Rev Drug Discov 2:347–360
87. Kabanov AV, Batrakova EV, Miller DW (2003) Pluronic block copolymers as modulators of drug efflux transporter activity in the blood–brain barrier. Adv Drug Deliv Rev 55:151–164
88. Nishiyama N, Kataoka K (2006) Current state, achievements, and future prospects of polymeric micelles as nanocarriers for drug and gene delivery. Pharmacol Ther 112:630–648
89. Aime S, Crich SG, Gianolio E, Giovenzana GB, Tei L, Terreno E (2006) High sensitivity lanthanide(III) based probes for MR-medical imaging. Coord Chem Rev 250:1562–1579

90. Wang YX, Hussain SM, Krestin GP (2001) Superparamagnetic iron oxide contrast agents: physicochemical characteristics and applications in MR imaging. Eur Radiol 11:2319–2331
91. Bremer C, Tung CH, Weissleder R (2001) In vivo molecular target assessment of matrix metalloproteinase inhibition. Nat Med 7:743–748
92. Min KH, Park K, Kim YS et al (2008) Hydrophobically modified glycol chitosan nanoparticles-encapsulated camptothecin enhance the drug stability and tumor targeting in cancer therapy. J Control Release 127:208–218
93. Weissleder R, Tung CH, Mahmood U, Bogdanov A Jr (1999) In vivo imaging of tumors with protease-activated near-infrared fluorescent probes. Nat Biotechnol 17:375–378
94. Massoud TF, Gambhir SS (2003) Molecular imaging in living subjects: seeing fundamental biological processes in a new light. Genes Dev 17:545–580
95. Blomley MJ, Cooke JC, Unger EC, Monaghan MJ, Cosgrove DO (2001) Microbubble contrast agents: a new era in ultrasound. Br Med J 322:1222–1225

Chapter 13
Multifunctional Nanoparticles for Personalized Medicine

Benjamin T. Roller, Kathleen M. McNeeley, and Ravi V. Bellamkonda

13.1 Introduction

Traditionally, physicians and surgeons have treated disease based on common symptoms with preestablished treatments and standards of care for a representative population. Therapies are generally developed, tested, and approved via screening processes that conclude in large randomized clinical trials, aiming to pick the safest, most effective treatment for the largest population of diseased individuals. Though this *modus operandi* is successful with common diseases, this approach has proven to be inefficient in the treatment of cancer where response to therapy is frustratingly unpredictable.

Though still in its infancy, personalized medicine holds promise for treating individuals with diseases such as cancer. For instance, breast cancers usually over-express one or more of the receptors for estrogen, progesterone, or epidermal growth factor (EGF). Tests such as Dako's HercepTest™, Ventana's Pathway™, Abbott's PathVysion™, and Genomic Health's Onco*Type* DX™ confirm the presence of overexpressed receptors or assess gene expression levels of breast cancer patients and can be used to predict whether the tumor will respond to treatment with one of the receptor targeted medications (i.e., Herceptin, Tamoxifen, etc.) [1, 2]. Similar evaluation is available for prostate cancer, including tests for p504S, hepsin, Pim-1, protease/KLK4, prostein, EHZ 2, GSTP1, and STEAP [3, 4]. However, these tests require invasive biopsies, and the application of nanoparticles could allow performance of similar diagnostics noninvasively.

B.T. Roller • K.M. McNeeley • R.V. Bellamkonda (✉)
Wallace H Coulter Department of Biomedical Engineering,
Georgia Institute of Technology and Emory School of Medicine,
Atlanta, GA 30332, USA
e-mail: ravi@gatech.edu

S. Svenson and R.K. Prud'homme (eds.), *Multifunctional Nanoparticles for Drug Delivery Applications: Imaging, Targeting, and Delivery*, Nanostructure Science and Technology,
DOI 10.1007/978-1-4614-2305-8_13,

Nanoparticles have potential for many applications that enable personalized medicine including imaging, diagnostics, and treatment. Large surface area, along with diverse surface chemistries and ease of modifications, allows nanoparticles to be readily adapted for specific applications. For example, the surface attachment of hydrophilic polymers is often utilized to alter the pharmacokinetics of encapsulated agents, while the conjugation of targeting moieties can facilitate homing of nanoparticles to specific biomarkers and receptors to treat tumors with high specificity. The ability to package large amounts of contrast agents or fluorescent/bioluminescent markers effectively increases sensitivity of imaging modalities to enable imaging at the molecular level, while packaging of therapeutic agents ensures high target area effectiveness while lessening or eliminating potentially detrimental off-target side effects. Nanoparticles can be utilized to diagnostically assess tumor vascular status, molecular profile, and treatment response, all of which influence treatment strategies. In addition, nanoparticles can be functionalized to facilitate targeting to specific disease sites, to enable triggered release of therapeutic agents, and to deliver DNA or RNA to target cells. These functions can be combined to create multifunctional nanoparticles and adapted for use in personalized medicine. This chapter will focus on examples of multifunctional nanoparticles and their application to personalized medicine.

13.2 Nanoparticles as Diagnostic Tools for Personalized Medicine

Perhaps one of the most essential areas of research for personalized medicine to move forward is the development of diagnostic tools to help assign particular treatments and treatment schedules to individuals with disease. Many of the current diagnostic tools in the area of cancer rely on blood or tissue samples. Evaluation of patient blood samples is typically an indirect method of evaluating disease status, and acquisition of tissue samples requires invasive surgeries to biopsy the tumor tissue for testing. Multifunctional nanoparticles might allow for performance of diagnostic assessments noninvasively.

13.2.1 Assessing Response to Chemotherapy in "Real Time" and Predicting Its Outcome

The current state of the art for personalized cancer treatments relies on drugs targeted to overexpressed cell surface receptors. Presently, there are diagnostic tools available that enable doctors to assess whether a patient's tumor will respond to treatment by one or more drugs based on tumor-specific cell surface receptor overexpression. However, many of these tests require homogeneous and representative tissue samples obtained through invasive surgery/biopsy that are often difficult to obtain due to the heterogeneity of most tumors, which often contain cancerous, benign, stromal, and cancer stem cells. Thus, it would be beneficial to have access to other diagnostic tools that are less invasive, less sensitive to the makeup of the tumor, and provide a more complete representation of the entire tumor accounting for spatial heterogeneity.

It is well known that tumor vasculature varies from tumor to tumor and from patient to patient, but it has also been shown to vary between identical tumor models as well as throughout an individual tumor [5–7]. This is because the tumor's complex microvasculature often grows in a rapid, disorganized fashion [8]. In addition, proliferating cells can cause the microvasculature to compress and/or collapse [9]. Thus, a major limiting factor in the effectiveness of nanoparticle chemotherapeutics is the variability in the inherent leakiness of the tumor vasculature. To date, there exists no standard method to determine vascular leakiness in a clinical setting. Current clinically approved liposomal chemotherapeutics are generally delivered once every 3–4 weeks [10]. For example, chemotherapeutics such as liposomal doxorubicin (Doxil®) and others have become an important part of successful cancer therapy [11, 12]. Having an *a priori* knowledge of whether or not a liposomal chemotherapeutic will accumulate in a tumor in large enough quantities to be effective before use would provide valuable information for a clinician to help tailor personalized treatment and would buy valuable time that a cancer patient does not often have.

Correctly sized (50–200 nm), long-circulating nanoparticles, when delivered systemically, preferentially accumulate in tumor tissue as a result of the enhanced permeability and retention (EPR) effect. Due to the presence of leaky vasculature within tumors, which bears gaps ranging from 100 to 800 nm, these nanoparticles are able to extravasate into cancerous tissue [13–15]. At the same time, the nanoparticles are too large to exit the vasculature present within healthy tissue, thus reducing off-target delivery. This phenomenon is critically dependent on the presence of a compromised endothelial lining on the tumor vasculature; however, not all tumors have vasculature that is leaky enough to allow sufficient accumulation of nanoparticles.

The elegantly straightforward solution proposed by Karathanasis et al. to determine tumor leakiness involves a multifunctional, long-circulating liposomal nanoparticle that encapsulated iodine for contrast enhanced mammography. The iodine contrast agent, Visipaque™ 320 (GE Healthcare, Milwaukee, WI), was encapsulated in 100-nm liposomes and delivered to rats that were inoculated with MAT B III breast tumors. Mammography was performed on a clinical mammography system (Senographe 2000D, GE Healthcare) for three consecutive days [16, 17]. Images were used to calculate the amount of enhancement and approximate accumulation rate of the iodine liposomes. Animals were then separated into two groups. Animals in the "good prognosis" group had rapid and intense iodine enhancement in their mammography images and were hypothesized to respond better to subsequent treatment with liposomal doxorubicin than those in the "bad prognosis" group, exhibiting slower, less pronounced iodine enhancement. Animals were then treated with liposomal doxorubicin, and tumor growth was monitored. Tumor growth rates proved to be much slower in the good prognosis group, resulting in longer survival times than the bad prognosis group. In a separate study, Karathanasis et al. demonstrated that the dual encapsulation of iodine and doxorubicin not only allowed one to track the accumulation of doxorubicin in the tumor in real time but also allowed one to track tumor response to the drug throughout treatment [17]. Prediction of effectiveness of treatment by this method could be used clinically to help personalize cancer therapy for individual patients. If enhancement is sufficient, then treatment can proceed utilizing the appropriate chemotherapeutic drugs encapsulated within similarly formulated liposomal nanoparticles. If the iodine liposomes do not accumulate in sufficient quantities to show that future liposomal therapy would be effective enough, then doctors could choose other treatment methods.

13.2.2 Determination of Tumor Malignancy

The malignancy of a tumor influences the personalized treatment strategies chosen by a physician. The more malignant the lesion, the more aggressive the cancer therapy should be to ensure best odds of success. While some of this information can be gleaned from traditional magnetic resonance imaging (MRI), computed tomography (CT), or mammography data, histological analysis of biopsied tissue samples still remains the primary method of tumor grading. Use of multifunctional nanoparticles could enable doctors to grade tumors noninvasively and choose personalized treatment strategies accordingly.

Turetschek et al. hypothesized that a poly(ethylene glycol) (PEG)-protected iron oxide nanoparticles could be used to quantitatively characterize tumor malignancy in vivo [18]. Nanoparticles were injected into rats with *N*-ethyl-*N*-nitrosourea (ENU)-induced mammary tumors. Subsequent MRI analysis enabled tumor grading in vivo by quantification of permeability of the tumors to the contrast agent. Vascularity is an indicator of malignancy in tumor grading; those tumors with higher vascularity tend to be more malignant, resulting in decreased patient survival times [19, 20]. Malignancy estimations via MRI after nanoparticle injection were confirmed using traditional biopsy and histology.

The molecular profile of a tumor is another good indicator of malignancy. In particular, the expression of the integrin $\alpha_V\beta_{III}$ on vascular endothelial cells has been shown to correlate with tumor malignancy and grade [21–27]. Thus, important information on prognosis could be obtained noninvasively if it was possible to detect and quantify $\alpha_V\beta_{III}$ in vivo. Molecular imaging with high sensitivity is needed to achieve $\alpha_V\beta_{III}$ quantification in vivo, and the use of nanoparticles in this respect is described in the following section.

13.2.3 Multifunctional Nanoparticles for Molecular Imaging

In order for personalized medicine to truly materialize, highly sensitive diagnostics must be developed to probe tumors and other diseases for patient-specific biomarkers. Several methods exist to obtain molecular information about tissue samples in the laboratory, but it would be much more beneficial if this information could be obtained noninvasively in vivo. Since multifunctional nanoparticles have the capacity to carry large payloads of contrast agents, they can be exploited to enable molecular imaging using clinically relevant imaging modalities, especially in the area of cancer.

As mentioned in the previous section, $\alpha_V\beta_{III}$ is an indicator of tumor grade and malignancy. Thus, it would be beneficial to be able to detect and quantify this molecule in vivo, allowing physicians to choose personalized treatment strategies based on the findings. Sipkins et al. used an antibody-targeted liposome containing gadolinium chelates to quantify expression of $\alpha_V\beta_{III}$ in a rabbit carcinoma model [28]. Treatment with targeted nanoparticles resulted in greater contrast compared to traditional gadolinium-enhanced MRI. As predicted, areas of concentrated targeted nanoparticle accumulation were found to be highly angiogenic, thus allowing non-invasive in vivo assessment of angiogenesis.

Regulatory molecules such as vascular endothelial growth factor (VEGF) and its receptor (VEGFR) play important rate-limiting roles in tumor angiogenesis. Therefore, it may be possible to image the angiogenic state of the tumor by quantifying the amount of VEGF or VEGFR that is present in the tumor. Karathanasis et al. used a multifunctional liposome to enable surrogate quantification of VEGF and VEGFR expression in tumors [29]. Liposomes filled with an iodine-based contrast agent and coated with PEG polymer were delivered to rats that had been inoculated with MAT B III breast tumors. Mammography was performed, and tumor enhancement was quantified and then compared to VEGF and VEGFR expression information obtained via qRT-PCR. It was found that tumor enhancement resulting from nanoparticle accumulation correlated strongly with VEGF and VEGFR expression. Though not directly imaging VEGF and VEGFR, information from this type of surrogate imaging could be used to personalize antiangiogenic therapies for patients with highly vascularized tumors.

Methods mentioned above involved clinically available technology. However, there are other examples of molecular imaging using nanoparticles, and though the technology used is not readily available in a clinical setting, it is realistic to think that the equipment could be available in the near future. One example is the use of gold nanoparticles and Raman spectroscopy. Cao et al. were able to perform multiplexed detection of six different Raman-labeled nanoparticles to distinguish six distinct DNA targets in vitro [30]. Detection limits were down to the range of 20 fM. This technology has the potential to be applied in vivo as well, though the depth of penetration of Raman scattering could be a limiting factor.

13.3 Nanoparticles as Cancer Therapeutic Carriers and Agents

Nanoparticles can package large quantities of drugs and contrast agents. The versatility of surface chemistries allows the design of nanoparticles that alter drug pharmacokinetics and specifically target cancer cell receptors. Nanoparticles can also be designed to be heat sensitive, pH sensitive, or enzyme sensitive to allow triggered release. Multifunctional nanoparticles for cancer therapy are presented below.

13.3.1 Packaging Chemotherapeutics

Traditional chemotherapy causes systemic side effects since intravenously delivered chemotherapeutics typically leak promiscuously into tissues throughout the body, including healthy, nontarget organs. Patients undergoing chemotherapy often become underweight, suffer from alopecia, nausea, and generally have a poor quality of life. As a result, treatments are restricted by the dose-limiting toxicities of chemotherapeutics, which ultimately lower the overall treatment efficacy and capability of completely eradicating the tumor. In addition, patient compliance to treatment is often an issue due to the severely unfavorable side effects associated with traditional chemotherapy. If chemotherapeutics could be delivered directly to the tumor while sparing healthy tissues of the body then, logically, the treatment should have less side effects, therapeutic efficacy could be increased, and patient quality of life could be improved.

Encapsulation of drugs within nanoparticles alters their pharmacokinetics. The size of nanoparticles can be designed to restrict entry into nontarget organs bearing intact vasculature lined with tight endothelial junctions. Since tumors bear disorganized, rapidly growing vasculature characterized by enhanced permeability, selective delivery to tumors can be achieved through nanoparticle packaging of a substance. In addition, circulation times of the nanoparticle-associated drug can be controlled by altering the external surface of the carrier (i.e., coating with a polymer such as PEG). This is particularly advantageous for drugs and contrast agents which exhibit short half-lives in circulation, limiting their ability to access and accumulate within tumors.

Some of the first nanoscale therapeutics approved for human use were liposomal nanoparticles that encapsulate cytotoxic, anthracycline-derived antitumor drugs. Liposomes are similar in structure to the lipid bilayer of cells and are thus highly biocompatible, making them quite safe for use in vivo. Myocet®, liposomal doxorubicin, and DaunoXome®, liposomal daunorubicin, are indicated for use as anticancer agents. Packaging chemotherapeutic drugs in liposome vessels allows for accumulation in the tumor, while reducing the deadly cardiotoxicity associated with freely delivered anthracyclines. However, these nanoparticles are not considered multifunctional since they merely encapsulate drugs. Newer formulations of these drugs, like Doxil®, are coated with poly(ethylene glycol) (PEG) to protect them from clearance by the reticuloendothelial system (RES), thereby prolonging circulation time in the bloodstream. So-called sterically stabilized or "Stealth" liposomes are able to evade opsonization and RES clearance because the hydrophilic PEG chains attract water, effectively creating a water barrier between the external surface of the liposomes and potential opsonizing proteins. For many patients with solid tumors, liposomal anthracyclines could be used. However, their efficacy relies on the specific vascular makeup of the tumor in question, which can vary from tumor to tumor as well as between the same tumor type in different individuals [5–7]. Having knowledge of the tumor vascular structure before treatment would allow doctors to assess feasibility of such a treatment. The work of Karathanasis et al., discussed above, demonstrates the potential use of nanoparticle contrast agents to noninvasively assess whether liposomal treatment would be beneficial before its application [16].

The versatility of nanoparticles makes them especially suited to package drugs for personalized medicine approaches. Packaging of agents within nanoparticles offers the potential to alter the drug pharmacokinetics as appropriate for a given tumor, allows for predictive and real-time assessment of treatment response, and enables triggered drug exposure and targeting to specific tumor types. Utilizing these approaches, treatment regimens can be designed specifically for an individual tumor rather than to the disease in general.

13.3.2 Nanoparticles as Targeting Agents

The versatility of nanoparticles lies not only in the ability to package small molecules but also in the many types of surface modifications that can be employed. As mentioned earlier, stealth liposomes, containing the hydrophilic polymer PEG on their surface, can be used to protect nanoparticles from RES clearance. Surface modifications allow one to make hydrophilic, hydrophobic, or amphipathic nanoparticles. Other surface modifications allow attachment of ligands such as proteins, antibodies, peptides, nucleic acids, and small molecules [31–33] that turn nanoparticles into cell-specific targeting vehicles. Molecular targeting of cancer cells would be invaluable for personalized medicine. One could envision a treatment strategy where tumor-specific surface receptors could be found via biopsy or noninvasively with the use of nanoparticles; then nanoparticles could be designed to target that specific array of cellular receptors.

One example of molecular targeting of nanoparticles is the use of surface-bound folate on PEGylated stealth liposomes to target 9L glioma, which inherently overexpress folate receptors [34]. Though these nanoparticles proved promising in vitro, their use in vivo did not produce the anticipated increased efficacy over nontargeted stealth liposomes [35]. The presence of folate on the surface of the liposomes counteracted the stealth nature of the PEGylated liposomes, allowing opsonization and accelerated clearance from the blood stream. McNeeley et al. proposed a solution using longer, cleavable PEG chains to mask the folate molecules until accumulation in the tumor had occurred [36]. PEG molecules with molecular weight of 5,000 Da were bound to the liposome surface via a cleavable disulfide bond. The longer PEG chains masked the folate-targeting molecules, which were attached to the nanoparticles via 2,000 Da PEG molecules. The liposomes were able to evade the RES and accumulate in the tumor via the EPR effect. Later, cysteine was delivered to cleave the disulfide bond and detach the PEG_{5000} coating. This exposed the folate molecules and allowed them to bind to the folate receptors of the target 9L glioma cells, resulting in increased uptake compared to control nontargeted stealth liposomes.

Kale and Tochilin have since engineered clever pH-sensitive liposomes that are able to penetrate tumors using cell-penetrating peptides [37]. These liposomes use a shielding method similar to that of McNeeley et al. [36], with the exception that the protective PEG molecules are cleaved via low pH-induced hydrolysis. When the liposomes extravasate into tumor or ischemic tissue, the lowered pH causes hydrolysis of protective PEG, allowing cell-penetrating peptide TATp to be exposed, which in turn induces cellular uptake.

The general trend of targeting tumor cells is to target a receptor that is overexpressed by a specific tumor. Target receptors, however, are often expressed on healthy cells as well, allowing targeted nanoparticles to cause side effects in these healthy cells. Saul et al. proposed a method to overcome this issue by creating a nanoparticle that targets more than one type of receptor overexpressed on a target tumor cell, thus increasing the targeting selectivity of the nanoparticles [38]. PEGylated stealth liposomes were designed carrying surface molecules targeted to the folate receptor as well as the EGF receptor. After carefully optimizing the targeting ligand density on the liposomal nanoparticles, experiments in vitro showed that only cells that overexpressed both receptors showed increased cytotoxicity in response to targeted treatment.

Other nanoparticles such as dendrimers have shown promise as targeted delivery vehicles. Dendrimers are multibranched polymer molecules that are generally sphere-shaped. The presence of many branches increases the surface area available for conjugation of targeting molecules as well as provides space for inclusion of chemotherapeutics and contrast agents. Folate molecules have been conjugated to a dendrimer that also contained fluorescein isothiocyanate imaging agent and the chemotherapeutic paclitaxel (39, see also chapter by Mullen et al.). These dendrimers showed increased toxicity and specificity to cancer cells overexpressing folate receptors in vitro. At optimal concentrations of folate molecules on the dendrimers, only cells overexpressing folate receptors showed increased cytotoxicity compared to controls. Similar results were seen using a folate-containing dendrimer conjugated with methotrexate chemotherapeutic agent [40]. Dendrimers containing shielded cell-penetrating proteins (CPPs) have also been devised that exposed the CPPs in response to matrix metalloproteinase (MMP) cleavage of the shielding moiety [41]. Though these particular dendrimers were formulated with gadolinium for MR imaging as well as fluorophores for in vivo fluorescent imaging, the next logical step would be to conjugate chemotherapeutic agents to the nanoparticles to allow targeted treatment as well.

Carbon nanotubes have also been used as targeted nanoparticles. Nanotubes modified with RGD peptides were used to target integrin-positive tumors in a mouse model using the U87MG human glioblastoma cell line. Nanotubes were coated with PEG molecules, which were then bound to the RGD peptide [42]. MicroPET images of tumor-bearing mice demonstrated selective delivery of targeted nanotubes to the tumor site. The use of nanotubes in humans is still questionable, however, since little is known about their long-term fate after injection.

13.3.3 Triggered Release

One of the biggest advantages—and drawbacks—of nanoparticles is their size. They are large enough to exclusively accumulate in areas of leaky vasculature, but too large to readily diffuse once they reach the targeted region, often traveling only as far as 30 μm away from the vessels from which they extravasate [15]. This drawback limits the effects of these particles to the few layers of perivascularly situated cells. If nanoparticles could be fabricated that would release their contents, chemotherapeutics for example, once they have reached their target destination, it would allow the chemotherapeutics to then diffuse within the target tissue and affect more cells. This improvement can be achieved through the use of multifunctional nanoparticles that can release their contents after a triggering effect caused by hyperthermia, pH changes, or enzymes present at the target location. With the addition of targeting molecules for specific tumors and/or the coencapsulation of trackable contrast agents as mentioned above, these nanoparticles could play an important role in the future of personalized medicine because they would allow the physician to control the drug bioavailability in a time-dependent manner that is specific for nanoparticle accumulation within a particular tumor.

13.3.3.1 Hyperthermia-Triggered Release

Several nanoparticles have been created that can be triggered to release their contents in response to physiologically safe increases in temperatures (greater than 37°C, but less than 42°C). Thermally sensitive liposomes for triggered release date back almost half a century, including those developed by Yatvin et al. [43]. These liposomes are stable at body temperature, but become unstable upon heating near or above their melting transition temperature (T_m), allowing their contents to diffuse out. Several studies have proven increased drug release and tumor response over Doxil® and free doxorubicin using different types of thermosensitive doxorubicin-containing liposomes [44–46]. In fact, Needham et al. reported full tumor regression in all 11 of 11 mice treated with doxorubicin-containing thermosensitive liposomes, accompanied by induced hyperthermia in a human xenograft mouse model [47].

Current examples of hyperthermia treatments like the one above require the patient to submerge the area of interest in a water bath to generate large regions of hyperthermia in the general vicinity of the target site [47, 48]. Though useful and easily translatable to a clinical setting, there are several disadvantages with this technique. First of all, one can only treat the regions on the body that are able to be submerged in hot water for an extended period of time. This rules out hyperthermia treatment for areas such as the torso (likely that the entire body would need to be submerged, negating selectivity) and areas that are highly temperature sensitive such as the brain. Also, this hyperthermia treatment is anything but local to the tumor area since large portions of the body around the tumor must be submerged, meaning that thermosensitive nanoparticle therapeutics in off-target areas could be released, causing undesirable side effects. Tumor-specific hyperthermia would be ideal when considering heat-triggered release and can be accomplished using iron oxide or gold nanoparticles.

Iron oxide nanoparticles, traditionally used as MRI contrast agents, can be stimulated by the same MRI magnetic fields to generate heat [49]. These nanoparticles generally are not present in high enough concentration to generate enough heat for cytotoxic hyperthermia but could be used in conjunction with thermally sensitive nanoparticles to induce triggered release. One advantage of using iron oxide nanoparticles for hyperthermia is the ability to image the particles to ensure proper accumulation in the target tissue. Unfortunately, production of such electromagnetic fields to cause heating is rather expensive and requires relatively expensive equipment.

Gold nanoparticles, on the other hand, can be heated using less expensive lasers that require less power and space, and therefore, result in lower operational/maintenance costs, though the frequency of light required for gold nanoparticle heating does not have the ability to penetrate tissue as deeply as the magnetic fields utilized to heat iron oxide nanoparticles. Gold nanorods and shells can be tuned via alteration of dimensional ratios to generate heat in response to stimulation by different frequencies of light [50, 51]. The most useful of these are rods and shells that respond to near-infrared (NIR) radiation. Near-infrared light can penetrate soft tissue up to 10 cm with little loss of energy due to minimal absorption and scattering by intrinsic pigments [52]. Thus, NIR light can be used to stimulate gold nanoparticles to generate local hyperthermia of several degrees, which could be utilized for initiating triggered release of encapsulated agents from thermosensitive nanoparticles. In fact, Paasonen et al. have already provided a proof of concept of this method [50]. Gold nanoparticles were incorporated into calcein-loaded liposomes, which were stable at 37°C. Upon illumination with near-infrared light, localized heating occurred, resulting in calcein release.

13.3.3.2 pH-Triggered Release

Nanoparticles can be engineered to release their contents based on changes in pH. Tumor interstitium, endosomes, and lysosomes are all known to have acidic pH. Thus, pH-sensitive nanoparticles can be used to passively accumulate in tumors based on size and longevity in circulation, then release their contents similarly to thermosensitive liposomes but without the need for an externally applied trigger.

Several types of pH-sensitive liposomes have been developed, again dating back to the mid- to late 1970s (reviewed in [53]). When first conceived, their in vivo use was limited due to the rapid clearance by the RES, and therefore, PEG was later added to the liposome surfaces to ensure longer circulation times [54]. Several examples of pH-sensitive liposomes have since been developed, including liposomes that carry cell-targeting moieties. Ishida et al. created a pH-sensitive doxorubicin liposome with a CD19-targeting moiety for targeting human B cell lymphoma [55]. This liposomal doxorubicin was stable at physiological pH (7.4), but was destabilized and released its contents at pH 5.5 in lysosomes. Despite its relatively rapid clearance from the blood stream, Ishida's multifunctional nanoparticle was more effective in killing target lymphoma cells than more traditional, long-circulating doxorubicin liposomes.

Carbon nanotubes have also been used to reversibly package drugs for delivery to tumors based on changes in pH [56]. Doxorubicin was bound to PEG-coated nanotubes in a pH-dependent manner. The binding was stable at physiological pH, but allowed release of the drug in acidic conditions like those present in endosomes or lysosomes. When treated with these nanoparticles, MCF-7 breast cancer cells and U87MG human glioblastoma cells showed high levels of cell death in vitro. In vivo use outside of a laboratory environment relies on the further investigation of the long-term effects of carbon nanotubes in the body.

13.3.3.3 Enzymatic-Triggered Release

Enzymatic-triggered release of nanoparticles takes advantage of enzymes located at the target site for the release of encapsulated agents. These nanoparticles could be utilized in personalized medicine by tailoring the type of enzyme cleavage site to the area of interest in a particular patient.

Matrix metalloproteinases (MMPs) are often overexpressed in tumors. The inclusion of MMP cleavage sites in the PEG coating on liposomes or in the membrane itself allows for enzyme-triggered release of contents. PEG is utilized to ensure RES evasion and long circulation times. Once the nanoparticles extravasate to the tumor site, cleavage of MMP sites in the PEG coating can expose targeting moieties or allow the liposomes to fuse with the target cell membranes, releasing their contents into the cytoplasm [57]. Lipoproteins can be included in the membrane that causes destabilization of the liposome bilayer when cleaved by MMPs, thus allowing release of contents into the extracellular space where they can diffuse throughout the tissue [58].

13.3.4 DNA/RNA Therapy

Recent research has uncovered an important link between cancer biomarkers and gene expression, and the pathology of tumors [59–61]. Thus, it would be advantageous to be able to control the expression of malicious genes through personalized or individualized gene therapy, either through DNA modifications or through RNA silencing. Several multifunctional nanoparticles are in development for this very task. High throughput methods exist to obtain gene expression and protein maps of tumors, including cDNA, RNA, and protein microarrays. These methods could be utilized to determine which gene expression to extinguish or silence. Nanoparticles could then be delivered to achieve the desired effect, enabling personalized gene therapy.

Delivery of DNA into mammalian cells is not a new concept. Various methods such as electroporation and use of viruses exist for transfecting mammalian cells. Multifunctional nanoparticles can also be exploited to deliver DNA. Legendre et al. used pH-sensitive nanoparticles to deliver DNA to mammalian cells, exploiting protective packaging of DNA and intracellular pH-triggered release to ensure transfection [62]. The group successfully transfected cells in vitro with a plasmid containing a luciferase reporter gene. Many other multifunctional nanoparticles have been used to achieve similar goals. Several examples are given below, with emphasis on their potential for use in personalized cancer therapy.

Tumors require rapid growth of vasculature to sustain their growth. Therefore, slowing or ceasing angiogenesis in a tumor could result in death of tumor cells and subsequent tumor regression. For example, Avastin® (bevacizumab) is a monoclonal antibody that binds VEGF and slows angiogenesis. An alternative approach to combat tumor angiogenesis is gene therapy. Hood et al. used a cationic nanoparticle, coupled with $\alpha_V\beta_{III}$ integrin targeting ligands, for targeted delivery of mutated Raf gene to tumor endothelial cells in a mouse model of M21 human melanoma [63]. Neovasculature endothelial cells overexpress $\alpha_V\beta_{III}$, making it an ideal target for delivery of therapeutics to these cells. Delivering mutated Raf to the neovasculature caused apoptosis of endothelial cells, which resulted in death of cancer cells near these vessels, and subsequent tumor regression.

Carbon nanotubes have been used effectively to deliver DNA in vivo. Folic acid-coated nanotubes were used to deliver small DNA segments into folate receptor overexpressing HeLa cells [64]. The cells were then exposed to pulses of NIR radiation to cause endosomal release of the DNA nanotubes, which allowed DNA to then enter the nucleus of the cells. Minimal toxicity of the nanotubes was observed, but the long-term effects of carbon nanotubes have yet to be discovered.

Reducing gene expression through RNA silencing has great appeal in the biomedical sciences due to its transient nature and safety compared to permanent alteration of cellular DNA. Thus, delivery of siRNA to tumor cells could be used for personalized cancer therapy. Schiffelers et al. achieved tumor-selective delivery of VEGFR expression silencing siRNA via an RGD-targeted polymer nanoparticle [65]. Intravenous injection of the nanoparticles in a mouse N2A tumor model resulted in successful targeted delivery, causing reduced angiogenesis and reduced tumor growth.

Carbon nanotubes have been used to deliver siRNA to silence telomerase reverse transcriptase (TERT) function in a murine tumor model [66]. Nanotubes with the anti-TERT siRNA induced tumor cell growth arrest, resulting in longer survival times for nude mice inoculated with HeLa tumors. Other uses for RNAi nanoparticles can be found in a recent review by Kedmi and Peer [67].

13.4 Future Perspectives

The near future of personalized medicine lies in the further development and mass production of trimodal nanoparticles that can not only deliver desired therapeutics to the specific area of interest through passive or active targeting but also treat and allow real-time imaging of treatment via clinically relevant imaging modalities such as MRI, CT, and mammography. An ideal treatment scheme would be one where nanoparticles could be used for molecular imaging to obtain a profile of a tumor without invasive biopsy. The molecular profile could be used to design multifunctional nanoparticles to ensure highly specific and efficient delivery to that patient's tumor.

These personalized trimodal nanoparticles with targeting ligands, chemotherapeutic, and imaging contrast agent would allow rapid assessment of treatment efficacy when compared to current strategies. Preliminary examples of this strategy have been developed using polymer nanoparticles [68], magnetic nanoparticles [69–72], gold nanoparticles [73], and hybrid nanoparticles [74], although long-term effects and toxicity of these nanoparticles must be assessed before safe use in humans can be considered.

One way to achieve clinically relevant personalized nanoparticle chemotherapy described above would be to design a "mix and match" nanoparticle system, in which a clinician would be able to manufacture personalized nanoparticles in house from a set of building blocks. Perhaps there would be a base set of nanoparticles containing different drugs and contrast agents, having specific surface modifications present that would enable the attachment of one or several different targeting ligands from a catalog of choices off the shelf. Clinicians would be able to design and manufacture the nanoparticle that would best apply to the individual patient's disease.

Though this chapter has focused on the application of personalized medicine to cancer, there are several other applications for this technology that are showing promise in the academic sector. Liposomes with $\alpha_V\beta_{III}$ targeting have been used to track progression of atherosclerotic plaques in vivo [75, 76]. Nanoparticles like these could be used to apply personalized treatment strategies based on plaque stability and other data. Cellular tracking using nanoparticles has enabled scientists to track stem cells, macrophages, and other cell types in the laboratory [77–80]. This tracking ability could have future implications in personalized cellular therapies, including stem cell therapies, among other applications.

13.5 Conclusions

Multifunctional nanoparticles show great promise for use in personalized medicine. Nanoparticles that target, treat, and allow a means of in vivo monitoring of treatment are being developed for diseases such as cancer. Nanoparticles such as liposomes will likely be swift to market because they mimic structures already in the body and several formulations of liposomes have already been approved for clinical use. Other molecules show promise but will need to undergo stringent testing for long-term toxicity and biocompatibility.

The future of personalized medicine, however, lies not only in the hands of researchers and academia who are developing novel diagnostics and treatments but also in the hands of industry, policy makers, and government agencies like the US Food and Drug Administration (FDA) and National Institutes of Health (NIH).

The FDA and NIH are expanding legislation, regulations, and guidelines to encourage growth of personalized medicine. The FDA is contributing by forming the Critical Paths Initiative (CPI), which aims to discover biomarkers and develop diagnostic tools, and the Voluntary Genomic Data Submission (VGDS) program. The NIH is also contributing to the growth by creating the Therapeutics for Rare and Neglected Diseases (TRND) program, which aims to expedite preclinical development of promising therapeutic compounds for personalized medicine. Together, these agencies aim to bring personalized medicine to patients as swiftly and safely as possible.

Acknowledgments Funding from the National Institutes of Health (NCI, CA 153229), National Science Foundation (CBET 0756567), Georgia Cancer Coalition, Coulter Foundation, and Ian's Friends Foundation is acknowledged.

References

1. Hinestrosa MC et al (2007) Shaping the future of biomarker research in breast cancer to ensure clinical relevance. Nat Rev Cancer 7(4):309–315
2. Levenson VV (2007) Biomarkers for early detection of breast cancer: what, when, and where? Biochim Biophys Acta 1770(6):847–856
3. Hopkins TG, Burns PA, Routledge MN (2007) DNA methylation of GSTP1 as biomarker in diagnosis of prostate cancer. Urology 69(1):11–16
4. Ballou B, Ernst LA, Waggoner AS (2005) Fluorescence imaging of tumors in vivo. Curr Med Chem 12(7):795–805
5. Fukumura D, Jain RK (2007) Tumor microenvironment abnormalities: causes, consequences, and strategies to normalize. J Cell Biochem 101(4):937–949
6. Hobbs SK et al (1998) Regulation of transport pathways in tumor vessels: role of tumor type and microenvironment. Proc Natl Acad Sci USA 95(8):4607–4612
7. Yuan F et al (1996) Time-dependent vascular regression and permeability changes in established human tumor xenografts induced by an anti-vascular endothelial growth factor/vascular permeability factor antibody. Proc Natl Acad Sci USA 93(25):14765–14770
8. Folkman J (1995) Angiogenesis in cancer, vascular, rheumatoid and other disease. Nat Med 1(1):27–31
9. Padera TP et al (2004) Pathology: cancer cells compress intratumour vessels. Nature 427:695
10. Wolff AC (2003) Liposomal anthracyclines and new treatment approaches for breast cancer. Oncologist 8(Suppl 2):25–30
11. Ferrari M (2005) Cancer nanotechnology: opportunities and challenges. Nat Rev Cancer 5(3):161–171
12. Service RF (2005) Materials and biology. Nanotechnology takes aim at cancer. Science 310:1132–1134
13. Maeda H (2001) SMANCS and polymer-conjugated macromolecular drugs: advantages in cancer chemotherapy. Adv Drug Deliv Rev 46(1–3):169–185
14. Maeda H et al (2000) Tumor vascular permeability and the EPR effect in macromolecular therapeutics: a review. J Control Release 65(1–2):271–284
15. Yuan F et al (1994) Microvascular permeability and interstitial penetration of sterically stabilized (stealth) liposomes in a human tumor xenograft. Cancer Res 54(13):3352–3356
16. Karathanasis E et al (2009) Imaging nanoprobe for prediction of outcome of nanoparticle chemotherapy by using mammography. Radiology 250(2):398–406
17. Karathanasis E et al (2008) Multifunctional nanocarriers for mammographic quantification of tumor dosing and prognosis of breast cancer therapy. Biomaterials 29(36):4815–4822
18. Turetschek K et al (2001) MR imaging characterization of microvessels in experimental breast tumors by using a particulate contrast agent with histopathologic correlation. Radiology 218(2):562–569

19. Elston C, Ellis I (1991) Pathological prognostic factors in breast cancer. I. The value of histological grade in breast cancer: experience from a large study with long-term follow-up. Histopathology 19(5):403–410
20. Weidner N et al (1991) Tumor angiogenesis and metastasis—correlation in invasive breast carcinoma. N Engl J Med 324(1):1–8
21. Horak E et al (1992) Angiogenesis, assessed by platelet/endothelial cell adhesion molecule antibodies, as indicator of node metastases and survival in breast cancer. Lancet 340:1120–1124
22. Carrau R et al (1995) Tumor angiogenesis as a predictor of tumor aggressiveness and metastatic potential in squamous cell carcinoma of the head and neck. Invasion Metastasis 15(5–6):197–202
23. Meitar D et al (1996) Tumor angiogenesis correlates with metastatic disease, N-myc amplification, and poor outcome in human neuroblastoma. J Clin Oncol 14(2):405–414
24. Takebayashi Y et al (1996) Angiogenesis as an unfavorable prognostic factor in human colorectal carcinoma. Cancer 78(2):226–231
25. Brooks P, Clark R, Cheresh D (1994) Requirement of vascular integrin alpha v beta 3 for angiogenesis. Science 264:569–571
26. Brooks P et al (1995) Antiintegrin alpha v beta 3 blocks human breast cancer growth and angiogenesis in human skin. J Clin Invest 96(4):1815–1822
27. Gladson C (1996) Expression of integrin alpha v beta 3 in small blood vessels of glioblastoma tumors. J Neuropathol Exp Neurol 55(11):1143–1149
28. Sipkins D et al (1998) Detection of tumor angiogenesis in vivo by alpha v beta 3-targeted magnetic resonance imaging. Nat Med 4(5):623–626
29. Karathanasis E et al (2009) Tumor vascular permeability to a nanoprobe correlates to tumor-specific expression levels of angiogenic markers. PLoS One 4(6):e5843
30. Cao Y, Jin R, Mirkin C (2002) Nanoparticles with Raman spectroscopic fingerprints for DNA and RNA detection. Science 297:1536–1540
31. Alivisatos P (2003) The use of nanocrystals in biological detection. Nat Biotechnol 22(1):47–52
32. Sinha R et al (2006) Nanotechnology in cancer therapeutics: bioconjugated nanoparticles for drug delivery. Mol Cancer Ther 5(8):1909–1917
33. Yezhelyev M et al (2006) Emerging use of nanoparticles in diagnosis and treatment of breast cancer. Lancet Oncol 7(8):657–667
34. Saul JM et al (2003) Controlled targeting of liposomal doxorubicin via the folate receptor in vitro. J Control Release 92(1–2):49–67
35. McNeeley K, Annapragada AV, Bellamkonda RV (2007) Decreased circulation time offsets increased efficacy of PEGylated nanocarriers targeting folate receptors of glioma. Nanotechnology 18:385101
36. McNeeley KM et al (2009) Masking and triggered unmasking of targeting ligands on nanocarriers to improve drug delivery to brain tumors. Biomaterials 30(23–24):3986–3995
37. Kale A, Torchilin V (2007) "Smart" drug carriers: PEGylated TATp-modified pH-sensitive liposomes. J Liposome Res 17(3–4):197–203
38. Saul JM, Annapragada AV, Bellamkonda RV (2006) A dual-ligand approach for enhancing targeting selectivity of therapeutic nanocarriers. J Control Release 114(3):277–287
39. Majoros I et al (2006) PAMAM dendrimer-based multifunctional conjugate for cancer therapy: Synthesis, characterization, and functionality. Biomacromolecules 7(2):572–579
40. Kukowska-Latallo J et al (2005) Nanoparticle targeting of anticancer drug improves therapeutic response in animal model of human epithelial cancer. Cancer Res 65(12):5317–5324
41. Olson ES et al (2010) Activatable cell penetrating peptides linked to nanoparticles as dual probes for in vivo fluorescence and MR imaging of proteases. Proc Natl Acad Sci USA 107(9):4311–4316
42. Liu Z et al (2007) In vivo biodistribution and highly efficient tumour targeting of carbon nanotubes in mice. Nat Nanotechnol 2(1):47–52

43. Yatvin M et al (1978) Design of liposomes for enhanced local release of drugs by hyperthermia. Science 202:1290–1293
44. Maruyama K et al (1993) Enhanced delivery of doxorubicin to tumor by long-circulating thermosensitive liposomes and local hyperthermia. Biochim Biophys Acta-Biomembranes 1149(2):209–216
45. Unezaki S et al (1994) Enhanced delivery and antitumor activity of doxorubicin using long-circulating thermosensitive liposomes containing amphipathic polyethylene glycol in combination with local hyperthermia. Pharm Res 11(8):1180–1185
46. Gaber M et al (1996) Thermosensitive liposomes: extravasation and release of contents in tumor microvascular networks. Int J Radiat Oncol Biol Phys 36(5):1177–1187
47. Needham D et al (2000) A new temperature-sensitive liposome for use with mild hyperthermia: characterization and testing in a human tumor xenograft model. Cancer Res 60(5):1197–1201
48. Li L et al (2010) Triggered content release from optimized stealth thermosensitive liposomes using mild hyperthermia. J Control Release 143(2):274–279
49. Johannsen M et al (2007) Morbidity and quality of life during thermotherapy using magnetic nanoparticles in locally recurrent prostate cancer: results of a prospective phase I trial. Int J Hyperthermia 23(3):315–323
50. Paasonen L et al (2007) Gold nanoparticles enable selective light-induced contents release from liposomes. J Control Release 122(1):86–93
51. Wu G et al (2008) Remotely triggered liposome release by near-infrared light absorption via hollow gold nanoshells. J Am Chem Soc 130(26):8175–8177
52. Huang X et al (2006) Cancer cell imaging and photothermal therapy in the near-infrared region by using gold nanorods. J Am Chem Soc 128(6):2115–2120
53. Drummond DC, Zignani M, Leroux JC (2000) Current status of pH-sensitive liposomes in drug delivery. Prog Lipid Res 39(5):409–460
54. Slepushkin V et al (1997) Sterically stabilized pH-sensitive liposomes. J Biol Chem 272(4):2382–2388
55. Ishida T et al (2001) Targeted delivery and triggered release of liposomal doxorubicin enhances cytotoxicity against human B lymphoma cells. Biochim Biophys Acta-Biomembranes 1515(2):144–158
56. Liu Z et al (2007) Supramolecular chemistry on water-soluble carbon nanotubes for drug loading and delivery. ACS Nano 1(1):50–56
57. Terada T et al (2006) Novel PEG-matrix metalloproteinase-2 cleavable peptide-lipid containing galactosylated liposomes for hepatocellular carcinoma-selective targeting. J Control Release 111(3):333–342
58. Elegbede A et al (2008) Mechanistic studies of the triggered release of liposomal contents by matrix metalloproteinase-9. J Am Chem Soc 130(32):10633–10642
59. Golub T et al (1999) Molecular classification of cancer: class discovery and class prediction by gene expression monitoring. Science 286:531–537
60. Ross D et al (2000) Systematic variation in gene expression patterns in human cancer cell lines. Nat Genet 24(3):227–235
61. Alizadeh A et al (2000) Distinct types of diffuse large B-cell lymphoma identified by gene expression profiling. Nature 403:503–511
62. Legendre J, Szoka F Jr (1992) Delivery of plasmid DNA into mammalian cell lines using pH-sensitive liposomes: comparison with cationic liposomes. Pharm Res 9(10):1235–1242
63. Hood J et al (2002) Tumor regression by targeted gene delivery to the neovasculature. Science 296:2404–2407
64. Kam N et al (2005) Carbon nanotubes as multifunctional biological transporters and near-infrared agents for selective cancer cell destruction. Proc Natl Acad Sci USA 102(33):11600–11605
65. Schiffelers R et al (2004) Cancer siRNA therapy by tumor selective delivery with ligand-targeted sterically stabilized nanoparticle. Nucleic Acids Res 32(19):e149
66. Zhang Z et al (2006) Delivery of telomerase reverse transcriptase small interfering RNA in complex with positively charged single-walled carbon nanotubes suppresses tumor growth. Clin Cancer Res 12(16):4933–4939

67. Kedmi R, Peer D (2009) RNAi nanoparticles in the service of personalized medicine. Nanomedicine (Lond) 4(8):853–855
68. Pridgen E, Langer R, Farokhzad O (2007) Biodegradable, polymeric nanoparticle delivery systems for cancer therapy. Nanomedicine 2(5):669–680
69. Kohler N et al (2005) Methotrexate-modified superparamagnetic nanoparticles and their intracellular uptake into human cancer cells. Langmuir 21(19):8858–8864
70. Kohler N et al (2006) Methotrexate immobilized poly (ethylene glycol) magnetic nanoparticles for MR imaging and drug delivery. Small 2(6):785–792
71. Yu M et al (2008) Drug loaded superparamagnetic iron oxide nanoparticles for combined cancer imaging and therapy in vivo. Angew Chem Int Ed 47(29):5362–5365
72. Wang A et al (2008) Superparamagnetic iron oxide nanoparticle-aptamer bioconjugates for combined prostate cancer imaging and therapy. ChemMedChem 3(9):1311–1315
73. Park H et al (2008) Multifunctional nanoparticles for photothermally controlled drug delivery and magnetic resonance imaging enhancement. Small 4(2):192–196
74. Yang J et al (2007) Multifunctional magneto polymeric nanohybrids for targeted detection and synergistic therapeutic effects on breast cancer. Angew Chem Int Ed 46(46):8836–8839
75. Winter P et al (2003) Molecular imaging of angiogenesis in nascent Vx-2 rabbit tumors using a novel 3-targeted nanoparticle and 1.5 Tesla magnetic resonance imaging. Cancer Res 63(18):5838–5843
76. Winter P et al (2006) Endothelial alpha v beta 3 integrin-targeted fumagillin nanoparticles inhibit angiogenesis in atherosclerosis. Arterioscler Thromb Vasc Biol 26(9):2103–2109
77. Alivisatos A, Gu W, Larabell C (2005) Quantum dots as cellular probes. Biomed Eng 7:55–76
78. Gao X et al (2005) In vivo molecular and cellular imaging with quantum dots. Curr Opin Biotechnol 16(1):63–72
79. Michalet X et al (2005) Quantum dots for live cells, in vivo imaging, and diagnostics. Science 307:538–544
80. Slotkin J et al (2007) Cellular magnetic resonance imaging: nanometer and micrometer size particles for noninvasive cell localization. Neurotherapeutics 4(3):428–433

Chapter 14
Dendrimer-Based Nanoparticle Therapies: Can Uniform Multifunctional Therapeutics Be Made with Current Chemical Approaches?

Douglas G. Mullen, Daniel Q. McNerny, Mark M. Banaszak Holl, and James R. Baker Jr.

14.1 Introduction

First introduced by Tomalia, Newkome, and Frechet in the 1970s, dendrimers have been investigated for a number of applications including photonics [1], antimicrobial agents [2–4], gene transfection [5–7], contrast imaging agents [8, 9], boron-neutron capture therapy [10], and targeted drug delivery [11, 12]. Dendrimers have a highly ordered branching architecture that results in low degrees of polydispersity compared to many other polymers.

One class of dendrimer, the poly(amidoamine) (PAMAM) dendrimer, has many characteristics that make the material well suited to be a platform for multifunctional therapeutics. PAMAM dendrimers are non-immunogenic and have good biocompatibility once their surface amines have been neutralized [13–18]. Their small size (~4.5 nm for a generation 5 dendrimer) is similar to many proteins in the body and this enables the particles to escape the vasculature and pass through the tissue matrix to access the desired target. The flexible nature of the dendrimer backbone also facilitates flow through vascular pores as well as promotes polyvalent interactions with the desired cellular target [19]. Since generations of PAMAM lower than 5 are smaller than the maximum particle size for renal filtration (~15 nm), these dendrimers are filtered through the kidney, therefore reducing the risk of long-term toxicity and removing the need for a degradable scaffold [20]. The hydrophilic nature of the PAMAM backbone maintains water solubility even when the dendrimer surface is joined to a large number of hydrophobic ligands. In fact, the PAMAM dendrimer can enhance therapeutic compound solubility at physiological conditions. In addition, the well-defined number of end groups on the dendrimer enables multivalent

D.G. Mullen • D.Q. McNerny • M.M.B. Holl • J.R. Baker Jr. (✉)
Michigan Nanotechnology Institute for Medicine and Biological Sciences,
University of Michigan, 1150 W Medical Center Dr., Ann Arbor, MI 48109, USA
e-mail: jbakerjr@umich.edu

S. Svenson and R.K. Prud'homme (eds.), *Multifunctional Nanoparticles for Drug Delivery Applications: Imaging, Targeting, and Delivery*, Nanostructure Science and Technology,
DOI 10.1007/978-1-4614-2305-8_14,

interactions with biological systems because multiple copies of a monovalent targeting ligand can be conjugated to each dendrimer. The existence of multiple end groups per dendrimer also provides the ability to conjugate several different functional molecules to a single dendrimer, thereby producing the multifunctional therapeutic.

These unique features have been leveraged to produce dendrimer therapeutics that combine a therapeutic agent, a targeting ligand, and in many cases, an imaging agent. It is hoped that the combination of these functionalities on the same particle will result in synergistic properties that include reduced nonspecific cytotoxicity vs. free drug, increased drug delivery to target cells, and the potential to overcome drug resistance in target cells. Multifunctional dendrimer conjugates have been shown to successfully target cells with a number of different membrane receptors including the EGF receptor [21], RGD-binding integrins [22, 23], HER2 receptor [24], PSMA receptor [25], and FAα receptor [26–28]. Therapeutic agents that have been successfully delivered to target cells in these systems include doxorubicin, methotrexate (MTX), and taxol. This chapter describes our efforts to translate one such multifunctional dendrimer, a generation 5 (G5) dendrimer that combines the targeting molecules folic acid (FA) and the therapeutic agent methotrexate.

14.2 Laboratory Development of a Multifunctional Dendrimer Therapeutic (G5-FA-MTX)

Of the multifunctional dendrimers synthesized to date, one of the most promising therapeutics has been a dendrimer that combined folic acid as a targeting ligand with the anti-proliferative agent methotrexate. Folic acid is an essential vitamin that is used by cells in DNA synthesis. This molecule binds to the FAα receptor, which is a cellular membrane-bound receptor, as well as the Reduced Folate Carrier (RFC), which is an ion channel-like molecule in the cell membrane. Due to the demand for a high rate of DNA synthesis by rapidly dividing cells, many types of cancer cells overexpress the FAα receptor. Tumor cells with a high-membrane concentration of the receptor include breast, ovary, endometrium, kidney, lung, head and neck, brain, and myeloid cancer cells [29–31]. A single FA molecule binds specifically to the FAα, but has relatively low affinity for this receptor [32]. However, when multiple copies of FA are conjugated to a dendrimer, the platform (targeting ligands and dendrimer) was found to be capable of high-avidity interactions with the membrane-bound FAα receptor. Quantitative Surface Plasmon Resonance (SPR) studies found the binding interaction of surface receptor with multiple FA molecules bound to a dendrimer is several orders of magnitude stronger than free folic acid [19]. This study also investigated the change in binding strength of folate-conjugated dendrimers based on the mean FA-dendrimer ratio. For dendrimers with mean numbers of FA ranging from 1 to 15, the optimal FA-dendrimer ratio was found to be approximately 4.

The second functional molecule coupled to the dendrimer, MTX, is a cytotoxic drug that functions by binding to dihydrofolate reductase (DHFR), an enzyme

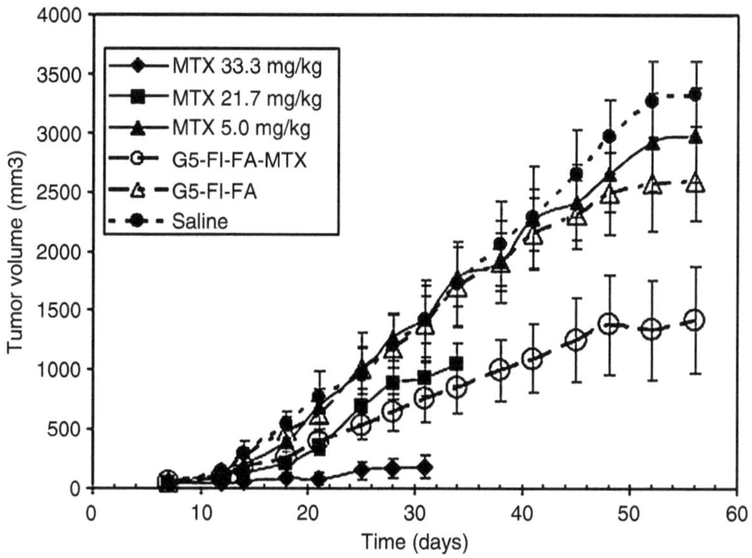

Fig. 14.1 In vivo antitumor activity of the G5-FA-MTX dendrimer synthesized by MNIMBS. The multifunctional dendrimer demonstrated an inhibition of tumor growth in a human epithelial tumor in SCID mice without the toxicity observed with similar concentrations of free MTX. Reproduced with permission from [28]

important in DNA synthesis. It has a higher affinity interaction with DHFR than FA, but inefficiently enters cells because it is only transported by the reduced folate carrier. Because of this, cells can develop resistance to MTX by overexpressing FAα to concentrate more FA in cancer cells, or by removing MTX from the cytoplasm via p-glycoprotein pumps [33].

In combining FA with MTX on the dendrimer, our hypothesis was that we could improve the therapeutic index of MTX by directing the drug with folate to FAα overexpressing cancer cells, while at the same time preventing MTX entry into normal cells through the reduced folate carrier by coupling to the macromolecular dendrimer. The prototype compound developed within MNIMBS had a mean ligand/dendrimer ratio of 4 FA and 5 MTX molecules per dendrimer [27]. Both in vitro and in vivo results of this compound showed that it did improve the therapeutic index for cancer therapy as it was highly successful at targeting and internalizing into cancer cells and then killing cells by inhibition of the DHFR [28]. Shown in Fig. 14.1 are the results of an in vitro and in vivo study that demonstrated the excellent antitumor properties of the G5-FA-MTX compound.

It is worth reviewing the somewhat complex synthetic path required to produce the G5-FA-MTX. A serial conjugation strategy was employed and is displayed in Fig. 14.2. Commercially supplied dendrimer was first purified by dialysis to remove small molecular weight impurities and improve the uniformity of the material. This material was then painstakingly characterized by nuclear magnetic resonance spectroscopy (NMR), gel permeation chromatography (GPC), and potentiometric

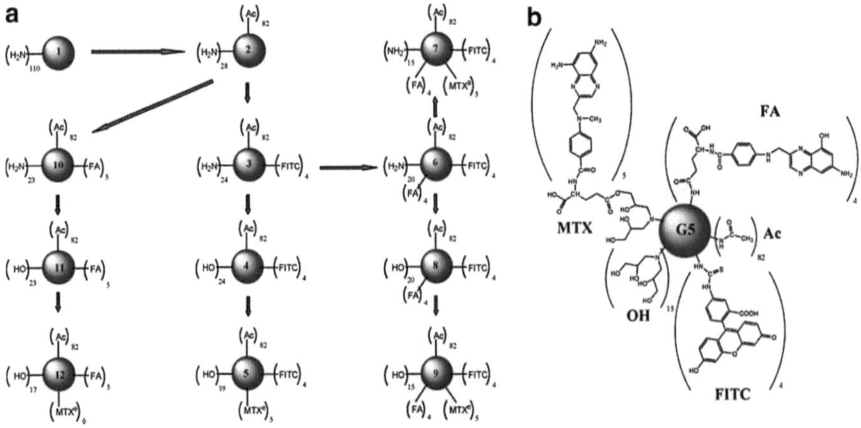

Fig. 14.2 The multistep synthetic process to synthesize the G5-FA-MTX dendrimer and controls. This particular version of the multifunctional dendrimer also incorporated the dye molecules FITC on the platform. Reproduced with permission from [27]

titration to establish the number average molecular weight (M_n) of the material as well as the average number of end groups per dendrimer. The dendrimer was then partially acetylated to reduce the number of surface amines on the dendrimer. Using activated-ester coupling chemistry (EDC), FA was then attached to the dendrimer through an amide bond through one (or the other) of its carboxylic acid groups. The remaining primary amines were then glycidolated in order to add MTX, which was also accomplished with EDC coupling through an ester bond.

14.3 Problems in Translation

The performance of the G5-FA-MTX dendrimer in ex vivo testing was extremely promising and a company called Avidimer Therapeutics was formed to continue development of the multifunctional dendrimer therapeutic and advance the G5-FA-MTX compound through clinical trials. Avidimer began preparing material for preclinical GLP (good laboratory practices) toxicity testing to support an Investigative New Drug (IND) filing with the U.S. Food and Drug Administration (FDA). Over five lots of the multifunctional dendrimer were synthesized and analyzed by the company to document the reproducibility of the production before synthesis was transferred to a contract manufacturer, who produced an additional five lots including one 200-g lot for GLP animal toxicity testing. This final lot of material successfully completed a 4-week repeated dose toxicity testing in rats and dogs.

As final preparation for Phase I clinical trials, a 2-kg lot of GMP (good manufacturing principals) material was produced, called ATI-101. Unfortunately, several factors that were unknown or not appreciated came into play to alter the product of this synthesis with significant consequences. The commercial supplier of the PAMAM dendrimer had altered its synthesis protocol to produce the starting material for ATI-101. While Avidimer was not originally aware of the changes in the protocol, it had been informed that the mean primary amine to dendrimer ratio was substantially lower than previous lots (90 vs. 112). In an attempt to compensate, Avidimer reduced its acetylation stoichiometry but was not aware that the PDI (polydispersity index) of the dendrimer was also significantly increased compared to prior lots. Another factor that was underappreciated was the substantial heterogeneity in the ratios of functional ligands (FA and MTX)-to-dendrimer particles, although defining this heterogeneity has become a major focus of our research efforts since ATI-101 was synthesized. A third unknown factor was the effects of the partial dendrimer acetylation reaction on the quality of mass transport and the subsequent acetyl-dendrimer distributions. As will be described in detail later in this chapter, changes in the mass transport can have major effects on the acetyl-dendrimer distribution and leave many dendrimers with very few primary amines available for conjugation with either FA or MTX.

The consequence of these factors resulted in ATI-101 having poor antitumor activity. Figure 14.3 shows the failure of ATI-101 to demonstrate the antitumor activity both in vitro and in vivo. Characterization of ATI-101 by both the MNIMBS analytical staff and the National Cancer Institute's Nanoparticle Characterization Laboratory (NCL) determined that only a small amount of FA had been conjugated to the dendrimer. This meant that the compound was no longer capable of the high avidity, multivalent targeting that was essential to the material's success. MNIMBS recognized that these problems were clearly preventing translation of the multifunctional dendrimer, and that a successful platform required that these hurdles be fully characterized, and overcome, to create a uniform, large-scale therapeutic.

14.4 Challenges to Current Synthetic Approaches of Dendrimer Therapeutics

In reviewing our experience with the G5-FA-MTX dendrimer and the failure of ATI-101, we identified four major challenges associated with the scale-up of the multifunctional dendrimer. The first challenge was controlling the heterogeneity of the starting dendrimer to minimize the adverse effects it had on the reproducible synthesis of the multifunctional therapeutic. The second challenge was developing analytical methods capable of determining the material composition, particularly the different ligand–dendrimer ratios, and thereby evaluating the consistency between batches. A third major challenge was developing synthetic strategies that reproducibly generated multifunctional dendrimer. The fourth and final challenge to successful translation was controlling the therapeutics' composition during scale-up.

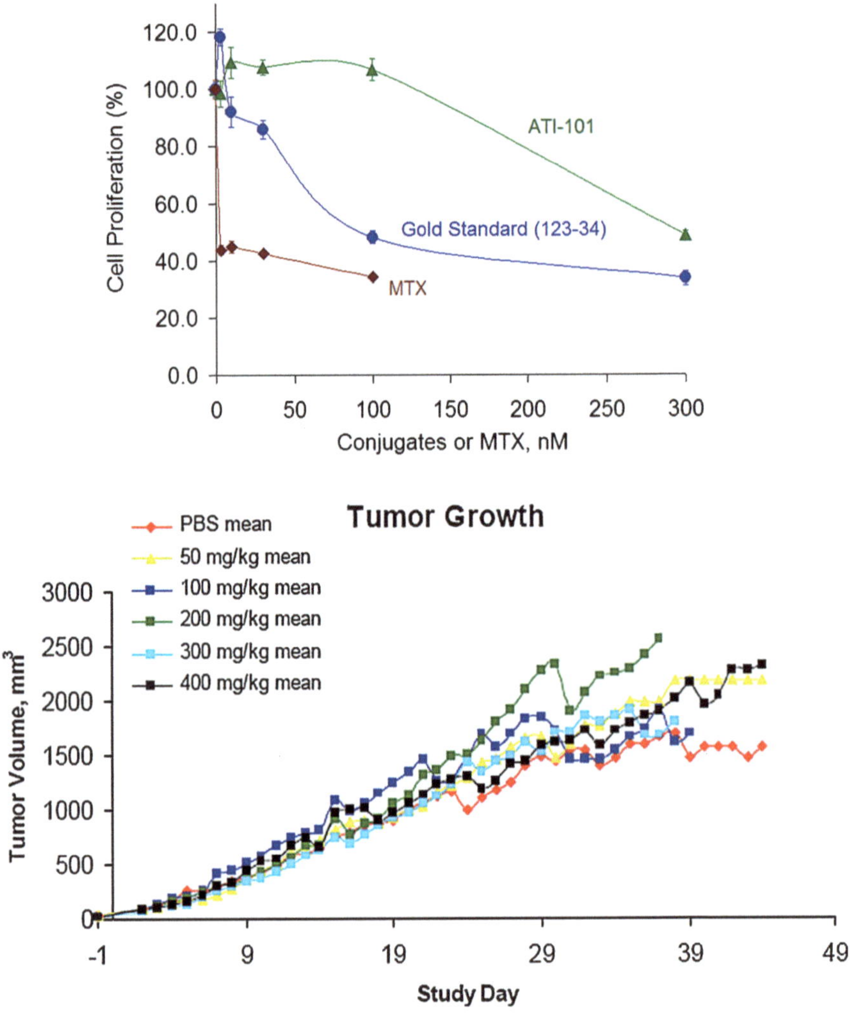

Fig. 14.3 The lack of antitumor activity of the 2-kg scale multifunctional dendrimer (ATI-101) compared with the antitumor capability of a 200-g scale lot of material (lot # 123-34). The lack of activity was found both in vitro (*top*) and in a SCID mice tumor model (*bottom*)

14.4.1 Variability and Heterogeneity of the Starting Dendrimer

Although dendrimers are among the least heterogeneous family of polymers, they are not monodisperse. This misconception seems to have led researchers to grossly underestimate the influence that dendrimer heterogeneity can have on the reproducibility of synthetic strategies. For PAMAM dendrimers, deviations from the

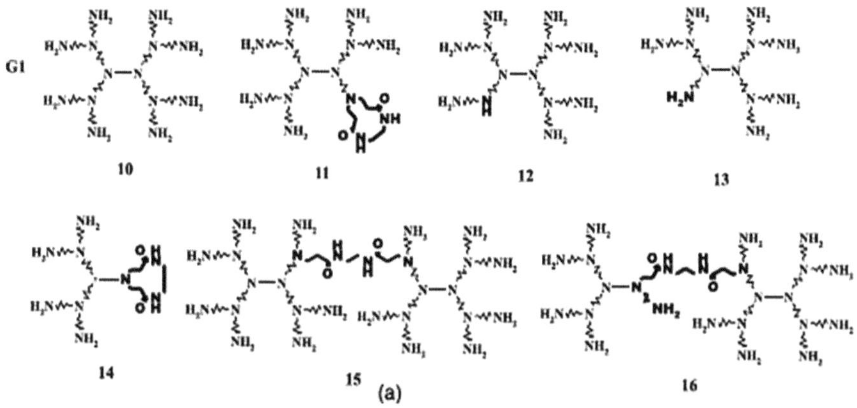

Fig. 14.4 Examples of different dendrimer defects for a generation 1 PAMAM dendrimer. Adapted from [36]

theoretical structure of these molecules are caused by undesired side reactions and site-blocking effects. A number of different defect structures in the PAMAM dendrimer architecture have been identified, including those shown for a generation 1 (G1) dendrimer in Fig. 14.4. These defects in the dendrimer architecture are the source of molecular weight heterogeneity in the starting material and are reflected in the material's polydispersity index. In our experience, the most optimal commercial lots of generation 5 dendrimer have a mean of ~110 primary surface amines per molecule instead of the theoretical 128 end groups per dendrimer. This number is obtained only after a post-purchase purification using dialysis that removes a portion of lower molecular weight defect structures [34].

The challenge that dendrimer heterogeneity presents to the successful translation of multifunctional dendrimer is that the relative amount and type of defect structures is different in each lot of material. If these changes are not well understood, reproducible conjugations of functional groups will become difficult to achieve due to improper reagent stoichiometry, and heterogeneity in the number of attachment sites per dendrimer. Furthermore, changes in the relative amount of defect structure impacts many of the platform's material properties related to biological function such as pharmacodynamics and receptor-ligand interactions (effected by arm flexibility).

The challenge of controlling the heterogeneity and variability of the starting PAMAM dendrimer is apparent in the failure of ATI-101. In this case, the inconsistency of the starting material caused the synthesis protocol to be modified. The change in the mean number of end groups was known (90 instead of 112); however, the change in the relative amount of defect structures was not. In retrospect, it may be that the previously developed synthetic conditions were no longer appropriate for this lot of starting material even when adjusted in a linear relationship to the mean number of end groups per dendrimer. It is also possible that the problems resulted from poorly controlled mass transport in the acetylation step (*vida infra*).

14.4.2 Analytical Methods that Verify Material Composition and Reproducibility

Control over the consistency of the starting material is a challenge that must be addressed in order to achieve reproducible batches of material. There are, however, fundamental consequences of the synthetic methods used to functionalize the dendrimer that must be realized in order to produce reproducible material. Essential to this realization are analytical methods that can actually identify the different ligand–dendrimer ratios that compose a batch of multifunctional dendrimer. Such analytical capabilities are important both to understand material-property relationships and to ensure the reproducibility between synthesis batches. This is a significant challenge for not only the multifunctional dendrimer field, but also the entire field of multifunctional nanoparticles.

The majority of synthetic reactions performed to functionalize dendrimers (partial acetylation, FA conjugation, MTX conjugation) occur where the molar amount of ligand is much smaller than the number of available conjugation sites per dendrimer. For example, in making the G5-FA-MTX dendrimer there was a mean of 35 primary amines per dendrimer and about 4 moles of FA added per dendrimer. Instead of generating material composed of dendrimer particles with exactly 4 conjugated FA, these conditions result in a distribution of FA-dendrimer ratios and a mean of 4 FA per dendrimer. We call this form of heterogeneity the "distribution of ligand–dendrimer components." Unfortunately, the extent of ligand heterogeneity was not initially appreciated when the synthesis of the G5-FA-MTX dendrimer was being developed. This misunderstanding is largely due to a reliance on standard analytical methods that failed to identify the ligand–dendrimer ratios that comprised the material. Many analytical techniques that are commonly used in multifunctional nanoparticle characterization (NMR spectroscopy, elemental analysis, UV-visible spectroscopy) are only capable of identifying the mean ligand/nanoparticle ratio. NMR spectroscopy was (and remains today) one of the primary characterization methods for multifunctional dendrimer. We have found, however, that the mean ligand–dendrimer ratio that NMR spectroscopy provides is insufficient to evaluate material reproducibility. Other techniques with potential to resolve component distributions (GPC, HPLC, MALDI-TOF) often produce unresolved single peaks and fail to provide sufficient information about the material. Figure 14.5 provides one such example in which both GPC (panel a) and MALDI (panel b) were unsuccessful at resolving the ligand–dendrimer ratios. Sample A in this figure was composed of three different ligand–dendrimer ratios, whereas sample D was composed of nine different ratios. Both techniques only produced single-peak signals that completely missed the significant differences in material composition [35]. This type of analytical result is clearly not sufficient to ensure the clinical translation of a multifunctional dendrimer.

Our understanding of actual ligand–dendrimer distributions since the failure of ATI-101 has been informed by the use of two different model ligands; the alkyne ligand (3-(4-(prop-2-ynyloxy)phenyl) propanoic acid) and the azide ligand

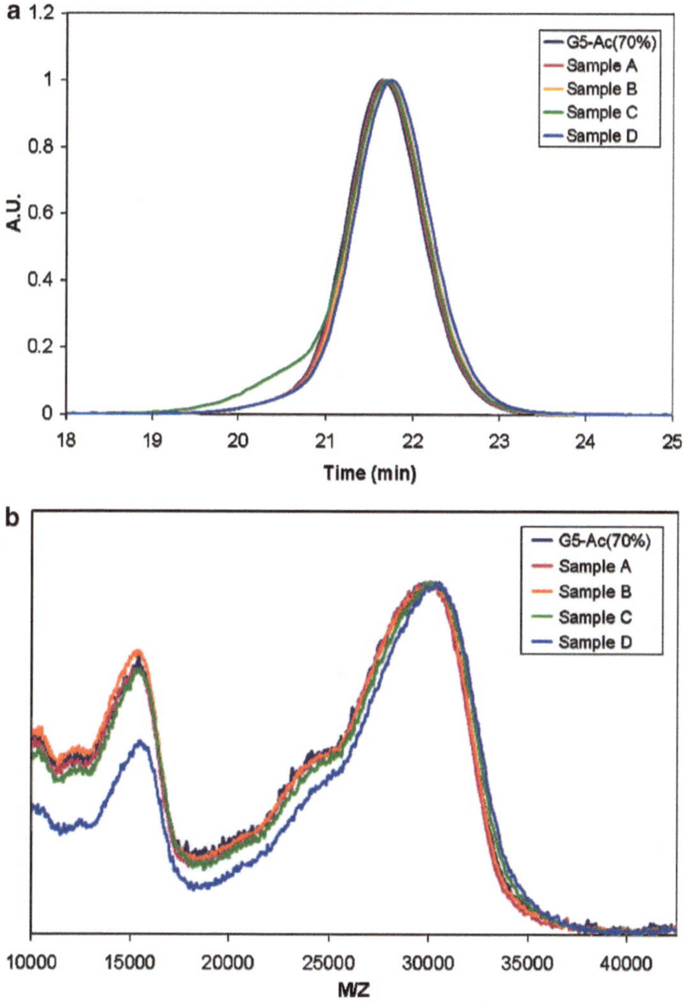

Fig. 14.5 GPC (**a**) and MALDI-TOF (**b**) characterization of four ligand–dendrimer samples. The number of components in samples A–D ranges from 3 to 9, yet both techniques failed to identify these differences. Reproduced with permission from [35]

3(-(4-(2-azidoethoxy)phenyl) propanoic acid) [34]. When either of these ligands was conjugated to the dendrimer, reverse-phase HPLC could be used to resolve and quantify dendrimers with specific amounts of each ligand within a sample. Examples of these HPLC traces can be found in Fig. 14.6. The ligand–dendrimer conjugate in this figure had mean ligand–dendrimer ratios ranging from 0.4 to 12.9. The starting dendrimer for the samples in panel (a) was a dendrimer with a mean of 112 end groups, whereas the dendrimer for samples in panel (b) was a partially acetylated dendrimer. The distinct peaks within each trace were determined to be composed of different numbers of ligand–dendrimer ratios [34–37].

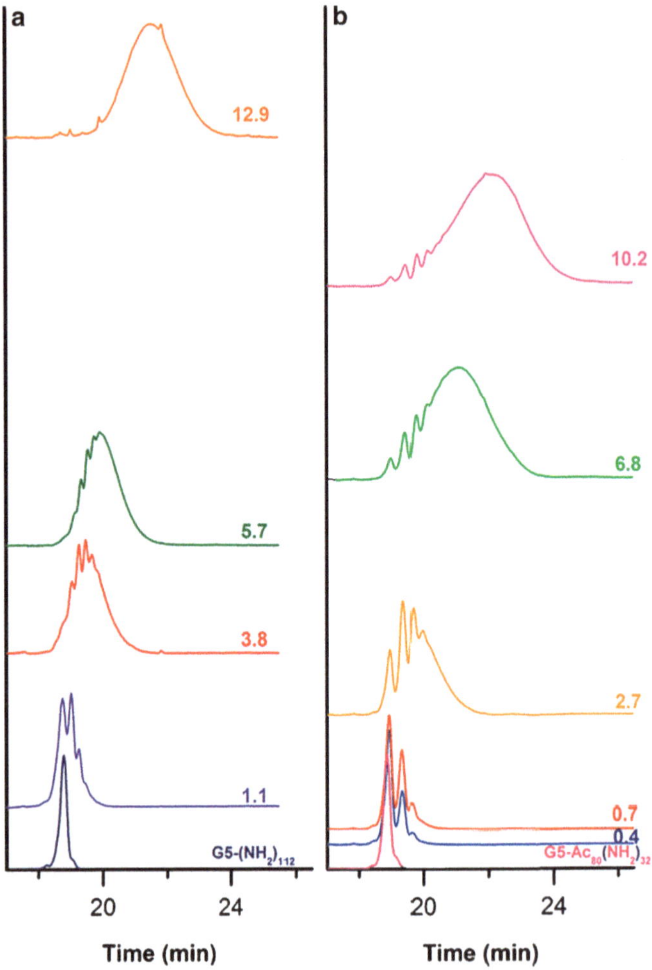

Fig. 14.6 Reverse phase HPLC traces for samples of dendrimers conjugated with the alkyne ligand. Traces are plotted with a vertical off-set that reflects the mean ligand–dendrimer ratio (ranging from 0.4 to 12.9). Two different starting dendrimers were used to make the samples: a dendrimer with a mean 112 primary amines (panel **a**) and a partially acetylated dendrimer with a mean of 32 primary amines (panel **b**). Reproduced with permission from [34]

The quantified distribution was determined by applying a peak-fitting method to each of the HPLC traces, and the results can be found in Fig. 14.7. Although these results are consistent with theoretical expectations [35], they are significantly more heterogeneous than common expectations. As an example, a dendrimer with a mean ligand–dendrimer ratio of 6.8 was composed of 18 different components. These components ranged from a dendrimer with 0 ligands to a dendrimer with 17 ligands. No component in this material made up more than 9% of the total population.

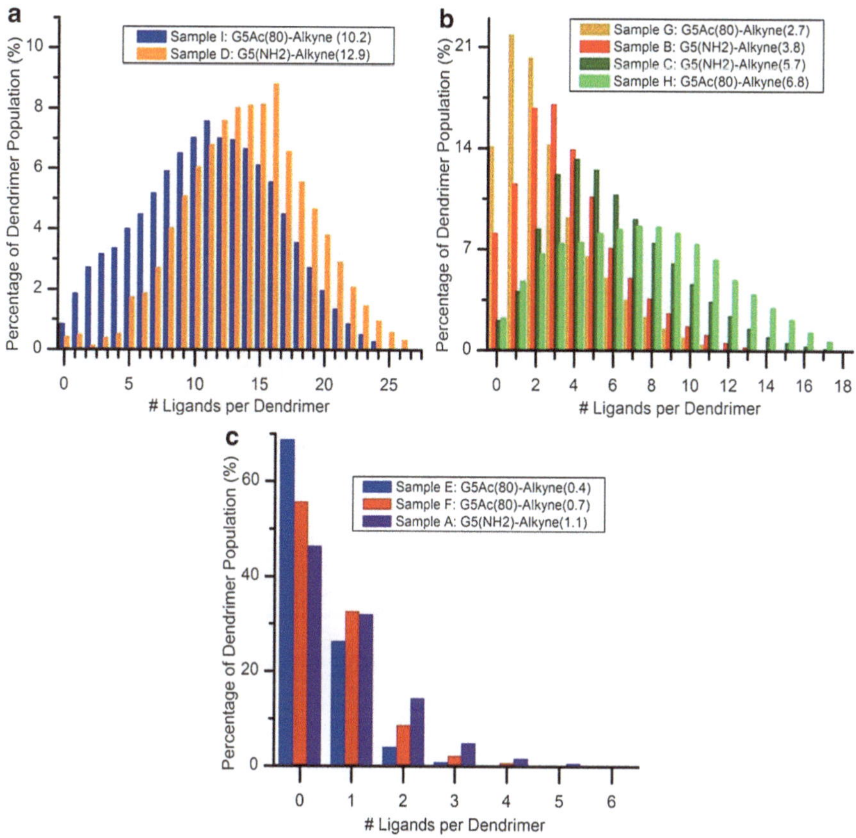

Fig. 14.7 Quantified dendrimer–ligand distributions based on the resolved HPLC traces in Fig. 14.5. Distributions are grouped based on the mean ligand–dendrimer ratio. Reproduced with permission from [34]

Possessing an analytical method that can actually identify the different ligand–dendrimer ratios has dramatically improved our understanding of the heterogeneous distributions generated by the functionalization methods. This new understanding has led to a fundamental change in the way we viewed the material content of the multifunctional dendrimers, and the approaches taken to generate the material. Quite simply, the number of different components in the G5-FA-MTX dendrimer was far greater than we had previously thought [34]. Our new experimental observations indicated that the ligand–dendrimer distributions were, in the most uniform scenario, approximately Poissonian. Figure 14.8, panel (a) shows Poisson distributions with means of 4 and 5. Recall that the G5-FA-MTX dendrimer that had promising in vitro and in vivo results had a mean of 4 FA molecules per dendrimer and 5 MTX molecules per dendrimer. Accounting for both distributions, Fig. 14.8, panel (b) shows the relative amount of different dendrimer components with both FA and

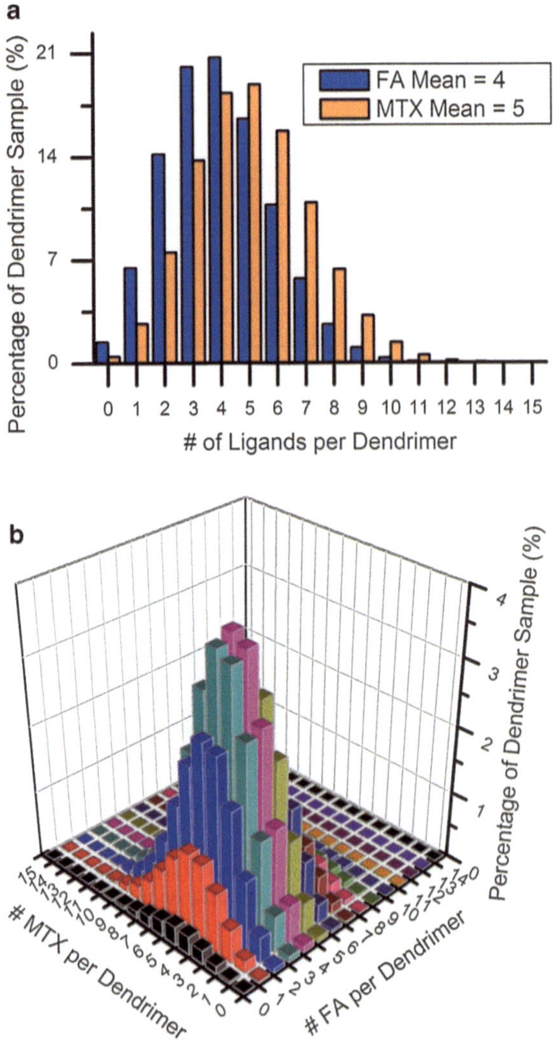

Fig. 14.8 Projection of the number and distribution of components in the G5-FA-MTX dendrimer, which had a mean of 4 FA and 5 MTX per dendrimer. (**a**) Poisson distributions with a means of 4 and 5. (**b**) The distribution of dendrimer components with different numbers of FA and MTX molecules. The dendrimer with 4 FA and 5 MTX ligands composed less than 4% of the total material. Reproduced with permission from [34]

MTX ligands. According to these projections, which are based on experimental data, the dendrimer component with exactly 4 FA and 5 MTX makes up less than 4% of the total material. Given that this component made up such a small portion of the entire population, we have concluded that it is insufficient to use the mean ligand–dendrimer ratio alone to describe the material and interpret material properties such as biological effects. Indeed, due to the nature of the two independent distributions, a substantial portion of the material will have large numbers of FA molecules and few MTX. This sub-population will have minimal therapeutic effect, and could actually cause growth of folate-dependent tumors. In contrast, a second sub-population will have large numbers of MTX molecules per dendrimer and few FA molecules. This material may be nonspecifically cytotoxic and could have a deleterious impact on the therapeutic index of the material. These concepts are in no way captured simply by examining the mean ligand–dendrimer ratio.

14.4.3 Synthetic Strategies that Can Produce Reproducible Batches of Material

One major implication of the complexity of ligand–dendrimer distributions is that it is a significant challenge to produce consistent batches of material using current synthetic approaches. Only after analytical methods have been developed to sufficiently identify the composition of a multifunctional material can this challenge be truly addressed. Our recent experimental results have indicated that the methods used to produce the G5-FA-MTX dendrimer are far more sensitive to slight variations in the synthetic conditions than we had originally anticipated. Specifically, we have found that two ligand–dendrimer conjugates can have the same mean number of ligands and yet have dramatically different distributions [38].

In the above example of two batches of ligand–dendrimer material having the same mean ratio and yet very different material compositions, we found that the major source of this discrepancy was the sensitivity of the partial acetylation to the quality of mass transport during the reaction. Similar to the alkyne, azide, FA, and MTX molecules, the acetic anhydride used to neutralize the dendrimer surface through acetylation is a small ligand. The nature of the partial acetylation reaction (excess of primary amines per dendrimer relative to the amount of acetic anhydride added) means that a distribution of dendrimers with different numbers of acetyl groups is generated. We have found that under poor mass transport conditions (low mixing, no dilution of acetic anhydride), the acetyl-dendrimer distribution is significantly more heterogeneous than a Poisson distribution. Essentially, the poor mass transport causes a sub-population of dendrimers to be nearly or completely acetylated, with the remaining dendrimers having fewer numbers of acetyl groups. This distribution meant that a subsequent ligand conjugation would occur in the presence of a highly heterogeneous distribution of attachment sites. Figure 14.9 demonstrates this phenomenon with the HPLC traces and quantified distributions for two batches of dendrimer conjugated with the azide ligand. The starting dendrimer for panels (a) and (c) was partially acetylated under optimal mass transport conditions prior to the azide ligand conjugation. In panels (b) and (d), the starting dendrimer was partially acetylated under poor mass transport conditions. Although the two dendrimer conjugates have the same mean ligand–dendrimer ratio (6.6 and 6.8), they have dramatically different distribution profiles. Importantly, although the two batches of partially acetylated dendrimer were significantly different, they could not be distinguished by either NMR or HPLC.

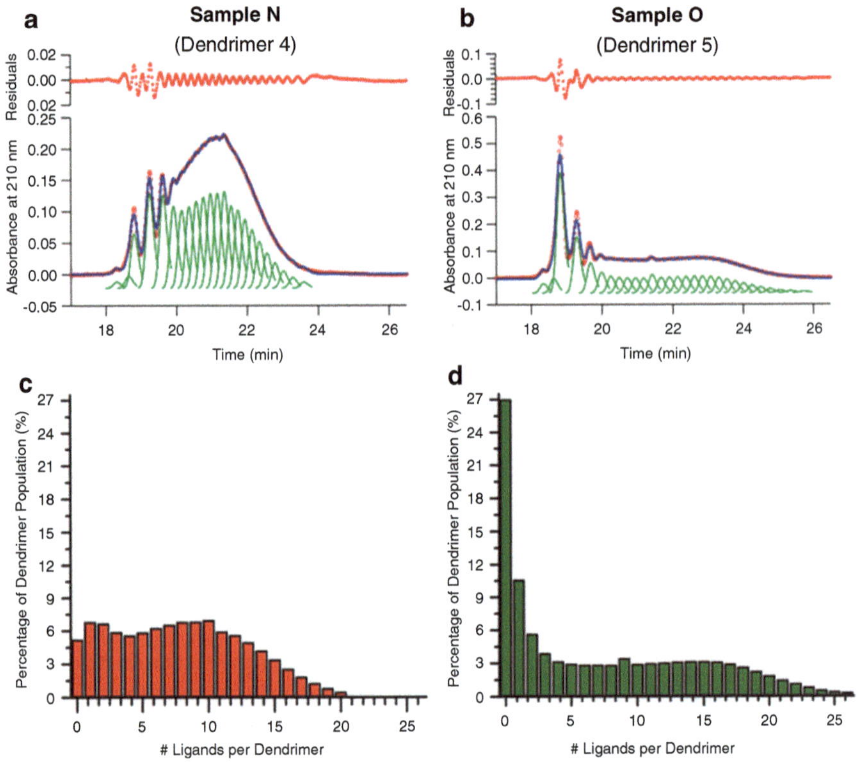

Fig. 14.9 The effect of acetylation mass transport quality on subsequent dendrimer–ligand distributions (Samples N and O). Sample N had a mean of 6.6 azide ligands per dendrimer. The parent dendrimer for Sample N was partially acetylated under optimal mass transport conditions. Sample O had a mean of 6.8 azide ligands per dendrimer. The parent dendrimer for Sample O was partially acetylated under poor mass transport conditions. Panels (**a**) and (**b**) contain the peak-fitted HPLC traces for Sample N and O. This fitted data provided the quantified distributions shown in panels (**c**) and (**d**). Reproduced with permission from [38]

14.4.4 Scale-Up

The final challenge that we identified is developing the capability for scale-up production of the material. The field of polymer processing has long been confronted with the challenge that many reactions and processes do not predictably scale. In the case of ATI-101, not only was there a change in the starting material (reduced mean end-group-dendrimer ratio and increase in heterogeneity), there was also an order of magnitude change in the reaction scale (200 g to 2 kg). Presumably, as the ligand-to-dendrimer amine ratio increases, the ability to maintain optimal mass transport is diminished. Both of these factors likely caused a significant sub-population of the dendrimer to become fully or close to fully acetylated, thereby reducing the number of FA that could be attached during subsequent conjugation reactions.

14.5 Solution Strategies

Given the challenges involved in the original synthetic process of the multifunctional dendrimer, we have focused significant efforts within MNIMBS to alter the design of the multifunctional dendrimer to improve the consistency, uniformity, and reproducibility of these products. Described below are five promising design strategies that improve upon the original multivalent dendrimer product. The first three design strategies are relatively minor modifications, while the final two detail different classes of dendritic platforms.

The first design strategy is to perform fewer functionalization reactions to the dendrimer. This goal is accomplished either by using fewer different types of functionalities per dendrimer, or by combination of functionalities prior to the conjugation to the dendrimer. Both approaches will not eliminate ligand–dendrimer distributions, but they will reduce the *heterogeneity* of the distribution. The strategy to perform only one conjugation to the dendrimer would still produce a distribution similar to a Poisson distribution; however, this is a significant reduction in heterogeneity from the combination distributions in Fig. 14.8b. More important, this approach insures that concentrations of both drug and targeting agent scale together, thus ensuring that the drug is only presented on targeted particles. A second strategy inspired by the material in Fig. 14.8 is to eliminate the partial acetylation step. This approach avoids the variability introduced by the partial acetylation reaction related to the quality of mass transport. It also reduces the heterogeneity of the ligand distribution, as the pre-existing acyl distribution increases the heterogeneity of subsequent ligand conjugations [34]. This is because the pre-existing ligand–dendrimer distribution creates variability in the number of attachment sites per dendrimer. A third strategy is to use alkyne or azide ligands as a quality control marker to monitor these reactions. When the distribution of ligand–dendrimer components cannot be resolved by HPLC, the alkyne or azide ligand can be conjugated and the product analyzed again by the HPLC to provide information about the material heterogeneity. The material in Fig. 14.9 is an excellent example of this strategy.

A fourth design strategy is to replace the dendrimer platform with a dendron [39]. Although the dendron's architecture is similar to a dendrimer, each particle has a single, chemically unique focal point. This focal point can be leveraged to reduce the heterogeneity of a multifunctional platform. If the focal point is used as an orthogonal reactive site, near 1:1 stoichiometry can be achieved between a functional ligand or drug and the dendron. This strategy has been successfully applied using an RGD-targeting ligand conjugated to the dendron's terminal arms, and either a dye molecule or the drug at the focal point (Fig. 14.10).

The final design strategy capitalizes on the fact that dendrimers conjugated with the azide ligand have different retention times on an HPLC column based on the number of ligands per dendrimer (Fig. 14.11). Using semi-preparative HPLC, individual dendrimer ligand components (with between 0 and 8 ligands) have been successfully isolated at purity levels greater than 80% [37]. These precision dendrimers are very interesting because functional ligands such as drugs, targeting

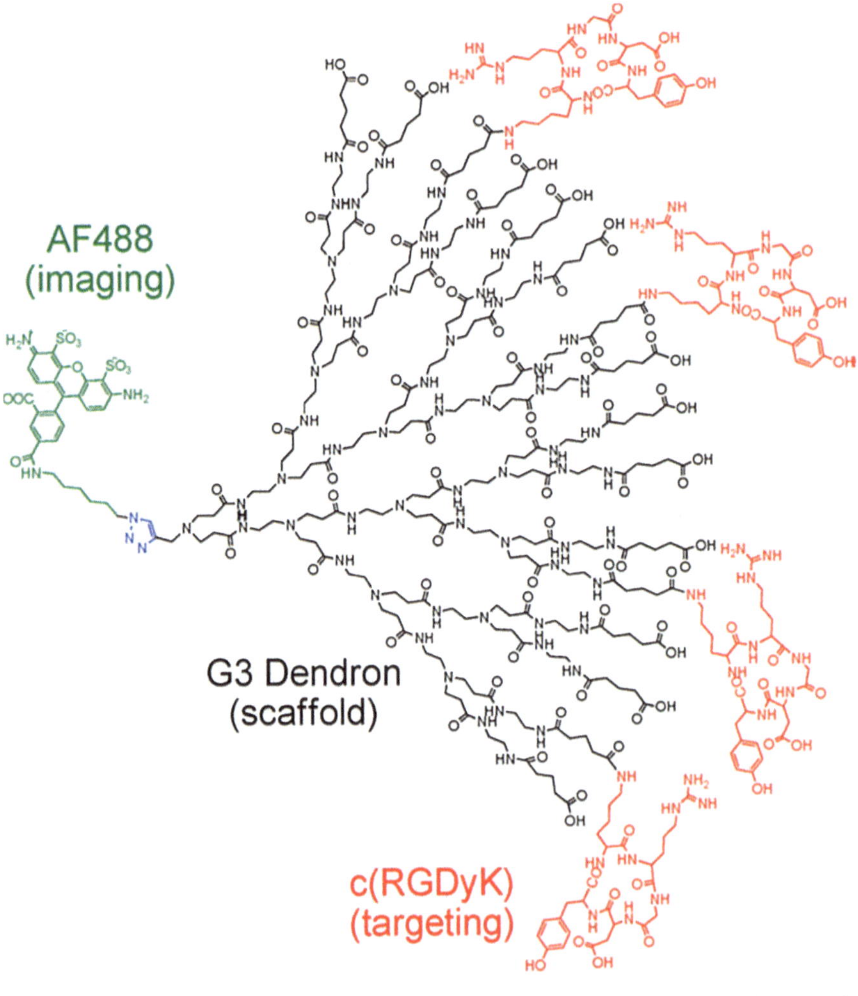

Fig. 14.10 PAMAM dendron strategy that leverages the unique focal point to reduce sample heterogeneity. Reproduced with permission from [39]

ligands or combinations thereof can be conjugated specifically with the azide ligands on the dendrimer using click chemistry. This design has the potential to probe biological activity of specific dendrimer ligand components to identify the reaction product with the optimal activity.

14.6 Conclusions

Multifunctional dendrimer platforms continue to be exciting materials with great promise. The lessons learned from the early synthetic experiences have been important to clarify the challenges that must be addressed in order to achieve uniform synthesis

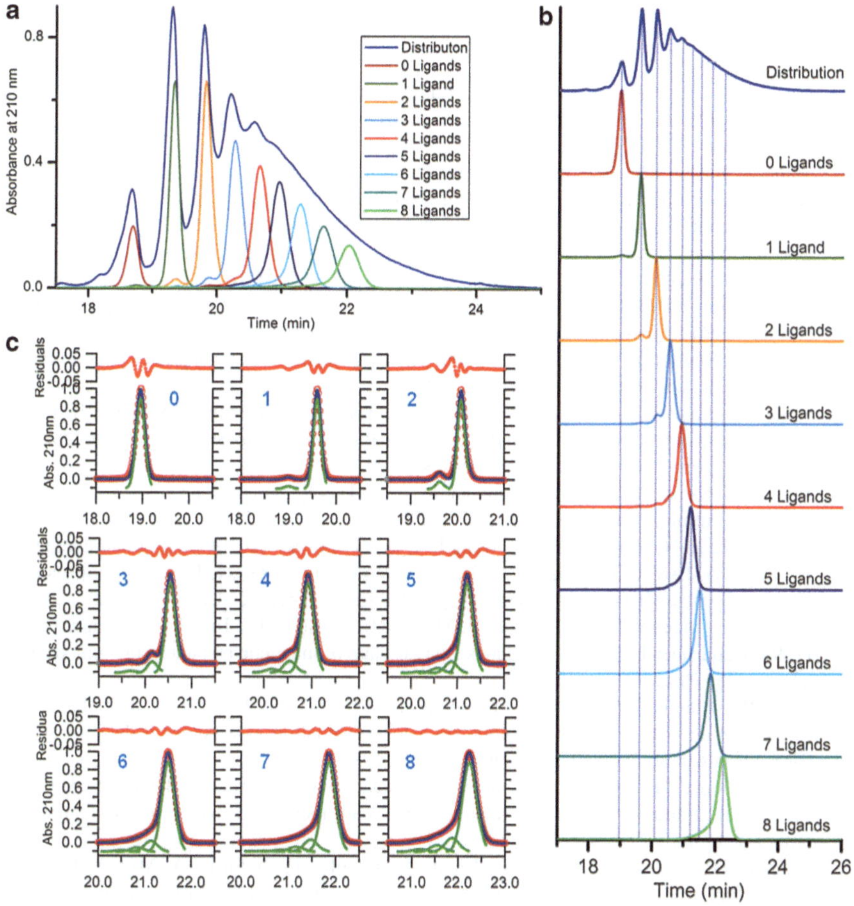

Fig. 14.11 Dendrimer with precise numbers of ligands isolated from the distribution of components. Dendrimer with 0–8 ligands were isolated with degrees of purity greater than 80%. Reproduced with permission from [37]

of functionalized nanoparticles for therapeutics. Indeed the challenges identified in this chapter—control of the starting material, control of ligand–dendrimer distributions, development of quality control methods, and control of scale-up—are applicable to many multifunctional nanoparticle systems. Strategies that provide solutions to these challenges with any type of nanoparticle will be critical to ensuring clinical success with these materials as therapeutics.

References

1. Ma H, Liu S, Luo JD, Suresh S, Liu L, Kang SH, Haller M, Sassa T, Dalton LR, Jen AKY (2002) Highly efficient and thermally stable electro-optical dendrimers for photonics. Adv Funct Mater 12:565–574
2. Wang B, Navath RS, Menjoge AR, Balakrishnan B, Bellair R, Dai H, Romero R, Kannan S, Kannan RM (2010) Inhibition of bacterial growth and intramniotic infection in a guinea pig model of chorioamnionitis using PAMAM dendrimers. Int J Pharm 395:298–308
3. Ghosh S, Yadav S, Vasanthan N, Sekosan G (2010) A study of antimicrobial property of textile fabric treated with modified dendrimers. J Appl Polym Sci 115:716–722
4. Polcyn P, Jurczak M, Rajnisz A, Solecka J, Urbanczyk-Lipkowska Z (2009) Design of antimicrobially active small amphiphilic peptide dendrimers. Molecules 14:3881–3905
5. Tang MX, Redemann CT, Szoka FC (1996) In vitro gene delivery by degraded polyamidoamine dendrimers. Bioconjug Chem 7:703–714
6. Kukowska-Latallo JF, Bielinska AU, Johnson J, Spindler R, Tomalia DA, Baker JR (1996) Efficient transfer of genetic material into mammalian cells using Starburst polyamidoamine dendrimers. Proc Natl Acad Sci USA 93:4897–4902
7. Haensler J, Szoka FC (1993) Polyamidoamine cascade polymers mediate efficient transfection of cells in culture. Bioconjug Chem 4:372–379
8. Zhang WL, Li N, Huang J, Yu JH, Wang DX, Li YP, Liu SY (2010) Gadolinium-conjugated FA-PEG-PAMAM-COOH nanoparticles as potential tumor-targeted circulation-prolonged macromolecular MRI contrast agents. J Appl Polym Sci 118:1805–1814
9. Wiener EC, Brechbiel MW, Brothers H, Magin RL, Gansow OA, Tomalia DA, Lauterbur PC (1994) Dendrimer-based metal-chelates—a new class of magnetic resonance imaging contrast agents. Mag Res Med 31:1–8
10. Barth RF, Adams DM, Soloway AH, Alam F, Darby MV (1994) Boronated Starburst dendrimer monoclonal-antibody immunoconjugates—evaluation as a potential delivery system for neutron-capture therapy. Bioconjug Chem 5:58–66
11. Myc A, Douce TB, Ahuja N, Kotlyar A, Kukowska-Latallo JF, Thomas TP, Baker JR (2008) Preclinical antitumor efficacy evaluation of dendrimer-based methotrexate conjugates. Anticancer Drugs 19:143–149
12. Thomas TP, Majoros IJ, Kotlyar A, Kukowska-Latallo JF, Bielinska A, Myc A, Baker JR (2005) Targeting and inhibition of cell growth by an engineered dendritic nanodevice. J Med Chem 48:3729–3735
13. Majoros IJ, Keszler B, Woehler S, Bull T, Baker JR (2003) Acetylation of poly(amidoamine) dendrimers. Macromolecules 36:5526–5529
14. Hong SP, Bielinska AU, Mecke A, Keszler B, Beals JL, Shi XY, Balogh L, Orr BG, Baker JR, Banaszak-Holl MM (2004) Interaction of poly(amidoamine) dendrimers with supported lipid bilayers and cells: hole formation and the relation to transport. Bioconjug Chem 15:774–782
15. Lee CC, MacKay JA, Frechet JMJ, Szoka FC (2005) Designing dendrimers for biological applications. Nat Biotechnol 23:1517–1526
16. Svenson S, Tomalia DA (2005) Commentary—dendrimers in biomedical applications—reflections on the field. Adv Drug Del Rev 57:2106–2129
17. Hong SP, Leroueil PR, Janus EK, Peters JL, Kober MM, Islam MT, Orr BG, Baker JR, Banaszak-Holl MM (2006) Interaction of polycationic polymers with supported lipid bilayers and cells: nanoscale hole formation and enhanced membrane permeability. Bioconjug Chem 17:728–734
18. Leroueil PR, Hong SY, Mecke A, Baker JR, Orr BG, Banaszak-Holl MM (2007) Nanoparticle interaction with biological membranes: does nanotechnology present a janus face? Acc Chem Res 40:335–342
19. Hong S, Leroueil PR, Majoros IJ, Orr BG, Baker JR, Banaszak-Holl MM (2007) The binding avidity of a nanoparticle-based multivalent targeted drug delivery platform. Chem Biol 14:107–115
20. Choi HS, Liu W, Liu F, Nasr K, Misra P, Bawendi MG, Frangioni JV (2010) Design considerations for tumour-targeted nanoparticles. Nat Nanotechnol 5:42–7

21. Wu G, Barth RF, Yang WL, Chatterjee M, Tjarks W, Ciesielski MJ, Fenstermaker RA (2004) Site-specific conjugation of boron-containing dendrimers to anti-EGF receptor monoclonal antibody cetuximab (IMC-C225) and its evaluation as a potential delivery agent for neutron capture therapy. Bioconjug Chem 15:185–194
22. Hill E, Shukla R, Park SS, Baker JR (2007) Synthetic PAMAM-RGD conjugates target and bind to odontoblast-like MDPC 23 cells and the predentin in tooth organ cultures. Bioconjug Chem 18:1756–1762
23. Shukla R, Thomas TP, Peters J, Kotlyar A, Myc A, Baker JR (2005) Tumor angiogenic vasculature targeting with PAMAM dendrimer-RGD conjugates. Chem Commun 5739–5741
24. Shukla R, Thomas TP, Peters JL, Desai AM, Kukowska-Latallo JF, Patri AK, Kotlyar A, Baker JR (2006) HER2 specific tumor targeting with dendrimer conjugated anti-HER2 mAb. Bioconjug Chem 17:1109–1115
25. Patri AK, Myc A, Beals J, Thomas TP, Bander NH, Baker JR (2004) Synthesis and in vitro testing of J591 antibody-dendrimer conjugates for targeted prostate cancer therapy. Bioconjug Chem 15:1174–1181
26. Majoros IJ, Myc A, Thomas TP, Mehta CB, Baker JR (2006) PAMAM dendrimer-based multifunctional conjugate for cancer therapy: synthesis, characterization, and functionality. Biomacromolecules 7:572–579
27. Majoros IJ, Thomas TP, Mehta CB, Baker JR (2005) Poly(amidoamine) dendrimer-based multifunctional engineered nanodevice for cancer therapy. J Med Chem 48:5892–5899
28. Kukowska-Latallo JF, Candido KA, Cao ZY, Nigavekar SS, Majoros IJ, Thomas TP, Balogh LP, Khan MK, Baker JR (2005) Nanoparticle targeting of anticancer drug improves therapeutic response in animal model of human epithelial cancer. Cancer Res 65:5317–5324
29. Campbell IG, Jones TA, Foulkes WD, Trowsdale J (1991) Folate-binding protein is a marker for ovarian cancer. Cancer Res 51:5329–5338
30. Weitman SD, Weinberg AG, Coney LR, Zurawski VR, Jennings DS, Kamen BA (1992) Cellular localization of the folate receptor: potential role in drug toxicity and folate homeostasis. Cancer Res 52:6708–6711
31. Ross JF, Chaudhuri PK, Ratnam M (1994) Differential regulation of folate receptor isoforms in normal and malignant tissues in vivo and in estabilshed cell lines. Physiologic and clinical implications. Cancer 73:2432–2443
32. Elnakat H, Ratnam M (2004) Distribution, functionality and gene regulation of folate receptor isoforms: implications in targeted therapy. Adv Drug Del Rev 56:1067–1084
33. Borst P, Evers R, Kool M, Wijnholds J (2000) A family of drug transporters: the multidrug resistance-associated proteins. J Natl Cancer Inst 92:1295–1302
34. Mullen DG, Fang M, Desai AM, Baker JR, Orr BG, Banaszak-Holl MM (2010) A quantitative assessment of nanoparticle-ligand distributions: implications for targeted drug and imaging delivery in dendrimer conjugates. ACS Nano 4:657–670
35. Mullen DG, Desai AM, Waddell JN, Cheng X-M, Kelly CV, McNerny DQ, Majoros IJ, Baker JR, Sander LM, Orr BG, Banaszak-Holl MM (2008) The implications of stochastic synthesis for the conjugation of functional groups to nanoparticles. Bioconjug Chem 19:1748–52
36. Peterson J, Allikmaa V, Subbi J, Pehk T, Lopp M (2003) Structural deviations in poly(amidoamine) dendrimers: a MALDI-TOF MS analysis. Eur Polym J 39:33–42
37. Mullen DG, Byrne EL, Desai A, van Dongen MA, Barash M, Cheng XM, Baker JR, Banaszak-Holl MM (2010) Isolation and characterization of dendrimer with precise numbers of functional groups. Chem Eur J 16:10675–10678
38. Mullen DG, Borgmeier EL, Fang M, McNerny DQ, Desai A, Baker JR, Orr BG, Banaszak-Holl MM (2010) Effect of mass transport in the synthesis of partially acetylated dendrimer: implications for functional ligand-nanoparticle distributions. Macromolecules 43:6577–6587
39. McNerny DQ, Kukowska-Latallo JF, Mullen DG, Wallace JM, Desai AM, Shukla R, Huang BH, Banaszak-Holl MM, Baker JR (2009) RGD dendron bodies: synthetic avidity agents with defined and potentially interchangeable effector sites that can substitute for antibodies. Bioconjug Chem 20:1853–1859v

Chapter 15
Multivalent Dendritic Architectures for Theranostics

Stephanie Reichert, Marcelo Calderón, Kai Licha, and Rainer Haag

15.1 Introduction

One of the biggest challenges in the field of medicinal chemistry is the distinction between healthy and malignant cells. One approach in cancer diagnostics is based on the fact that there is a high density of membrane proteins on the surface of tumor cells which perform specific cellular functions [1]. Such tumor markers can be made to act as receptors for modular ligands that can then be linked to a contrast agent. For example, fluorescent dyes with near infrared absorbance (~770 nm) are used as contrast agents for in vivo analysis in animal models and to identify target tumor cells. In nature, cell recognition processes and the resulting transduction of signals underlie multivalent interactions. Antibodies, for example, can bind to cells via multivalent interaction and thereby initiate cell death. Synthetic multivalent ligands can be used to determine the surfaces of different cell types, for example, bacteria, cancer cells, and antigens. Multivalent interactions play an important role in many biological systems. A great advantage of multivalent drugs, which are bridged by polymeric spacers, is attributed to higher entropy loss and hence higher binding constant (Fig. 15.1) [2, 3].

The particles in multivalent interactions can be small molecules, oligosaccharides, proteins, nucleic acids, lipids, viruses, bacteria, cells, and polymers. As a result, new synthetic strategies can be developed for drugs, which are based on multiple interactions of ligands conjugated to, for instance, a polymer which interacts

S. Reichert • M. Calderón • R. Haag (✉)
Department of Chemistry and Biochemistry, Organic and Macromolecular Chemistry, Freie Universität Berlin, Takustr. 3, 14195 Berlin, Germany
e-mail: haag@chemie.fu-berlin.de

K. Licha
Mivenion GmbH, Robert-Koch-Platz 4, 10115 Berlin, Germany

S. Svenson and R.K. Prud'homme (eds.), *Multifunctional Nanoparticles for Drug Delivery Applications: Imaging, Targeting, and Delivery*, Nanostructure Science and Technology, DOI 10.1007/978-1-4614-2305-8_15,

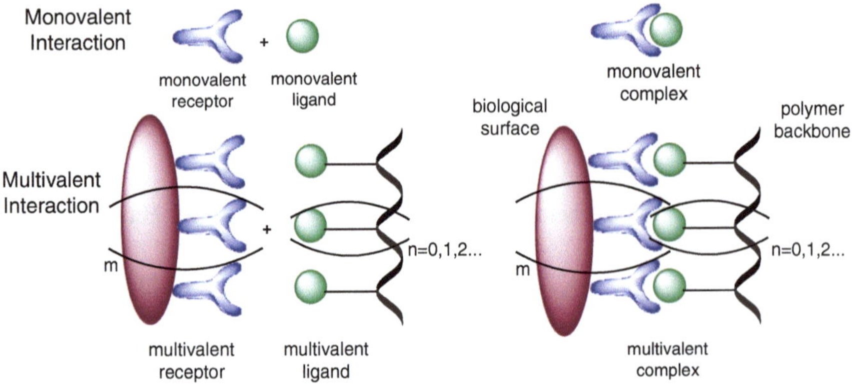

Fig. 15.1 Comparison of monovalent and multivalent interactions

with multiple receptor sites in protein complexes and simultaneously multiple receptors on the cell surface. Therefore the scientific and industrial communities are interested in identifying and developing materials that take advantage of such features on the cellular level. In this context, compounds with a high density of functional groups and sizes on the nanometer scale (5–200 nm) have emerged as optimal candidates in biological, multivalency-related applications. Especially the highly branched, multivalent nature of dendritic polymers makes them ideal candidates for a variety of tissue-engineering applications, for example, as crosslinking agents, modulators of surface charge and surface chemistry, and as primary components in scaffolds that mimic natural extracellular matrices [4].

15.2 Dendrimers and Dendritic Architectures

Many types of macromolecular architectures, namely linear, crosslinked, star-like, cyclic, and branched have evolved from Staudinger's work on macromolecular chemistry from the 1920s (Fig. 15.2). For around 30 years, dendritic molecules have been used in a variety of subfields of chemistry, biology, medicine, and materials science. Dendrimers are highly branched macromolecules with a defined mass which are advantageous for several reasons [5]. Their high water solubility, chemical flexibility, low viscosity in solution, and multivalency [3] make them attractive molecules for many applications. Nowadays, they are used in sensing [6, 7], catalysis [8, 9], light harvesting systems [10], and in biological and medical applications [11, 12]. Most applications of dendrimers have been based mainly on the high number of functional groups.

Dendrimers have a well-defined homogeneous structure and can be nearly perfect, monodisperse molecules [13, 14] consisting of tree-like arms or branches. So far, two fundamentally different synthesis methods have been developed. The divergent

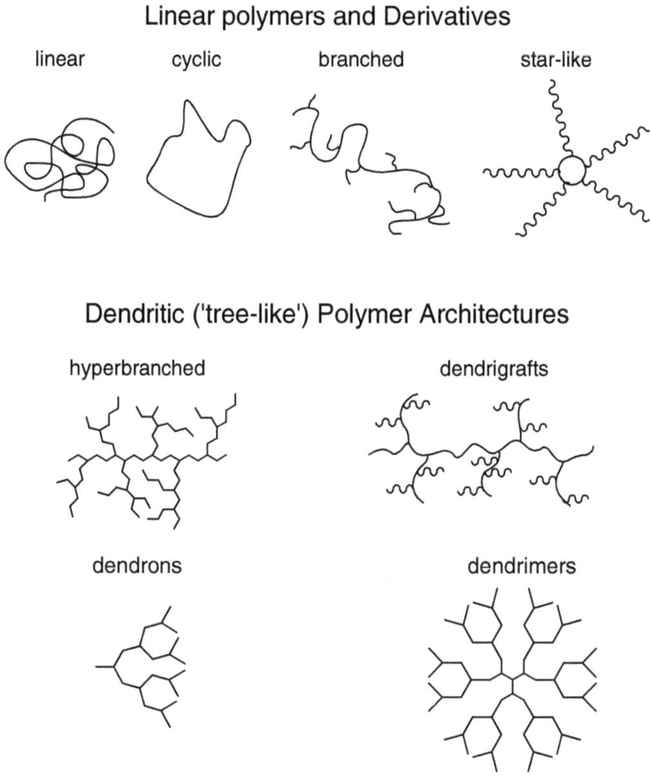

Fig. 15.2 Examples of macromolecular architectures

approach initiated by Tomalia et al. [15] and Newkome et al. [16], and the convergent approach developed by Hawker and Fréchet and coworkers [17, 18]. It is relatively easy to precisely control sizes, composition, and chemical reactivity of dendrimers. Depending on the type of end group, the properties of dendrimers vary in form, stability, solubility, flexibility, and viscosity. With increasing generations (G), the number of end groups increases which can result in intensification of definite properties such as light collector effects [19] and signal amplification [20]. In contrast to linear polymers, the intrinsic viscosity of dendrimers does not grow linearly with its molar mass but rather reaches a maximum at a specific generation, beyond which it decreases again with higher generations.

Synthesis of these dendrimers, however, requires a high purity of the used educts and high yields of the individual steps. As repeating reactions tend to be costly, hyperbranched polymers have emerged as an alternative. In this context, the hyperbranched polymer shown in Fig. 15.3 may not be perfectly branched but can be prepared conveniently in one-step procedures via polyaddition, polycondensation, and radical polymerization, etc., on the kilogram scale.

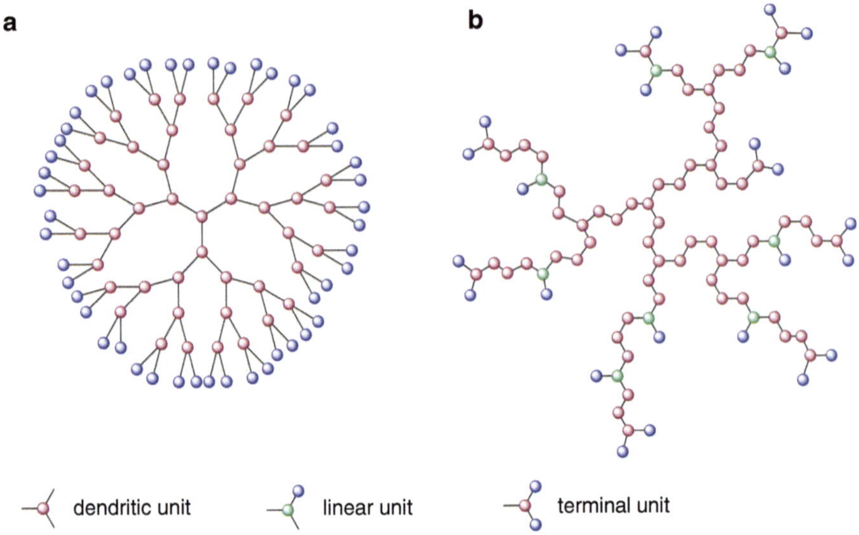

Fig. 15.3 General Structure of (**a**) a perfect dendrimer with only dendritic and terminal units, and (**b**) a hyperbranched polymer with dendritic, linear, and terminal units

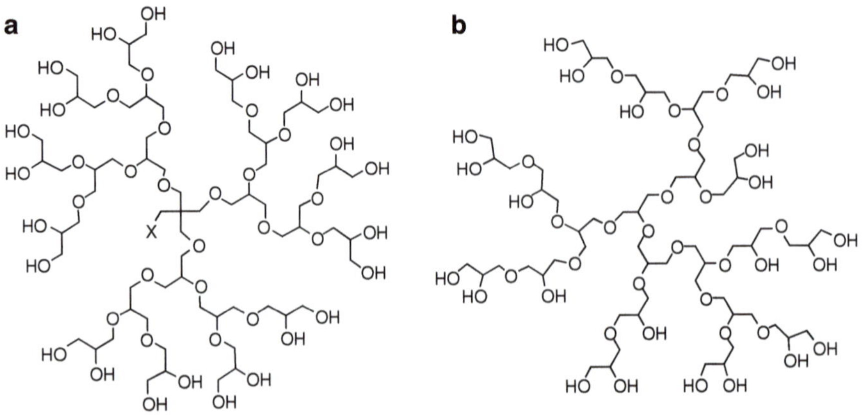

Fig. 15.4 (**a**) Perfect [G3] polyglycerol dendrimer, and (**b**) hyperbranched polyglycerol. The depicted polymer structure represents only one possible isomer and a small part of the polyglycerol scaffold

Hyperbranched and perfect dendrimers exhibit similar chemical and physical properties such as high functionality due to their large number of end groups, and a high degree of branching that prevents crystallization, and causes low melting point and low solution viscosity.

Our group has focused its attention on the study of dendritic polyglycerols (PGs) as shown in Fig. 15.4, which are glycerol-based macromolecular architectures with a high degree of branching and end-group functionalities. They have compact, well-defined, dendrimer-mimicking structures that show appreciable aqueous solubility, can be prepared in facile one-pot synthesis at large scale, and possess substantially low tissue toxicity. Due to their low degree of molecular weight variation, flexible design, and biocompatibility profile, dendritic polyglycerols have also found a broad range of potential applications in medicine and pharmacology [21–25].

15.3 Biological Scenario

The surface decoration of dendritic nanostructures with solubilizing agents and targeting moieties, along with imaging and therapeutic modalities, and the inherent charge profile of the dendritic polymer, confer structural benefits such as faster cellular entry, reduced macrophage uptake, targetability, and easier passage across biological barriers by transcytosis [26]. In addition, the branched nature of dendritic nanostructures has been shown to improve their in vivo application profile in comparison to linear polymeric analogs. For instance, increasing the number of branches or arms for polymers with similar molecular weight (MW) and chemistry has increased the blood circulation half-life ($t_{1/2)}$. A systematic study with a library of PEGylated polyester "bow tie" dendrimers established the relationship between branching and blood-circulation time [27, 28]. For a series of bow ties with equivalent MW (~40 kDa), there was an increase in $t_{1/2}$, from 1.4 ± 0.4 h for the two-arm dendrimer, essentially a linear polymer, to 26 ± 6 h for the four-arm dendrimer, and finally to 31 ± 2 h for the eight-arm dendrimer. Corresponding biodistribution studies in healthy mice showed no significant variation in tissue uptake among the three polymers; however, decreasing polymer excretion in the urine was observed with increasing branching. This polymeric drug carrier, which was studied in C26 colon carcinoma cells-bearing mice, showed long blood-circulation times and remarkable efficacy in delivering the chemotherapeutic drug doxorubicin (DOX) to tumors. The increased efficiency of bow tie-DOX conjugates led to complete tumor cell disappearance in sharp contrast to free DOX, which was ineffective.

The aforementioned example illustrates one of the many environmental biological challenges that dendritic polymers face. In following sections, a brief description of the biological scenario will be considered with regard to the design of therapeutic and diagnostic approaches using dendrimers.

15.3.1 Particular Features of Malignant Tissues

The term tumor is generally used for an uncontrolled growth of the body's own tissue. Malignant tumors grow into surrounding tissue and destroy local structures. By means of the blood stream or lymphatic fluid tumors can spread in every body region, which is called metastasis formation. Tumors often occur after surgeries and

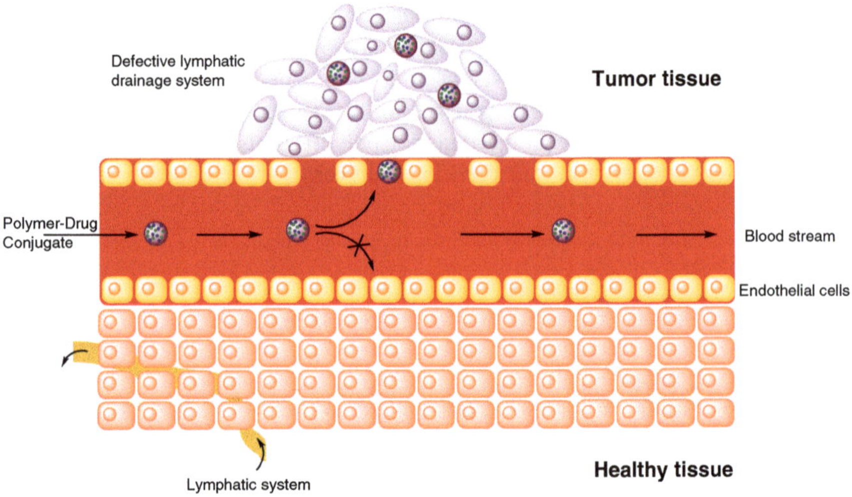

Fig. 15.5 Schematic representation of the EPR effect

cannot be removed completely. Therefore, in most cases, chemotherapy and permanent controls are necessary. For this reason, several working groups investigate in the development of multivalent active agents. Structure-defined multivalent and dendritic polymers allow the production of customized self-organized systems or polymer therapeutics [58]. The aim is the right therapy for the right patient at the right time or, in other words, diagnostics is accompanied by therapy which is expressed by the term *theranostics* (*thera*py and diag*nostics*).

15.3.2 Passive and Active Targeting

The targetability of dendritic polymers to malignant cells and tissues can be achieved by adopting either of the two approaches [1] passive targeting and [2] active targeting [29]. In the passive targeting approach, the dendritic system delivers bioactives directly to the cell/tissue as a consequence of the environmental conditions in the malignant tissue. It is well recognized that tumor microvascular endothelium is leakier than healthy tissue, which results in enhanced permeability for macromolecules [30]. Furthermore, tumor tissue is characterized by inefficient lymphatic drainage [31]. The combination of these two characteristics, along with the hypervascularization evident in the tumor microenvironment, leads to an accumulation of low molecular weight drugs coupled to high molecular weight nanocarriers in tumors. This so-called enhanced permeation and retention (EPR) effect [31–34] is depicted schematically in Fig. 15.5. In this type of passive targeting, the macromolecules accumulate in tumor tissue to a greater extent than in healthy tissue [35]. The EPR effect is predominantly observed for biocompatible macromolecules, macromolecular drugs, and lipids in solid tumors [36].

The size of the macromolecule is a critical factor with respect to tumor uptake. Following the pioneering work of Maeda, many other groups [37–39] were able to show that polymeric drugs with a molecular weight above the renal threshold (i.e., >40 kDa) accumulate in tumor tissues for prolonged time periods following intravenous (i.v.) injection. Liposomes, micelles, and polymeric nanocarriers are the most extensively studied drug carriers and possess the most suitable characteristics for encapsulation of many drugs and diagnostic (imaging) agents.

Active targeting of a dendritic system is usually achieved by coupling a targeting ligand to a dendritic polymer that provides preferential accumulation of the entire drug delivery system in the malignant tissue [40]. The active targeting approach is based on the interactions between the ligand and its cognate receptor or between specific biological pairs (e.g., avidin-biotin, antibody-antigen, and sialic acid-carbohydrate) [41]. In most cases, a targeting moiety in a dendritic polymer is focused on the specific receptor or antigen overexpressed in the plasma membrane or intracellular membrane in the malignant cells.

A recent example showing why dendritic polymers are good candidates for targeted delivery has been reported by Hashida et al. [42]. They synthesized a sixth-generation lysine dendrimer (KG6) and two PEGylated derivatives for tumor-selective targeting after i.v. injection in tumor-bearing mice. The study has shown that the PEGylated KG6 conjugates effectively accumulated in tumor tissue by the EPR effect. Many other examples will be explored in the following sections.

15.3.3 Cellular Entry (Endocytosis)

In recent years, a number of research groups have demonstrated the potential of dendrimers to enhance the cellular delivery of drugs through endocytosis [43, 44]. Endocytic pathways are subdivided into pinocytosis, the uptake of fluid and solutes, and phagocytosis, the uptake of large particles. Phagocytosis is the process of engulfing and destroying extracellularly derived material by a phagocytic cell, such as a macrophage, neutrophil, or amoeba. Pinocytosis includes four basic mechanisms: macropinocytosis, clathrin-mediated endocytosis, caveolae-mediated endocytosis, and clathrin- and caveolae-independent endocytosis (Fig. 15.6) [45, 46].

Cells overcome the lack of essential nutrients in the environment by expressing high-affinity receptors or binding sites on the membrane surface. In the case of the receptor-mediated process of phagocytosis, the receptors function as adhesive elements that bind the plasma membrane to the particle. As a result, surface membranes contain actin-binding proteins that link the phagocytic receptor to the actin cytoskeleton of the cell. Invagination produces a vesicle called phagosome, which usually fuses with early, then late endosomes, and finally with lysosomes to yield a phagolysosome [47]. In the case of pinocytosis, macropinocytosis is a cell-type specific and receptor-independent endocytic pathway. The function of receptor-mediated endocytosis is diverse. It is widely used for the specific uptake of certain substances required by the cell. The best studied endocytosis mechanism is

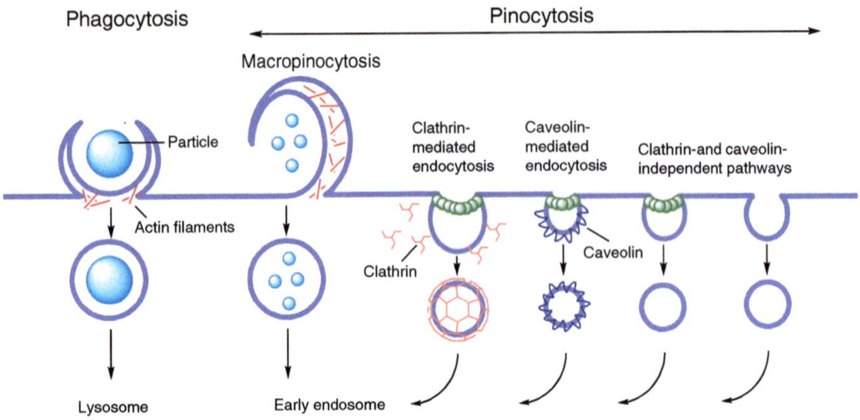

Fig. 15.6 Entry of material into cells via phagocytosis for engulfing large particles or via different ways of pinocytosis for small particles, which includes macropinocytosis, clathrin- and caveolin-mediated endocytosis, and clathrin- and caveolin-independent pathways. Adapted from [43]

clathrin-mediated endocytosis [48–50], which is the uptake of receptors, membrane, and cargo at the cell surface through a process that specifically involves the coat protein clathrin. Clathrin-coated vesicles are found in virtually all cells and form domains of the plasma membrane, termed clathrin-coated pits. Caveolae-mediated endocytosis is a form of uptake at the plasma membrane that involves the protein caveolin [51]. A third pathway, which is both clathrin and caveolae independent, may constitute a specialized high capacity endocytic pathway for lipids and fluid [52].

Dendrimers are a good model for use as a carrier/delivery system. Duncan et al. reported that poly(amido amine) (PAMAM) dendrimers generation 2.5 and 3.5 had particularly rapid serosal transfer rates and a low tissue deposition, indicating a very efficient transport pathway [53]. D'Emanuele et al. demonstrated that G3 PAMAM dendrimers and lauroyl-G3 PAMAM dendrimer conjugates could be visualized in individual endocytotic vesicles at the apical domain of the cell, and their association with multivesicular bodies in the cell interior was seen as well [54].

15.3.4 Blood Compatibility

Good blood compatibility and low cytotoxicity are important concerns in potential applications of dendrimers and other polymers as drug carriers. Higher molecular weight polymers are less prone to excretion through the kidneys and stay in the bloodstream for longer. During this circulation time, an immune reaction should be avoided, or the polymer–cargo complex might be quickly eliminated from the body. The use of a poly(ethylene glycol) (PEG) coating of surfaces in contact with biological fluids has been shown to reduce toxicological adverse reactions [36, 55]. Recently, Brooks et al. tested hyperbranched polyglycerols in blood compatibility studies for significant effects in complement activation, platelet activation, coagulation, erythrocyte aggregation, and hemolysis, and found them to be highly blood compatible [56, 57]. In addition, the blood compatibility of hyperbranched PG-based polymers containing multivalent cationic sites was investigated [56]. In comparison to standard cationic polymers such as poly(ethylene imine) (PEI), PG-based polymers showed much lower cytotoxicity. These observations indicate that hyperbranched polyglycerols are potential candidates for polymer therapeutics.

15.4 Dendritic Architectures for Therapeutic Applications

Dendritic polymers, as described above, have been applied in various fields such as gene expression [58], immunodiagnostics [59], and controlled [60] and targeted delivery [61]. The synergy between their multivalency and nanoscale size provides a range of options for chemical "smartness" along their molecular scaffold to achieve environmentally sensitive modalities [62].

Tomalia-type PAMAM dendrimers (Fig. 15.7), which are commercially available, represent the most widely investigated dendrimer family for biomedical applicability [63]. Besides PAMAM, there are numerous alternative types of dendrimers such as polylysines, polyesters, polyglycerols (PG), poly(propylene imines) (PPI), poly(ethylene imines) (PEI), and triazines that have been introduced for biomedical applications to molecularly amplify or multiply pathopharmacological effects [62]. These polymers show nontoxic behavior sufficient for in vivo applications, high transport capacity, multivalent charge, and ligand display for targeting of biological cells and tissue. These features are advantages for life-science applications, especially in the treatment of cancer, virus-related illness, and inflammation.

15.4.1 Dendritic Polymers as Drug or Gene Vectors

Several research groups have compared dendritic and linear polymers and found the dendritic architecture to be advantageous for delivery applications [64]. For example, the multivalency of dendrimers can be utilized to encapsulate or conjugate similar or different drug molecules, while adding targeting and/or solubilizing modalities to the same construct in a controllable pattern. In addition, the low polydispersity of these dendrimers is expected to provide a more reproducible pharmacokinetic behavior than linear polymers.

The application of dendrimers as nanocarriers follows two different mechanisms, namely the supramolecular or encapsulation approach and conjugation. The first approach makes use of the supramolecular voids within the dendritic structure to encapsulate guest molecules. Pioneered by the work of Maciejewski [65], this concept led the way toward dendritic nanocarriers, where a wide range of drug molecules are noncovalently entrapped within the dendritic core [66]. The so-called

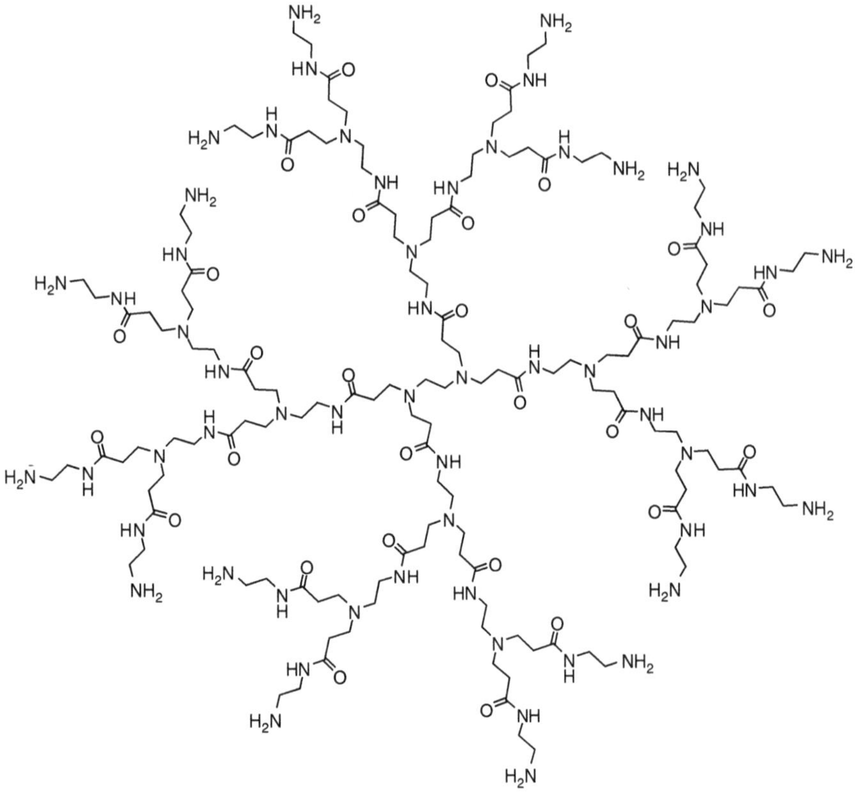

Fig. 15.7 Generation 2 PAMAM dendrimer

"unimolecular micelle" or "dendritic box" type architecture has been demonstrated with PAMAM dendrimers [63, 67]. Following the noncovalent interaction approach, a wide range of dendritic structures have been used for the delivery of therapeutic agents to intracellular target sites for disease management via oral, topical, ocular, and transdermal routes [24].

The other form of dendrimer application in medicine involves Ringsdorf's use of drug-polymer conjugation to the dendritic scaffold [2]. The idea involves the attachment of bioactive molecules directly or via spacer molecules to dendrimer terminal groups. The attachments are cellularly hydrolysable, employing ester or amide bonds in most cases. To improve therapeutic efficiency, targeting fragments or marker proteins can also be attached to the multiple functional group features of dendrimers [68].

Our group recently reviewed different mechanisms of transport, passive or active targeting, and smart delivery and release of bioactives at the site of action, performed by dendritic polymers [62]. An extensive description of the different cleavage modalities is presented, along with examples illustrating the underlying mechanisms.

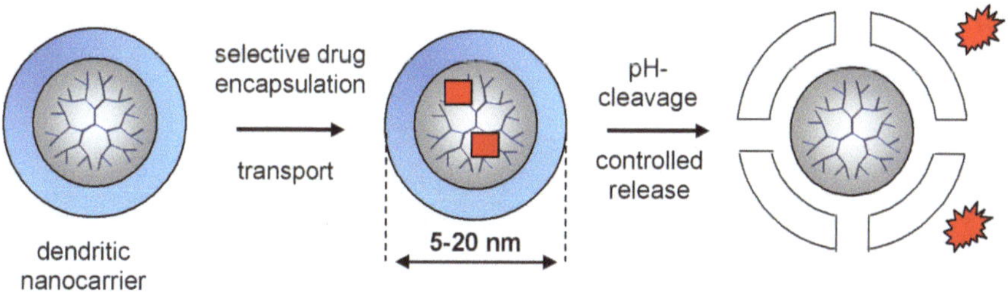

Fig. 15.8 Unimolecular dendritic nanocarriers for encapsulation of biologically active compounds. Controlled release after triggered shell cleavage (e.g., pH-controlled release)

15.4.2 Macromolecular or Encapsulation Approach

Physical entrapment of drugs or other bioactive molecules within polymeric networks by non-specific, non-covalent interaction between complementary functional modalities of the involved species is an attractive way to design nanotransport systems for drug delivery. One of the major problems faced by the pharmaceutical industry is the poor water solubility of many existing and novel bioactive species (e.g., drugs and imaging probes). Many potential candidates fail in preclinical studies because of limited solubility, stability, and toxicity, because of the hydrophobic character of the concerned species [2]. Thus, a number of nanocarriers have been designed and developed to overcome the solubility issue. This includes physical aggregates of amphiphilic molecules such as polymeric micelles as well as stable unimolecular micelles, which have been considered to be powerful nanocarriers in the dawning era of polymer therapeutics.

Initial studies of dendrimers as potential delivery systems focused on their use as unimolecular micelles for the non-covalent encapsulation of bioactive agents [69]. For example, in early studies, DNA was complexed with PAMAM dendrimers for gene delivery applications [70], and hydrophobic drugs and dye molecules were incorporated into various dendritic cores [69, 71]. A major drawback of these delivery systems is the lack of controlled drug release kinetics because most systems release their payload quickly over the course of several hours. An alternative approach is the covalent modification of dendritic macromolecules with an appropriate shell that results in stable, micelle-type structures that are suitable for non-covalent encapsulation of guest molecules. Dendritic polymers can be chemically modified either at the core (to increase hydrophobicity) or at the shell (to increase hydrophilicity), thereby tailoring the solubility profile of such nanotransport systems (Fig. 15.8).

Based on this concept, a simple and general method for the generation of core-shell type architectures from readily accessible, hyperbranched polymers was extensively explored by our group [72–75]. Universal nanotransporters (Fig. 15.9a) as well as several pH-sensitive nanocarriers were prepared by attaching pH-sensitive shells

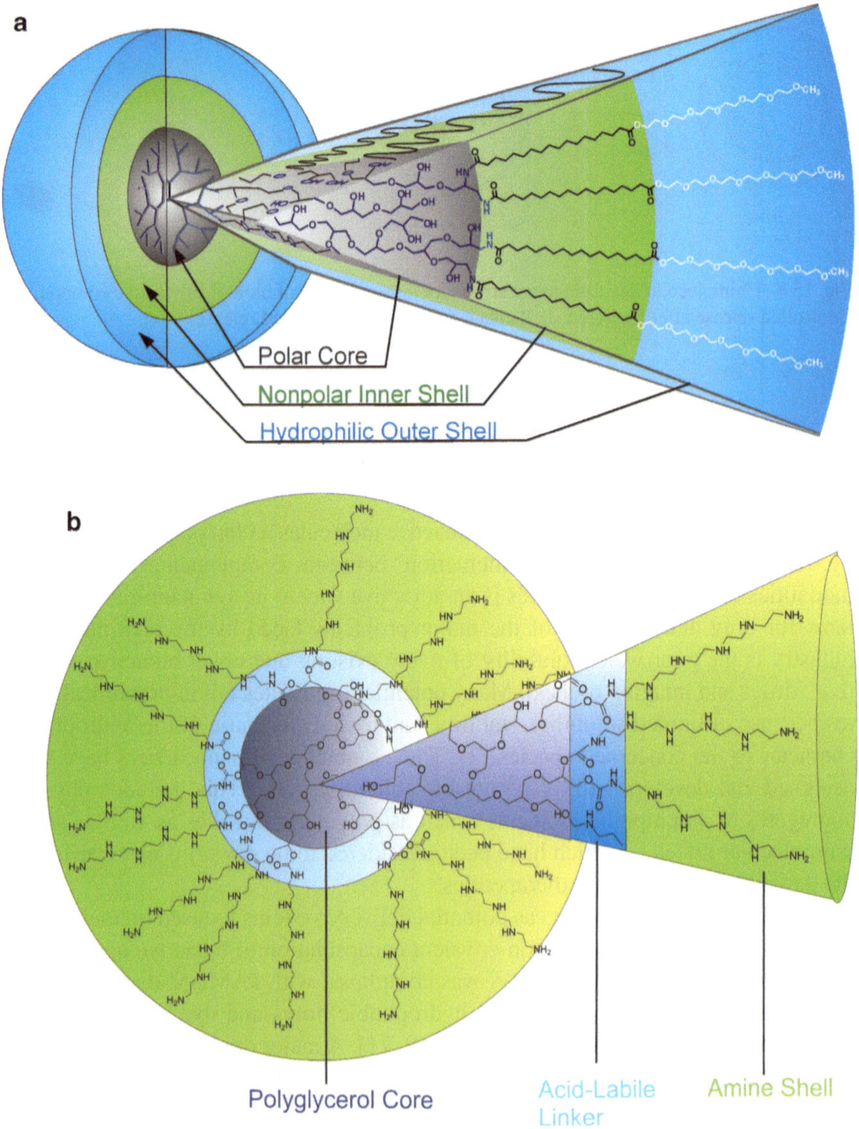

Fig. 15.9 Core-shell architectures based on dendritic polyglycerols. (**a**) Multishell universal nanocarrier. Reproduced with permission from [76]. (**b**) Amine-bearing, responsive gene carrier. Reproduced with permission from [77]

through acetal or imine bonds to commercially available dendritic core structures such as polyglycerol and poly(ethylene imine) [76]. In some cases, the pH-responsive nanocarriers showed a very high transport capacity, which is an important criterion for efficient drug delivery. Various guest molecules such as polar dyes, oligonucleotides, and anticancer drugs have been encapsulated inside these den-

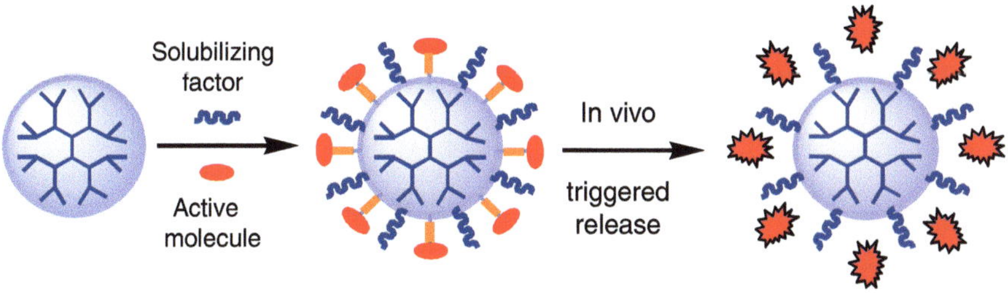

Fig. 15.10 Conjugation of biologically active molecules to dendritic polymers. Triggered release by pH drop, enzyme activity, reductive environment, or other biological signals

dritic core-shell architectures. Furthermore, an optimal release behavior was observed: fast release at pH 5–6 and slow release at pH 7.4 [72]. This approach was not only used to transport imaging and cytostatic agents, but also for the successful complexation, stabilization, and intracellular and intratumor delivery of genes (Fig. 15.9b) [77, 78].

15.4.3 Conjugation Approach

The covalent attachment of bioactive molecules to dendritic scaffolds is a promising route for controlling the loading and release of active molecules. Chemical conjugation to a dendritic scaffold allows covalent attachment of different kinds of active molecules in a controlled ratio. The approach is to some extent superior to physical, non-specific encapsulation within polymeric networks since strict control over active pay load can be imparted. The loading as well as the release can be tuned by incorporating cleavable bonds, which can be degraded under specific conditions present at the site of action (endogeneous stimuli, e.g., acidic pH, overexpression of specific enzymes, or reductive conditions, as well as exogeneous stimuli, e.g., light, salt concentration, or electrochemical potential) as shown in Fig. 15.10 [62].

The drug loading can be tuned by varying the generation number of the dendrimer, and release kinetics can be controlled by incorporating degradable linkages between the drug and the dendrimer. For example, Duncan and coworkers pioneered this field by preparing conjugates of PAMAM dendrimers with cisplatin, a potent anticancer drug with nonspecific toxicity and poor water solubility [79, 80]. The conjugates showed increased solubility, decreased systemic toxicity, and selective accumulation in solid tumors. Several other examples have been extensively described by us and others in the literature [2, 26, 81–83].

15.4.4 Dendritic Polymers as Effector Molecules

The utilization of the amplified effects derived from multiple interactions has paved the way to a new field in nanomedine with a broad field of applications, ranging from preparation of antifouling surfaces to antiviral agents. These particular features

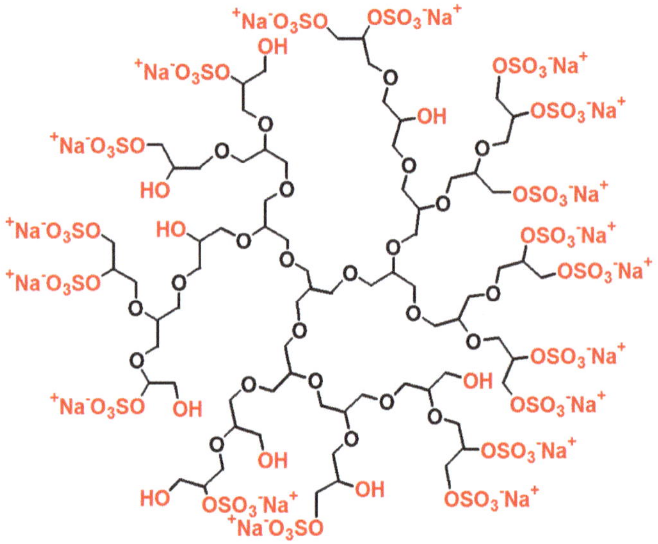

Fig. 15.11 Dendritic polyglycerol sulphate. The depicted polymer structure represents only one possible isomer and a small part of the polyglycerol scaffold

make dendritic polymers interesting as potential molecular entities because of their inherent highly dense functionality, which fulfills all the requirements for multivalent approaches. Multivalency describes the binding of two or more entities and involves the simultaneous interaction between multiple and complementary functionalities. These interactions are considerably stronger than the individual bonding of a corresponding number of monovalent ligands to a multivalent receptor and are the most predominant phenomena in biological systems, particularly for recognition and attachment, signal transduction, and numerous cellular interactions [2].

Several examples of dendritic polymers as effector molecules that take advantage of their multiple functionalities are described in literature. A common characteristic is that multivalency plays an essential role in their therapeutic activity. Recently Gajbhiye et al. have reviewed the applications of dendritic polymers as therapeutic agents, with activity against prion diseases, Alzheimer's disease, inflammation, human immunodeficiency virus (HIV), herpes simplex virus (HSV), bacteria, and cancer. They have also shown that dendrimers were efficacious in tissue repair and prevention of scar tissue, and had potential for neutralizing toxins and removing drug and metal overdose from the body. Use of these nanostructures as bioenzymes and biosensors has also been reviewed [64].

It is important to mention the dendrimer-based product "VivaGel™" from Starpharma, which is the first of this kind of dendrimers to enter into Phase II human clinical trials. Vivagel is the first dendrimer-based product to have received Fast Track Status from the U.S. Food and Drug Administration (FDA) under an investigational new drug application for the prevention of genital herpes. A G4 poly(L-lysine)-based dendrimer with naphthalene disulfonic acid surface groups (i.e., SPL7013) is the active ingredient in Vivagel. The dendrimer binds to surface proteins on HIV, making it impossible for the viruses to attach to the binding sites on their cellular targets.

Another outstanding example is dendritic polyglycerol sulfate, primarily reported as a heparin analog by our group [84], which has emerged as a potential new anti-inflammatory agent based on a dendritic polymer (Fig. 15.11). This structure has been found to prolong the time of activated partial thromboplastin as thrombin, and to inhibit both the classical and alternative complement activation more effectively than heparin itself. The biocompatible and well-tolerated PG sulfate acts as multivalent selectin ligand mimetic and efficiently blocks leukocyte migration [85].

15.5 Dendritic Architectures for Biomedical In Vivo Imaging

15.5.1 Imaging Modalities and Design Approaches for Dendritic Architectures

Dendritic architectures have been employed for in vivo imaging purposes. A broad variety of probes have been synthesized for magnetic resonance imaging (MRI), X-ray imaging, ultrasound, radiodiagnostics, and optical imaging, as well as multimodal techniques combining these diagnostic techniques.

There are two major design principles that combine dendritic structures and a diagnostic entity. The first design principle is that dendrimers can covalently carry one or a multitude of signaling molecules on their periphery, thereby introducing physical detectability into a polymeric or dendritic conjugate. The dose size of contrast agents required for the less sensitive MRI and X-ray techniques can only be achieved through multimerization of the signaling molecules, for example, gadolinium complexes or triiodinated benzenes [26]. On the other hand, the more sensitive molecular imaging techniques radio imaging and fluorescence require not more than one signaling molecule, for example, one chelator for ^{99m}Tc or ^{111}In in Single Photon Emission Computed Tomography (SPECT), or one chelator for ^{64}Cu or ^{68}Ga in Positron Emission Tomography (PET), or one fluorescent dye in a given macromolecular dendritic targeting entity [86] as illustrated in Fig. 15.12a–c.

The other design principle employs a complementary class of imaging probes, i.e., inorganic nanoparticles combined with dendritic structures, which impart stability and biocompatibility to the nanoparticles and provide multivalent surface architecture. The most studied particles have been semiconductor quantum dots (QD) for fluorescence detection [87], gold nanoparticles and nanorods suited for optoacoustic/photoacoustic detection [88], and iron oxide particles for MRI [89] (Fig. 15.12d, e). Additionally, both design approaches can incorporate a multitude of targeting moieties in order to achieve multivalent target-binding properties, for example, by using low molecular weight peptides, glycans, or other biological or synthetic targeting moieties. Furthermore, a combination of different signaling units permits multimodal imaging applications, for example, by doping a quantum dot surface with a radioisotope or an iron oxide particle with a fluorescent dye [90].

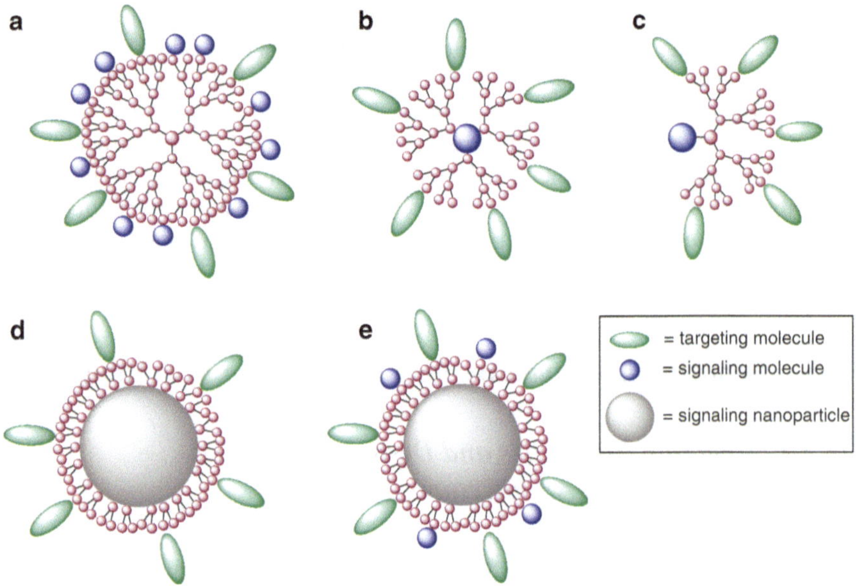

Fig. 15.12 Different architectures employing diagnostic signaling molecules with one or several signaling molecules, placed either centrally or at the periphery. Solid nanoparticles serve as a central entity, covered with targeting moieties and additional complementary signaling molecules for multimodality purposes

15.5.2 Imaging Conjugates Based on Organic Polymers and Dendrimers

One of the first applications of dendrimers in diagnostic imaging has been in MRI. Dendritic molecules have been applied as scaffold to multimerize paramagnetic gadolinium (Gd) complexes for contrast enhancement, tissue retention, and improved clearance characteristics. Most prominently, PAMAM, poly(lysine), and poly(ester) dendrimers of different generations/molecular weights have shown great utility for the design of paramagnetic MRI contrast agents [84]. For example, the gadolinium dendrimer Gadomer-17, consisting of a trimesic acid central moiety with a polylysine cascade that bears 24 DOTA (1,4,7,10-tetraazacyclododecane-1,4,7,10-tetraacetic acid) chelating groups, has been introduced as one of the first clinically tested dendrimer-based MRI imaging agents (Fig. 15.13). Gadomer 17 is a novel type of macromolecular contrast agent, optimized for blood pool imaging similar to the known linear Gd-DTPA (diethylenetriamine pentaacetic acid)-polylysine systems, but with superior pharmacokinetic and toxicological profile [84].

Subsequently, approaches have been expanded toward multivalent conjugates employing different ways of combining signaling entities with targeting units. The major purpose of covalent labeling with contrasting agents is the in vivo monitoring of carrier distribution, degradation, and excretion. One example is the use of folate/

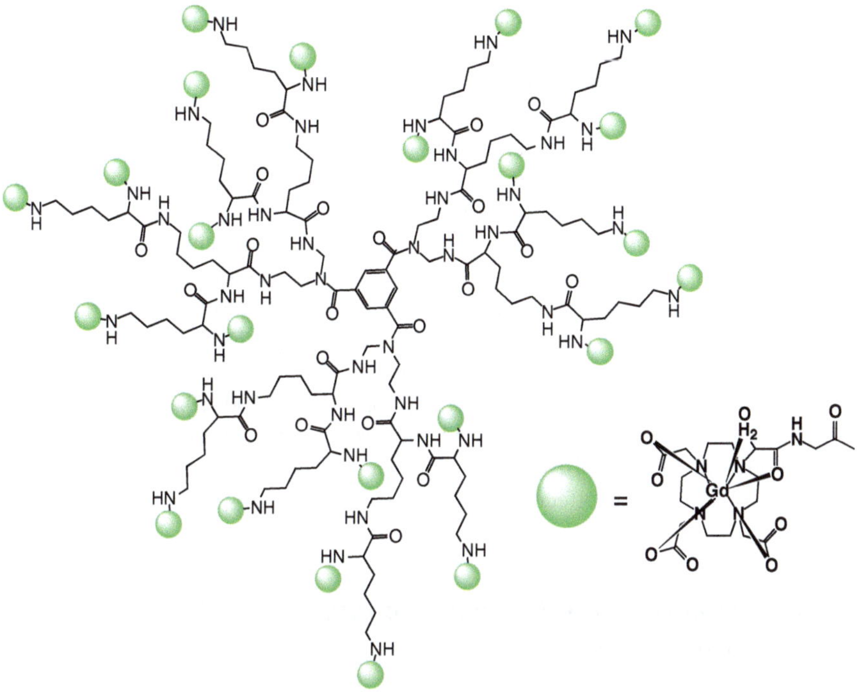

Fig. 15.13 Gd-bearing MRI contrast agent based on poly(lysine) dendrimer (Gadomer-17)

folic acid to enhance the specificity of dendrimers for the folic acid receptor, involving MRI complexes, fluorescent dyes, or a combination of both for dual imaging purposes [91, 92]. Imaging studies revealed an eightfold increase in tumor uptake of the targeted dendrimer compared to a non-targeted control dendrimer [91]. With a PEGylated PAMAM-folic acid conjugate, quantitative biodistribution studies were conducted after SPECT radiolabeling via a ^{99m}Tc-chelator [93].

Specific targeting nanoarchitectures also include low molecular weight peptides used in combination with signaling units. For instance, RGD (Arg-Gly-Asp)-peptide constructs, targeting integrins, have been multimerized on PAMAM surfaces and combined with DTPA, which served as chelator for either Gd^{3+} for MRI or ^{111}In for radioactive biodistribution studies. Additionally, the dendrimer was linked to a fluorescent dye to enable cellular microscopic detection. The design approach shown in Fig. 15.12c was chosen by Dijkgraaf et al. who used a central aromatic core linked to one DOTA for radiolabeling and four RGD peptide moieties, attached via "click" chemistry, for targeting [94]. Incorporating the capability of PET imaging, Almutairi et al. designed a biodegradable nanocarrier based on pentaerythritol and 2,2-bis(hydroxymethyl) propanoic acid to afford a dendrimer with flexible PEG chains and eight branching points for RGD peptide attachment. Each branching linkage contained a tyrosine, which allowed labeling with the PET isotope ^{76}Br or the SPECT isotope ^{125}I [95]. Multivalent RGD binding affinity to $\alpha_V\beta_3$ integrin receptors was improved 50-fold to 0.18 nM.

These and many further studies, employing either radioimaging [96] or fluorescence detection [97, 98], revealed the multivalency effect of dendrimer-RGD constructs targeting the $\alpha_V\beta_3$ integrin. Similarly, targeting of somatostatin receptors was achieved with dimeric and tetrameric [Tyr^3]-octreotide labeled with ^{111}In-DOTA. Here, the multimerization was reported to encounter complications, as the tetramer showed decreased receptor affinity and tumor uptake compared to monomer and dimer, probably due to unfavorable lipophilicity resulting from the synthetic design approach [97]. Imaging conjugates employing a central fluorescent dye, multivalently decorated with targeting residues (Fig. 15.12b) were described by the group of Achilefu et al. who synthesized a heptmethine dye with a dendritic polycarboxy array for further conjugation, for example, to glucosamins [99] or RGD motifs [96].

Targeted dendrimers employing antibodies, proteins, and oligonucleotides as targeting vehicles were described in several approaches [100–102]. Thomas et al. described a fluorescein-labeled PAMAM dendrimer carrying approximately 2 epidermal growth factor (EGF) proteins per dendrimer, critically highlighting its observed superagonistic effect above monomeric EGF itself [95].

15.5.3 Dendrimers in Combination with Inorganic Nanoparticle Cores

Semiconductor quantum dots have been widely employed for diagnostic purposes due to their bright and tunable fluorescence. The ability of multiplex labeling, ligand interaction studies based on quenching and Fluorescence Resonance Energy Transfer (FRET), and in vivo targeting have yielded a fundamental database of applications in biology and biomedicine in the past few years [96]. Surface modification and stabilization have involved dendrimers and dendritic polymers in order to render such particles target specific. The use of PAMAM in combination with PEG-folate building blocks grafted to the surface of quantum dots through direct ligand-exchange reactions resulted in increased cellular internalization [103]. It was shown that a single G7 PAMAM dendrimer was able to encapsulate a mercury telluride (HgTe) quantum dot particle, whereas several G5 dendrimers were required for this purpose [104].

Gold layers and nanoparticles are well established materials in biomedical research due to their applicability not only for ligand–ligand interaction studies with Surface Plasmon Resonance (SPR) [105], but also as fluorescent or photoacoustic signaling entities [87]. The application of gold nanorods for the photothermal treatment of tumors was demonstrated by covering the gold surface with partially thiolated PAMAM dendrimer building blocks to which RGD peptide moieties were coupled. The constructs exhibited cell-destructive effects upon irradiation with near-infrared light [106]. The comparison of PAMAM dendrimer cores with gold nanoparticle cores of identical surface properties showed similar targeting behavior when equipped with folic acid [107], thus demonstrating a certain general applicability of particle-based multivalent targeting.

Iron oxide nanoparticles are well-established MRI agents. Dendrimers can be employed to stabilize and modify the particle surface, either by growing particles in

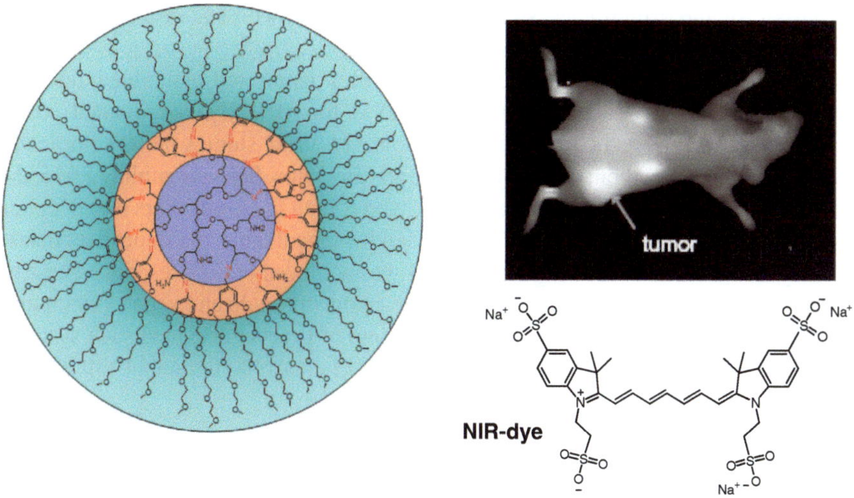

Fig. 15.14 (*Left*) Core-shell architecture based on PEI with a PEG shell, and (*right*) near-infrared dye for in vitro and in vivo studies

the presence of dendrimers or by post synthetic interventions [88]. Multifunctional, superparamagnetic iron oxides conjugated to peptides and antibodies have been shown to exhibit multivalent binding properties and improved cellular and in vivo uptake, such as integrin-targeted systems [108] or bombesin-conjugated systems [109]. Along with gold particles and quantum dots, iron oxides particles can only be fabricated with comparably broad particle size distribution, and therefore, attempts toward structurally defined surface modification with dendrimers have been rather limited to date.

15.5.4 Noncovalently Encapsulated Signaling Molecules and Self-Assembling Architectures

The noncovalent encapsulation of different low molecular weight signaling molecules such as fluorescent dyes, metal complexes, and radioactive entities is a principal alternative route for generating nanosized imaging agents. Encapsulated signaling molecules are primarily useful for the study of drug delivery and release capabilities of nanocarriers, with the dye or radiotracer serving as a model payload for an otherwise difficult-to-detect therapeutic drug molecule.

Dendritic multi-shell architectures based on polyglycerol and PEG have been shown to encapsulate different guest molecules, such as Nile Red, cyanine dye, doxorubicin, and methotrexate (Fig. 15.14). With the near-infrared cyanine dye, in vivo imaging of tumors based on passive macromolecular targeting was demonstrated [110]. Using pH-responsive dendrimers, which carried a pH-cleavable PEG shell conjugated via imine moieties, in vivo tumor uptake was demonstrated with fluorescence imaging, and cell toxicity and drug release studies were conducted with encapsulated doxorubicin [72].

Non-covalent transport of imaging molecules using self-assembled micellar nanostructures included gadolinium complexes for MRI [111], iodinated X-ray contrast agents, and radioactive chelates; however, these studies mostly employed block copolymers without distinct dendritic architectures. Self-assembling PAMAM systems were described by Criscione et al., who forced the dendrimers into micellar assemblies after partial fluorination of the dendrimer subunits. pH-dependent disassembly and in vivo monitoring by ^{19}F-MRI was achieved [112]. Responsiveness to external stimuli, for example, a pH change upon internalization, can principally improve drug delivery and selectivity. The group of Fréchet designed such systems based on PEGylated polyester dendrons and acid-cleavable units that self-assembled into micelles, containing a dye payload [113].

15.6 Dendritic Architectures in Theranostics

A combination of the above-described approaches provides the basis for a novel "theranostic" design. For many authors, September 25, 1998 is considered the birth of "theranostics" [114] as the FDA granted the approval for both Genentech's Herceptin® for the treatment of Stage IV breast cancer and Dako's HercepTest® for diagnosis of Her2 overexpression. But these rudimentary examples of first theranostics are still a long way from an early diagnosis of diseases on the molecular level. The goal of theranostics is to enhance treatment decisions by providing information to clinicians and their patients. The theranostic concept of "find, fight, and follow" includes early diagnosis with specific agents for an image of individual cells, the supply of active substances, and then the therapy [115]. The main elements of the diagnosis in theranostics include determining the genetic predisposition, characterization of the disease stage, and monitoring the healing progress [116]. Theranostic applications of dendrimers require external binding of a drug and a fluorescent probe (one-package system), and the ability to interact with the target such as cell walls and/or proteins via a targeting device (Fig. 15.15).

Dendritic polymer versatility and compatibility with nanoscale building blocks such as genetic material, bioactive molecules, and fluorescent probes, make them promising candidates for theranostic applications [117]. Comprehensive in vivo research is necessary in order to evaluate the true value of dendrimers as multifunctional packages. It would be especially attractive to investigate their adsorption and interaction with biological surfaces, and to determine the associated uptake mechanisms, long-term effects, and bio-elimination.

Backer et al. reported a boronated fifth-generation PAMAM dendrimer conjugated to a vascular endothelial growth factor (VEGF) ligand and tagged with the near-IR Cy5 dye to allow for near-IR fluorescent imaging of the bioconjugate in vitro and in vivo [118]. They reported an accumulation of the conjugate in 4T1 breast carcinoma cells with a toxin-VEGF fusion protein that selectively killed

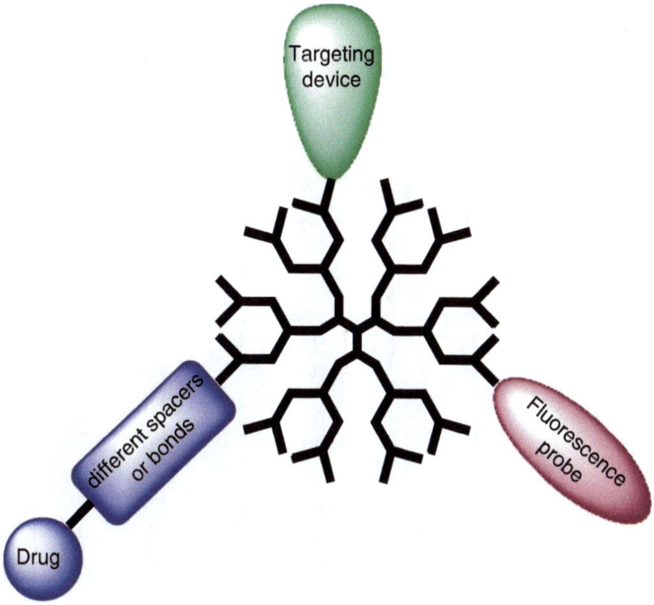

Fig. 15.15 Dendrimer properties for theranostic applications

VEGFR-2-overexpressing endothelial cells. Extensive investigation has been performed by the same research group combining dendritic scaffold with targeting modalities (e.g., folic acid, riboflavin, etc.), anticancer drugs (methotrexate (MTX), Taxol®) and fluorescent probes. These data provide the groundwork for future studies using these compounds as potential theranostics.

Baker et al. synthesized PAMAM dendrimer-based multifunctional cancer therapeutic conjugates [119]. They conjugated generation-five PAMAM to fluorescein isothiocyanate (FITC, an imaging agent), folic acid (FA, targets overexpressed folate receptors on specific cancer cells), and paclitaxel (Taxol, a chemotherapeutic drug) for in vitro and in vivo studies of a targeted delivery of chemotherapeutic and imaging agents to specific cancer cells. They characterized their conjugates with gel permeation chromatography (GPC), nuclear magnetic resonance spectroscopy (NMR), potentiometric titration, high-performance liquid chromatography (HPLC), and UV spectroscopy. The results indicated that the conjugate can be a promising nano-platform for a combination of therapeutic use with an imaging modality.

Pei et al. have linked different PEGylated PAMAM dendrimers with doxorubicin (DOX) via acid-sensitive *cis*-aconityl linkage and acid-insensitive succinic acid linkage to produce PEG-PAMAM-*cis*-aconityl-DOX (PPCD) and PEG-PAMAM-succinic-DOX (PPSD) conjugates [120]. PPSD internalized into SKOV-3 cells via clathrin-mediated and adsorptive endocytosis, respectively. They have shown that DOX diffused into the nuclei after release from PPCD in acidic lysosomes. In vivo fluorescence imaging studies demonstrated that PPCD with the highest PEGylation degree accumulated in tumor sites most efficiently (Fig. 15.16).

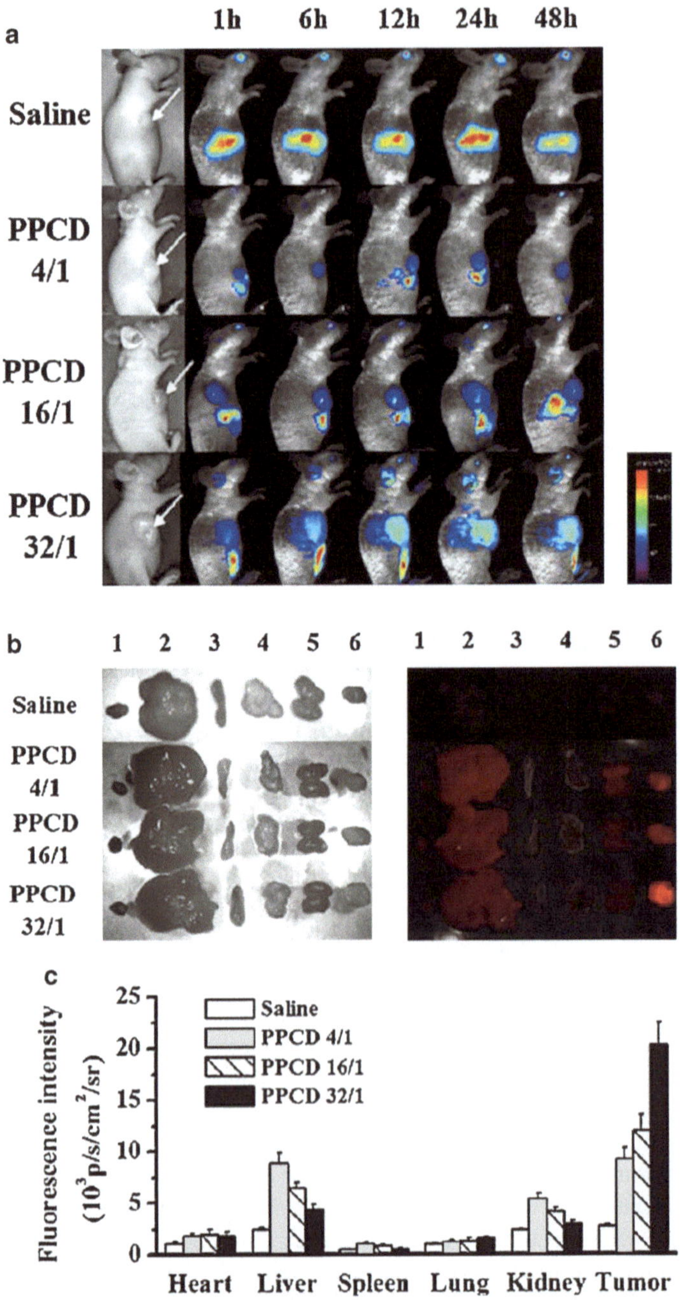

Fig. 15.16 (**a**) Intravenous injection of PPCD conjugates for in vivo imaging of subcutaneous SKOV-3 ovarian carcinoma tumor-bearing nude mice. *Arrow*: position of the tumor. (**b**) Ex vivo evaluation of excised tissues and tumors (1: heart, 2: liver, 3: spleen, 4: lung, 5: kidney, 6: tumor). (**c**) Quantification of the fluorescence intensity. Reproduced with permission from [119]

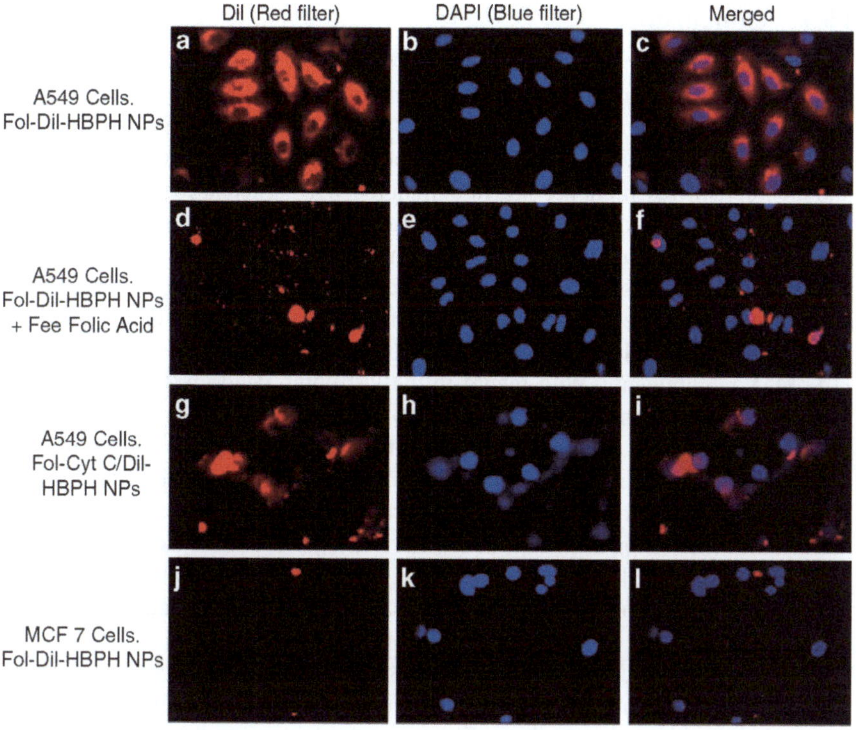

Fig. 15.17 Fluorescence images of A549 lung epithelial cells and breast adenocarcinoma MCF 7 cells, incubated with folate-HBPH nanoparticles. Enhanced fluorescence was observed in the cytoplasm of A549 cells (**a–c**), whereas no significant internalization was observed in MCF 7 cells (**j–l**). A549 cells pre-incubated with free folic acid showed minimal uptake (**d–f**); A549 cells incubated with folate-functionalized HBPH nanoparticles containing both Cyt C and DiI (Fol-Cyt C/DiI-HBPH NP) induced significant cell death (**g–i**). Reproduced with permission from [120]

A good example employing hyperbranched polymers has been reported by Santra et al. [121]. This group synthesized a new water-soluble, hyperbranched polyhydroxyl (HBPH) nanoparticle and formulated this nanoparticle for the simultaneous encapsulation of cytochrome *c* (Cyt *c*) and the near-infrared dye indocyanine green (ICG), and folic acid for targeting to folate-expressing cancer cells (Fig. 15.17). It was reported that this HBPH conjugate could serve as a targeted transmembrane carrier, delivering Cyt C to cancer cells that overexpressed the folate receptor and induced apoptosis. This novel class of polymers exhibited unique physicochemical and biological properties, which have great potential use in therapeutic and diagnostic applications.

The above-mentioned example of dendritic polyglycerol sulphate [83] is a good example of a potential theranostic that does not need an external effector moiety once it is fluorescently labeled. Ongoing in vivo research has demonstrated the feasibility of such a system to accumulate in inflamed tissue, which allows the diagnosis of several diseases, including rheumatism.

15.7 Conclusions

Dendrimers and dendritic polymers have emerging as a powerful, multifunctional nanotherapeutic platform for imaging and therapeutic applications. Looking into the future, there are a number of medical needs where multifunctional dendrimers and their derivatives might become important tools, in particular for early detection, diagnosis, and personalized treatment of diseases. It is important to investigate the adsorption and interaction of new multifunctional nanoparticles with malignant materials (such as tissue specimens and circulating tumor cells) to improve their biocompatibility, to test long-term effects, and to standardize the dendrimers and their derivatives for clinical applications, which must be done in compliance with U.S. FDA requirements. With regard to the last point, dendrimer-based products need further improvement on a high level of productivity with an excellent cost–benefit ratio. VivaGel, Starpharma's dendrimer-based microbicide for the prevention of HIV and HSV injection, is currently in Phase II clinical trials. These are the first human dendrimer pharmaceutical clinical trials. Baker and coworkers synthesized a fifth-generation PAMAM dendrimer conjugated to fluorescein isothiocyanate for imaging [118]. In preclinical studies they have used recombinant fibroblast growth factor-1 for tumor targeting. These examples show that dendritic polymers will emerge and enter clinical trials in the coming years.

In conclusion, dendrimers may be used as tools for theranostics in areas such as the detection and treatment of cardiovascular, cancer, and inflammatory diseases due to their multifunctional nature. The property of imaging, detection, and treatment of diseases at the same time is interesting for clinical trials. Aside from these advantages, issues of safety and complexity must be investigated.

References

1. Chabner BA, Roberts TG (2005) Timeline: chemotherapy and the war on cancer. Nat Rev Cancer 5:65–72
2. Haag R, Kratz F (2006) Polymer therapeutics: concepts and applications. Angew Chem Int Ed 45:1198–1215
3. Mammen M, Chio SK, Whitesides GM (1998) Polyvalent interactions in biological systems: implications for design and use of multivalent ligands and inhibitors. Angew Chem Int Ed 37:2755–2794; Angew Chem 110:2908–2963
4. Joshi N, Grinstaff M (2008) Applications of dendrimers in tissue engineering. Curr Top Med Chem 7:1225–1236
5. Klajnert B, Bryszewska M (2001) Dendrimers: properties and applications. Acta Biochim Pol 48:199–208
6. Ong W, Gomez-Kaifer M (2004) Dendrimers as guests in molecular recognition phenomena. Chem Commun 15:1677–1683
7. Kaifer AE (2007) Electron transfer and molecular recognition in metallocene-containing dendrimers. Eur J Inorg Chem 32:5015–5027
8. Van Heerbeek R, Kamer PCJ, Van Leeuwen PWNM, Reek JNH (2002) Dendrimers as support for recoverable catalysts and reagents. Chem Rev 102:3717–3756

9. Andres R, De Jesus E, Flores JC (2007) Catalysts based on palladium dendrimers. New J Chem 31:1161–1191
10. Flomenboma O, Amir RJ, Shabat D, Klafter J (2005) Some new aspects of dendrimer applications. J Lumin 111:315–325
11. Aulenta F, Hayes W, Rannard S (2003) Dendrimers: a new class of nanoscopic containers and delivery devices. Eur Polym J 39:1741–1771
12. Boas U, Heegaard PMH (2004) Dendrimers in drug research. Chem Soc Rev 33:43–63
13. Hawker CJ, Fréchet JMJ (1991) One-step synthesis of hyperbranched dendritic polyesters. J Am Chem Soc 113:4583–4588
14. Vögtle F, Richardt G, Werner N (2007) Dendritische Moleküle: Konzepte, Synthesen, Eigenschaften, Anwendungen. Teubner, Wiesbaden, Germany, pp 17–28
15. Tomalia DA, Baker H, Dewald J, Hall M, Kallos G, Martin S, Roeck J, Ryder J, Smith P (1985) A new class of polymers: starburst-dendritic macromolecules. Polym J 17:117–132
16. Newkome GR, Zhong-qi Y, Baker GR, Gupta VK (1985) Micelles. Part 1. Cascade molecules: a new approach to micelles. A [27]-arborol. J Org Chem 50:2003–2004
17. Hawker CJ, Fréchet JMJ (1990) A new convergent approach to monodisperse dendritic macromolecules. J Chem Soc Chem Commun 15:1010–1013
18. Hawker CJ, Fréchet JMJ (1990) Preparation of polymers with controlled molecular architecture. A new convergent approach to dendritic macromolecules. J Am Chem Soc 112:7638–7647
19. Adronov A, Fréchet JMJ (2002) Light-harvesting dendrimers. Chem Commun 1701–1710
20. Stiriba SE, Frey H, Haag R (2002) Dendritic polymers in biomedical applications: from potential to clinical use in diagnostics and therapy. Angew Chem Int Ed 41:1329–1334
21. Gao C, Yan D (2004) Hyperbranched polymers: from synthesis to applications. Prog Polym Sci 29:183–275
22. Sunder A, Mülhaupt R, Haag R, Frey H (2000) Hyperbranched polyether polyols: a modular approach to new materials with complex polymer architectures. Adv Mater 12:235–239
23. Sunder A, Hanselmann R, Frey H, Mülhaupt R (1999) Controlled synthesis of hyperbranched polyglycerols by ring-opening multibranching polymerization. Macromolecules 32:4240–4246
24. Lee CC, MacKay JA, Fréchet JMJ, Szoka FC (2005) Designing dendrimers for biological applications. Nat Biotechnol 23:1517–1526
25. Kolhe P, Khandare J, Pillai O, Kannan S, Lieh-Lai M, Kannan RM (2006) Preparation, cellular transport, and activity of polyamidoamine-based dendritic nanodevices with a high drug payload. Biomaterials 27:660–669
26. Menjoge AR, Kannan RM, Tomalia DA (2010) Dendrimer-based drug and imaging conjugates: design considerations for nanomedical applications. Drug Discov Today 15:171–185
27. Gillies ER, Fréchet JMJ (2002) Designing macromolecules for therapeutic applications: polyester dendrimer poly(ethylene oxide) "bow-tie" hybrids with tunable molecular weight and architecture. J Am Chem Soc 124:14137–14146
28. Gillies ER, Fréchet JMJ, Szoka FC (2005) Biological evaluation of polyester dendrimer—poly(ethylene oxide) "bow-tie" hybrids with tunable molecular weight and architecture. Mol Pharm 2:129–138
29. Minko T, Dharap SS, Pakunlu RI, Wang Y (2004) Molecular targeting of drug delivery systems to cancer. Curr Drug Targets 5:389–406
30. Luo Y, Prestwich GD (2002) Cancer-targeted polymeric drugs. Curr Cancer Drug Targets 2:209–226
31. Matsumura Y, Maeda H (1986) A new concept for macromolecular therapeutics in cancer chemotherapy: mechanism of tumoritropic accumulation of proteins and the antitumor agent Smancs. Cancer Res 46:6387–6392
32. Maeda H, Wu J, Sawa T, Matsumura Y, Hori K (2000) Tumor vascular permeability and the EPR effect in macromolecular therapeutics: a review. J Control Release 65:271–284
33. Jain RK (1987) Transport of molecules across tumor vasculature. Cancer Metathesis Rev 6:559–593
34. Jain RK (1987) Transport of molecules in the tumor interstitium: a review. Cancer Res 47:3039–3051

35. Fox ME, Szoka FC, Fréchet JMJ (2009) Soluble polymer carriers for the treatment of cancer: the importance of molecular architecture. Acc Chem Res 42:1141–1151
36. Knop K, Hoogenboom R, Fischer D, Schubert US (2010) Anwendung von Poly(ethylenglycol) beim Wirkstoff-Transport: Vorteile, Nachteile und Alternativen. Angew Chem 122:6430–6452
37. Yazawa K, Fujimori M, Amano J, Kano Y, Taniguchi S (2000) *Bifidobacterium longum* as a delivery system for cancer gene therapy: selective localization and growth in hypoxic tumors. Cancer Gene Ther 2:269–274
38. Stroh M, Zimmer JP, Duda DG, Levchenko TS, Cohen KS, Brown EB, Scadden DT, Torchilin VP, Bawendi MG, Fukumura D, Jain RK (2005) Quantum dots spectrally distinguish multiple species within the tumor milieu in vivo. Nat Med 6:678–682
39. Noguchi Y, Wu J, Duncan R, Strohalm J, Ulbrich K, Akaike T, Maeda H (1998) Early phase tumor accumulation of macromolecules: a great difference in clearance rate between tumor and normal tissues. Jpn J Cancer Res 89:307–314
40. Minko T, Khandare J, Jayant S (2007) In Matyjaszewski K, Gnanou Y, Leibler L (eds) Macromolecular engineering: from precise macromolecular synthesis to macroscopic material properties and application. Wiley-VCH Verlag GmbH & Co.KGaA, Weinheim
41. Jayant S, Khandare J, Wang Y, Singh AP, Vorsa N, Minko T (2007) Targeted sialic acid-doxorubicin prodrugs for intracellular delivery and cancer treatment: Polymeric Drugs 24:2541–2595
42. Okuda T, Kawakami S, Akimoto N, Niidome T, Yamashita F, Hashida M (2006) PEGylated lysine dendrimers for tumor-selective targeting after intravenous injection in tumor-bearing mice. J Control Release 116:330–336
43. Kitchens KM, Kolhatkar RB, Swaan PW, Ghandehari H (2008) Endocytosis inhibitors prevent poly(amidoamine) dendrimer internalization and permeability across $CaCo_2$ cells. Mol Pharm 5:364–369
44. Najlah M, D'Emanuele A (2006) Crossing cellular barriers using dendrimer nanotechnologies. Curr Opin Pharmacol 6:522–527
45. Conner SD, Schmid SL (2003) Regulated portals of entry into the cell. Nature 422:37–44
46. Mayor S, Pagano RE (2007) Pathways of clathrin-independent endocytosis. Nat Rev Mol Cell Biol 8:603–612
47. Lopes L, Godoy LMF, de Oliveira CC, Gabardo J, Schadeck RJG, de Freitas Buchi D (2006) Phagocytosis, endosomal/lysosomal system and other cellular aspects of macrophage activation by canova medication. Micron 37:277–287
48. Schmid EM, McMahon HAT (2007) Integrating molecular and network biology to decode endocytosis. Nature 448:883–888
49. Benmerah A, Lamaze C (2007) Clathrin-coated pits: vive la différence? Traffic 8:970–982
50. Ungewickell EJ, Hinrichsen L (2007) Endocytosis: clathrin-mediated membrane budding. Curr Opin Cell Biol 19:417–425
51. Nabi IR, Phuong ULJ (2003) Caveolae/raft-dependent endocytosis. Cell Biol 161:673–677
52. Gold S, Monaghan P, Mertens P, Jackson T (2010) A clathrin independent macropinocytosis-like entry mechanism used by bluetongue virus-1 during infection of BHK cells. PLoS One 5(6):e11360
53. Wiwattanapatapee R, Carreno-Gomez B, Malik N, Duncan R (2000) Anionic PAMAM dendrimers rapidly cross adult rat intestine in vitro: a potential oral delivery system? Pharm Res 17:991–998
54. Jevprasesphant R, Penny J, Attwood D, D'Emanuele A (2004) Transport of dendrimer nanocarriers through epithelial cells via the transcellular route. J Control Release 97:259–267
55. Mosbah IB, Franco-Go R, Abdennebi HB, Hernandez R, Escolar G, Saidane D, Rosello-Catafau J, Peralta C (2006) Effects of polyethylene glycol and hydroxyethyl starch in University of Wisconsin preservation solution on human red blood cell aggregation and viscosity. Transplant Proc 38:1229–1235
56. Kainthan RK, Gnanamani M, Ganguli M, Ghosh T, Brooks DE, Maiti S, Kizhakkedathu JN (2006) Blood compatibility of novel water soluble hyperbranched polyglycerol-based multivalent cationic polymers and their interaction with DNA. Biomaterials 27:5377–5390

57. Fernandes EGR, De Queiroz AAA, Abraham GA, San Roma J (2006) Antithrombogenic properties of bioconjugate streptokinase-polyglycerol dendrimers. J Mater Sci Mater Med 17:105–111
58. Schatzlein AG, Zinselmeyer BH, Elouzi A, Dufes C, Chim YTA, Roberts CJ, Davies MC, Munro A, Gray AI, Uchegbu IF (2005) Preferential liver gene expression with polypropylenimine dendrimers. J Control Release 101:247–258
59. Singh P (1996) StarburstTM dendrimers: a novel matrix for multifunctional reagents in immunoassays. Clin Chem 42:1567–1569
60. Kojima C, Tsumura S, Harada A, Kono K (2009) A collagen-mimic dendrimer capable of controlled release. J Am Chem Soc 131:6052–6053
61. Myc A, Majoros IJ, Thomas TP, Baker JR Jr (2007) Dendrimer-based targeted delivery of an apoptotic sensor in cancer cells. Biomacromolecules 8:13–18
62. Calderón M, Quadir MA, Strumia M, Haag R (2010) Functional dendritic polymer architectures as stimuli-responsive nanocarriers. Biochimie 92:1242–1251
63. Tomalia DA, Naylor AM, Goddard WA (1990) Starburst dendrimers: molecular-level control of size, shape, surface chemistry, topology, and flexibility from atoms to macroscopic matter. Angew Chem Int Ed 29:138–175
64. Gajbhiye V, Palanirajan VK, Tekade RK, Jain NK (2009) Dendrimers as therapeutic agents: a systematic review. J Pharm Pharmacol 61:989–1003
65. Maciejewski M (1982) Concepts of trapping topologically by shell molecules. J Macromol Sci Chem A17:689–703
66. D'Emanuele A, Attwood D (2005) Dendrimer-drug interactions. Adv Drug Deliv Rev 57:2147–2162
67. Svenson S, Tomalia DA (2005) Dendrimers in biomedical applications—reflections on the field. Adv Drug Deliv Rev 57:2106–2129
68. Quadir MA, Calderón M, Haag R (2011) Drug Delivery in Oncology: From Basic Research to Cancer Therapy. Dendritic Polymers in Oncology: Facts, Features, and Applications, Weinheim 513–551
69. Jansen JF, De Brabander-Van Den Berg EMM, Meijer EW (1994) Encapsulation of guest molecules into a dendritic box. Science 266:1226–1229
70. Haensler J, Szoka FC Jr (1993) Polyamidoamine cascade polymers mediate efficient transfection of cells in culture. Bioconjug Chem 4:372–379
71. Newkome GR, Moorefield CN, Baker GR, Saunders MJ, Grossman SH (1991) Unimolecular micelles. Angew Chem Int Ed 30:1178–1180
72. Xu S, Luo Y, Gräser R, Warnecke A, Kratz F, Hauff P, Licha K, Haag R (2009) Development of pH-responsive core-shell nanocarriers for delivery of therapeutic and diagnostic Agents. Bioorg Med Chem Lett 19:1030–1034
73. Xu S, Luo Y, Haag R (2007) Water-soluble pH-responsive dendritic core-shell nanocarriers for polar dyes based on poly(ethylene imine). Macromol Biosci 7:968–974
74. Xu S, Krämer M, Haag R (2006) pH-responsive dendritic core-shell architectures as amphiphilic nanocarriers for polar drugs. J Drug Target 14:367–374
75. Krämer M, Stumbé JF, Türk H, Krause S, Komp A, Delineau L, Prokhorova S, Kautz H, Haag R (2002) pH-responsive molecular nanocarriers based on dendritic core-shell architectures. Angew Chem Int Ed 41:4252–4256
76. Radowski MR, Shukla A, v. Berlepsch H, Böttcher C, Pickaert G, Rehage H, Haag R (2007) Supramolecular aggregates of dendritic multishell architectures as universal nanocarriers. Angew Chem Int Ed 46:1265–1292
77. Fischer W, Calderón M, Schulz A, Andreou I, Weber M, Haag R (2010) Dendritic polyglycerols with oligoamine shells show low toxicity and high transfection efficiency in vitro. Bioconj Chem 21:1744–1752
78. Ofek P, Fischer W, Calderón M, Haag R, Satchi-Fainaro R (2010) In vivo delivery of siRNA to tumors and their vasculature by novel dendritic nanocarriers. FASEB J 24: 3122–3134
79. Malik N, Evagorou EG, Duncan R (1999) Dendrimer-platinate: a novel approach to cancer chemotherapy. Anticancer Drugs 10:767–776

80. Duncan R, Malik N (1996) Dendrimers: biocompatibility and potential for delivery of anticancer agents. Proc Int Symp Control Release Bioact Mater 23:105–106
81. Gillies ER, Fréchet JMJ (2005) Dendrimers and dendritic polymers in drug delivery. Drug Discov Today 10:35–43
82. Kratz F, Müller IA, Ryppa C, Warnecke A (2008) Prodrug strategies in anticancer chemotherapy. ChemMedChem 3:20–53
83. Calderón M, Welker P, Licha K, Fichtner I, Graeser R, Haag R, Kratz F (2011) Development of efficient acid cleavable multifunctional prodrugs derived from dendritic polyglycerol with a poly(ethylene glycol) shell. J Control Release 151(3):295–301
84. Türk H, Haag R, Alban S (2004) Dendritic polyglycerol sulfates as new heparin analogues and potent inhibitors of the complement system. Bioconjug Chem 15:162–167
85. Dernedde J, Rausch A, Weinhart M, Enders S, Tauber R, Licha K, Schirner M, Zügel U, von Bonin A, Haag R (2010) Dendritic polyglycerol sulfates as multivalent inhibitors of inflammation. Proc Natl Acad Sci USA 107:19679–19684
86. Hamoudeh M, Kamleh MA, Diab R, Fessi H (2008) Radionuclides delivery systems for nuclear imaging and radiotherapy of cancer. Adv Drug Deliv 60:1329–1346
87. Zhou M, Ghosh I (2007) Quantum dots and peptides: a bright future together. Biopolymers 8:325–339
88. Wei A, Leonov AP, Wei Q (2010) Gold nanorods: multifunctional agents for cancer imaging and therapy. Methods Mol Biol 624:119–130
89. Shen M, Shi X (2010) Dendrimer-based organic/inorganic hybrid nanoparticles in biomedical applications. Nanoscale 2:1596–1610
90. Licha K, Schirner M (2008) Emerging optical imaging technologies: contrast agents. In: Azar FS, Intes X (eds) Translational multimodality optical imaging. Artech House, Boston, pp 327–337
91. Swanson SD, Kukowska-Latallo JF, Patri AK, Chen C, Ge S, Cao Z, Kotlyar A, East AT, Baker JR (2008) Targeted gadolinium-loaded dendrimer nanoparticles for tumor-specific magnetic resonance contrast enhancement. Int J Nanomedicine 3:201–210
92. Thomas TP, Majoros IJ, Kotlyar A, Kukowska-Latallo JF, Bielinksa A (2005) Targeting and inhibition of cell growth by an engineered dendritic synthetic nanoscale bioconjugate. J Med Chem 48:3729–3735
93. Zhang Y, Sun Y, Xu X, Zhang X, Zhu H, Huang L, Qi Y, Shen YM (2010) Synthesis, biodistribution, and microsingle photon emission computed tomography (SPECT) imaging study of technetium-99m labeled PEGylated dendrimer poly(amidoamine) (PAMAM)-folic acid conjugates. J Med Chem 53:3262–3272
94. Dijkgraaf I, Rijnders AY, Soede A, Dechesne AC, van Esse GW, Brouwer AJ, Corstens FH, Boerman OC, Rijkers DT, Liskamp RM (2007) Synthesis of DOTA-conjugated multivalent cyclic-RGD peptide dendrimers via 1,3-dipolar cycloaddition and their biological evaluation: implications for tumor targeting and tumor imaging purposes. Org Biomol Chem 21:935–944
95. Almutairi A, Rossin R, Shokeen M, Hagooly A, Ananth A, Capoccia B, Guillaudeu S, Abendschein D, Anderson DJ, Welch MJ, Fréchet JMJ (2009) Biodegradable dendritic positron-emitting nanoprobes for the noninvasive imaging of angiogenesis. Proc Natl Acad Sci USA 106:685–690
96. Liu S (2006) Radiolabeled multimeric cyclic RGD peptides as integrin alpha-v-beta-3 targeted radiotracers for tumor imaging. Mol Pharm 3:472–487
97. Ye Y, Bloch S, Xu B, Achilefu S (2006) Design, synthesis, and evaluation of near infrared luorescent multimeric RGD peptides for targeting tumors. J Med Chem 49:2268–2275
98. Yim CB, Dijkgraaf I, Merkx R, Versluis C, Eek A, Mulder GE, Rijkers DT, Boerman OC, Liskamp RS (2010) Synthesis of DOTA-conjugated multimeric [Tyr3]octreotide peptides via a combination of Cu(I)-catalyzed "click" cycloaddition and thio acid/sulfonyl azide "sulfo-click" amidation and their in vivo evaluation. J Med Chem 53:3944–3953
99. Ye Y, Bloch S, Kao J, Achilefu S (2005) Multivalent carbocyanine molecular probes: synthesis and applications. Bioconjug Chem 16:51–61

100. Kobayashi H, Sato N, Saga T, Nakamoto Y, Ishimori T, Toyama S, Togashi K, Konishi J, Brechbiel MW (2000) Monoclonal antibody-dendrimer conjugates enable radiolabeling of antibody with markedly high specific activity with minimal loss of immunoreactivity. Eur J Nucl Med 27:1334–1339
101. Sato N, Kobayashi H, Saga T, Nakamoto Y, Ishimori T, Togashi K, Fujibayashi Y, Konishi J, Brechbiel MW (2001) Tumor targeting and imaging of intraperitoneal tumors by use of antisense oligo-DNA complexed with dendrimers and/or avidin in mice. Clin Cancer Res 7:3606–3612
102. Thomas TP, Shukla R, Kotlyar A, Liang B, Ye JY, Norris TB, Baker JR Jr (2008) Dendrimer-epidermal growth factor conjugate displays superagonist activity. Biomacromolecules 9:603–609
103. Zhao Y, Liu S, Li Y, Jiang W, Chang Y, Pan S, Fang X, Wang YA, Wang J (2010) Synthesis and grafting of folate-PEG-PAMAM conjugates onto quantum dots for selective targeting of folate-receptor-positive tumor cells. J Colloid Interface Sci 350:44–50
104. Priyam A, Blumling DE, Knappenberger KL Jr (2010) Synthesis, characterization, and self-organization of dendrimer-encapsulated HgTe quantum dots. Langmuir 26:10636–10644
105. Dernedde J, Enders S, Reissig HU, Roskamp M, Schlecht S, Yekta S (2009) Inhibition of selectin binding by colloidal gold with functionalized shells. Chem Commun 28:932–934
106. Li Z, Huang P, Zhang X, Lin J, Yang S, Liu B, Gao F, Xi P, Ren Q, Cui D (2010) RGD-conjugated dendrimer-modified gold nanorods for in vivo tumor targeting and photothermal therapy. Mol Pharm 7:94–104
107. Shi X, Wang SH, Lee I, Shen M, Baker JR Jr (2009) Comparison of the internalization of targeted dendrimers and dendrimer-entrapped gold nanoparticles into cancer cells. Biopolymers 91:936–942
108. Yu MK, Park J, Jeong YY, Moon WK, Jon S (2010) Integrin-targeting thermally cross-linked superparamagnetic iron oxide nanoparticles for combined cancer imaging and drug delivery. Nanotechnology 21:415102
109. Lee CM, Jeong HJ, Cheong SJ, Kim EM, Kim DW, Lim ST, Sohn MH (2010) Prostate cancer-targeted imaging using magnetofluorescent polymeric nanoparticles functionalized with bombesin. Pharm Res 27:712–721
110. Quadir MA, Radowski MR, Kratz F, Licha K, Hauff P, Haag R (2010) Dendritic multishell architectures for drug and dye transport. J Control Release 132:289–294
111. Nakamura E, Makino K, Okano T, Yamamoto T, Yokoyama M (2006) A polymeric micelle MRI contrast agent with changeable relaxivity. J Control Release 114:325–333
112. Criscione JM, Le BL, Stern E, Brennan M, Rahner C, Papademetris X, Fahmy TM (2009) Self-assembly of pH-responsive fluorinated dendrimer-based particulates for drug delivery and noninvasive imaging. Biomaterials 23–24:3946–3955
113. Gillies ER, Jonsson TB, Fréchet JMJ (2004) Stimuli-responsive supramolecular assemblies of linear-dendritic copolymers. J Am Chem Soc 126:11936–11943
114. Gilham I (2002) Theranostics: an emerging tool in drug discovery and commercialisation. Drug Discov World Fall 1:17–23
115. Riehemann K, Schneider SW, Luger TA, Godin B, Ferrari M, Fuchs H (2009) Nanomedicine: challenge and oppurtunities. Angew Chem Int Ed 48:872–897
116. Wagner V, Wechsler D (2004) Technologiefrüherkennung, Nanobiotechnologie II: Anwendungen in der Medizin und Pharmazie, Band 50. Düsseldorf, pp 43–45
117. Oliveira JM, Salgado AJ, Sousa N, Mano JF, Reis RL (2010) Dendrimers and derivatives as a potential therapeutic tool in regenerative medicine strategies—a review. Science 35:1163–1194
118. Backer MV, Gaynutdinov TI, Patel V, Bandyopadhyaya AK, Thirumamagal BTS, Tjarks W, Barth RF, Claffey K, Backer JM (2005) Vascular endothelial growth factor selectively targets boronated dendrimers to tumor vasculature. Mol Cancer Ther 4:1423–1429
119. Majoros IJ, Myc A, Thomas T, Mehta CB, Baker JR Jr (2006) PAMAM dendrimer-based multifunctional conjugate for cancer therapy: synthesis, characterization, and functionality. Biomacromolecules 7:572–579

120. Zhu S, Hong M, Zhang L, Tang G, Jiang Y, Pei Y (2009) Erratum to: PEGylated PAMAM dendrimer-doxorubicin conjugates: in vitro evaluation and in vivo tumor accumulation. Pharm Res 27:161–174
121. Santra S, Kaittanis C, Perez JM (2010) Cytochrome *c* encapsulating theranostic nanoparticles: a novel bifunctional system for targeted delivery of therapeutic membrane-impermeable proteins to tumors and imaging of cancer therapy. Mol Pharm 7:1209–1222

Chapter 16
Design and Engineering of Multifunctional Quantum Dot-Based Nanoparticles for Simultaneous Therapeutic-Diagnostic Applications

Bowen Tian, Wafa' T. Al-Jamal, Jeroen Van den Bossche, and Kostas Kostarelos

16.1 Introduction

Quantum dots (QD) have many unique properties which render their application advantageous over more traditional technologies. One such use of QD is as a novel imaging agent, exhibiting certain characteristics which are superior to current organic dyes, such as an increased fluorescence lifetime. However, as with many nanoscale materials, both the lay public and the scientific community are apprehensive about the safety of such materials in clinical applications. One possible way to alleviate this concern is through the incorporation of QD in existing clinically used drug delivery systems, resulting in a more sophisticated "hybrid nanomaterial." The hope is that the resulting hybrid system will maintain the pharmacokinetic, biodistribution, and toxicological profiles of the existing drug delivery systems. In addition, therapeutic moieties transported along with QD can result in a theranostic device, delivering simultaneous diagnosis and therapy. Ideally, this would result in a sophisticated construct that accumulates specifically at the disease site, is taken up by the cells of interest, and is delivered into the cytosol, exerting the desired therapeutic effect. This process would allow for real-time imaging of the treatment, while maintaining minimal cytotoxicity to nontarget cells.

B. Tian • W.T. Al-Jamal • J. Van den Bossche • K. Kostarelos (✉)
Nanomedicine Laboratory, Centre for Drug Delivery Research, The School of Pharmacy, University of London, 29-39 Brunswick Square, London WC1N 1AX, UK
e-mail: jrwvdb@gmail.com

S. Svenson and R.K. Prud'homme (eds.), *Multifunctional Nanoparticles for Drug Delivery Applications: Imaging, Targeting, and Delivery*, Nanostructure Science and Technology, DOI 10.1007/978-1-4614-2305-8_16, © Springer Science+Business Media, LLC 2012

16.2 Quantum Dots

16.2.1 Fluorescence Properties of Quantum Dots

Semiconductor nanocrystals (also known as quantum dots or QD) have been extensively investigated as fluorescent probes for a variety of fluorescence-based imaging applications in biological studies both in vitro and in vivo [1–7]. Their unique fluorescent properties such as long-term fluorescence, diameter-dependent color, and photostability [8] have established QD as one of the most promising fluorescent probes, revolutionizing the fluorescence-based imaging field. By simply adjusting the QD size, the emission spectra can be tuned from the UV to the mid-infrared range [9–12]. The size-dependent fluorescence can be explained by the fact that QD have a band gap (quantum confinement) that can be tuned according to QD diameter [8]. In comparison to organic dyes, QD are approximately 10–100 times brighter and about 100–1,000 times more stable against photobleaching. This makes them excellent candidates for long-term imaging [13, 14]. The narrower emission spectra and broader absorption spectra that QD possess enable simultaneous detection of QD with different emission wavelengths, using a single excitation source [9–12, 15, 16]. In addition, a passivation shell on the QD surface (e.g., ZnS) can further enhance their photoluminescence efficiency by up to 90% [8] and their stability against oxidative photobleaching [8, 17, 18]. Due to the broad excitation spectra of QD, simultaneous detection of QD in different colors has enabled multiplexed imaging for tracking cancer cell metastasis [19] and differentiating tumor tissues [20] in vivo. Their bright fluorescence and pronounced photostability allow for broad applications in various biological investigations, e.g., cell labeling [1, 7], bioassays [21, 22], tissue specimens in vitro [23, 24], as well as tumor vessels [25] lymph nodes [26, 27], and solid tumors [3, 28] in vivo.

16.2.2 QD Chemical Composition and Synthesis

In 1982, semiconductor microcrystals were first synthesized in glassy dielectric matrices by Efros and Ekimov [29, 30]. In 1993, the first monodispersed semiconductor nanocrystals (CdS, CdSe, and CdTe) that exhibited size-dependent fluorescence were prepared by Bawendi and coworkers [31]. Following these discoveries, the synthesis of CdSe QD has advanced rapidly, and as a result, they are currently some of the most widely used QD in biological applications [6]. CdSe QD are typically prepared using the following process: a selenium precursor (commonly trioctylphosphine selenide or tributylphosphine selenide) and a cadmium precursor (dimethyl cadmium or cadmium oleate) are dissolved in trioctylphosphine and injected into a high-temperature (300°C) solution containing coordinating molecules such as trioctylphosphine oxide or hexadecylamine under inert conditions (nitrogen or argon) [17, 31, 32]. The selenium and cadmium precursors react fast to

Table 16.1 Core composition, size, and emission wavelength of QD

Core composition	Size (nm)	Emission range (nm)	References
CdSe	1–10	Visible (450–650)	[31, 32, 34–36]
CdS	1–6	UV, visible (320–450)	[31, 36, 37]
CdTe	2–8	Visible, near IR (500–750)	[31, 36]
ZnSe	4.3–6.0	UV, visible (350–420)	[37–39]
InAs	2.8–6.0	Near IR (850–1,400)	[40]
InP	2.6–4.6	Visible (650–720)	[41]
PbSe	3–12	Near IR (740–990)	[12, 42]
PbS	2–6.5	Near IR (850–1,500)	[11, 43]
PbTe	2.6–8.3	Near/mid-IR (1,020–2,070)	[44]
ZnSe:Mn	2.7–6.3	UV, visible (380–420)	[45]
$Zn_xCd_{1-x}S$	2.4–4.0	Visible (391–474)	[9]
$CdSe_{0.34}Te_{0.66}$	3.5–6.5	Near IR (741–800)	[10]

form the CdSe core, after which the core is allowed to grow at a lower temperature (230–260°C). The size of the QD cores can be quantified through absorption spectroscopy, and their growth is halted at the desired size [33]. ZnS capping of the core can be done after QD core synthesis through addition of bis(trimethylsilyl) selenide and diethyl zinc, where the shell thickness is modified through reaction time and temperature. In most of these syntheses, the solvent is a mixture of trioctylphosphine and trioctylphosphine oxide (TOP/TOPO), leading to capping of the resulting QD core with coordinating ligands attached on the surface. These ligands aid in maintaining the colloidal stability of the QD. Due to the coating, the QD resulting from this procedure are hydrophobic and require further engineering to be useful for biological applications.

Until now, QD have been synthesized in aqueous solution, high-temperature organic solvents, and solid substrates [6]. Many types of fluorescent QD have been synthesized with various compositions, mainly from II to IV and III to V group semiconductor materials, with emission wavelengths spreading from UV to mid-infrared (Table 16.1). Very recently, alloyed QD that allow tuning of the emission wavelength by manipulating compositions (but keeping the same size) have been developed [9, 10].

16.2.3 QD Solubilization and Functionalization

As mentioned above, further engineering of QD particles is required to obtain QD suitable for medical applications. Two strategies have been developed to address this challenge: ligand exchange and amphiphilic polymer coating (Fig. 16.1). Ligand exchange involves the use of thiolated ligands to replace the TOPO coating from the QD surface, resulting from the chelation of the QD core by sulfur atoms. This technique allows QD to be modified with different thiol-containing molecules, such as

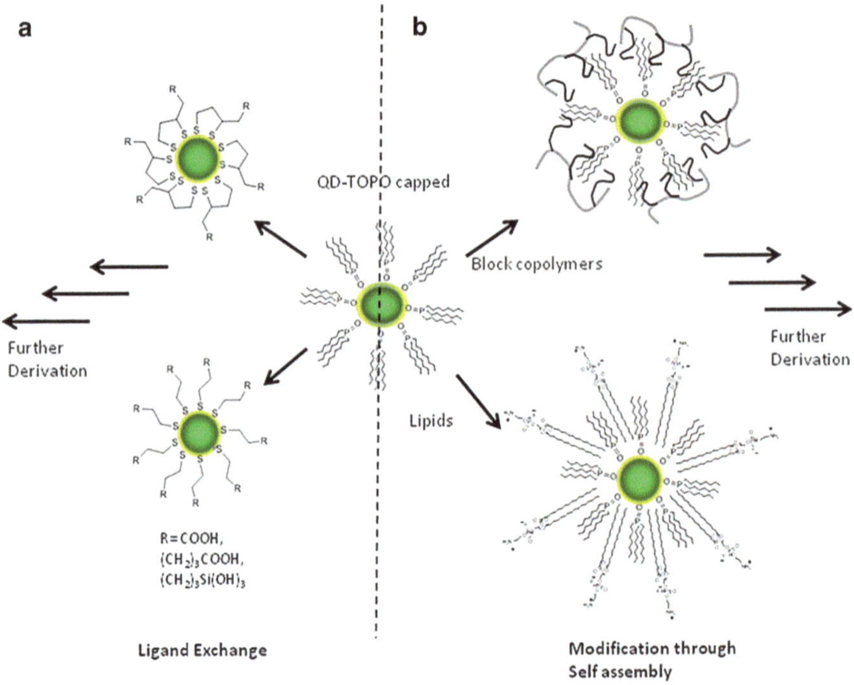

Fig. 16.1 QD water solubilization. Schematic diagram represents TOPO-capped QD (*middle*) solubilized into aqueous solutions by (**a**) exchanging TOPO surface coat with thiol-containing molecules or (**b**) self-assembly within phospholipid and diblock copolymers [22]

mercaptoacetic acid (MAA), dihydrolipoic acid, and mercaptopropyltris (methoxy) silane (MPS) [14, 34, 46]. As a result, these modifications introduce different functional groups such as carboxylic acids and silanols onto the QD surface, which allows for easy further modification with peptides [47], dendrons [48], etc. It should be noted that, while ligand exchange has successfully produced water-soluble QD, negative effects have been found on the QD fluorescence and stability as a result of the replacement of TOPO [34, 46, 49].

Therefore, amphiphilic polymer coatings have been used as an alternative method to TOPO replacement. The hydrophobic parts of amphiphilic polymers need to allow for interactions with TOPO. Phospholipid micelles, triblock copolymers, and amphiphilic diblock copolymers have all been successful in QD coating. The hydrophilic part of the polymers will be exposed on the outside of the QD, resulting in a hydrophilic material [3, 7, 50, 51]. As this method avoids TOPO displacement, it typically maintains the QD fluorescence properties and stability for a longer time. However, amphiphilic polymer coating has been shown to significantly increase the size of the construct when compared to ligand exchange methods [6]. Both QD solubilization strategies have resulted in various water-soluble QD constructs functionalized with antibodies [3, 14, 52, 53], peptides [25, 54–56], endosome-disruptive polymers [57], aptamers [58, 59], radionuclides [60–62], and magnetic resonance imaging (MRI) agents [56]. It has also been shown that QD coated with polyethylene glycol (PEG) exhibited prolonged blood circulation in vivo and minimized immunogenicity and cytotoxicity [27, 63, 64]. The flexibility of QD functionalization allows for the construction of customized, multifunctional QD constructs tailored toward many applications.

16.2.4 QD Biodistribution and Pharmacokinetics In Vivo

The effects of surface chemistry, size, and administration routes on QD biodistribution have been investigated using various techniques, such as fluorescence imaging, cadmium analysis, and radiolabeling techniques. In 2004, Ballou et al. first investigated the effect of surface chemistry modifications on QD blood circulation and organ biodistribution following intravenous administration [65]. This study found that a poly(acrylic acid)-coated QD (PAA-QD), surface-modified with PEG_{750} and PEG_{3400}, exhibited very short blood circulation half-lives ($t_{1/2} < 12$ min) when compared to PEG_{5000}-modified QD ($t_{1/2} = 140$ min) as measured by noninvasive fluorescence imaging [65]. In addition, PAA-QD modified with PEG_{750} was predominantly taken up by the liver, spleen, lymph nodes, and bone marrow 24-h postinjection, while considerably less liver and lymph node uptake was observed from PEG_{5000}-modified QD [27]. Consistently long blood circulation half-lives were reported by Akerman et al. and Gao et al. using PEG_{5000}-conjugated QD (15–20 nm in diameter) [3, 28]. Recently, Choi and coworkers systematically studied the size effect of PEG on the biodistribution and clearance of ^{99m}Tc-labeled infrared QD (InAs/ZnS) [66]. They found that QD conjugated with tetraethylene glycol primarily accumulated in the bladder and kidney, in marked contrast to PEG14 (14 ethylene glycol repeat units) which showed high uptake in the liver and intestine at 4-h postinjection. Furthermore, manipulation of the PEG length led to preferential accumulation of QD in the liver (diethylene glycol modified) and the pancreas (octaethylene glycol modified), excretion via renal clearance (triethylene glycol and tetraethylene glycol modified), or prolonged circulation in the vasculature (PEG22-modified QD) [66]. These studies highlighted the critical effect of surface chemistry and PEG length on the in vivo biodistribution of QD.

Fischer et al. described the first quantitative biodistribution study of non-PEGylated CdSe/ZnS QD by cadmium analysis in Sprague–Dawley rats following intravenous injection [67]. Two types of QD were selected in the study, 25 nm QD-LM (coated with mercaptoundecanoic acid and cross-linked with lysine) and 80 nm QD-BSA (bovine serum albumin conjugated to the QD-LM surface). The study found that QD-LM and QD-BSA exhibited different blood circulation half-lives, respectively, $t_{1/2}$ of 58.5 ± 17 min and 38.7 ± 3.5 min. Interestingly, major differences were also observed in organ distribution, with a liver uptake of 90% from QD-BSA compared to 40% from QD-LM at 90-min postinjection. This study indicated that the differences in QD biodistribution can be a result of both surface chemistry and the hydrodynamic diameter of the studied QD [67]. Choi et al. reported that zwitterionic (cysteine coated) or neutral (PEGylated dihydrolipoic acid coated) QD showed size-dependent biodistribution and clearance rates when ^{99m}Tc-labeled QD were intravenously injected into CD-1 mice. That study found renal excretion for QD smaller than 5.5 nm in hydrodynamic diameter, while no renal clearance was observed at 4-h postinjection for larger QD; however, liver uptake was high in this case [68]. On the other hand, micro-PET imaging showed that ^{64}Cu-labeled PEGylated QD (21 and 12 nm in diameter) exhibited rapid liver uptake without renal clearance [60].

The fate of QD following different routes of administration has also been studied. Polymer-coated QD with an average diameter between 15 and 20 nm were found to migrate rapidly to the sentinel lymph nodes (SLN) after subcutaneous, intradermal, or intraparenchymal injection in living animals [4, 26, 69–72]. QD migration to the lymph nodes occurred within 1–5-min postinjection and was found to be selective for the first lymph node. Similar behavior was observed after injecting QD of different size and surface charge in M21 human melanoma and MH-15 mouse teratocarcinoma xenograft models [27].

Overall, it appears that QD biodistribution in living animals is dependent on many factors, such as the hydrodynamic diameter, surface charge, PEG length, PEG density, and the route of administration. Furthermore, it has been shown that only very small neutral and zwitterionic QD (below the renal filtration threshold) can be excreted [66, 68], while larger QD have a tendency to accumulate in the body for extended periods of time [67, 73]. This observation requires further investigation to identify the long-term toxicity of QD before embarking on their clinical use.

16.2.5 Toxicity Profiles of Nonfunctionalized QD

The concern over QD toxicity is mainly derived from their intrinsic core composition, which in general, is composed of CdSe or CdTe. The correlation between cytotoxicity and free cadmium (Cd^{2+}) ions has been established [74–76], with the occurrence of significant cell death in the range of 100–400 μM Cd^{2+} ions [27]. Derfus et al. reported that CdSe QD can be toxic because of the release of cadmium ions by photolysis and/or oxidation, evidenced by the blueshift in QD absorbance spectra due to size reduction and subsequent release of Cd^{2+} [74]. Furthermore, the process of producing Cd^{2+} ions has been found to be accompanied by the formation of reactive oxygen species (ROS), such as singlet oxygen (O_2^-), due to QD electron donation to oxygen [77–79]. Cho et al. observed significant lysosomal damage due to the presence of both Cd^{2+} ions and ROS [79]. Interestingly, both Derfus et al. [74] and Cho et al. [79] found that ZnS coating (CdSe/ZnS and CdTe/ZnS) can protect the QD core from oxidation and minimize Cd^{2+} leakage, subsequently reducing the QD-induced cytotoxicity [75, 79, 80]. In addition, the use of antioxidants such as

N-acetylcysteine (NAC) has been shown to be effective in reducing QD cytotoxicity as they can act as ROS quenchers [78, 81].

QD-induced cell dysfunction is often accompanied by apoptotic and necrotic biochemical changes, including morphological alterations in the plasma membrane, mitochondrial and nucleic damage [81], lysosomal enlargement [79], reduction in cytochrome C concentration [76, 79, 82], loss of mitochondrial membrane potential [82], and upregulation of peroxidized lipids [83]. In addition, QD-induced cytotoxicity dramatically increased in the case of QD exposure to oxygen or ultraviolet (UV) light [74, 81]. To date, the most cytotoxic QD (usually without ZnS coating) have been solubilized by the ligand exchange method, such as mercaptopropionic acid (MPA-QD) [78, 80, 81, 84], mercaptoacetic acid (MAA-QD) [74], mercaptoundecanoic acid (MUA-QD) [84], cysteamine (QD-NH_2) [79, 80], or thioglycerol (QD-OH) [76]. This toxicity is caused by ligand detachment from the QD surface as a result of weak interactions between QD surface and the ligands [85, 86], especially under unfavorable conditions such as seen in the endosomal compartment [34]. Exposure of the QD core was correlated to cell death in a QD concentration-dependent manner, similar to the cytotoxicity caused by Cd^{2+} ions [74, 76, 78, 79, 81].

There have been ways proposed to alleviate QD cytotoxicity. Hoshino et al. reported that there is no evidence of Cd^{2+}-induced cytotoxicity after QD coating with ZnS [80]. Furthermore, Kirchner et al. showed that toxicity profiles of QD can be further improved after conjugation of PEG to CdSe/ZnS QD [75]. This study also found that PEG-conjugated, silica-coated QD were nontoxic up to high Cd^{2+} surface concentrations. It has to be noted that the QD core was coated with a shell of cross-linked silica molecules, which was subsequently conjugated to PEG. These QD were highly resistant to chemical and metabolic degradation [87] and showed low toxicity after nuclear translocation [88]. In addition, such type of QD exhibited low genotoxicity [63].

Further information about QD toxicity can be obtained by alternative models of more complex organisms. *Xenopus* embryos [50] and zebrafish embryos [92] are some of the most sensitive models in which QD toxicity has been studied. The embryos were microinjected with 1×10^8 QD/cell and 2×10^9 QD/cell, respectively, and did not exhibit any sign of toxicity. However, both *Xenopus* embryos and zebrafish exhibited abnormalities when the doses were increased to 2×10^8 QD/cell and 5×10^9 QD/cell, respectively. This change in biocompatibility was attributed to either the intrinsic toxicity of QD or the osmotic equilibrium changes [50].

Several groups have injected QD in animals for targeting and imaging purposes; however, very few studies reported QD toxicity in living animals. QD injected systemically (via tail vein or jugular vein) in mice and rats (pmol to nmol range) have shown no apparent toxicity several months postinjection [20, 27, 67, 93]. Other studies have shown that when 200–400 pmol of near-infrared (NIR) QD were injected locally into Yorkshire pigs to map their sentinel lymph nodes (SLN), no changes in the heart rate, blood pressure, and oxygen level were observed during the experimental procedure or after several hours [4, 26, 69, 71]. Recently, a study indicated no in vivo toxicity over longer periods of time (>80 days), even after in vivo breakdown of the QD over time. It has been previously described that QD nanoparticles of certain sizes could be inhaled and deposited in the respiratory tract [89, 90]. Furthermore, QD have been found to be capable of penetrating intact skin [91]. These preliminary studies suggest that extra care should be taken during QD handling and especially during QD manufacturing, as systemic toxicity can occur via QD inhalation and direct skin contact.

Overall, the cytotoxicity studies have shown that the key determinants of QD toxicity are their chemical composition and functionalization. However, other factors such as cell type [84], QD size [78], and QD exposure to oxygen and UV light [74] were also found to influence QD cytotoxicity.

16.3 Quantum Dot Hybrid Nanomaterials

As has been described earlier, the pharmacological profile of QD can be significantly altered through the surface modification of these materials. Apart from covalent surface modifications, several studies have shown that the QD tissue uptake profiles and pharmacokinetics can be further modified through their incorporation into other nanoscale materials such as liposomes [94–99], (polymeric) micelles [50, 100–108], or carbon nanotubes [109, 110]. Co-delivery of therapeutic agents with QD is an additional motivation for using other delivery systems. In addition, the latter can be easily tailored to exhibit desirable characteristics such as responsiveness to both internal as external triggers and showing higher specificity and recognition by functionalizing their surface with targeting ligands.

Among many nanostructured drug delivery systems, liposomes are currently the most extensively clinically studied. Liposomes are vesicles that are typically comprised of naturally occurring phospholipids that enclose an aqueous compartment by lipid bilayers. These vesicular structures allow the incorporation and delivery of both hydrophobic (within the lipid bilayer) and hydrophilic drugs (within the aqueous compartment) [111]. The physicochemical characteristics of these drug delivery systems vary with their composition and can be easily modified through formulation of different lipids. It is possible to add cell or tissue specificity by conjugation or association with targeting moieties such as peptides, antibodies, or small molecules [111]. Biocompatibility issues can be solved through the incorporation of PEGylated lipids or through the incorporation of acid-sensitive lipids that can result in pH-responsive materials [112].

Such attractive liposome features have initiated investigations into the possibility of structurally incorporating quantum dots that would result in hybrid QD-lipid vesicle systems (H-QD-V). Currently, these H-QD-V materials have been engineered following different methodologies. The first H-QD-V systems were constructed through complexation of anionic PEG-QD with cationic liposomes, purely relying on electrostatic interactions to drive the self-assembly of these materials. It was observed that in comparison to other cationic delivery systems (e.g., peptides or dendrimers), the cationic liposomes were the most efficient delivery systems in vitro, attributed to them following similar uptake pathways to cationic liposome–DNA lipoplexes [113].

An alternative approach was later proposed independently by the Vogel and Kostarelos groups, taking advantage of the hydrophobic nature of TOPO-capped QD [96, 99]. The hydrophobic QD materials were shown to be easily embedded within the bilayer of liposomes when formulated through the lipid film hydration method. It should be noted that in order to incorporate QD within a bilayer, the QD diameter should be comparable to the dimensions of the bilayers (approx. 4 nm for most lipid bilayers). Clearly, the quantum confinement effect restricts the fluorescence properties of the QD that can be incorporated within lipid bilayers, whereas a wide variety of lipid formulations can be used for successful incorporation of QD. Incorporation of QD within lipid bilayers neither affected the liposome size nor altered the QD optical properties [96]. Both studies showed that the constructs could be efficiently internalized in vitro by tumor cells [96, 99]. In vivo intratumoral injection of H-QD-V showed that tumor retention was feasible; however, this was dependent on the bilayer characteristics of the liposome [96]. Cationic liposomes lead to efficient cellular binding and uptake over 24 h, whereas zwitterionic formulations exhibited tumor clearance immediately after administration. Further studies were conducted using PEGylated QD as these were shown to have superior in vivo pharmacokinetics. These hydrophilic QD were incorporated within the aqueous compartment of the liposomes. Similar lipid formulations could be used as for hydrophobic QD; however, as the size restriction is removed, a wider variety of QD can be used. These structures were shown to exhibit similar properties, in vitro and in vivo to H-QD-V with hydrophobic QD [94]. Furthermore, following systemic administration, these hybrids showed faster tumor accumulation compared to PEGylated QD [95]. Very recently, Sigot et al. showed dual QD-labeled H-QD-V, with QD both within the aqueous environment and bound to the liposome bilayer using the streptavidin–biotin interaction. These dual QD-labeled hybrids efficiently targeted to EGFR-expressing cells and showed no photobleaching during live imaging over 2 h [114].

It appears that once QD are encapsulated or conjugated to liposomes, the construction of multifunctional drug delivery systems is possible and that the use of such H-QD-V results in decreased cytotoxicity as compared to unmodified QD. Hence, the encapsulation of QD within drug delivery systems can be considered a valid approach to alleviate some of the toxicity issues associated with QD and allow better control of their in vivo fate [115].

16.4 Quantum Dot Theranostics

The ultimate goal of theranostic devices is to achieve simultaneous therapy and diagnosis. Recent research efforts have been devoted toward the construction of such devices by combining QD with therapeutic agents such as small chemotherapeutic drugs and nucleic acids, as will be discussed below.

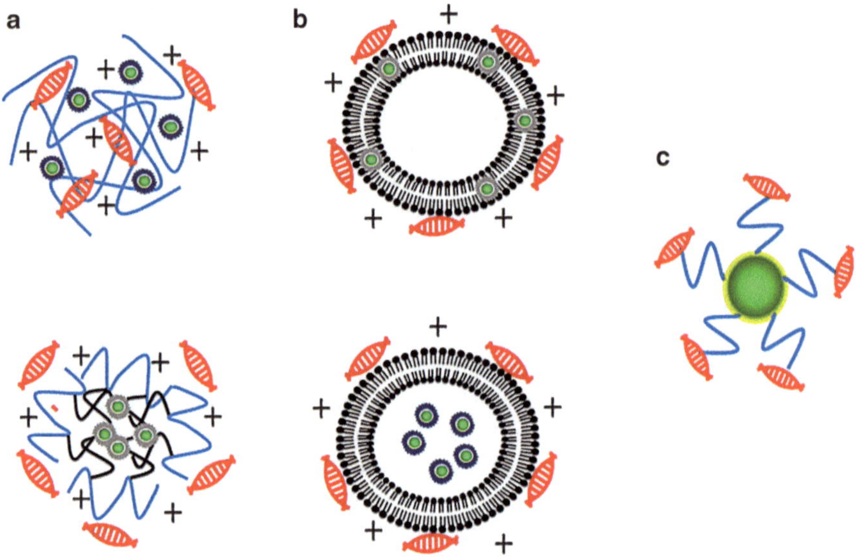

Fig. 16.2 Alternative strategies for the design of QD-nucleic acid theranostics (**a**) QD-nucleic acid polymers, (**b**) QD-nucleic acid liposomes, (**c**) Direct ligation of nucleic acids to QD

16.4.1 Quantum Dot Nucleic Acid Delivery

The delivery of genetic material into cells, also named gene therapy, is one of the most promising strategies for the treatment of debilitating diseases, including cancer. The barriers to successful gene delivery are numerous [116]. Understanding of such barriers can be achieved through effective fluorescence labeling of either the cargo or the delivery system. Therefore, there is a need for fluorescence probes that will resist photobleaching and allow dynamic imaging over longer periods of time, which make QD ideal candidates [117]. Labeling of nucleic acids or delivery vectors with QD has been achieved in the past. A powerful approach that can be used to investigate the release of plasmid from nonviral vectors is via fluorescence resonance energy transfer (FRET). This phenomenon can be achieved if two fluorophores (a donor and acceptor) have similar fluorescence properties and are within a small distance (called the Förster radius) of each other. In the case of QD-FRET, the QD is the donor material and gets excited and the acceptor molecule is typically a small organic dye. As this dye is not excited by the laser source, it is less likely to undergo significant photobleaching [115].

Figure 16.2 represents some alternative QD-nucleic acid theranostic hybrid systems reported that could enable tracking of the gene delivery system in vitro and in vivo. The first attempt to visualize plasmid DNA was through the use of micelle-dispersed QD. Maleimide lipids were incorporated within the micellar containers, allowing for pDNA strand conjugation (Fig. 16.2c) and the construction of pDNA-QD hybrids. These hybrids were visualized efficiently using AFM and confocal microscopy over 24 h without any significant photobleaching. In addition, these constructs exhibited a very low cytotoxicity profile [118]. An alternative strategy used complexation of QD with Cy5-labeled chitosan to construct a FRET pair. The complexes could be visualized by TEM, and the kinetics of decomplexation and release process could be modeled as a reversible first-order reaction. It was shown that the use of FRET constructs enabled highly sensitive and quantitative monitoring of polyplexes during intracellular transport [119, 120]. It should be noted that photoactivation of QD can also lead to the generation of reactive oxygen intermediates (ROI), which has been used as a strategy for photodynamic therapy (PDT). Hence, if the nucleic acid is in close proximity to the QD, significant DNA strand breakages may occur [121]. This possibility should be considered in the construction of DNA-QD theranostic devices.

A common issue with nucleic acid delivery is associated with the efficiency in decomplexation and release from its carrier system. In the case of QD-pDNA constructs, this has been achieved in vitro through exchange of the cationic thiol-capping reagent with glutathione (GSH). This process results in a decrease of the cationic charge of the QD hybrid, resulting in DNA release from the complex [122]. Very recently, QD-pDNA systems were able to illustrate dynamic DNA release in transdermal applications [123]. An in vivo study has also been recently reported using anionic QD to track pDNA after complexation with poly(ethylene imine) (PEI). Upon systemic administration, the QD alone accumulated mainly in the liver and were rapidly eliminated from the blood stream. If complexed, the hybrid materials were rapidly accumulated in the lung and liver, similar to the biodistribution of unmodified PEI polyplexes, and it was observed that this accumulation did not change over several hours [124].

An alternative system for gene therapy applications are QD-siRNA constructs. Where pDNA delivery aims to incorporate a gene, siRNA delivery aims to achieve a therapeutic effect through inhibition of the translation of target proteins. In comparison to plasmid DNA gene delivery, much less genetic material needs to be delivered to the cytoplasm. However, efficient and specific delivery of siRNA and protection from nuclease activity is still required [125]. As with plasmid DNA delivery, attempts have been made to use QD to visualize and determine the release of the nucleic acid. QD imaging would be of significant interest to the siRNA delivery field as release and gene knockdown will take place (>24h postadministration) over a period during which organic dyes would be subject to significant photobleaching [117]. This challenge can be addressed in vitro through the simultaneous delivery of a fluorescent protein plasmid.

An initial approach used to achieve QD-siRNA theranostics was through complexation of the QD with cationic liposomes, without affecting the nucleic acid delivery. The resulting QD-siRNA/liposomes were rather large in size, proving problematic for systemic delivery [126]. An alternative approach used PEGylated QD that were covalently modified with siRNA through both cleavable and non-cleavable linkers. Attachment of F3-targeting peptides showed that increasing amounts of targeting peptide decreased the amount of siRNA that could be attached and vice versa. Although that conjugation did not affect gene knockdown when compared to naked siRNA, cleavable QD-siRNA constructs (disulfide linked) did lead to an improvement compared to the noncleavable materials [117]. Chitosan-modified QD could be complexed with siRNA and induced green fluorescence protein (GFP) gene silencing. The materials obtained this way showed relative good stability over 48 h, maintaining its size at 80 nm in diameter. Their overall surface charge was negative; so to obtain sufficient cellular uptake, monoclonal antibodies were attached to achieve uptake through receptor-mediated endocytosis. These theranostic systems were further developed through the generation of RGD- and Tat-targeted QD-siRNA constructs. Higher gene knockdown values were observed after 24 h when compared to naked siRNA. The effect was still present over longer time points (48 and 96 h) but less pronounced. This theranostic system allowed a therapeutic knockdown of EGFvIII protein, which is a growth factor exclusive to cancer cells [127, 128].

Alternatively, it has been possible to conjugate siRNA to QD through peptide-modified QD surfaces. Cationic peptides can also assist in improved cellular uptake and allow for siRNA complexation through electrostatic interactions. It was shown that such constructs exhibited good stability and are taken up through energy-dependent endocytosis via multiple pathways, depending on the cell type, while maintaining minimal cytotoxicity [129]. One of the major problems with these types of systems is the dissociation of the complex after addition of chloroquine to the cells, which is needed to stimulate release of siRNA. The FRET effect was also recently exploited to monitor the dissociation of QD-siRNA-polymer complexes. Here, the complexation reagent of choice was PEI as this is known to induce a proton sponge effect, which can aid in decomplexation. Conjugation with the cell-penetrating peptide (CPP), Hph-1 enabled faster internalization of the polyplexes. In the absence of the CPP, the siRNA followed a typical endocytic and subcellular trafficking process. In the presence of the CPP, perinuclear targeting was observed within 30 min [130]. Alternative targeting of QD-siRNA-PEI complexes was achieved through PEI-hyaluronic acid (HA) constructs. These constructs were shown to increase gene knockdown as compared to nontargeted controls, and in vivo biodistribution showed increased uptake to tumor, liver, and kidney [131]. Despite such achievements, the dissociation of the siRNA from its carrier agent remains a major problem. Amphipol-modified QD were developed to solve this issue. Amphipol is an amphipathic linear polymer, and was shown to facilitate cytoplasmic siRNA internalization while protecting its activity. The resulting particles were relatively small (approximately 16 nm in diameter). Uptake could be achieved within 1 h; after 5 h, the siRNA was observed throughout the cells, indicating endosomal escape. In terms of gene knockdown efficacy, similar activity to Lipofectamine® was obtained with minimal cytotoxicity [132].

16.4.2 Quantum Dot Photodynamic Therapy

Photodynamic therapy (PDT) is a technique in which light-activated therapeutic agents, photosensitizers (PS), are used. Light activation leads to the generation of singlet oxygen, the main intermediate in photosensitized cell toxicity, while the photosensitizer should exhibit only low toxicity in the dark. By definition, this technique can be classified as theranostic since the photosensitizers function simultaneously as imaging and cytotoxic agents [133, 134]. QD can generate singlet oxygen, which has previously resulted in DNA cleavage when nucleic acid materials were bound to QD. It should be noted that in the case of UV irradiation of QD, other reactive oxygen and nitrogen species can be generated, leading to phototoxic effects. Further proof was obtained through addition of antioxidant scavengers such as *N*-acetylcysteine, which resulted in significant reduction of DNA damage [78, 81].

However, as the phototoxicity of QD is relatively low, QD-activated PDT has been explored. It has been possible to combine photosensitizers with QD to access excitation wavelengths at which the photosensitizer alone does not absorb. The presence of the QD facilitates the photosensitizer excitation through a FRET-related process, leading to higher quantum yields for the combination of QD and PS compared to the PS alone. Hence, it is more common to use QD function as the donor in conjunction with conventional photosensitizers [134]. The major limiting factor in this approach is that the PS needs to be bound within a small distance of the QD (the Förster radius). QD offer several advantages to PDT over PS: (1) large surface area that enables binding of numerous PS molecules, (2) better photostability, (3) tuning of the emission spectrum of the QD to the desired PS absorption according to dimensions, and (4) extinction coefficients that exceed the unmodified PS [134]. The first reported system suffered from low solubility in an aqueous environment, and the uncoated QD used could lead to cytotoxic side effects [135]. To solve this issue, peptide-coated QD have been used and were successful in singlet oxygen generation [136]. Blanco et al. have recently generated a proof of concept study involving a new, promising way to obtain phototoxicity through the formation of a Pt(IV) complex with CDSe–ZnS QD [137]. Pd(IV) complexes are chemically inert and can accumulate at tumor sites; however, upon QD irradiation, photoreduction can take place, leading to Pt(II) generation, which has a well-reported antitumor effect (cf., cisplatin).

16.4.3 Quantum Dot-Based Chemotherapeutic Devices

The construction of QD-based theranostic devices in which a standard chemotherapeutic small molecule such as anthracycline or paclitaxel is being used as the therapeutic moiety is currently an area of intense research. The most successful approach to date involving QD as the diagnostic element of such a system has been the construction of aptamer-QD-doxorubicin hybrids. This approach made use of a dual FRET modality; in the nonactive state, DOX fluorescence was quenched through intercalation with the aptamer, while the QD fluorescence was quenched by the DOX. After the release of DOX, both fluorescent signals could be measured and used for imaging purposes. This system was able to deliver DOX to prostate cancer cell lines in vitro and was able to differentiate between PSMA+ and PSMA− cell lines (the chosen aptamer was specific to this PMSA receptor) [58].

An alternative approach used DOX-loaded liposomes, which were targeted toward HER2 using monoclonal antibodies (mAbs). QD were conjugated to the outside of the liposomes. It was shown that this coupling to QD had little effect on the in vitro cytotoxicity of DOX. The in vivo blood circulation half-life decreased when compared to liposome controls without QD conjugation. However, a significant increase in the QD half-life was observed, shifting from 10 min to about 3 h. After 24 h, both targeted as well as nontargeted systems showed approximately 14% of the total body fluorescence in the tumor [138]. DOX-QD have also been incorporated into lipid-PEG micelles together with Fe_2O_3 nanoparticles, resulting in a cytotoxic construct capable of MRI and fluorescence imaging. Without a targeting ligand, these materials did not show cellular uptake by fluorescence, whereas with T3 targeting, these materials showed significantly increase in the MRI contrast (in vitro) and an improved cytotoxic response compared to free DOX [139]. Recently, it was shown that the multidrug resistance of cell lines can be reduced through complexation of daunorubicin with anionic QD [140]. Paclitaxel has also been incorporated into polymeric micelles, together with QD and the anti-HIF-1 antibody as targeting ligand. Epifluorescence imaging revealed that this system was targeted specifically to stomach cancer cells and resulted in significant increase in the cytotoxic activity of the drug [141].

16.5 Conclusions

Many studies have demonstrated that QD are superior imaging agents and can aid in the construction of theranostic devices. Recent investigations have shown that it is possible to modify QD to become more biologically and environmentally benign. At the same time, it was shown that QD can be successfully incorporated into existing delivery systems such as liposomes, polymeric micelles, and polymeric conjugates. It appears that the characteristics of these hybrid materials for both in vitro and in vivo applications are mainly determined by the respective delivery system. The resulting theranostic devices have been extensively studied in vitro and have shown great promise as dual imaging/therapeutic agents. Despite the advantages that QD may offer, more systems should be developed and studied in vivo to appreciate the feasibility of QD-based theranostic devices for clinical applications.

References

1. Alivisatos AP, Gu W, Larabell C (2005) Quantum dots as cellular probes. Annu Rev Biomed Eng 7:55–76
2. Alivisatos AP (2004) The use of nanocrystals in biological detection. Nat Biotechnol 22(1):47–52
3. Gao X, Cui Y, Levenson RM, Chung LW, Nie S (2004) In vivo cancer targeting and imaging with semiconductor quantum dots. Nat Biotechnol 22(8):969–976

4. Kim S, Lim YT, Soltesz EG, De Grand AM, Lee J, Nakayama A, Parker JA, Mihaljevic T, Laurence RG, Dor DM, Cohn LH, Bawendi MG, Frangioni JV (2004) Near-infrared fluorescent type II quantum dots for sentinel lymph node mapping. Nat Biotechnol 22(1):93–97
5. Rhyner MN, Smith AM, Gao X, Mao H, Yang L, Nie S (2006) Quantum dots and multifunctional nanoparticles: new contrast agents for tumor imaging. Nanomedicine 1(2):209–217
6. Smith AM, Duan H, Mohs AM, Nie S (2008) Bioconjugated quantum dots for in vivo molecular and cellular imaging. Adv Drug Del Rev 60(11):1226–1240
7. Wu X, Liu H, Liu J, Haley KN, Treadway JA, Larson JP, Ge N, Peale F, Bruchez MP (2003) Immunofluorescent labeling of cancer marker Her2 and other cellular targets with semiconductor quantum dots. Nat Biotechnol 21(1):41–46
8. Alivisatos AP (1996) Semiconductor clusters, nanocrystals, and quantum dots. Science 271:933–937
9. Zhong X, Feng Y, Knoll W, Han M (2003) Alloyed Zn(x)Cd(1-x)S nanocrystals with highly narrow luminescence spectral width. J Am Chem Soc 125(44):13559–13563
10. Bailey RE, Nie S (2003) Alloyed semiconductor quantum dots: tuning the optical properties without changing the particle size. J Am Chem Soc 125(23):7100–7106
11. Hines MA, Scholes GD (2003) Colloidal PbS nanocrystals with size-tunable near-infrared emission: observation of post-synthesis self-narrowing of the particle size distribution. Adv Mater 15(21):1844–1849
12. Pietryga JM, Schaller RD, Werder D, Stewart MH, Klimov VI, Hollingsworth JA (2004) Pushing the band gap envelope: mid-infrared emitting colloidal PbSe quantum dots. J Am Chem Soc 126(38):11752–11753
13. Medintz IL, Uyeda HT, Goldman ER, Mattoussi H (2005) Quantum dot bioconjugates for imaging, labelling and sensing. Nat Mater 4(6):435–446
14. Jaiswal JK, Mattoussi H, Mauro JM, Simon SM (2003) Long-term multiple color imaging of live cells using quantum dot bioconjugates. Nat Biotechnol 21(1):47–51
15. Kim S, Fisher B, Eisler HJ, Bawendi M (2003) Type-II quantum dots: CdTe/CdSe(core/shell) and CdSe/ZnTe(core/shell) heterostructures. J Am Chem Soc 125(38):11466–11467
16. Qu L, Peng X (2002) Control of photoluminescence properties of CdSe nanocrystals in growth. J Am Chem Soc 124(9):2049–2055
17. Dabbousi BO, RodriguezViejo J, Mikulec FV, Heine JR, Mattoussi H, Ober R, Jensen KF, Bawendi MG (1997) (CdSe)ZnS core-shell quantum dots: synthesis and characterization of a size series of highly luminescent nanocrystallites. J Phys Chem B 101(46):9463–9475
18. Hines MA, Guyot-Sionnest P (1996) Synthesis and characterization of strongly luminescing ZnS-Capped CdSe nanocrystals. J Phys Chem 100(2):468–471
19. Voura EB, Jaiswal JK, Mattoussi H, Simon SM (2004) Tracking metastatic tumor cell extravasation with quantum dot nanocrystals and fluorescence emission-scanning microscopy. Nat Med 10(9):993–998
20. Stroh M, Zimmer JP, Duda DG, Levchenko TS, Cohen KS, Brown EB, Scadden DT, Torchilin VP, Bawendi MG, Fukumura D, Jain RK (2005) Quantum dots spectrally distinguish multiple species within the tumor milieu in vivo. Nat Med 11(6):678–682
21. Smith RA, Giorgio TD (2009) Quantitative measurement of multifunctional quantum dot binding to cellular targets using flow cytometry. Cytom Part A 75A(5):465–474
22. Smith AM, Dave S, Nie SM, True L, Gao XH (2006) Multicolor quantum dots for molecular diagnostics of cancer. Expert Rev Mol Diagn 6(2):231–244
23. Fountaine TJ, Wincovitch SM, Geho DH, Garfield SH, Pittaluga S (2006) Multispectral imaging of clinically relevant cellular targets in tonsil and lymphoid tissue using semiconductor quantum dots. Mod Pathol 19(9):1181–1191
24. Ferrara DE, Weiss D, Carnell PH, Vito RP, Vega D, Gao XH, Nie SM, Taylor WR (2006) Quantitative 3D fluorescence technique for the analysis of en face preparations of arterial walls using quantum dot nanocrystals and two-photon excitation laser scanning microscopy. Am J Physiol-Regul Integr Comp Physiol 290(1):R114–R123
25. Cai WB, Shin DW, Chen K, Gheysens O, Cao QZ, Wang SX, Gambhir SS, Chen XY (2006) Peptide-labeled near-infrared quantum dots for imaging tumor vasculature in living subjects. Nano Lett 6(4):669–676

26. Parungo CP, Ohnishi S, Kim SW, Kim S, Laurence RG, Soltesz EG, Chen FY, Colson YL, Cohn LH, Bawendi MG, Frangioni JV (2005) Intraoperative identification of esophageal sentinel lymph nodes with near-infrared fluorescence imaging. J Thorac Cardiovasc Surg 129(4):844–850
27. Ballou B, Ernst LA, Andreko S, Harper T, Fitzpatrick JA, Waggoner AS, Bruchez MP (2007) Sentinel lymph node imaging using quantum dots in mouse tumor models. Bioconjug Chem 18(2):389–396
28. Akerman ME, Chan WC, Laakkonen P, Bhatia SN, Ruoslahti E (2002) Nanocrystal targeting in vivo. Proc Natl Acad Sci USA 99(20):12617–12621
29. Efros AIL, Efros AL (1982) Interband absorption of light in a semiconductor sphere. Sov Phys Semicond USSR 16:772–775
30. Ekimov AI, Onushchenko AA (1981) Quantum size effect in the optical-spectra of semiconductor microcrystals. Sov Phys Semicond USSR 16:775–778
31. Murray CB, Norris DJ, Bawendi MG (1993) Synthesis and characterization of nearly monodisperse CdE (E=S, Se, Te) semiconductor nanocrystallites. J Am Chem Soc 115(19):8706–8715
32. Talapin DVR, Rogach AL, Kornowski A, Haase M, Weller H (2001) Highly luminescent monodisperse CdSe and CdSe/ZnS nanocrystals synthesized in a hexadecylamine-trioctylphosphine oxide-trioctylphosphine mixture. Nano Lett 1(4):207–211
33. Yu WW, Qu LH, Guo WZ, Peng XG (2003) Experimental determination of the extinction coefficient of CdTe, CdSe, and CdS nanocrystals. Chem Mater 15(14):2854–2860
34. Bruchez M Jr, Moronne M, Gin P, Weiss S, Alivisatos AP (1998) Semiconductor nanocrystals as fluorescent biological labels. Science 281:2013–2016
35. Qu LP, Peng ZA, Peng XG (2001) Alternative routes toward high quality CdSe nanocrystals. Nano Lett 1:333–337
36. Peng ZA, Peng XG (2001) Formation of high-quality CdTe, CdSe, and CdS nanocrystals using CdO as precursor. J Am Chem Soc 123(1):183–184
37. Yu WW, Peng XG (2002) Formation of high-quality CdS and other II-VI semiconductor nanocrystals in noncoordinating solvents: tunable reactivity of monomers. Angew Chem Int Ed 41(13):2368–2371
38. Hines MA, Guyot-Sionnest P (1998) Bright UV-blue luminescent colloidal ZnSe nanocrystals. J Phys Chem B 102(19):3655–3657
39. Dawood F, Schaak RE (2009) ZnO-templated synthesis of wurtzite-type ZnS and ZnSe nanoparticles. J Am Chem Soc 131(2):424–425
40. Peng XG, Wickham J, Alivisatos AP (1998) Kinetics of II-VI and III-V colloidal semiconductor nanocrystal growth: "focusing" of size distributions. J Am Chem Soc 120(21):5343–5344
41. Guzelian AA, Katari JEB, Kadavanich AV, Banin U, Hamad K, Juban E, Alivisatos AP, Wolters RH, Arnold CC, Heath JR (1996) Synthesis of size-selected, surface-passivated InP nanocrystals. J Phys Chem 100(17):7212–7219
42. Du H, Chen CL, Krishnan R, Krauss TD, Harbold JM, Wise FW, Thomas MG, Silcox J (2002) Optical properties of colloidal PbSe nanocrystals. Nano Lett 2(11):1321–1324
43. Zhao XS, Gan JQ, Liu GH, Chen AM (2008) One-step synthesis and optical properties of PbS quantum dots. Acta Chim Sinica 66(16):1869–1872
44. Murphy JE, Beard MC, Norman AG, Ahrenkiel SP, Johnson JC, Yu PR, Micic OI, Ellingson RJ, Nozik AJ (2006) PbTe colloidal nanocrystals: synthesis, characterization, and multiple exciton generation. J Am Chem Soc 128(10):3241–3247
45. Norris DJ, Yao N, Charnock FT, Kennedy TA (2001) High-quality manganese-doped ZnSe nanocrystals. Nano Lett 1(1):3–7
46. Chan WCW, Nie SM (1998) Quantum dot bioconjugates for ultrasensitive nonisotopic detection. Science 281:2016–2018
47. Pinaud F, King D, Moore HP, Weiss S (2004) Bioactivation and cell targeting of semiconductor CdSe/ZnS nanocrystals with phytochelatin-related peptides. J Am Chem Soc 126(19):6115–6123

48. Huang BH, Tomalia DA (2005) Dendronization of gold and CdSe/cdS (core-shell) quantum functionalized dendrons dots with tomalia type, thiol core, poly(amidoamine) (PAMAM) dendrons. J Lumin 111(4):215–223
49. Gerion D, Pinaud F, Williams SC, Parak WJ, Zanchet D, Weiss S, Alivisatos AP (2001) Synthesis and properties of biocompatible water-soluble silica-coated CdSe/ZnS semiconductor quantum dots. J Phys Chem B 105(37):8861–8871
50. Dubertret B, Skourides P, Norris DJ, Noireaux V, Brivanlou AH, Libchaber A (2002) In vivo imaging of quantum dots encapsulated in phospholipid micelles. Science 298:1759–1762
51. Larson DR, Zipfel WR, Williams RM, Clark SW, Bruchez MP, Wise FW, Webb WW (2003) Water-soluble quantum dots for multiphoton fluorescence imaging in vivo. Science 300:1434–1436
52. Liu TC, Zhang HL, Wang JH, Wang HQ, Zhang ZH, Hua XF, Cao YC, Luo QM, Zhao YD (2008) Study on molecular interactions between proteins on live cell membranes using quantum dot-based fluorescence resonance energy transfer. Anal Bioanal Chem 391(8):2819–2824
53. Xing Y, Chaudry Q, Shen C, Kong KY, Zhau HE, Chung LW, Petros JA, O'Regan RM, Yezhelyev MV, Simons JW, Wang MD, Nie S (2007) Bioconjugated quantum dots for multiplexed and quantitative immunohistochemistry. Nat Protoc 2(5):1152–1165
54. Anas A, Okuda T, Kawashima N, Nakayama K, Itoh T, Ishikawa M, Biju V (2009) Clathrin-mediated endocytosis of quantum dot-peptide conjugates in living cells. ACS Nano 3(8):2419–2429
55. Zhang Y, So MK, Rao JH (2006) Protease-modulated cellular uptake of quantum dots. Nano Lett 6(9):1988–1992
56. Mulder WJM, Koole R, Brandwijk RJ, Storm G, Chin PTK, Strijkers GJ, Donega CD, Nicolay K, Griffioen AW (2006) Quantum dots with a paramagnetic coating as a bimodal molecular imaging probe. Nano Lett 6(1):1–6
57. Duan HW, Nie SM (2007) Cell-penetrating quantum dots based on multivalent and endosome-disrupting surface coatings. J Am Chem Soc 129(11):3333–3338
58. Bagalkot V, Zhang L, Levy-Nissenbaum E, Jon S, Kantoff PW, Langer R, Farokhzad OC (2007) Quantum dot-aptamer conjugates for synchronous cancer imaging, therapy, and sensing of drug delivery based on bi-fluorescence resonance energy transfer. Nano Lett 7(10):3065–3070
59. Chen XC, Deng YL, Lin Y, Pang DW, Qing H, Qu F, Xie HY (2008) Quantum dot-labeled aptamer nanoprobes specifically targeting glioma cells. Nanotechnology 19:235105
60. Schipper ML, Cheng Z, Lee SW, Bentolila LA, Iyer G, Rao JH, Chen XY, Wul AM, Weiss S, Gambhirl SS (2007) MicroPET-based biodistribution of quantum dots in living mice. J Nucl Med 48(9):1511–1518
61. Cai WB, Chen K, Li ZB, Gambhir SS, Chen XY (2007) Dual-function probe for PET and near-infrared fluorescence imaging of tumor vasculature. J Nucl Med 48(11):1862–1870
62. Patt M, Schildan A, Habermann B, Mishchenko O, Patt JT, Sabri O (2010) F-18- and C-11-labelling of quantum dots with n.c.a. [F-18]fluoroethyltosylate and [C-11]methyliodide: a feasibility study. J Radioanal Nucl Chem 283(2):487–491
63. Zhang TT, Stilwell JL, Gerion D, Ding LH, Elboudwarej O, Cooke PA, Gray JW, Alivisatos AP, Chen FF (2006) Cellular effect of high doses of silica-coated quantum dot profiled with high throughput gene expression analysis and high content cellomics measurements. Nano Lett 6(4):800–808
64. Ryman-Rasmussen JP, Riviere JE, Monteiro-Riviere NA (2007) Surface coatings determine cytotoxicity and irritation potential of quantum dot nanoparticles in epidermal keratinocytes. J Invest Dermatol 127(1):143–153
65. Ballou B, Lagerholm BC, Ernst LA, Bruchez MP, Waggoner AS (2004) Noninvasive imaging of quantum dots in mice. Bioconjug Chem 15(1):79–86
66. Choi HS, Ipe BI, Misra P, Lee JH, Bawendi MG, Frangioni JV (2009) Tissue- and organ-selective biodistribution of NIR fluorescent quantum dots. Nano Lett 9(6):2354–2359

67. Fischer HC, Liu LC, Pang KS, Chan WCW (2006) Pharmacokinetics of nanoscale quantum dots: in vivo distribution, sequestration, and clearance in the rat. Adv Funct Mater 16(10):1299–1305
68. Choi HS, Liu W, Misra P, Tanaka E, Zimmer JP, Ipe BI, Bawendi MG, Frangioni JV (2007) Renal clearance of quantum dots. Nat Biotechnol 25(10):1165–1170
69. Soltesz EG, Kim S, Kim SW, Laurence RG, De Grand AM, Parungo CP, Cohn LH, Bawendi MG, Frangioni JV (2006) Sentinel lymph node mapping of the gastrointestinal tract by using invisible light. Ann Surg Oncol 13(3):386–396
70. Parungo CP, Colson YL, Kim SW, Kim S, Cohn LH, Bawendi MG, Frangioni JV (2005) Sentinel lymph node mapping of the pleural space. Chest 127(5):1799–1804
71. Soltesz EG, Kim S, Laurence RG, DeGrand AM, Parungo CP, Dor DM, Cohn LH, Bawendi MG, Frangioni JV, Mihaljevic T (2005) Intraoperative sentinel lymph node mapping of the lung using near-infrared fluorescent quantum dots. Ann Thorac Surg 79(1):269–277
72. Gopee NV, Roberts DW, Webb P, Cozart CR, Siitonen PH, Warbritton AR, Yu WW, Colvin VL, Walker NJ, Howard PC (2007) Migration of intradermally injected quantum dots to sentinel organs in mice. Toxicol Sci 98(1):249–257
73. Yang RS, Chang LW, Wu JP, Tsai MH, Wang HJ, Kuo YC, Yeh TK, Yang CS, Lin P (2007) Persistent tissue kinetics and redistribution of nanoparticles, quantum dot 705, in mice: ICP-MS quantitative assessment. Environ Health Perspect 115(9):1339–1343
74. Derfus AM, Chan WC, Bhatia SN (2003) Probing the cytotoxicity of semiconductor quantum dots. Nano Lett 4(1):11–18
75. Kirchner C, Javier AM, Susha AS, Rogach AL, Kreft O, Sukhorukov GB, Parak WJ (2005) Cytotoxicity of nanoparticle-loaded polymer capsules. Talanta 67(3):486–491
76. Kirchner C, Liedl T, Kudera S, Pellegrino T, Munoz Javier A, Gaub HE, Stolzle S, Fertig N, Parak WJ (2005) Cytotoxicity of colloidal CdSe and CdSe/ZnS nanoparticles. Nano Lett 5(2):331–338
77. Clarke SJ, Hollmann CA, Zhang Z, Suffern D, Bradforth SE, Dimitrijevic NM, Minarik WG, Nadeau JL (2006) Photophysics of dopamine-modified quantum dots and effects on biological systems. Nat Mater 5(5):409–417
78. Lovric J, Bazzi HS, Cuie Y, Fortin GR, Winnik FM, Maysinger D (2005) Differences in subcellular distribution and toxicity of green and red emitting CdTe quantum dots. J Mol Med 83(5):377–385
79. Cho SJ, Maysinger D, Jain M, Roder B, Hackbarth S, Winnik FM (2007) Long-term exposure to CdTe quantum dots causes functional impairments in live cells. Langmuir 23(4):1974–1980
80. Hoshino A, Fujioka K, Oku T, Suga M, Sasaki YF, Ohta T, Yasuhara M, Suzuki K, Yamamoto K (2004) Physicochemical properties and cellular toxicity of nanocrystal quantum dots depend on their surface modification. Nano Lett 4(11):2163–2169
81. Lovric J, Cho SJ, Winnik FM, Maysinger D (2005) Unmodified cadmium telluride quantum dots induce reactive oxygen species formation leading to multiple organelle damage and cell death. Chem Biol 12(11):1227–1234
82. Chan WH, Shiao NH, Lu PZ (2006) CdSe quantum dots induce apoptosis in human neuroblastoma cells via mitochondrial-dependent pathways and inhibition of survival signals. Toxicol Lett 167(3):191–200
83. Choi AO, Cho SJ, Desbarats J, Lovric J, Maysinger D (2007) Quantum dot-induced cell death involves Fas upregulation and lipid peroxidation in human neuroblastoma cells. J Nanobiotechnol 5:1. doi:10.1186/1477-3155-5-1
84. Shiohara A, Hoshino A, Hanaki K, Suzuki K, Yamamoto K (2004) On the cytotoxicity caused by quantum dots. Microbiol Immunol 48(9):669–675
85. Boldt K, Bruns OT, Gaponik N, Eychmuller A (2006) Comparative examination of the stability of semiconductor quantum dots in various biochemical buffers. J Phys Chem B 110(5):1959–1963
86. Dollefeld H, Hoppe K, Kolny J, Schilling K, Weller H, Eychmuller A (2002) Investigations on the stability of thiol stabilized semiconductor nanoparticles. Phys Chem Chem Phys 4(19):4747–4753

87. Pellegrino T, Manna L, Kudera S, Liedl T, Koktysh D, Rogach AL, Keller S, Radler J, Natile G, Parak WJ (2004) Hydrophobic nanocrystals coated with an amphiphilic polymer shell: a general route to water soluble nanocrystals. Nano Lett 4(4):703–707
88. Chen FQ, Gerion D (2004) Fluorescent CdSe/ZnS nanocrystal-peptide conjugates for long-term, nontoxic imaging and nuclear targeting in living cells. Nano Lett 4(10):1827–1832
89. Hardman R (2006) A toxicologic review of quantum dots: toxicity depends on physicochemical and environmental factors. Environ Health Perspect 114(2):165–172
90. Oberdörster G, Maynard A, Donaldson K, Castranova V, Fitzpatrick J, Ausman K, Carter J, Karn B, Kreyling W, Lai D, Olin S, Monteiro-Riviere N, Warheit D, Yang H (2005) Principles for characterizing the potential human health effects from exposure to nanomaterials: elements of a screening strategy. Part Fibre Toxicol 2:8. doi:10.1186/1743-8977-2-8
91. Upadhyay P (2006) Enhanced transdermal-immunization with diptheria-toxoid using local hyperthermia. Vaccine 24(27–28):5593–5598
92. Rieger S, Kulkarni RP, Darcy D, Fraser SE, Koster RW (2005) Quantum dots are powerful multipurpose vital labeling agents in zebrafish embryos. Dev Dyn 234(3):670–681
93. Manabe N, Hoshino A, Liang YQ, Goto T, Kato N, Yamamoto K (2006) Quantum dot as a drug tracer in vivo. IEEE Trans Nanobioscience 5(4):263–267
94. Al-Jamal WT, Al-Jamal KT, Bomans PH, Frederik PM, Kostarelos K (2008) Functionalized-quantum-dot-liposome hybrids as multimodal nanoparticles for cancer. Small 4(9):1406–1415
95. Al-Jamal WT, Al-Jamal KT, Cakebread A, Halket JM, Kostarelos K (2009) Blood circulation and tissue biodistribution of lipid-quantum dot (L-QD) hybrid vesicles intravenously administered in mice. Bioconjug Chem 20(9):1696–1702
96. Al-Jamal WT, Al-Jamal KT, Tian B, Lacerda L, Bornans PH, Frederik PM, Kostarelos K (2008) Lipid-quantum dot bilayer vesicles enhance tumor cell uptake and retention in vitro and in vivo. ACS Nano 2(3):408–418
97. Al-Jamal WT, Al-Jamal KT, Tian B, Cakebread A, Halket JM, Kostarelos K (2009) Tumor targeting of functionalized quantum dot-liposome hybrids by intravenous administration. Mol Pharm 6(2):520–530
98. Al-Jamal WT, Kostarelos K (2007) Liposome-nanoparticle hybrids for multimodal diagnostic and therapeutic applications. Nanomedicine 2(1):85–98
99. Gopalakrishnan G, Danelon C, Izewska P, Prummer M, Bolinger PY, Geissbuhler I, Demurtas D, Dubochet J, Vogel H (2006) Multifunctional lipid/quantum dot hybrid nanocontainers for controlled targeting of live cells. Angew Chem Int Ed 45(33):5478–5483
100. Erogbogbo F, Yong KT, Hu R, Law WC, Ding H, Chang CW, Prasad PN, Swihart MT (2010) Biocompatible magnetofluorescent probes: luminescent silicon quantum dots coupled with superparamagnetic iron(III) oxide. ACS Nano 4(9):5131–5138
101. Yong KT, Ding H, Roy I, Law WC, Bergey EJ, Maitra A, Prasad PN (2009) Imaging pancreatic cancer using bioconjugated InP quantum dots. ACS Nano 3(3):502–510
102. Niebling T, Zhang F, Ali Z, Parak WJ, Heimbrodt W (2009) Excitation dynamics in polymer-coated semiconductor quantum dots with integrated dye molecules: the role of reabsorption. J Appl Phys 106(10):104701
103. Law WC, Yong KT, Roy I, Xu G, Ding H, Bergey EJ, Zeng H, Prasad PN (2008) Optically and magnetically doped organically modified silica nanoparticles as efficient magnetically guided biomarkers for two-photon imaging of live cancer cells. J Phys Chem C 112(21):7972–7977
104. Schabas G, Wang CW, Oskooei A, Yusuf H, Moffitt MG, Sinton D (2008) Formation and shear-induced processing of quantum dot colloidal assemblies in a multiphase microfluidic chip. Langmuir 24(19):10596–10603
105. Wang CW, Oskooei A, Sinton D, Moffitt MG (2010) Controlled self-assembly of quantum dot-block copolymer colloids in multiphase microfluidic reactors. Langmuir 26(2):716–723
106. Guo Y, Moffitt MG (2007) Semiconductor quantum dots with environmentally responsive mixed polystyrene/poly(methyl methacrylate) brush layers. Macromolecules 40(16):5868–5878
107. Hu R, Yong KT, Roy I, Ding H, Law WC, Cai HX, Zhang XH, Vathy LA, Bergey EJ, Prasad PN (2010) Functionalized near-infrared quantum dots for in vivo tumor vasculature imaging. Nanotechnology 21:145105

108. Law WC, Yong KT, Roy I, Ding H, Hu R, Zhao WW, Prasad PN (2009) Aqueous-phase synthesis of highly luminescent CdTe/ZnTe core/shell quantum dots optimized for targeted bioimaging. Small 5(11):1302–1310
109. Guo Y, Shi DL, Cho HS, Dong ZY, Kulkarni A, Pauletti GM, Wang W, Lian J, Liu W, Ren L, Zhang QQ, Liu GK, Huth C, Wang LM, Ewing RC (2008) In vivo imaging and drug storage by quantum-dot-conjugated carbon nanotubes. Adv Funct Mater 18(17):2489–2497
110. Shi DL, Cho HS, Huth C, Wang F, Dong ZY, Pauletti GM, Lian J, Wang W, Liu GK, Bud'ko SL, Wang LM, Ewing RC (2009) Conjugation of quantum dots and Fe3O4 on carbon nanotubes for medical diagnosis and treatment. Appl Phys Lett 95(22):223702. doi:10.163/1.3268469
111. Torchilin VP (2005) Recent advances with liposomes as pharmaceutical carriers. Nat Rev Drug Discov 4(2):145–160
112. Gerasimov OV, Boomer JA, Qualls MM, Thompson DH (1999) Cytosolic drug delivery using pH- and light-sensitive liposomes. Adv Drug Del Rev 38(3):317–338
113. Derfus AM, Chan WCW, Bhatia SN (2004) Intracellular delivery of quantum dots for live cell labeling and organelle tracking. Adv Mater 16(12):961–966
114. Sigot V, Arndt-Jovin DJ, Jovin TM (2010) Targeted cellular delivery of quantum dots loaded on and in biotinylated liposomes. Bioconjug Chem 21(8):1465–1472
115. Ho YP, Leong KW (2010) Quantum dot-based theranostics. Nanoscale 2(1):60–68
116. Kostarelos K, Miller AD (2005) Synthetic, self-assembly ABCD nanoparticles: a structural paradigm for viable synthetic non-viral vectors. Chem Soc Rev 34(11):970–994
117. Derfus AM, Chen AA, Min DH, Ruoslahti E, Bhatia SN (2007) Targeted quantum dot conjugates for siRNA delivery. Bioconjug Chem 18(5):1391–1396
118. Srinivasan C, Lee J, Papadimitrakopoulos F, Silbart LK, Zhao MH, Burgess DJ (2006) Labeling and intracellular tracking of functionally active plasmid DNA with semiconductor quantum dots. Mol Ther 14(2):192–201
119. Ho YP, Chen HH, Leong KW, Wang TH (2006) Evaluating the intracellular stability and unpacking of DNA nanocomplexes by quantum dots-FRET. J Control Release 116(1):83–89
120. Chen HH, Ho YP, Jiang X, Mao HQ, Wang TH, Leong KW (2008) Quantitative comparison of intracellular unpacking kinetics of polyplexes by a model constructed from quantum Dot-FRET. Mol Ther 16(2):324–332
121. Anas A, Akita H, Harashima H, Itoh T, Ishikawa M, Biju V (2008) Photosensitized breakage and damage of DNA by CdSe-ZnS quantum dots. J Phys Chem B 112(32):10005–10011
122. Li D, Li GP, Guo WW, Li PC, Wang EK, Wang J (2008) Glutathione-mediated release of functional plasmid DNA from positively charged quantum dots. Biomaterials 29(18):2776–2782
123. Lee PW, Hsu SH, Tsai JS, Chen FR, Huang PJ, Ke CJ, Liao ZX, Hsiao CW, Lin HJ, Sung HW (2010) Multifunctional core-shell polymeric nanoparticles for transdermal DNA delivery and epidermal Langerhans cells tracking. Biomaterials 31(8):2425–2434
124. Zintchenko A, Susha AS, Concia M, Feldmann J, Wagner E, Rogach AL, Ogris M (2009) Drug nanocarriers labeled with near-infrared-emitting quantum dots (quantoplexes): imaging fast dynamics of distribution in living animals. Mol Ther 17(11):1849–1856
125. David S, Pitard B, Benoit JP, Passirani C (2010) Non-viral nanosystems for systemic siRNA delivery. Pharmacol Res 62(2):100–114
126. Chen AA, Derfus AM, Khetani SR, Bhatia SN (2005) Quantum dots to monitor RNAi delivery and improve gene silencing. Nucleic Acids Res 33(22):e190. doi:10.1093/nar/gni188
127. Tan WB, Jiang S, Zhang Y (2007) Quantum-dot based nanoparticles for targeted silencing of HER2/neu gene via RNA interference. Biomaterials 28(8):1565–1571
128. Jung JJ, Solanki A, Memoli KA, Kamei K, Kim H, Drahl MA, Williams LJ, Tseng HR, Lee K (2010) Selective inhibition of human brain tumor cells through multifunctional quantum-dot-based siRNA delivery. Angew Chem Int Ed 49(1):103–107
129. Walther C, Meyer K, Rennert R, Neundorf I (2008) Quantum dot-carrier peptide conjugates suitable for imaging and delivery applications. Bioconjug Chem 19(12):2346–2356
130. Yezhelyev MV, Qi LF, O'Regan RM, Nie S, Gao XH (2008) Proton-sponge coated quantum dots for siRNA delivery and intracellular imaging. J Am Chem Soc 130(28):9006–9012

131. Jiang G, Park K, Kim J, Kim KS, Hahn SK (2009) Target specific intracellular delivery of siRNA/PEI-HA complex by receptor mediated endocytosis. Mol Pharm 6(3):727–737
132. Qi L, Gao X (2008) Quantum dot-amphipol nanocomplex for intracellular delivery and real-time imaging of siRNA. ACS Nano 2(7):1403–1410
133. Juzenas P, Chen W, Sun YP, Coelho MAN, Generalov R, Generalova N, Christensen IL (2008) Quantum dots and nanoparticles for photodynamic and radiation therapies of cancer. Adv Drug Del Rev 60(15):1600–1614
134. Yaghini E, Seifalian AM, MacRobert AJ (2009) Quantum dots and their potential biomedical applications in photosensitization for photodynamic therapy. Nanomedicine 4(3):353–363
135. Dayal S, Lou YB, Samia ACS, Berlin JC, Kenney ME, Burda C (2006) Observation of non-Förster-type energy-transfer behavior in quantum dot-phthalocyanine conjugates. J Am Chem Soc 128(43):13974–13975
136. Shi LX, Hernandez B, Selke M (2006) Singlet oxygen generation from water-soluble quantum dot-organic dye nanocomposites. J Am Chem Soc 128(19):6278–6279
137. Blanco NG, Maldonado CR, Mareque-Rivas JC (2009) Effective photoreduction of a Pt(IV) complex with quantum dots: a feasible new light-induced method of releasing anticancer Pt(II) drugs. Chem. Commun 5257–5259
138. Weng KC, Noble CO, Papahadjopoulos-Sternberg B, Chen FF, Drummond DC, Kirpotin DB, Wang DH, Hom YK, Hann B, Park JW (2008) Targeted tumor cell internalization and imaging of multifunctional quantum dot-conjugated immunoliposomes in vitro and in vivo. Nano Lett 8(9):2851–2857
139. Park JH, von Maltzahn G, Ruoslahti E, Bhatia SN, Sailor MJ (2008) Micellar hybrid nanoparticles for simultaneous magnetofluorescent imaging and drug delivery. Angew Chem Int Ed 47(38):7284–7288
140. Zhou YY, Shi LX, Li QN, Jiang H, Lv G, Zhao J, Wu CH, Selke M, Wang XM (2010) Imaging and inhibition of multidrug resistance in cancer cells via specific association with negatively charged CdTe quantum dots. Biomaterials 31(18):4958–4963
141. Song H, He R, Wang K, Ruan J, Bao CC, Li N, Ji JJ, Cui DX (2010) Anti-HIF-1 alpha antibody-conjugated pluronic triblock copolymers encapsulated with Paclitaxel for tumor targeting therapy. Biomaterials 31(8):2302–2312

Index

S. Svenson and R. Prud'homme (eds.), *Multifunctional Nanoparticles for Drug Delivery Applications: Imaging, Targeting, and Delivery*, Nanostructure Science and Technology, DOI 10.1007/978-1-4614-2305-8, © Springer Science+Business Media, LLC 2012

MIX
Papier aus verantwortungsvollen Quellen
Paper from responsible sources
FSC® C105338

If you have any concerns about our products,
you can contact us on
ProductSafety@springernature.com

In case Publisher is established outside the EU,
the EU authorized representative is:
Springer Nature Customer Service Center GmbH
Europaplatz 3, 69115 Heidelberg, Germany

Printed by Libri Plureos GmbH
in Hamburg, Germany